Mathematische Optimierung und Wirtschaftsmathematik | Mathematical Optimization and Economathematics

Reihe herausgegeben von
Ralf Werner, Augsburg, Deutschland
Tobias Harks, Augsburg, Deutschland
Vladimir Shikhman, Chemnitz, Deutschland

In der Reihe werden Arbeiten zu aktuellen Themen der mathematischen Optimierung und der Wirtschaftsmathematik publiziert. Hierbei werden sowohl Themen aus Grundlagen, Theorie und Anwendung der Wirtschafts-, Finanz- und Versicherungsmathematik als auch der Optimierung und des Operations Research behandelt. Die Publikationen sollen insbesondere neue Impulse für weitergehende Forschungsfragen liefern, oder auch zu offenen Problemstellungen aus der Anwendung Lösungsansätze anbieten. Die Reihe leistet damit einen Beitrag der Bündelung der Forschung der Optimierung und der Wirtschaftsmathematik und der sich ergebenden Perspektiven.

Weitere Bände in der Reihe http://www.springer.com/series/15822

Ole Martin

High-Frequency Statistics with Asynchronous and Irregular Data

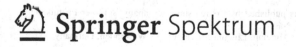

Springer Spektrum

Ole Martin
Department of Mathematics
Kiel University (CAU)
Kiel, Germany

Dissertation, Kiel University (CAU), 2018

ISSN 2523-7926　　　　　　　　ISSN 2523-7934　(electronic)
Mathematische Optimierung und Wirtschaftsmathematik | Mathematical Optimization and Economathematics
ISBN 978-3-658-28417-6　　　　ISBN 978-3-658-28418-3　(eBook)
https://doi.org/10.1007/978-3-658-28418-3

This Springer Spektrum imprint is published by the registered company Springer Fachmedien Wiesbaden GmbH part of Springer Nature.
The registered company address is: Abraham-Lincoln-Str. 46, 65189 Wiesbaden, Germany

Preface

Over the past two decades the field of high-frequency statistics especially with applications to financial data experienced a rapid growth[1]. This growth was fueled by the increasing availability of high-frequency data generated by electronic trading platforms and the rising importance of high-frequency traders. Various mathematical methods have therefore been designed to help practitioners as well as academics to investigate such data. The setting of high-frequency statistics is characterized as follows: We assume that we observe (multiple) stochastic process(es) $(X_t^{(l)})_{t\geq0}$, $l = 1,\ldots,d$, which exist in continuous time over discrete grids of time points $t_{i,n}^{(l)}$, $i \in \mathbb{N}$. The goal then is to infer on properties of the continuous-time model based on the discrete observations

$$X_{t_{i,n}^{(l)}}^{(l)}, \quad i \in \mathbb{N}.$$

To this end we consider a fixed time interval $[0,T]$ and investigate the asymptotics where the mesh of the observation times $t_{i,n}^{(l)}$, $i \in \mathbb{N}$, (i.e. the maximal distance of consecutive observation times) tends to zero. Although the number of observations cannot be controlled by the statistician and is in practice always finite, we assume that their density in $[0,T]$ is high enough such that asymptotic results approximate the finite case reasonably well. Indeed in financial applications, financial processes are often observed over intervals as short as seconds or even milliseconds.

Quite naturally people first started to work with simple and regular models for the data. But as the field grew and practical applications presented further challenges the models also evolved. So people moved from models for continuous processes to models incorporating jumps, started to model market frictions by including micro-structure noise and also began to consider more complex models for the observation times. While there exists a vast amount of literature on the situation of equidistant and synchronous observation times $t_{i,n}^{(l)} = i/n$, see the monographs [30] and [3] for an overview, the situation of irregular and asynchronous observation times has been investigated less often because of its higher complexity. Due to the lack of methods based on irregularly observed data often methods which were designed for equidistant and synchronous data are also applied to real data which usually comes in irregular and asynchronous form. Here, these methods can't be applied to the data directly, but the data has to be artificially synchronized beforehand. This means that not all available observations

---

[1]compare Page XVII in [3]

are used but instead only the observations closest to the fixed marks $i\Delta_n$, $i \in \mathbb{N}$, enter the estimation. Here the new grid width $\Delta_n \geq 0$ is usually much larger than the length of the average observation interval in the original dataset. The newly obtained dataset is then almost synchronous and the observation times are almost equidistant with mesh Δ_n. This technique allows to use the methods developed for equidistant and synchronous observations also in the irregular and asynchronous setting as the error originating from the fact, that the modified sampling scheme is only almost equidistant and synchronous, is usually dominated by other approximation errors. The big disadvantage of this method, however, is that we lose a lot of data as we only use very few selected observations. Although there exist methods which synchronize the data more efficiently, compare the concept of "refresh times" times used e.g. in [5] and [1], a synchronization step always leads to a reduction of the number of used data points. Thereby the effective sample size is diminished and hence estimators have a higher variance, asymptotic distributions are less accurate proxies for finite sample distributions and the power of tests is usually smaller than in the setting where we observe the processes equally often in an equidistant and synchronous manner. Further, it has been shown both empirically in [14] and demonstrated mathematically in [22] that this artificial synchronization of the observations of multiple processes may lead to estimation bias. For these reasons, it seems superior to work with methods based directly on all irregular and asynchronous observations. Such methods for the estimation of the quadratic covariation were first presented in [22] and have been further investigated in [23], [24] and [20]. These methods have been extended to the discontinuous case where the processes are allowed to have jumps in the paper [8] for the estimation of the quadratic covariation including jumps and in [35], [34] for the estimation of the continuous part of the quadratic covariation under the presence of jumps. In this work, we generalize the methods developed in these papers for the estimation of the quadratic covariation to obtain asymptotic results for more general power-variation type functionals. These results are then used to construct statistical tests which allow to decide whether (common) jumps are present in irregularly and asynchronously observed processes. Further, we present a new bootstrap technique which allows to estimate asymptotic variances in the central limit theorems for statistics based on irregular and asynchronous observations.

We find that in general the behaviour of functionals based on irregular and asynchronous observations is not only more complicated but sometimes even fundamentally different compared to the simpler setting of equidistant and synchronous observation times; see especially the results in Chapter 2. Further we will see that it becomes in general infeasible to use certain central limit theorems in the setting of irregular observations as the asymptotic variance can in general not be estimated from the data anymore; see Chapters 3 and 4. To circumvent this problem additional assumptions have to be made on the observation scheme such that a bootstrap method can be applied. The simulation results in Chapters 8 and 9 show that the finite sample performance of our methods based on irregular and asynchronous observations is comparable to the finite sample performance of similar methods based on equidistant and synchronous observations. This is a remarkable result as in practice data usually comes in irregular and asynchronous form and therefore for the application of methods designed for equidistant and synchronous observations the data has to be artificially made equidistant and synchronous as described before. Hence, our methods perform as well as methods based on equidistant and synchronous observations for similar sample sizes but the effective sample size is usually much larger when using methods designed for irregular and asynchronous data. We conclude that although the methods become more complicated it is worthwhile to work with procedures based directly on irregular and asynchronous observations as these methods yield much better results in practical applications.

This work is divided into two parts: Part I contains general theoretical results which in similar form already have proven to be useful in the context of high-frequency statistics and which will be used in the applications in Part II. In Chapter 1 we characterize the framework and especially the general model for the stochastic processes and the observation times in which we are going to derive results. In Chapter 2 generalizations of functionals to the setting of irregular and asynchronous observations are investigated which regularly occur in the context of high-frequency statistics. In Chapters 3 and 4 we discuss how to obtain central limit theorems for the functionals introduced in Chapter 2 and how to estimate the asymptotic variances in these central limit theorems. Although Chapters 3 and 4 contain only results for specific functionals we use these as an example to introduce general methods which then later are also used in the applications part for more elaborate statistics. Throughout Chapters 2–4 we try to highlight differences in the results for irregular and asynchronous observation times compared to existing results in the simpler setting of equidistant and synchronous observations. In Chapter 5 we collect some results which are necessary to prove that certain specific observation schemes, most importantly the one where observation times are generated by independent Poisson processes, fulfil the conditions which were made in Chapters 2–4. Despite the use of these results already in the previous chapters, I decided to gather them in a separate chapter as the results are rather technical and partially also of individual interest. Part II contains applications of the methods developed in Part I for the estimation of certain quantities and for the testing of specific hypotheses which are of direct interest to practitioners. Chapter 6 is devoted to the estimation of spot volatilities and correlations. Albeit these estimators are already used in Chapter 4 I decided to discuss them in Part II separately. Their investigation to me seems to fit better into Part II because their estimation has been discussed in the literature in several contexts individually which makes the topic of their estimation worthy of an own chapter. In Chapter 7 we discuss the estimation of realized quadratic variation and covariation processes which is probably the quantity which has received the most attention in the high-frequency statistics literature. Chapters 8 and 9 contain tests for the presence of (common) jumps in paths of observed processes.

This work is mainly based on the research published in the three papers [36], [38] and [37]. It aims at presenting the methods used and results found therein in a unifying context. Thereby I try to draw a comprehensive picture of our approach in addressing the challenges arising from the use of irregular and asynchronous data in the field of high-frequency statistics. In particular Sections 2.1 and 2.2 are almost entirely adopted from the paper [37]. Further, Section 9.1 is based on [36] and Chapter 8 and Section 9.2 contain the results presented in [38]. Additionally, parts of the results and proofs which make up the content of the remaining chapters and sections are also modified versions of results and proofs which are included in the papers mentioned above.

I am very thankful to my doctoral advisor Prof. Dr. Mathias Vetter who gave me the opportunity to write this thesis and also for giving me a lot of freedom in choosing the direction of my research. Further, I would like to thank my colleagues from the working groups stochastics and financial mathematics at the CAU Kiel for providing a welcoming working environment and the participants of the „Oberseminar Stochastik und Finanzmathematik" for listening to presentations on earlier stages of this work and for giving helpful comments. I am thankful to the organizers of the DynStoch meeting 2017 held in Siegen, to the organizers of the DynStoch meeting 2018 in Porto and to the organizing committee of the 13th German Probability and Statistics Days 2018 in Freiburg i. Br. for giving me the opportunity to present my work in front of experienced researchers. Additionally, I would like to thank two anonymous referees whose comments improved the paper [36] and also further research. Further, I am grateful to my parents Ulla and Ulrich who from an early childhood age on encouraged my interest in the field of mathematics and who have always supported me. Finally, I am deeply indebted to my girlfriend Kathrin who is always there for me.

Kiel, Germany   Ole Martin

Contents

Notation

Basic mathematical notation

$\mathbb{R}$, $\mathbb{R}_{\geq 0}$, $\mathbb{R}^d$	The real numbers, the non-negative real numbers and the space of d-dimensional vectors with real entries
$\mathbb{N}$, $\mathbb{N}_0$	The natural numbers without 0, the natural numbers including 0
$\|x\|$, $\|y\|$	The absolute value of $x \in \mathbb{R}^d$, the Euclidean norm of $y \in \mathbb{R}^d$
$\|A\|$	The Lebesgue-measure of a set $A \subset \mathbb{R}$
v^*	The transpose of a vector $v \in \mathbb{R}^d$
$\lfloor x \rfloor$, $\lceil x \rceil$	The largest (smallest) natural number less (greater) or equal than $x \in \mathbb{R}$
$x \wedge y$, $x \vee y$	The minimum, maximum of $x, y \in \mathbb{R}$
x^+	The positive part $x \vee 0$ of $x \in \mathbb{R}$
$A \cup B$, $A \cap B$, $A \setminus B$	The union, intersection and difference of two sets A, B
A^c	The complement of a set A
$\mathbb{1}_A$	The indicator function of a set A
$\xrightarrow{\mathbb{P}}$	Convergence in probability
$\xrightarrow{\mathcal{L}}$, $\overset{\mathcal{L}}{=}$	Convergence in law, identity in law
$\xrightarrow{\mathcal{L}-s}$	Stable convergence in law, see Appendix A
$[X, Y]_t$	The covariation process of the stochastic processes X and Y
X_{t-}	The left limit $\lim_{s \uparrow t} X_s$ of X at time t if it exists
ΔX_t	The jump $X_t - X_{t-}$ of a process X at time t
$\mathcal{N}(0, \Sigma)$	The d-dimensional centered normal distribution with covariance matrix $\Sigma \in \mathbb{R}^{d \times d}$
$\mathcal{U}[a, b]$	The uniform distribution over the interval $[a, b]$
$\text{Exp}(\lambda)$	The exponential distribution with parameter $\lambda > 0$
$f(x) = o(g(x))$	It holds $f(x)/g(x) \to 0$ as $x \to x_0$
$f(x) = O(g(x))$	It holds $\limsup_{x \to x_0} |f(x)/g(x)| < \infty$
$X_n = o_{\mathbb{P}}(Y_n)$	It holds $X_n/Y_n \xrightarrow{\mathbb{P}} 0$
$X_n = O_{\mathbb{P}}(Y_n)$	The sequence X_n/Y_n is bounded in probability

Repeatedly used variables in alphabetical order

$\gamma(z)$, Γ_t	Function γ and locally bounded proces Γ_t such that $\|\delta(t,z)\| \leq \Gamma_t \gamma(z)$, see Condition 1.3
$\delta(s,z)$	The function relating the jump measure μ to the jump sizes of $\mathcal{X}$, see (1.1)
$\Delta_{i,n}^{(l)} Y$, $\Delta_{i,k,n}^{(l)} Y$	The increments of a process Y over the intervals $\mathcal{I}_{i,n}^{(l)}$, $\mathcal{I}_{i,k,n}^{(l)}$; see (1.6) and (2.87)
μ, ν	The Poisson random measure driving the jump part of X and its compensator, see (1.1)
π_n	The set of all observation times at stage n, see (1.3)
$|\pi_n|_t$	The mesh of the observation times, see (1.4)
ρ	The correlation process of the continuous martingale parts of $X^{(1)}$ and $X^{(2)}$, see (1.2)
$\tilde{\rho}_n(s,-)$, $\hat{\rho}_n(s,-)$, $\tilde{\rho}_n(s,+)$, $\hat{\rho}_n(s,+)$	Estimators for the spot correlation of $C^{(1)}$ and $C^{(2)}$ at time s, see Chapter 6
σ, $\sigma^{(l)}$	The volatility matrix process of X, the volatility process of $X^{(l)}$, see (1.2)
$\tilde{\sigma}_n^{(l)}(s,-)$, $\hat{\sigma}_n^{(l)}(s,-)$, $\tilde{\sigma}_n^{(l)}(s,+)$, $\hat{\sigma}_n^{(l)}(s,+)$	Estimators for the spot volatility of $X^{(l)}$ at time s, see Chapter 6
$\tau_{n,-}^{(l)}(s)$, $\tau_{n,+}^{(l)}(s)$	The observation times of $X^{(l)}$ at stage n immediately before and after s, see (2.28)
$(\Omega, \mathcal{F}, \mathbb{P})$	Probability space on which the process X and the observation times are defined
$(\widetilde{\Omega}, \widetilde{\mathcal{F}}, \widetilde{\mathbb{P}})$	Extended probability space on which limits of stable limit theorems can be defined
$B(q) = (B^{(1)}(q), B^{(2)}(q))^*$	The continuous part of finite variation in the decomposition of X, see (1.7)
$B(f)_T$, $B^*(f)_T$, $B^{(l)}(g)_T$	Sums of function evaluations at the jumps of X, $X^{(l)}$; see (2.5)
c_s	The spot covariance matrix $c_s = \sigma_s \sigma_s^*$ of $C^{(1)}$, $C^{(2)}$
$C = (C^{(1)}, C^{(2)})^*$	The continuous martingale part in the decomposition of X, see (1.7)
$G_p^{(l),n}(t)$, $G_{p_1,p_2}^n(t)$, $H_{k,m,p}^n(t)$	Sums of products of observation interval lengths occuring in the limits of the normalized functionals, see (2.39)

$G_p^{[k],(l),n}(t)$, $G_{p_1,p_2}^{[k],n}(t)$, $H_{l,m,p}^{[k],n}(t)$	Sums of products of cummulative lengths of multiple observation intervals, see (2.89)
$i_n^{(l)}(s)$	The index of the observation interval characterized by $s \in \mathcal{I}_{i_n^{(l)}(s),n}^{(l)}$
$\mathcal{I}_{i,n}^{(l)}$, $\mathcal{I}_{i,k,n}^{(l)}$	Observation interval $(t_{i-1,n}^{(l)}, t_{i,n}^{(l)}]$ of $X^{(l)}$, interval $(t_{i-k,n}^{(l)}, t_{i,n}^{(l)}]$; see (1.5) and (2.86)
K, K_q, K_p	Generic constants (depending on q respectively p)
$m_\Sigma(h)$	Moment $\mathbb{E}[h(Z)]$ for $Z \sim \mathcal{N}(0,\Sigma)$, see (2.38)
$M(q) = (M^{(1)}(q), M^{(2)}(q))^*$	The martingale part containing small jumps in the decomposition of X, see (1.7)
$N(q) = (N^{(1)}(q), N^{(2)}(q))^*$	The part containing big jumps in the decomposition of X, see (1.7)
$\mathcal{S}$	The σ-algebra generated by the observation times, see Definition 1.2
$t_{i,n}^{(l)}$	The i-th observation time of $X^{(l)}$ at stage n
T	The time horizon $T \geq 0$
$V(f,\pi_n)_T$, $V^{(l)}(g,\pi_n)_T$	Non-normalized functionals, see Section 2.1
$\overline{V}(p,f,\pi_n)_T$, $\overline{V}^{(l)}(p,g,\pi_n)_T$	Normalized functionals, see Section 2.2
$V_+(f,\pi_n,(\beta,\varpi))_T$, $\overline{V}_-(p,f,\pi_n,(\beta,\varpi))_T$	Functionals of truncated increments, see Section 2.3, also: $V_+^{(l)}(g,\pi_n,(\beta,\varpi))_T$, $\overline{V}_-^{(l)}(p,g,\pi_n,(\beta,\varpi))_T$
$V(f,[k],\pi_n)_T$, $\overline{V}(p,f,[k],\pi_n)_T$	Functionals of increments over multiple observation intervals, see Section 2.4, also: $V^{(l)}(g,[k],\pi_n)_T$, $\overline{V}^{(l)}(p,g,[k],\pi_n)_T$
$X = (X^{(1)}, X^{(2)})^*$	The bivariate observed process
$\mathcal{X}$	The σ-algebra generated by the process X and its components, see Definition 1.2

Part I

Theory

1 Framework

In this chapter, we specify the mathematical framework within we will derive theoretical results and develop statistical procedures.

First, we characterize the stochastic processes which we observe. To this end let $X_t = (X_t^{(1)}, X_t^{(2)})^*$, $t \geq 0$, be a stochastic process in continuous time. We restrict ourselves to bivariate processes as the case $d = 2$ is sufficient to study most effects of asynchronous observations and because a lot of questions on the dependence of multiple processes can be answered by investigating pairs of processes. Additionally, it is already in the bivariate case challenging to derive certain results. A common class of processes used to model dynamics in continuous time are semimartingales i.e. processes X which can be written as $X_t = A_t + M_t$ where A is a process with paths of finite variation and M is a martingale. Semimartingales form the largest class of processes with respect to which a nice integration theory can be defined, compare page 35 in [3]. Thereby they naturally occur as solutions to stochastic differential equations. Further it has been shown that in financial mathematics price processes under certain assumptions have to be semimartingales; compare [13]. In this work we will not derive a theory for the whole class of semimartingales but only for the subclass of *Itô semimartingales*. Itô semimartingales can be understood as a generalization of Lévy processes where the Lévy-Khintchine triplet is allowed to be time dependent. Itô semimartingales X distinguish themselves from general semimartingales by the property that it holds $\mathbb{P}(\Delta X_t \neq 0) = 0$ for any $t \geq 0$ where $\Delta X_t = X_t - X_{t-}$, $X_{t-} = \lim_{s \uparrow t} X_s$, denotes the jump of X at time t, compare Section 1.4.1 of [3]. This property implies that Itô semimartingales have no fixed jump times. All Itô semimartingales can be written in the so-called Grigelionis representation, compare Section 1.4.3 in [3], which we use in the following to give a more precise definition of the class of Itô semimartingales. We denote by $(\Omega, (\mathcal{F}_t)_{t \geq 0}, \mathbb{P})$ the filtered probability space on which the upcoming random variables will be defined.

Definition 1.1. *We call a process $X = (X^{(1)}, X^{(2)})^*$ a two-dimensional Itô semimartingale if it can be written in the form*

$$X_t = X_0 + \int_0^t b_s ds + \int_0^t \sigma_s dW_s + \int_0^t \int_{\mathbb{R}^2} \delta(s, z) \mathbb{1}_{\{\|\delta(s,z)\| \leq 1\}} (\mu - \nu)(ds, dz)$$

© Springer Fachmedien Wiesbaden GmbH, part of Springer Nature 2019
O. Martin, *High-Frequency Statistics with Asynchronous and Irregular Data*,
Mathematische Optimierung und Wirtschaftsmathematik | Mathematical Optimization
and Economathematics, https://doi.org/10.1007/978-3-658-28418-3_1

$$+ \int\limits_0^t \int\limits_{\mathbb{R}^2} \delta(s,z) \mathbb{1}_{\{\|\delta(s,z)\|>1\}} \mu(ds,dz), \quad (1.1)$$

where W is a two-dimensional standard Brownian motion, μ is a Poisson random measure on $(0,\infty) \times \mathbb{R}^2$ whose predictable compensator satisfies the identity $\nu(ds,dz) = ds \otimes \lambda(dz)$ for some σ-finite measure λ on $\mathbb{R}^2$ endowed with the Borelian σ-algebra. b is a two-dimensional adapted process, σ is a 2×2 adapted process and δ is a two-dimensional predictable function on $\Omega \times (0,\infty) \times \mathbb{R}^2$. $\square$

For a more detailed definition of the components of (1.1) and a detailed characterization of semimartingales and Itô semimartingales in general we refer to Chapter 1 of [3]. As in Section 4 of [8] we further assume that σ is of the form

$$\sigma_s = \begin{pmatrix} \sigma_s^{(1)} & 0 \\ \rho_s \sigma_s^{(2)} & \sqrt{1-\rho_s^2}\sigma_s^{(2)} \end{pmatrix} \quad (1.2)$$

for non-negative adapted processes $\sigma_s^{(1)}$, $\sigma_s^{(2)}$ and an adapted process ρ_s with values in the interval $[-1,1]$. This assumption is no additional restriction on the model because the law of (1.1) only depends on $\sigma_t(\sigma_t)^*$ and not on σ_t itself.

If we are mainly interested in applications in finance, the decision to work only with Itô semimartingales instead of general semimartingales is not very restrictive because according to [3] (page 1975, second paragraph) the assumption to work only with Itô semimartingales is „a mild structural assumption that is satisfied in all continuous-time models with stochastic volatility used in finance, at least as long as one wants to rule out arbitrage opportunities."

Next we characterize the observation times: As data is usually aquired over time and previous states of a system normally can not be recovered later if the system has evolved over time it seems reasonable to assume that the observation times $t_{i,n}^{(l)}$ are stopping times with respect to the filtration $(\mathcal{F}_t)_{t \geq 0}$ to which the process X is adapted. At stage n we therefore assume that the process $X^{(l)}$, $l = 1,2$, is observed at the stopping times $t_{i,n}^{(l)}$, $i \in \mathbb{N}_0$. Further, $\left(t_{i,n}^{(l)}\right)_{i \in \mathbb{N}_0}$, $l = 1,2$, are for each $n \in \mathbb{N}$ increasing sequences of stopping times and it holds $t_{0,n}^{(l)} = 0$. We denote by

$$\pi_n = \left\{ \left(t_{i,n}^{(1)}\right)_{i \in \mathbb{N}_0}, \left(t_{i,n}^{(2)}\right)_{i \in \mathbb{N}_0} \right\} \quad (1.3)$$

the collection of all observation times at stage $n \in \mathbb{N}$ which we will also call *observation scheme* and by

$$|\pi_n|_T = \sup \left\{ t_{i,n}^{(l)} \wedge T - t_{i-1,n}^{(l)} \wedge T \big| i \geq 1, \ l = 1,2 \right\} \quad (1.4)$$

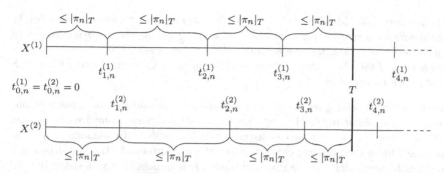

Figure 1.1: A realization of the observation scheme π_n restricted to $[0, T]$.

the mesh of the observation times up to T.

Based on the observation times we introduce some additional notation. By

$$\mathcal{I}_{i,n}^{(l)} = (t_{i-1,n}^{(l)}, t_{i,n}^{(l)}] \tag{1.5}$$

we denote the i-th observation interval at stage n corresponding to the process $X^{(l)}$. Further we define by

$$\Delta_{i,n}^{(l)} Y = Y_{t_{i,n}^{(l)}} - Y_{t_{i-1,n}^{(l)}} \tag{1.6}$$

the increment of an arbitrary adapted process Y over the observation interval $\mathcal{I}_{i,n}^{(l)} = (t_{i-1,n}^{(l)}, t_{i,n}^{(l)}]$. By $|A|$ we denote the Lebesgue measure of a set $A \subset [0, \infty)$ e.g. $|\mathcal{I}_{i,n}^{(l)}|$ is equal to the length of the observation interval $\mathcal{I}_{i,n}^{(l)}$. The random variable $i_n^{(l)}(s)$ is the index of the observation interval $\mathcal{I}_{i,n}^{(l)}$ which contains s i.e. $i_n^{(l)}(s)$ is characterized via $s \in \mathcal{I}_{i_n^{(l)}(s),n}^{(l)}$. Throughout this book n is an unobservable variable governing the observations and the asymptotics.

If we allow the observation times to be arbitrary stopping times which may depend on the process X in an unspecified way it is very difficult to derive certain results. For this reason we will restrict ourselves very often to the setting of *exogenous* observation times i.e. observation times that are independent of the process X. Although this assumption seems to be rarely justified in practical applications it covers various interesting models and is far more general than the setting of equidistant and synchronous observation times usually considered in the literature. A more precise definition for exogeneity of the observation scheme is stated in the following.

Definition 1.2. *Let* $\mathcal{S} = \sigma(\{\pi_n : n \in \mathbb{N}\})$ *denote the σ-algebra generated by the observation scheme and* $\mathcal{X} = \sigma(X, b, \sigma, \delta, W, \mu)$ *denote the σ-algebra generated by the process X and its components. We call an observation scheme* $(\pi_n)_{n \in \mathbb{N}}$ *exogenous if the observation scheme and the process X are independent, i.e. if* $\mathcal{S}$ *and* $\mathcal{X}$ *are independent.*   □

To derive the upcoming results we have to impose some mild structural assumptions on the Itô semimartingale (1.1) and the observation scheme, compare assumption (H) in [30] or [32], and we have to assume that the mesh $|\pi_n|_T$ vanishes as $n \to \infty$ because this is the property characterizing the asymptotics in the high-frequency setting. In particular if $|\pi_n|_T$ does not vanish it is impossible to consistently infer on properties of the model like the jump behaviour which only become visible when considering the whole path in continuous time. These assumptions are summarized in the following condition.

Condition 1.3. *The process* $(b_t)_{t\geq0}$ *is locally bounded and the processes* $(\sigma_t^{(1)})_{t\geq0}$, $(\sigma_t^{(2)})_{t\geq0}$, $(\rho_t)_{t\geq0}$ *are càdlàg. Furthermore, there exists a locally bounded process* Γ_t *with* $\|\delta(t,z)\| \leq \Gamma_t \gamma(z)$ *almost surely for some deterministic bounded function* γ *which satisfies* $\int (\gamma(z)^2 \wedge 1)\lambda(dz) < \infty$. *The sequence of observation schemes* $(\pi_n)_{n\in\mathbb{N}}$ *fulfils*

$$|\pi_n|_T \xrightarrow{\mathbb{P}} 0 \quad (n \to \infty).$$

In the decomposition (1.1) used in the definition of X small jumps and large jumps are distinguished by the property of whether or not their absolute size is less or equal respectively larger than 1. For $q > 0$ we additionally introduce the decomposition $X_t = X_0 + B(q)_t + C_t + M(q)_t + N(q)_t$ of X, compare Appendix A in [8], where

$$B(q)_t = \int_0^t \left(b_s - \int_{\mathbb{R}^2} (\delta(s,z)\mathbb{1}_{\{\|\delta(s,z)\|\leq1\}} - \delta(s,z)\mathbb{1}_{\{\gamma(z)\leq1/q\}})\lambda(dz)\right)ds,$$

$$C_t = \int_0^t \sigma_s dW_s,$$

$$M(q)_t = \int_0^t \int_{\mathbb{R}^2} \delta(s,z)\mathbb{1}_{\{\gamma(z)\leq1/q\}}(\mu - \nu)(ds, dz), \tag{1.7}$$

$$N(q)_t = \int_0^t \int_{\mathbb{R}^2} \delta(s,z)\mathbb{1}_{\{\gamma(z)>1/q\}}\mu(ds, dz).$$

Here q is a parameter which controls whether jumps are classified as small jumps or big jumps. If the parameter q becomes larger, then less jumps are classified as small jumps and more jumps are classified as large jumps. Here, the process $N(q)$ has almost surely only finitely many jumps in each compact time interval $[0,t]$, $t \geq 0$; compare Lemma 2.1.7 a) in [30]. A lot of results are easier to prove for processes of finite jump activity. Therefore a common strategy in the upcoming chapters will be to first prove the results where we only consider $N(q)$ instead of $M(q) + N(q)$ and then in a second step to show that the asymptotic contribution of $M(q)$ vanishes as $q \to \infty$.

A key observation which is especially important when disentangling asymptotic contributions of the continuous part of X and its jump part is that moments of the increments of $B(q)$, C, $M(q)$, $N(q)$ scale differently with the corresponding interval length. Their specific behaviour is summarized in the following lemma which will be used repeatedly in the upcoming proofs.

Lemma 1.4. *If Condition 1.3 is fulfilled and the processes b_t, σ_t and Γ_t are bounded there exist constants $K_p, K_{p'}, K_{p,q}, \widetilde{K}_{p,q}, e_q \geq 0$ such that*

$$\|B(q)_{s+t} - B(q)_s\|^p \leq K_{p,q} t^p, \tag{1.8}$$

$$\mathbb{E}\big[\|C_{s+t} - C_s\|^p\big|\mathcal{F}_s\big] \leq K_p t^{\frac{p}{2}}, \tag{1.9}$$

$$\mathbb{E}\big[\|M(q)_{s+t} - M(q)_s\|^p\big|\mathcal{F}_s\big] \leq K_p t^{\frac{p}{2}\wedge 1}(e_q)^{\frac{p}{2}\wedge 1}, \tag{1.10}$$

$$\mathbb{E}\big[\|N(q)_{s+t} - N(q)_s\|^{p'}\big|\mathcal{F}_s\big] \leq \widetilde{K}_{p',q} t + K_{p',q} t^{p'}, \tag{1.11}$$

$$\mathbb{E}\big[\|X_{s+t} - X_s\|^p\big|\mathcal{F}_s\big] \leq K_p t^{\frac{p}{2}\wedge 1}, \tag{1.12}$$

for all $s, t \geq 0$ with $s + t \leq T$ and all $q > 0$, $p \geq 0$, $p' \geq 1$. Here, e_q can be chosen such that $e_q \to 0$ for $q \to \infty$. For $p' \geq 2$ the constant $\widetilde{K}_{p',q}$ may be chosen independently of q.

Throughout the proofs in this book to simplify notation we denote by K and K_a generic constants, the latter dependent on some variable a. This means that e.g. statements like $K = 2K$ or $K = K^2$ may occur. In fact the numeric value of these constants will never be of importance.

The proof of Lemma 1.4 will be given in Appendix A. Lemma 1.4 can be used to bound moments of increments over the observation intervals $\mathcal{I}_{i,n}^{(l)}$ if the observation scheme is deterministic and also if the observation times are exogenous. In the second case we can work conditionally on $\mathcal{S}$ and then apply the inequalities above for conditional expectations with respect to the σ-algebra $\sigma(\mathcal{F}_s, \mathcal{S})$. If we consider endogenous observation times we require more general results. These are stated in the following Lemma which will be proved in Appendix A as well.

Lemma 1.5. *If Condition 1.3 is fulfilled and the processes b_t, σ_t and Γ_t are bounded there exist constants $K, K_p, K_{p,q}$ such that*

$$\sup_{t\in(S,S']} \|B(q)_t - B(q)_S\|^p \leq K_{p,q}(S' - S)^p, \tag{1.13}$$

$$\mathbb{E}\big[\sup_{t\in(S,S']} \|C_t - C_S\|^p\big|\mathcal{F}_S\big] \leq K_p\mathbb{E}[(S' - S)^{p/2}|\mathcal{F}_S], \tag{1.14}$$

$$\mathbb{E}\big[\sup_{t\in(S,S']} \|M(q)_t - M(q)_S\|^2 \big| \mathcal{F}_S\big] \leq Ke_q \mathbb{E}[(S'-S)|\mathcal{F}_S], \qquad (1.15)$$

$$\mathbb{E}\big[\sup_{t\in(S,S']} \|X_t - X_S\|^2 \big| \mathcal{F}_S\big] \leq K\mathbb{E}[(S'-S)|\mathcal{F}_S], \qquad (1.16)$$

for all stopping times $0 \leq S \leq S'$ and all $q > 0$, $p \geq 1$. Here, e_q can be chosen such that $e_q \to 0$ for $q \to \infty$.

Throughout the proofs in the upcoming chapters we will without further notice assume that the processes b_t, $\sigma_t^{(1)}$, $\sigma_t^{(2)}$, ρ_t and Γ_t are bounded on $[0, T]$. This assumption e.g. allows to directly apply the above lemmata. The processes b_t, $\sigma_t^{(1)}$, $\sigma_t^{(2)}$, ρ_t and Γ_t are all locally bounded by Condition 1.3 and a localization procedure then shows that the results for bounded processes can be carried over to the case of locally bounded processes. See Section 4.4.1 in [30] for a detailed proof of the validity of this argument.

2 Laws of Large Numbers

In the setting of high-frequency statistics for stochastic processes, the information contained in the observed data $\{X^{(l)}_{t^{(l)}_{i,n}} : t^{(l)}_{i,n} \leq T\}$, $l = 1, 2$, is also (almost) equivalently stored in the increments $\{\Delta^{(l)}_{i,n} X^{(l)} : t^{(l)}_{i,n} \leq T\}$, $l = 1, 2$. Note that to fully recover the observed data from the increments additionally only the starting values $X^{(l)}_{t^{(l)}_{0,n}}$, $l = 1, 2$, would be needed. However in many applications like volatility estimation or the examination of the jump behaviour the properties of interest are invariant under constant shifts of the processes $X^{(l)}$, $l = 1, 2$. Hence the information contained in the starting values $X^{(l)}_{t^{(l)}_{0,n}}$, $l = 1, 2$, is often not relevant and it is sufficient to work with statistics which are based on the increments $\{\Delta^{(l)}_{i,n} X^{(l)} : t^{(l)}_{i,n} \leq T\}$, $l = 1, 2$.

A very prominent class of such statistics is given by sums of certain transformations of the increments. Suppose for the moment that the observation scheme is synchronous and denote

$$t_{i,n} := t^{(1)}_{i,n} = t^{(2)}_{i,n}, \quad \Delta_{i,n} X := X_{t_{i,n}} - X_{t_{i-1,n}}. \tag{2.1}$$

It is a classic result in stochastic analysis, compare e.g. Theorem 23 in [41], that

$$\sum_{i: t_{i,n} \leq T} \Delta_{i,n} X^{(k)} \Delta_{i,n} X^{(l)} \xrightarrow{\mathbb{P}} [X^{(k)}, X^{(l)}]_T, \quad k, l \in \{1, 2\}. \tag{2.2}$$

As a generalization of (2.2) the asymptotics of statistics of the form

$$\sum_{i: t_{i,n} \leq T} f(\Delta_{i,n} X) \tag{2.3}$$

for various functions $f \colon \mathbb{R}^2 \to \mathbb{R}$ have been investigated; compare e.g. Chapter 3 of [30]. Applications of such statistics include the construction of tests for the presence of (common) jumps in [2] and [32].

In this chapter, we discuss how statistics similar to (2.3) can be generalized to the setting of asynchronous observation times. Further we investigate the asymptotics of these generalized statistics and we will develop (weak) laws of large numbers for them if possible. We will compare those results to the results obtained in the setting of synchronous observation times, learn where the results differ and what

© Springer Fachmedien Wiesbaden GmbH, part of Springer Nature 2019
O. Martin, *High-Frequency Statistics with Asynchronous and Irregular Data*,
Mathematische Optimierung und Wirtschaftsmathematik | Mathematical Optimization
and Economathematics, https://doi.org/10.1007/978-3-658-28418-3_2

kind of extra assumptions have to be made in the case of asynchronous observations to obtain results which match those in the setting of synchronous observation times.

In Section 2.1 we investigate the asymptotics of non-normalized functionals, i.e. statistics of the form (2.3), and in Section 2.2 we discuss normalized functionals, i.e. statistics of the form $n^{p-1} \sum_{t_{i,n} \leq T} f(\Delta_{i,n} X)$ for some $p \geq 0$ which is related to f. While the limits of the non-normalized functionals usually only depend on the jump part of X the limits of normalized functionals usually only depend on the continuous part of X. In Sections 2.3 and 2.4 we will discuss further extensions of the non-normalized and normalized functionals discussed in Sections 2.1 and 2.2.

2.1 Non-Normalized functionals

First note that when considering functionals of the form (2.3) in the setting of asynchronous observation times it is not straightforward anymore for which pairs of increments $(\Delta_{i,n}^{(1)} X^{(1)}, \Delta_{j,n}^{(2)} X^{(2)})$ the evaluation of the function f should be included in the sum. An idea utilized by [22] is to include $f(\Delta_{i,n}^{(1)} X^{(1)}, \Delta_{j,n}^{(2)} X^{(2)})$ if and only if the observation intervals $\mathcal{I}_{i,n}^{(1)}, \mathcal{I}_{j,n}^{(2)}$ overlap. In this case a consistent estimator for the quadratic covariation of $[X^{(1)}, X^{(2)}]_T$ is obtained by using the function $f(x_1, x_2) = x_1 x_2$ also in the setting of asynchronous observations, i.e. they showed

$$\sum_{i,j:t_{i,n}^{(1)} \vee t_{j,n}^{(2)} \leq T} \Delta_{i,n}^{(1)} X^{(1)} \Delta_{j,n}^{(2)} X^{(2)} \mathbb{1}_{\{\mathcal{I}_{i,n}^{(1)} \cap \mathcal{I}_{j,n}^{(2)} \neq \emptyset\}} \xrightarrow{\mathbb{P}} [X^{(1)}, X^{(2)}]_T \quad (2.4)$$

in the case of a continuous Itô semimartingale X and for exogenous observation times. A corresponding result which also holds for endogenous observations was later given in [21] and the extension to processes including jumps has been derived in [8]. The structure of the sum is illustrated in Figure 2.1.

In the style of this famous *Hayashi-Yoshida estimator for the quadratic covariation* we define

$$V(f, \pi_n)_T = \sum_{i,j:t_{i,n}^{(1)} \vee t_{j,n}^{(2)} \leq T} f(\Delta_{i,n}^{(1)} X^{(1)}, \Delta_{j,n}^{(2)} X^{(2)}) \mathbb{1}_{\{\mathcal{I}_{i,n}^{(1)} \cap \mathcal{I}_{j,n}^{(2)} \neq \emptyset\}}$$

for functions $f: \mathbb{R}^2 \to \mathbb{R}$. We will see that these functionals converge to similar limits as the functionals (2.3) in the setting of synchronous observation times for a large class of functions f, and not only for $f(x_1, x_2) = x_1 x_2$ as in the case of the Hayashi-Yoshida estimator.

Further we define

$$V^{(l)}(g, \pi_n)_T = \sum_{i:t_{i,n}^{(l)} \leq T} g(\Delta_{i,n}^{(l)} X^{(l)}), \quad l = 1, 2,$$

for functions $g\colon \mathbb{R} \to \mathbb{R}$. We will also state an asymptotic result for $V^{(l)}(g, \pi_n)_T$ to compare the results in the setting of asynchronously observed bivariate processes to those in simpler settings and also because such a result will be used in Chapter 8.

To describe the limits of the functionals $V(f, \pi_n)_T$ and $V^{(l)}(g, \pi_n)_T$ we denote

$$B(f)_T = \sum_{s \leq T} f(\Delta X_s^{(1)}, \Delta X_s^{(2)}),$$

$$B^*(f)_T = \sum_{s \leq T} f(\Delta X_s^{(1)}, \Delta X_s^{(2)}) \mathbb{1}_{\{\Delta X_s^{(1)} \Delta X_s^{(2)} \neq 0\}}, \tag{2.5}$$

$$B^{(l)}(g)_T = \sum_{s \leq T} g(\Delta X_s^{(l)}), \quad l = 1, 2,$$

for functions $f\colon \mathbb{R}^2 \to \mathbb{R}$ and $g\colon \mathbb{R} \to \mathbb{R}$ for which the sums are well-defined.

2.1.1 The Results

In the setting of synchronous observation times (2.1) the functional $V(f, \pi_n)_T$ coincides with the classical statistic (2.3) and hence the convergence of $V(f, \pi_n)_T$ in this situation follows from Theorem 3.3.1 of [30].

Theorem 2.1. *Suppose that the observation scheme is synchronous and that Condition 1.3 holds. Then we have*

$$V(f, \pi_n)_T \xrightarrow{\mathbb{P}} B(f)_T \tag{2.6}$$

for all continuous functions $f\colon \mathbb{R}^2 \to \mathbb{R}$ with $f(x) = o(\|x\|^2)$ as $x \to 0$.

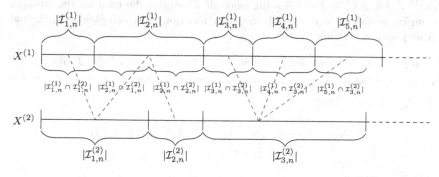

Figure 2.1: All products of increments of $X^{(1)}$ and $X^{(2)}$ over intersecting intervals enter the estimation of $[X^{(1)}, X^{(2)}]_T$.

Actually the statement in Theorem 3.3.1 of [30] holds for general d-dimensional Itô semimartingales and all functions $f\colon \mathbb{R}^d \to \mathbb{R}$ with the property $f(x) = o(\|x\|^2)$ for any $d \in \mathbb{N}$. The case $d = 1$ then yields the convergence for the functionals $V^{(l)}(g, \pi_n)_T$ stated in the following corollary.

Corollary 2.2. *Under Condition 1.3 we have*

$$V^{(l)}(g, \pi_n)_T \xrightarrow{\mathbb{P}} B^{(l)}(g)_T$$

for all continuous functions $g\colon \mathbb{R} \to \mathbb{R}$ with $g(x) = o(x^2)$ as $x \to 0$.

The following theorem states the most general result which can be obtained if the convergence in (2.6) is supposed to hold for arbitrary Itô semimartingales and any asynchronous observation scheme.

Theorem 2.3. *Under Condition 1.3 we have*

$$V(f, \pi_n)_T \xrightarrow{\mathbb{P}} B^*(f)_T \tag{2.7}$$

for all continuous functions $f\colon \mathbb{R}^2 \to \mathbb{R}$ with $f(x, y) = O(x^2 y^2)$ as $|xy| \to 0$.

As for the convergence in Theorem 2.1 in the setting of synchronous observation times we need that $f(x, y)$ vanishes as $(x, y) \to 0$ also in the setting of asynchronous observation times. However in the asynchronous setting we further need $f(x_k, y_k) \to 0$ also for sequences $(x_k, y_k)_{k \in \mathbb{N}}$ which do not converge to zero, but which fulfil $|x_k y_k| \to 0$. Hence the condition on f needed to obtain convergence of $V(f, \pi_n)_T$ in the asynchronous setting is stronger compared to the corresponding condition in the synchronous setting. Further we observe that in the asynchronous setting the limit only consists of common jumps of $X^{(1)}$ and $X^{(2)}$ i.e. jumps with $\Delta X_s^{(1)} \neq 0 \neq \Delta X_s^{(2)}$. The following example illustrates the need for the stronger condition as well as why we only consider functions f which yield a limit that consists only of common jumps.

Example 2.4. *Consider the function $f_{3,0}(x, y) = x^3$, which fulfils $f_{3,0}(x, y) \to 0$ as $(x, y) \to 0$ but not as $|xy| \to 0$, and the observation scheme given by $t_{i,n}^{(1)} = i/n$ and $t_{i,n}^{(2)} = i/(2n)$. Then*

$$V(f_{3,0}, \pi_n)_T = 2 \sum_{i: t_{i,n}^{(1)} \leq T} (\Delta_{i,n}^{(1)} X^{(1)})^3 \xrightarrow{\mathbb{P}} 2 B(f_{3,0})_T$$

where the convergence is due to Corollary 2.2. However for the standard synchronous observation scheme $t_{i,n}^{(1)} = t_{i,n}^{(2)} = i/n$ we have $V(f_{3,0}, \pi_n)_T \xrightarrow{\mathbb{P}} B(f_{3,0})_T$ also *due to Corollary 2.2. Hence the limit here depends on the observation scheme. If we further consider the observation scheme with $t_{i,n}^{(1)} = i/n$ and*

$$t_{i,n}^{(2)} = \begin{cases} i/n, & n \text{ even,} \\ i/(2n), & n \text{ odd,} \end{cases}$$

then $V(f_{3,0}, \pi_n)_T$ does not converge at all unless $B(f_{3,0})_T = 0$, as one subsequence converges to $B(f_{3,0})_T$ and the other one to $2B(f_{3,0})_T$. Hence there cannot exist a convergence result for $V(f_{3,0}, \pi_n)_T$ which holds for any Itô semimartingale X and any sequence of observation schemes π_n, $n \in \mathbb{N}$.

If we consider instead a function $f(x,y)$ that vanishes as $|xy| \to 0$ such a behaviour cannot occur because idiosyncratic jumps do not contribute in the limit as e.g. for $\Delta X_s^{(1)} \neq 0$ and $\Delta X_s^{(2)} = 0$ we have

$$\sup_{(i,j):s\in\mathcal{I}_{i,n}^{(1)},\,\mathcal{I}_{i,n}^{(1)}\cap\mathcal{I}_{j,n}^{(2)}\neq\emptyset} |\Delta_{i,n}^{(1)} X^{(1)} \Delta_{j,n}^{(2)} X^{(2)}| \xrightarrow{\mathbb{P}} 0.$$

If on the other hand there is a common jump at time s there is only one summand $f(\Delta_{i,n}^{(1)} X^{(1)}, \Delta_{j,n}^{(2)} X^{(2)})$ such that $s \in \mathcal{I}_{i,n}^{(1)}$ and $s \in \mathcal{I}_{j,n}^{(2)}$. Hence common jumps only enter the limit once.

This example shows that the assumption that $f(x,y)$ vanishes as $|xy| \to 0$ is needed to filter out the contribution of idiosyncratic jumps. These jumps may enter $V(f, \pi_n)_T$ multiple times, where the multiplicity by which they occur may depend on n and ω and therefore may prevent $V(f, \pi_n)_T$ from converging. □

Let us now consider the order by which the function $f(x,y)$ has to decrease as $(x,y) \to 0$ or, respectively, $|xy| \to 0$. We observe that in the asynchronous setting the function f has to decrease quadratically in both x and y while in the synchronous setting it only has to decrease quadratically in (x,y). Adding this condition to the requirement that $f(x_k, y_k)$ has to vanish for any sequence with $|x_k y_k| \to 0$ further diminishes the class of functions f for which $V(f, \pi_n)_T$ converges in the asynchronous setting compared to the synchronous one. The need for this stronger condition on f is due to the fact that the lengths of the observation intervals of $X^{(1)}$ and $X^{(2)}$ may decrease with different rates in the asynchronous setting which is illustrated in the following example.

Example 2.5. *Let $X_t^{(1)} = \mathbb{1}_{\{t \geq U\}}$ for $U \sim \mathcal{U}[0,1]$ and $X^{(2)}$ be a standard Brownian motion independent of U. The observation schemes are given by*

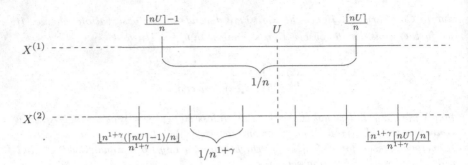

Figure 2.2: Observation times of $X^{(1)}$ and $X^{(2)}$ around the jump time U.

$t_{i,n}^{(1)} = i/n$ and $t_{i,n}^{(2)} = i/n^{1+\gamma}$ with $\gamma > 0$. Then for $f(x,y) = |x|^{p_1}|y|^{p_2}$ as illustrated in Figure 2.2 we have

$$V(f,\pi_n)_1 = \sum_{i=\lfloor n^{1+\gamma}(\lceil nU \rceil-1)/n \rfloor/n^{1+\gamma}+1}^{\lceil n^{1+\gamma}\lceil nU \rceil/n \rceil/n^{1+\gamma}} \left| X_{i/n^{1+\gamma}}^{(2)} - X_{(i-1)/n^{1+\gamma}}^{(2)} \right|^{p_2}$$

$$\geq \sum_{i=1}^{\lceil n^{\gamma}\rceil} \left| n^{-(1+\gamma)/2} Z_i^n \right|^{p_2}$$

$$= n^{-(1+\gamma)p_2/2+\gamma} \left(n^{-\gamma} \sum_{i=1}^{\lceil n^{\gamma}\rceil} |Z_i^n|^{p_2} \right)$$

where $Z_i^n := n^{(1+\gamma)/2} \Delta_{\lfloor n^{1+\gamma}(\lceil nU \rceil-1)/n \rfloor/n^{1+\gamma}+i/n^{1+\gamma},n}^{(2)} X^{(2)}$, $i = 1,\ldots\lceil n^{\gamma}\rceil$, are i.i.d. standard normal random variables for each $n \in \mathbb{N}$. Hence $V(f,\pi_n)_1$ diverges for $p_2 < 2$ if

$$\gamma > \frac{p_2}{2-p_2}$$

because the expression in parantheses converges in probability to $\mathbb{E}[|Z|^{p_2}]$ for some $Z \sim \mathcal{N}(0,1)$, by the law of large numbers. Here we are able to find a suitably large γ explicitly because the p_2-variations of a Brownian motion are infinite for $p_2 < 2$. But we also have $B^*(f)_1 = 0$ in this setting because $X^{(2)}$ is continuous. Hence (2.7) cannot hold for $f(x,y) = |x|^{p_1}|y|^{p_2}$, any Itô semimartingale of the form (1.1) and any observation scheme which fulfils Condition 1.3 if $p_1 \wedge p_2 < 2$. □

Example 2.5 shows that the convergence (2.7) fails for functions $f(x,y) = |x|^{p_1}|y|^{p_2}$ with $p_1 \wedge p_2 < 2$ in combination with observation schemes where the observation frequency for one process increases much faster as $n \to \infty$ than the observation frequency of the other process. If we consider only observation schemes where such a behaviour is prohibited, we can also obtain the convergence in (2.7) for functions $f(x,y) = |x|^{p_1}|y|^{p_2}$ with $p_1 \wedge p_2 < 2$.

First, we state a result in the case of exogenous observation times introduced in Definition 1.2, i.e. random observation times that do not depend on the process X or its components.

Theorem 2.6. *Assume that Condition 1.3 holds and that the observation scheme is exogenous. Further let $p_1, p_2 > 0$ with $p_1 + p_2 \geq 2$. If we have*

$$\sum_{i,j:t_{i,n}^{(1)} \vee t_{j,n}^{(2)} \leq T} \left|\mathcal{I}_{i,n}^{(1)}\right|^{\frac{p_1}{2} \wedge 1} \left|\mathcal{I}_{j,n}^{(2)}\right|^{\frac{p_2}{2} \wedge 1} \mathbb{1}_{\{\mathcal{I}_{i,n}^{(1)} \cap \mathcal{I}_{j,n}^{(2)} \neq \emptyset\}} = O_{\mathbb{P}}(1) \qquad (2.8)$$

as $n \to \infty$ it holds

$$V(f, \pi_n)_T \xrightarrow{\mathbb{P}} B^*(f)_T \qquad (2.9)$$

for all continuous functions $f \colon \mathbb{R}^2 \to \mathbb{R}$ with $f(x,y) = o(|x|^{p_1}|y|^{p_2})$ as $|xy| \to 0$.

In the boundary case $p_1 + p_2 = 2$ in Theorem 2.6 we achieve convergence for all functions f that are for $|xy| \to 0$ dominated by the function $|x|^{p_1}|y|^{p_2}$ which is of order $p_1 + p_2 = 2$. Hence Theorem 2.6 allows to achieve the convergence in (2.9) for functions f which are dominated by functions of the same order as the dominating function $\|(x,y)\|^2$ in the synchronous case in Theorem 2.1. However, the requirement that $f(x,y)$ vanishes as $|xy| \to 0$ cannot be relaxed because this assumption is as illustrated in Example 2.4 fundamentally necessary due to the asynchronous nature of the observation scheme.

Like in the synchronous setting we cannot have the general convergence in (2.9) for functions f which do not fulfil $f(x,y) = O(\|(x,y)\|^2)$ as $(x,y) \to 0$, because in this case $B^*(f)$ might not be well defined. An indication for this fact is also given by the observation that (2.8) can never be fulfilled if $p_1 + p_2 < 2$ and $|\pi_n|_T \to 0$ as shown in the following remark.

Remark 2.7. *Suppose that we have $p_1 + p_2 < 2$ and $|\pi_n|_T \to 0$. In this situation we obtain the following estimate for the left-hand side of (2.8), using $(p_l/2) \wedge 1 = p_l/2$, $l = 1, 2$, and the inequality $\sum_{i=1}^{N} a_i^p \geq (\sum_{i=1}^{N} a_i)^p$, which holds for all $N \in \mathbb{N}$, $a_i \geq 0$, $p \in [0, 1)$:*

$$\sum_{i,j:t_{i,n}^{(1)} \vee t_{j,n}^{(2)} \leq T} \left|\mathcal{I}_{i,n}^{(1)}\right|^{\frac{p_1}{2} \wedge 1} \left|\mathcal{I}_{j,n}^{(2)}\right|^{\frac{p_2}{2} \wedge 1} \mathbb{1}_{\{\mathcal{I}_{i,n}^{(1)} \cap \mathcal{I}_{j,n}^{(2)} \neq \emptyset\}}$$

$$\geq \sum_{i:t_{i,n}^{(1)} \leq T} \left|\mathcal{I}_{i,n}^{(1)}\right|^{\frac{p_1}{2}} \left(\sum_{j:t_{j,n}^{(2)} \leq T} \left|\mathcal{I}_{j,n}^{(2)}\right| \mathbb{1}_{\{\mathcal{I}_{i,n}^{(1)} \cap \mathcal{I}_{j,n}^{(2)} \neq \emptyset\}} \right)^{\frac{p_2}{2}}$$

$$\geq \sum_{i:t_{i,n}^{(1)}\leq T} |\mathcal{I}_{i,n}^{(1)}|^{\frac{p_1}{2}} |\mathcal{I}_{i,n}^{(1)}|^{\frac{p_2}{2}} - O((|\pi_n|_T)^{\frac{p_1+p_2}{2}})$$

$$\geq (|\pi_n|_T)^{\frac{p_1+p_2}{2}-1} T - O((|\pi_n|_T)^{\frac{p_1+p_2}{2}}).$$

Here the expression in the last line converges in probability to infinity due to $p_1 + p_2 < 2$ and $|\pi_n|_T \overset{\mathbb{P}}{\longrightarrow} 0$.

On the other hand the condition (2.8) is always fulfilled if $p_1 \wedge p_2 \geq 2$ holds. Indeed in that case we obtain

$$\sum_{i,j:t_{i,n}^{(1)}\vee t_{j,n}^{(2)}\leq T} |\mathcal{I}_{i,n}^{(1)}|^{\frac{p_1}{2}\wedge 1} |\mathcal{I}_{j,n}^{(2)}|^{\frac{p_2}{2}\wedge 1} \mathbb{1}_{\{\mathcal{I}_{i,n}^{(1)}\cap\mathcal{I}_{j,n}^{(2)}\neq\emptyset\}}$$

$$\leq K \sum_{i,j:t_{i,n}^{(1)}\vee t_{j,n}^{(2)}\leq T} |\mathcal{I}_{i,n}^{(1)}||\mathcal{I}_{j,n}^{(2)}| \mathbb{1}_{\{\mathcal{I}_{i,n}^{(1)}\cap\mathcal{I}_{j,n}^{(2)}\neq\emptyset\}} \leq 3K|\pi_n|_T T.$$

Hence condition (2.8) is by Condition 1.3 always fulfilled in the setting of Theorem 2.3.   □

Example 2.8. *Let $p \in [1,2)$ and consider the deterministic sampling scheme given by $t_{i,n}^{(1)} = i/n$ and $t_{i,n}^{(2)} = i/n^{1+\gamma}$ from Example 2.5 with $\gamma = \gamma(p) = \frac{2p-2}{2-p}$. In this case it holds*

$$\sum_{i,j:t_{i,n}^{(1)}\vee t_{j,n}^{(2)}\leq T} (|\mathcal{I}_{i,n}^{(1)}||\mathcal{I}_{j,n}^{(2)}|)^{\frac{p}{2}} \mathbb{1}_{\{\mathcal{I}_{i,n}^{(1)}\cap\mathcal{I}_{j,n}^{(2)}\neq\emptyset\}} = Tn^{1+\gamma(p)}\left(\frac{1}{n}\frac{1}{n^{1+\gamma(p)}}\right)^{\frac{p}{2}}(1+o(1))$$

$$= Tn^{1+\gamma(p)-(2+\gamma(p))\frac{p}{2}}(1+o(1)) = O(1).$$

Hence if we want (2.9) to hold for all functions f with $f(x,y) = o(|xy|^p)$ we can allow for observation schemes where the observation frequencies differ by a factor of up to $n^{\gamma(p)}$ where $\gamma(p)$ increases in p. For $p = 1$ we have $\gamma(1) = 0$ and for $p \to 2$ we have $\gamma(p) \to \infty$.   □

In general we observe that if the $o(|x|^{p_1}|y|^{p_2})$-restriction on f is less restrictive, then the restriction (2.8) on the observation scheme has to be more restrictive, and vice versa. Here the abstract criterion (2.8) characterizing the allowed classes of observation schemes can be related, as illustrated in Example 2.8, to the asymptotics of the ratio of the observation frequencies of the two processes. Hence if the observation frequency of one process increases much faster than the observation frequency of the other process we obtain the convergence in (2.9) only for a small class of functions f.

Example 2.9. *Condition (2.8) is fulfilled for any $p_1, p_2 \geq 0$ with $p_1 + p_2 = 1$ in the case where the observation times $\{t_{i,n}^{(l)} : i \in \mathbb{N}\}$, $l = 1, 2$, are given by the jump* times of two independent time-homogeneous Poisson processes with intensities $n\lambda_1, n\lambda_2$. Indeed in that situation Corollary 5.6 yields

$$\sum_{i,j:t_{i,n}^{(1)}\vee t_{j,n}^{(2)}\leq t} \left|\mathcal{I}_{i,n}^{(1)}\right|^{\frac{p_1}{2}} \left|\mathcal{I}_{j,n}^{(2)}\right|^{\frac{p_2}{2}} \mathbb{1}_{\{\mathcal{I}_{i,n}^{(1)}\cap\mathcal{I}_{j,n}^{(2)}\neq\emptyset\}} \xrightarrow{\mathbb{P}} ct, \quad t\geq 0,$$

for some positive real number $c > 0$. Further note that if (2.8) is fulfilled for $p_1, p_2 \geq 0$ it is clearly also fulfilled for any $p_1' \geq p_1, p_2' \geq p_2$. $\qquad\square$

Next, we give a result that may also be applied in a setting with endogenous observation times.

Theorem 2.10. *Assume that Condition 1.3 holds. If for all $\varepsilon > 0$ there exists some $N_\varepsilon \in \mathbb{N}$ with*

$$\limsup_{n\to\infty} \mathbb{P}\left(\sup_{i:t_{i,n}^{(l)}\leq T} \sum_{j\in\mathbb{N}} \mathbb{1}_{\{\mathcal{I}_{i,n}^{(l)}\cap\mathcal{I}_{j,n}^{(3-l)}\neq\emptyset\}} > N_\varepsilon, \ l=1,2 \right) < \varepsilon, \qquad (2.10)$$

then it holds

$$V(f,\pi_n)_T \xrightarrow{\mathbb{P}} B^*(f)_T$$

for all continuous functions $f: \mathbb{R}^2 \to \mathbb{R}$ such that $f(x,y) = o(|x|^{p_1}|y|^{p_2})$ as $|xy| \to 0$ for some $p_1, p_2 \geq 0$ with $p_1 + p_2 = 2$.

Here (2.10) ensures that as n tends to infinity the maximal number of observations of the process $X^{(3-l)}$ during one observation interval of $X^{(l)}$ is bounded. This yields that the ratio of the observation frequencies of the two processes is also bounded as $n \to \infty$ and cannot tend to infinity as in Example 2.5.

Example 2.11. *Consider the case where $X^{(1)}$ and $X^{(2)}$ are observed alternately. In this case we have*

$$\sum_{j\in\mathbb{N}} \mathbb{1}_{\{\mathcal{I}_{i,n}^{(l)}\cap\mathcal{I}_{j,n}^{(3-l)}\}} \leq 2$$

for $l = 1, 2$ and all i, n.

Note that, although this goes along with a data reduction, the statistician may always use only a subset of all available observations and hence is able to turn the real observation scheme for example into an obervation scheme where the processes are observed alternately. One way to achieve this is the following: Start with the first observation time $t_{i,n}^{(1)}$ of $X^{(1)}$ and set $\tilde{t}_{1,n}^{(1)} = t_{1,n}^{(1)}$, then take the smallest observation time of $X^{(2)}$ larger than $\tilde{t}_{i,n}^{(1)}$ and set $t_{1,n}^{(2)} = \inf\{t_{i,n}^{(2)}|t_{i,n}^{(2)} > \tilde{t}_{1,n}^{(1)}\}$. Further set $\tilde{t}_{2,n}^{(1)} = \inf\{t_{i,n}^{(1)}|t_{i,n}^{(1)} > \tilde{t}_{1,n}^{(2)}\}$ and define recursively the new observation scheme $\tilde{\pi}_n$ by continuing this procedure in the natural way. $\qquad\square$

Remark 2.12. *It can be shown that Theorems 2.1, 2.3, 2.6 and 2.10 do not just hold for continuous functions $f\colon \mathbb{R}^2 \to \mathbb{R}$ but also for functions $f\colon \mathbb{R}^2 \to \mathbb{R}$ which are only discontinuous at points $(x, y) \in \mathbb{R}^2$ for which almost surely no jump ΔX_s is realized in $[0, T]$ with $\Delta X_s = (x, y)$. Hence if the jump measure admits a Lebesgue-density the convergences in Theorems 2.1, 2.3, 2.6 and 2.10 remain valid for all functions f where the set of all discontinuity points is a Lebesgue null set. For a more precise formulation of the above claim see Theorem 3.3.5 in [30]. In this section, we restricted ourselves to the case of continuous f because all functions for which these results will be applied in Part II are continuous and to keep the the notation and the proofs clearer.* □

2.1.2 The Proofs

As a preparation for the proof of Theorem 2.3 we prove (2.7) for functions f that vanish in a neighbourhood of the two axes $\{(x, y) \in \mathbb{R}^2 | xy = 0\}$ in Lemma 2.13 and for the function $f^*\colon \mathbb{R}^2 \to \mathbb{R}, (x, y) \mapsto x^2 y^2$ in Lemma 2.14.

Lemma 2.13. *Under Condition 1.3 we have*

$$V(f, \pi_n)_T \xrightarrow{\mathrm{P}} B^*(f)_T$$

for all continuous functions $f\colon \mathbb{R}^2 \to \mathbb{R}$ for which some $\rho > 0$ exists such that $f(x, y) = 0$ whenever $(x', y') \in \{(x, y) \in \mathbb{R}^2 : |x'y'| < \rho\}$.

Proof. The following arguments are similar to the proof of Lemma 3.3.7 in [30]: Note that as X is càdlàg there can only exist countably many jump times $s \geq 0$ with $|\Delta X_s^{(1)} \Delta X_s^{(2)}| \geq \rho/2$ and in each compact time interval there are only finitely many such jumps. Denote by $(S_p)_{p \in \mathbb{N}}$ an enumeration of those jump times and let

$$\widetilde{X}_t = X_t - \int_0^t \int_{\mathbb{R}^2} \delta(s, z) \mathbb{1}_{\{|\delta^{(1)}(s,z)\delta^{(2)}(s,z)| \geq \rho/2\}} \mu(ds, dz)$$

denote the process X without those jumps. This yields $|\Delta \widetilde{X}_s^{(1)} \Delta \widetilde{X}_s^{(2)}| < \rho/2$ for all $s \in [0, T]$. Hence

$$\limsup_{\theta \to 0} \sup_{0 \leq s_l \leq t_l \leq T - |\pi_n|_T, t_l - s_l \leq \theta, (s_1, t_1] \cap (s_2, t_2] \neq \emptyset} |(\widetilde{X}_{t_1}^{(1)}(\omega) - \widetilde{X}_{s_1}^{(1)}(\omega))(\widetilde{X}_{t_2}^{(2)}(\omega) - \widetilde{X}_{s_2}^{(2)}(\omega))| < \frac{\rho}{2}$$

for all $\omega \in \Omega$. Then there exists $\theta' \colon \Omega \to (0, \infty)$ such that

$$\sup_{0 \leq s_l \leq t_l \leq T, t_l - s_l \leq \theta'(\omega), (s_1, t_1] \cap (s_2, t_2] \neq \emptyset} |(\widetilde{X}_{t_1}^{(1)}(\omega) - \widetilde{X}_{s_1}^{(1)}(\omega))(\widetilde{X}_{t_2}^{(2)}(\omega) - \widetilde{X}_{s_2}^{(2)}(\omega))| < \rho.$$

Denote by $\Omega(n)$ the subset of Ω which is defined as the intersetion of the set $\{|\pi_n|_T \leq \theta'\}$ and the set on which any two different jump times $S_p \neq S_{p'}$ with

$S_p, S_{p'} \leq T$ satisfy $|S_{p'} - S_p| > 2|\pi_n|_T$ and on which $|T - S_p| > |\pi_n|_T$ for any $S_p \leq T$. Then we have

$$V(f, \pi_n)_T \mathbb{1}_{\Omega(n)} = \sum_{p: S_p \leq T - |\pi_n|_T} f\big(\Delta^{(1)}_{i_n^{(1)}(S_p), n} X^{(1)}, \Delta^{(2)}_{i_n^{(2)}(S_p), n} X^{(2)}\big) \mathbb{1}_{\Omega(n)}$$

$$(2.11)$$

where $i_n^{(l)}(s)$ denotes the index of the interval characterized by $s \in \mathcal{I}^{(l)}_{i_n^{(l)}(s), n}$. Further we get from Condition 1.3

$$f\big(\Delta^{(1)}_{i_n^{(1)}(S_p), n} X^{(1)}, \Delta^{(2)}_{i_n^{(2)}(S_p), n} X^{(2)}\big) \mathbb{1}_{\{S_p \leq T\}} \xrightarrow{\mathbb{P}} f\big(\Delta X^{(1)}_{S_p}, \Delta X^{(2)}_{S_p}\big) \mathbb{1}_{\{S_p \leq T\}}$$

for any $p \in \mathbb{N}$ because X is càdlàg and f is continuous. Using this convergence, the fact that there exist almost surely only finitely many $p \in \mathbb{N}$ with $S_p \leq T$ and $\mathbb{P}(\Delta X_T = 0) = 1$ we obtain

$$\sum_{p: S_p \leq T - |\pi_n|_T} f\big(\Delta^{(1)}_{i_n^{(1)}(S_p), n} X^{(1)}, \Delta^{(2)}_{i_n^{(2)}(S_p), n} X^{(2)}\big)$$

$$\xrightarrow{\mathbb{P}} \sum_{s \leq T} f\big(\Delta X^{(1)}_s, \Delta X^{(2)}_s\big) \mathbb{1}_{\{|\Delta X^{(1)}_s \Delta X^{(2)}_s| \geq \rho/2\}} = B^*(f)_T,$$

where the last equality holds because of $f(x, y) = 0$ for $|xy| < \rho$. This yields the claim because of (2.11) and $\mathbb{P}(\Omega(n)) \to 1$ as $n \to \infty$. $\qquad \square$

Lemma 2.14. *Under Condition 1.3 we have*

$$V(f^*, \pi_n)_T \xrightarrow{\mathbb{P}} B^*(f^*)_T \qquad (2.12)$$

for $f^*: \mathbb{R}^2 \to \mathbb{R}, (x, y) \mapsto x^2 y^2$.

Proof. The convergence (2.12) follows from

$$\lim_{q \to \infty} \limsup_{n \to \infty} \mathbb{P}\big(\big| \sum_{i,j: t_{i,n}^{(1)} \vee t_{j,n}^{(2)} \leq T} (\Delta^{(1)}_{i,n} N^{(1)}(q) \Delta^{(2)}_{j,n} N^{(2)}(q))^2 \mathbb{1}_{\{\mathcal{I}^{(1)}_{i,n} \cap \mathcal{I}^{(2)}_{j,n} \neq \emptyset\}} - B^*(f^*)_T \big| > \varepsilon\big) \to 0$$

$$(2.13)$$

and

$$\lim_{q \to \infty} \limsup_{n \to \infty} \mathbb{P}\big(\big| V(f^*, \pi_n)_T - \sum_{i,j: t_{i,n}^{(1)} \vee t_{j,n}^{(2)} \leq T} (\Delta^{(1)}_{i,n} N^{(1)}(q) \Delta^{(2)}_{j,n} N^{(2)}(q))^2 \mathbb{1}_{\{\mathcal{I}^{(1)}_{i,n} \cap \mathcal{I}^{(2)}_{j,n} \neq \emptyset\}} \big| > \varepsilon\big) \to 0$$

$$(2.14)$$

for any $\varepsilon > 0$. We will prove (2.13) and (2.14) in the following.

For proving (2.13) we denote by $\Omega(n, q)$ the set on which two different jumps of $N(q)$ are further apart than $2|\pi_n|_T$. On $\Omega(n, q)$ we have

$$\sum_{i,j:t_{i,n}^{(1)} \vee t_{j,n}^{(2)} \leq T} \left(\Delta_{i,n}^{(1)} N^{(1)}(q) \Delta_{j,n}^{(2)} N^{(2)}(q)\right)^2 \mathbf{1}_{\{\mathcal{I}_{i,n}^{(1)} \cap \mathcal{I}_{j,n}^{(2)} \neq \emptyset\}} \mathbf{1}_{\Omega(n,q)}$$

$$= \sum_{s \leq T} \left(\Delta N^{(1)}(q)_s\right)^2 \left(\Delta N^{(2)}(q)_s\right)^2 \mathbf{1}_{\Omega(n,q)}. \tag{2.15}$$

Note that the sum without the indicator in (2.15) converges to $B_T(f^*)$ for $q \to \infty$. Thus, (2.13) follows since $\mathbb{P}(\Omega(n, q)) \to 1$ as $n \to \infty$ for any $q > 0$.

For proving (2.14) we apply inequality (2.46) for the function $f(x_1, x_2) = x_1^2 x_2^2$ and

$$x_1 = \Delta_{i,n}^{(1)} N^{(1)}(q), \quad y_1 = \Delta_{i,n}^{(1)} B^{(1)}(q) + \Delta_{i,n}^{(1)} C^{(1)} + \Delta_{i,n}^{(1)} M^{(1)}(q),$$

$$x_2 = \Delta_{j,n}^{(2)} N^{(2)}(q), \quad y_2 = \Delta_{j,n}^{(2)} B^{(2)}(q) + \Delta_{j,n}^{(2)} C^{(2)} + \Delta_{j,n}^{(2)} M^{(2)}(q)$$

which for arbitrary $\varepsilon' > 0$ yields

$$\left| V(f^*, \pi_n)_T - \sum_{i,j:t_{i,n}^{(1)} \vee t_{j,n}^{(2)} \leq T} \left(\Delta_{i,n}^{(1)} N^{(1)}(q) \Delta_{j,n}^{(2)} N^{(2)}(q)\right)^2 \mathbf{1}_{\{\mathcal{I}_{i,n}^{(1)} \cap \mathcal{I}_{j,n}^{(2)} \neq \emptyset\}} \right|$$

$$\leq \theta(\varepsilon') \sum_{i,j:t_{i,n}^{(1)} \vee t_{j,n}^{(2)} \leq T} \left(\Delta_{i,n}^{(1)} N^{(1)}(q) \Delta_{j,n}^{(2)} N^{(2)}(q)\right)^2 \mathbf{1}_{\{\mathcal{I}_{i,n}^{(1)} \cap \mathcal{I}_{j,n}^{(2)} \neq \emptyset\}} \tag{2.16}$$

$$+ K_{\varepsilon'} \sum_{l=1,2} \sum_{i,j:t_{i,n}^{(l)} \vee t_{j,n}^{(3-l)} \leq T} \left(\left(\Delta_{i,n}^{(l)} B^{(l)}(q)\right)^2 + \left(\Delta_{i,n}^{(l)} M^{(l)}(q)\right)^2\right) \mathbf{1}_{\{\mathcal{I}_{i,n}^{(l)} \cap \mathcal{I}_{j,n}^{(3-l)} \neq \emptyset\}}$$

$$\times \left(\left(\Delta_{j,n}^{(3-l)} B^{(3-l)}(q)\right)^2 + \left(\Delta_{j,n}^{(3-l)} C^{(3-l)}\right)^2 + \left(\Delta_{j,n}^{(3-l)} M^{(l)}(q)\right)^2 \right.$$

$$\left. + \left(\Delta_{j,n}^{(3-l)} N^{(3-l)}(q)\right)^2\right) \tag{2.17}$$

$$+ K_{\varepsilon'} \sum_{l=1,2} \sum_{i,j:t_{i,n}^{(3-l)} \vee t_{j,n}^{(l)} \leq T} \left(\Delta_{i,n}^{(l)} C^{(l)}\right)^2 \left(\left(\Delta_{j,n}^{(3-l)} C^{(3-l)}\right)^2 + \left(\Delta_{j,n}^{(l)} N^{(3-l)}(q)\right)^2\right)$$

$$\times \mathbf{1}_{\{\mathcal{I}_{i,n}^{(l)} \cap \mathcal{I}_{j,n}^{(3-l)} \neq \emptyset\}}. \tag{2.18}$$

First note that for (2.16) we obtain

$$\lim_{\varepsilon' \to 0} \lim_{q \to \infty} \limsup_{n \to \infty} \mathbb{P}\left(\theta(\varepsilon') \sum_{i,j:t_{i,n}^{(1)} \vee t_{j,n}^{(2)} \leq T} \left(\Delta_{i,n}^{(1)} N^{(1)}(q) \Delta_{j,n}^{(2)} N^{(2)}(q)\right)^2 \mathbf{1}_{\{\mathcal{I}_{i,n}^{(1)} \cap \mathcal{I}_{j,n}^{(2)} \neq \emptyset\}} > \varepsilon\right) = 0$$

for any $\varepsilon > 0$ using (2.13). Furthermore for (2.17) it holds

$$K_{\varepsilon'} \sum_{i,j:t_{i,n}^{(l)} \vee t_{j,n}^{(l)} \leq T} \left((\Delta_{i,n}^{(l)} B^{(l)}(q))^2 + (\Delta_{i,n}^{(l)} M^{(l)}(q))^2 \right) \left[(\Delta_{j,n}^{(3-l)} B^{(3-l)}(q))^2 \right.$$

$$+ (\Delta_{j,n}^{(3-l)} C^{(3-l)})^2 + (\Delta_{j,n}^{(3-l)} M^{(3-l)}(q))^2 + (\Delta_{j,n}^{(3-l)} N^{(3-l)}(q))^2 \right]$$

$$\times \mathbb{1}_{\{\mathcal{I}_{i,n}^{(l)} \cap \mathcal{I}_{j,n}^{(3-l)} \neq \emptyset\}}$$

$$\leq K_{\varepsilon'} \Big(\sum_{i:t_{i,n}^{(l)} \leq T} \left((\Delta_{i,n}^{(l)} B^{(l)}(q))^2 + (\Delta_{i,n}^{(l)} M^{(l)}(q))^2 \right) \Big)$$

$$\times \Big(\sum_{j:t_{j,n}^{(3-l)} \leq T} \left[(\Delta_{j,n}^{(3-l)} B^{(3-l)}(q))^2 + (\Delta_{j,n}^{(3-l)} C^{(3-l)})^2 \right.$$

$$+ (\Delta_{j,n}^{(3-l)} M^{(3-l)}(q))^2 + (\Delta_{j,n}^{(3-l)} N^{(3-l)}(q))^2 \right] \Big)$$

$$\xrightarrow{\mathbb{P}} K_{\varepsilon'} \left([B^{(l)}(q), B^{(l)}(q)]_T + [M^{(l)}(q), M^{(l)}(q)]_T \right) \left[X^{(3-l)}, X^{(3-l)} \right]_T$$

which tends to zero for $q \to \infty$ as $[B^{(l)}(q), B^{(l)}(q)]_T = 0$ for any $q > 0$. For the treatment of the remaining terms (2.18) we set

$$K_{n,T}(l,\rho) = \sup_{m,m' \in \mathbb{N}, m \leq m', |t_{m',n}^{(l)} - t_{m-1,n}^{(l)}| \leq \rho, t_{m',n}^{(l)} \leq T} \sum_{k=m}^{m'} \left(C_{t_{k,n}^{(l)}}^{(l)} - C_{t_{k-1,n}^{(l)}}^{(l)} \right)^2,$$

$l = 1, 2$. Then we have

$$\lim_{\rho \to 0} \limsup_{n \to \infty} \mathbb{P}(K_{n,T}(l,\rho) > \delta) = 0 \qquad (2.19)$$

for any $\delta > 0$ due to the u.c.p. convergence of realized volatility to the quadratic variation; see Theorem II.22 in [41]. In fact on the set $\{K_{n,T}(l,\rho) > \delta\}$ it holds

$$\sup_{0 \leq s \leq T} \left| \sum_{i=1}^{\infty} \left(C_{t_{i,n}^{(l)} \wedge s}^{(l)} - C_{t_{i-1,n}^{(l)} \wedge s}^{(l)} \right)^2 - \int_0^s (\sigma_u^{(l)})^2 du \right| \mathbb{1}_{\{K_{n,T}(l,\rho) > \delta\}} \geq \frac{\delta - K^2 \rho}{2}$$

where $\|\sigma_s\| \leq K$. This is illustrated in Figure 2.3 and yields (2.19). Using the fact that the total length of the observation intervals of one process which overlap with a specific observation interval of the other process is at most $3|\pi_n|_T$, we get on the set $\{3|\pi_n|_T \leq \rho\}$

$$K_{\varepsilon'} \sum_{i,j:t_{i,n}^{(l)} \vee t_{j,n}^{(3-l)} \leq T} (\Delta_{i,n}^{(l)} C^{(l)})^2 \left[(\Delta_{j,n}^{(l)} C^{(3-l)})^2 + (\Delta_{j,n}^{(3-l)} N^{(3-l)}(q))^2 \right]$$

$$\times \mathbb{1}_{\{\mathcal{I}_{i,n}^{(l)} \cap \mathcal{I}_{j,n}^{(3-l)} \neq \emptyset\}} \mathbb{1}_{\{3|\pi_n|_T \leq \rho\}}$$

$$\leq K_{\varepsilon'} K_{n,T}(l,\rho) \sum_{j:t_{j,n}^{(3-l)} \leq T} \left((\Delta_{j,n}^{(3-l)} C^{(3-l)})^2 + (\Delta_{j,n}^{(3-l)} N^{(3-l)}(q))^2 \right) \mathbb{1}_{\{3|\pi_n|_T \leq \rho\}}.$$

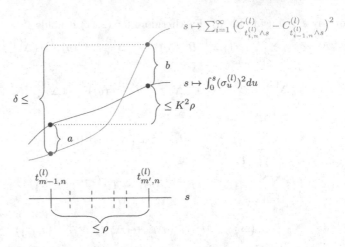

Figure 2.3: Either a or b has to be larger than $(\delta - K^2\rho)/2$.

As the latter sum converges to the quadratic variation of $X^{(l)}$ as first $n \to \infty$ and then $q \to \infty$, we obtain that the sum (2.18) vanishes by (2.19) and because of $\mathbb{P}(3|\pi_n|_T \leq \rho) \to 1$ as $n \to \infty$ for any fixed $\rho > 0$. $\qquad\square$

Proof of Theorem 2.3. Define $f_\rho(x,y) = f(x,y)\psi(|xy|/\rho)$ where $\psi : [0,\infty) \to [0,1]$ is continuous with $\psi(u) = 0$ for $u \leq 1/2$ and $\psi(u) = 1$ for $u \geq 1$. By Lemma 2.13 we have

$$V(f_\rho, \pi_n)_T \xrightarrow{\ \mathbb{P}\ } B^*(f_\rho)_T \tag{2.20}$$

for all $\rho > 0$. Because of $f(x,y) = O(x^2y^2)$ as $|xy| \to 0$ there exist constants $K_\rho > 0$ with $|f_\rho(x,y)| \leq |f(x,y)| \leq K_\rho|x|^2|y|^2$ for all (x,y) with $|xy| < \rho$ and $\rho \mapsto K_\rho$ is non-increasing as $\rho \to 0$. Hence it holds

$$|B^*(f)_T - B^*(f_\rho)_T| \leq 2 \sum_{s \leq T} K_\rho |\Delta X_s^{(1)}|^2 |\Delta X_s^{(2)}|^2 \mathbb{1}_{\{|\Delta X_s^{(1)}\Delta X_s^{(2)}|<\rho\}} \to 0 \tag{2.21}$$

as $\rho \to 0$. To conclude (2.7) we have to show

$$\lim_{\rho \to 0} \limsup_{n \to \infty} \mathbb{P}(|V(f,\pi_n)_T - V(f_\rho, \pi_n)_T| > \delta) = 0 \tag{2.22}$$

for any $\delta > 0$ in addition to (2.20) and (2.21). To this end we consider the following inequality which is obtained like (2.21)

$$|V(f, \pi_n)_T - V(f_\rho, \pi_n)_T| \leq \sum_{i,j: t_{i,n}^{(1)} \vee t_{j,n}^{(2)} \leq T} 2K_\rho |\Delta_{i,n}^{(1)} X^{(1)}|^2 |\Delta_{j,n}^{(2)} X^{(2)}|^2$$

$$\times \mathbb{1}_{\{|\Delta_{i,n}^{(1)} X^{(1)} \Delta_{j,n}^{(2)} X^{(2)}| < \rho\}} \mathbb{1}_{\{\mathcal{I}_{i,n}^{(1)} \cap \mathcal{I}_{j,n}^{(2)} \neq \emptyset\}}. \quad (2.23)$$

Looking at the proof of (2.14) we conclude that (2.23) vanishes in probability as first $n \to \infty$ and then $\rho \to 0$ if

$$\lim_{\rho \to 0} \lim_{q \to \infty} \limsup_{n \to \infty} \mathbb{P}\Big(\sum_{i,j: t_{i,n}^{(1)} \vee t_{j,n}^{(2)} \leq T} 2K_\rho |\Delta_{i,n}^{(1)} N^{(1)}(q)|^2 |\Delta_{j,n}^{(2)} N^{(2)}(q)|^2$$

$$\times \mathbb{1}_{\{|\Delta_{i,n}^{(1)} X^{(1)} \Delta_{j,n}^{(2)} X^{(2)}| < \rho\}} \mathbb{1}_{\{\mathcal{I}_{i,n}^{(1)} \cap \mathcal{I}_{j,n}^{(2)} \neq \emptyset\}} > \delta \Big) = 0 \quad (2.24)$$

holds for any $\delta > 0$. Denote by $S_{q,p}$ the jump times of $N(q)$ and let $\Omega(n, q, \rho)$ be the set on which no two jump times $S_{q,p}, S_{q,p'} \leq T$ fulfil $|S_{q,p} - S_{q,p'}| \leq 2|\pi_n|_T$, it holds $S_{q,p} \leq T - |\pi_n|_T$ and

$$|\Delta N^{(1)}(q)_{S_{q,p}} \Delta N^{(2)}(q)_{S_{q,p}} - \Delta_{i_n^{(1)}(S_{q,p}),n} X^{(1)} \Delta_{i_n^{(2)}(S_{q,p}),n} X^{(2)}| \leq \rho$$

for any $S_{q,p} \leq T$. Further we denote by $\widetilde{S}_{q,p}^\rho$ the jump times of $N(q)$ with

$$|\Delta N^{(1)}(q)_{\widetilde{S}_{q,p}^\rho} \Delta N^{(2)}(q)_{\widetilde{S}_{q,p}^\rho}| < 2\rho.$$

Then it holds

$$\sum_{i,j: t_{i,n}^{(1)} \vee t_{j,n}^{(2)} \leq T} 2K_\rho |\Delta_{i,n}^{(1)} N^{(1)}(q)|^2 |\Delta_{j,n}^{(2)} N^{(2)}(q)|^2$$

$$\times \mathbb{1}_{\{|\Delta_{i,n}^{(1)} X^{(1)} \Delta_{j,n}^{(2)} X^{(2)}| < \rho\}} \mathbb{1}_{\{\mathcal{I}_{i,n}^{(1)} \cap \mathcal{I}_{j,n}^{(2)} \neq \emptyset\}} \mathbb{1}_{\Omega(n,q,\rho)}$$

$$\leq 2K_\rho \sum_{p: \widetilde{S}_{q,p}^\rho \leq T} |\Delta N^{(1)}(q)_{\widetilde{S}_{q,p}^\rho}|^2 |\Delta N^{(2)}(q)_{\widetilde{S}_{q,p}^\rho}|^2$$

$$\times \mathbb{1}_{\{|\Delta_{i_n^{(1)}(\widetilde{S}_{q,p}^\rho),n} X^{(1)} \Delta_{i_n^{(2)}(\widetilde{S}_{q,p}^\rho),n} X^{(2)}| < \rho\}} \mathbb{1}_{\Omega(n,q,\rho)}$$

$$\leq 2K_\rho \sum_{p: \widetilde{S}_{q,p}^\rho \leq T} |\Delta N^{(1)}(q)_{\widetilde{S}_{q,p}^\rho}|^2 |\Delta N^{(2)}(q)_{\widetilde{S}_{q,p}^\rho}|^2$$

$$\times \mathbb{1}_{\{|\Delta N^{(1)}(q)_{\widetilde{S}_{q,p}^\rho} \Delta N^{(2)}(q)_{\widetilde{S}_{q,p}^\rho}| < 2\rho\}} \mathbb{1}_{\Omega(n,q,\rho)}$$

$$\leq 2K_\rho \sum_{s \leq T} |\Delta X_s^{(1)}|^2 |\Delta X_s^{(2)}|^2 \mathbb{1}_{\{|\Delta X_s^{(1)} \Delta X_s^{(2)}| < 2\rho\}}$$

where the expression in the last line vanishes as $\rho \to 0$. Using $\mathbb{P}(\Omega(n,q,\rho)) \to 1$ as $n \to \infty$ for any $q, \rho > 0$ this yields (2.24) and hence (2.22).   □

The following elementary lemma states that if we want to show that a sequence of random variables vanishes in probability it is sufficient to show that their conditional expectations with respect to some sub-σ-algebra vanish in probability. We will use this lemma frequently in the upcoming proofs in this chapter as well as in later chapters. Mostly in the exogenous setting, we will apply this lemma for conditional expectations with respect to the σ-algebra $\mathcal{S}$ generated by the observation scheme.

Lemma 2.15. *Let $(Y_n)_{n \in \mathbb{N}}$ be a sequence of real-valued integrable random variables on $(\Omega, \mathcal{F}, \mathbb{P})$ and $\mathcal{G} \subset \mathcal{F}$ a sub-σ-algebra. Then it holds*

$$\mathbb{E}\big[|Y_n| \big| \mathcal{G}\big] \xrightarrow{\mathbb{P}} 0 \Rightarrow Y_n \xrightarrow{\mathbb{P}} 0$$

Proof. For $\varepsilon < 1$ it holds by Markov's inequality and Jensen's inequality

$$\mathbb{P}(|Y_n| > \varepsilon) = \mathbb{P}(|Y_n| \wedge 1 > \varepsilon) \leq \frac{\mathbb{E}[|Y_n| \wedge 1]}{\varepsilon} = \frac{\mathbb{E}\big[\mathbb{E}[|Y_n| \wedge 1 | \mathcal{G}]\big]}{\varepsilon} \leq \frac{\mathbb{E}\big[\mathbb{E}[|Y_n| | \mathcal{G}] \wedge 1\big]}{\varepsilon}.$$

This estimate yields the claim as a sequence of random variables $(\widetilde{Y}_n)_{n \in \mathbb{N}}$ converges in probability to zero if and only if $\mathbb{E}[|\widetilde{Y}_n| \wedge c] \to 0$ for any $c > 0$; compare Theorem 6.7 in [33].   □

Proof of Theorem 2.6. Set $f_\rho(x,y) = f(x,y)\psi(|xy|/\rho)$ like in the proof of Theorem 2.3. As in the proof of Theorem 2.3 we obtain $V(f_\rho, \pi_n)_T \xrightarrow{\mathbb{P}} B^*(f_\rho)_T$ and

$$\lim_{\rho \to 0} |B^*(f)_T - B^*(f_\rho)_T| \leq \lim_{\rho \to 0} \sum_{s \leq T} 2K_\rho |\Delta X_s^{(1)}|^{p_1} |\Delta X_s^{(2)}|^{p_2} \mathbb{1}_{\{|\Delta X_s^{(1)} \Delta X_s^{(2)}| < \rho\}}$$

$$\leq \lim_{\rho \to 0} K_\rho \sum_{s \leq T} \big(|\Delta X_s^{(1)}|^{p_1+p_2} + |\Delta X_s^{(2)}|^{p_1+p_2}\big) \mathbb{1}_{\{|\Delta X_s^{(1)} \Delta X_s^{(2)}| < \rho\}}$$

$$= 0 \tag{2.25}$$

because of $p_1 + p_2 \geq 2$ where we used Muirhead's inequality, compare Theorem 45 of [19], which yields $2a^{p_1}b^{p_2} \leq a^{p_1+p_2} + b^{p_1+p_2}$ for any $a, b \geq 0$. Hence it remains to show

$$\lim_{\rho \searrow 0} \limsup_{n \to \infty} \mathbb{P}(|V(f, \pi_n)_T - V(f_\rho, \pi_n)_T| > \varepsilon) = 0 \quad \forall \varepsilon > 0. \tag{2.26}$$

Because of $f(x,y) = o(|x|^{p_1}|y|^{p_2})$ as $|xy| \to 0$ we obtain as in (2.23)

$$|V(f, \pi_n)_T - V(f_\rho, \pi_n)_T| \leq 2K_\rho \sum_{i,j : t_{i,n}^{(1)} \vee t_{j,n}^{(2)} \leq T} |\Delta_{i,n}^{(1)} X^{(1)}|^{p_1} |\Delta_{j,n}^{(2)} X^{(2)}|^{p_2} \mathbb{1}_{\{\mathcal{I}_{i,n}^{(1)} \cap \mathcal{I}_{j,n}^{(2)} \neq \emptyset\}}$$

$$\tag{2.27}$$

with $K_\rho \to 0$ as $\rho \to 0$. Define stopping times $(T_k^n)_{k \in \mathbb{N}_0}$ via $T_0^n = 0$ and

$$T_k^n = \inf\{t_{i,n}^{(l)} > T_{k-1}^n | i \in \mathbb{N}_0, l = 1, 2\}, \quad k \geq 1.$$

Hence the T_k^n mark the times, where at least one of the processes $X^{(1)}$ or $X^{(2)}$ is newly observed. Further we set

$$
\begin{aligned}
\tau_{n,-}^{(l)}(s) &= \sup\{t_{i,n}^{(l)} \leq s | i \in \mathbb{N}_0\}, \\
\tau_{n,+}^{(l)}(s) &= \inf\{t_{i,n}^{(l)} \geq s | i \in \mathbb{N}_0\}
\end{aligned}
\tag{2.28}
$$

for the observation times of $X^{(l)}$, $l = 1, 2$, immediately before and after s. Then we denote

$$
\begin{aligned}
\Delta_k^n X^{(l)} &= X_{T_k}^{(l)} - X_{T_{k-1}}^{(l)}, \\
\Delta_k^{n,l,-} X^{(l)} &= X_{T_{k-1}}^{(l)} - X_{\tau_{n,-}^{(l)}(T_{k-1}^n)}^{(l)}, \\
\Delta_k^{n,l,+} X^{(l)} &= X_{\tau_{n,+}^{(l)}(T_k^n)}^{(l)} - X_{T_k}^{(l)}
\end{aligned}
\tag{2.29}
$$

for $l = 1, 2$. Using this notation we obtain

$$
\sum_{i,j:t_{i,n}^{(1)} \vee t_{j,n}^{(2)} \leq T} |\Delta_{i,n}^{(1)} X^{(1)}|^{p_1} |\Delta_{j,n}^{(2)} X^{(2)}|^{p_2} \mathbb{1}_{\{\mathcal{I}_{i,n}^{(1)} \cap \mathcal{I}_{j,n}^{(2)} \neq \emptyset\}}
$$

$$
\leq \sum_{k:T_k^n \leq T} \prod_{l=1}^{2} |\Delta_k^{n,l,-} X^{(l)} + \Delta_k^n X^{(l)} + \Delta_k^{n,l,+} X^{(l)}|^{p_l}
$$

$$
\leq \sum_{k:T_k^n \leq T} \prod_{l=1}^{2} K_{p_l} \left(|\Delta_k^{n,l,-} X^{(l)}|^{p_l} + |\Delta_k^n X^{(l)}|^{p_l} + |\Delta_k^{n,l,+} X^{(l)}|^{p_l} \right).
$$

The $\mathcal{S}$-conditional expectation of this quantity is bounded by

$$
\sum_{k:T_k^n \leq T} \mathbb{E}[\prod_{l=1}^{2} K_{p_l} \left(|\Delta_k^{n,l,-} X^{(l)}|^{p_l} + |\Delta_k^n X^{(l)}|^{p_l} + |\Delta_k^{n,l,+} X^{(l)}|^{p_l} \right) | \mathcal{S}]
$$

$$
= K_{p_1} K_{p_2} \sum_{k:T_k^n \leq T} \mathbb{E}[|\Delta_k^n X^{(1)}|^{p_1} |\Delta_k^n X^{(2)}|^{p_2} | \mathcal{S}]
$$

$$
+ K_{p_1} K_{p_2} \sum_{l=1,2} \sum_{k:T_k^n \leq T} \mathbb{E}\big[|\Delta_k^{n,l,-} X^{(l)}|^{p_l}
$$

$$
\times \mathbb{E}[|\Delta_k^n X^{(3-l)}|^{p_{3-l}} + |\Delta_k^{n,3-l,+} X^{(3-l)}|^{p_{3-l}} | \sigma(\mathcal{F}_{T_{k-1}^n}, \mathcal{S})] | \mathcal{S}\big]
$$

$$
+ K_{p_1} K_{p_2} \sum_{l=1,2} \sum_{k:T_k^n \leq T} \mathbb{E}\big[|\Delta_k^n X^{(l)}|^{p_l} \mathbb{E}[|\Delta_k^{n,3-l,+} X^{(3-l)}|^{p_{3-l}} | \sigma(\mathcal{F}_{T_k^n}, \mathcal{S})] | \mathcal{S}\big]
$$

$$
\tag{2.30}
$$

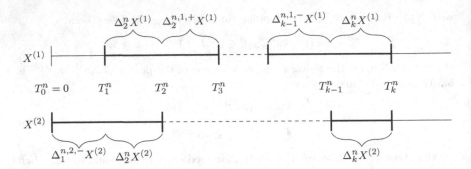

Figure 2.4: Merged observation times and the corresponding intervals $\Delta_k^{n,k,-}X^{(l)}$, $\Delta_k^n X^{(l)}$ and $\Delta_k^{n,k,+}X^{(l)}$.

where we used

$$|\Delta_k^{n,1,-}X^{(1)}|^{p_1}|\Delta_k^{n,2,-}X^{(2)}|^{p_2} = |\Delta_k^{n,1,+}X^{(1)}|^{p_1}|\Delta_k^{n,2,+}X^{(2)}|^{p_2} = 0$$

which holds because one of each two increments is always zero. Further, using

$$|\Delta_k^n X^{(1)}|^{p_1}|\Delta_k^n X^{(2)}|^{p_2} \leq 2\|\Delta_k^n X\|^{p_1+p_2},$$

which we obtain as in (2.25), and inequality (1.12), (2.30) is bounded by

$$K_{p_1}K_{p_2}\sum_{k:T_k^n\leq T}\mathbb{E}[\|\Delta_k^n X\|^{p_1+p_2}|\mathcal{S}]$$

$$+ K_{p_1}K_{p_2}\sum_{l=1,2}\sum_{k:T_k^n\leq T}\mathbb{E}[|\Delta_k^{n,l,-}X^{(l)}|^{p_l}K(\tau_+^{(3-l)}(T_k^n)-T_{k-1}^n)^{\frac{p_{3-l}}{2}\wedge 1}|\mathcal{S}]$$

$$+ K_{p_1}K_{p_2}\sum_{l=1,2}\sum_{k:T_k^n\leq T}\mathbb{E}[|\Delta_k^n X^{(l)}|^{p_l}K(\tau_+^{(3-l)}(T_k^n)-T_k^n)^{\frac{p_{3-l}}{2}\wedge 1}|\mathcal{S}]$$

$$\leq K_{p_1}K_{p_2}\sum_{k:T_k^n\leq T}K(T_k^n-T_{k-1}^n)$$

$$+ 4K_{p_1}K_{p_2}\sum_{k:T_k^n\leq T}\prod_{l=1,2}(\tau_+^{(l)}(T_k^n)-\tau_-^{(l)}(T_{k-1}^n))^{\frac{p_l}{2}\wedge 1}$$

$$\leq K_{p_1,p_2}T + K_{p_1,p_2}\sum_{i,j:t_{i,n}^{(1)}\vee t_{j,n}^{(2)}\leq T}K|\mathcal{I}_{i,n}^{(1)}|^{\frac{p_1}{2}\wedge 1}|\mathcal{I}_{j,n}^{(2)}|^{\frac{p_2}{2}\wedge 1}\mathbb{1}_{\{\mathcal{I}_{i,n}^{(1)}\cap\mathcal{I}_{j,n}^{(2)}\neq\emptyset\}}.$$

This expression is bounded in probability by condition (2.8) and hence the right-hand side of (2.27) vanishes for $\rho \to 0$ due to $K_\rho \to 0$ as $\rho \to 0$ which yields (2.26). $\qquad\square$

Proof of Theorem 2.10. Comparing the proof of Theorem 2.6 it is sufficient to show that

$$Y(n) = \sum_{i,j:t_{i,n}^{(1)} \vee t_{j,n}^{(2)} \leq T} |\Delta_{i,n}^{(1)} X^{(1)}|^{p_1} |\Delta_{j,n}^{(2)} X^{(2)}|^{p_2} \mathbb{1}_{\{\mathcal{I}_{i,n}^{(1)} \cap \mathcal{I}_{j,n}^{(2)} \neq \emptyset\}}$$

is bounded in probability as $n \to \infty$. To this end fix $\varepsilon > 0$ and define

$$\Omega(n,\varepsilon) = \{ \sup_{i:t_{i,n}^{(l)} \leq T} \sum_{j \in \mathbb{N}} \mathbb{1}_{\{\mathcal{I}_{i,n}^{(l)} \cap \mathcal{I}_{j,n}^{(3-l)} \neq \emptyset\}} > N_\varepsilon, \ l = 1,2\}$$

with N_ε as in (2.10). We then obtain using Muirhead's inequality like in (2.25)

$$Y(n)\mathbb{1}_{\Omega(n,\varepsilon)^c}$$
$$\leq \sum_{i,j:t_{i,n}^{(1)} \vee t_{j,n}^{(2)} \leq T} \left((\Delta_{i,n}^{(1)} X^{(1)})^2 + (\Delta_{j,n}^{(2)} X^{(2)})^2 \right) \mathbb{1}_{\{\mathcal{I}_{i,n}^{(1)} \cap \mathcal{I}_{j,n}^{(2)} \neq \emptyset\}} \mathbb{1}_{\Omega(n,\varepsilon)^c}$$
$$\leq N_\varepsilon \sum_{l=1,2} \sum_{i:t_{i,n}^{(l)} \leq T} (\Delta_{i,n}^{(l)} X^{(l)})^2$$

where the sums in the last line converge to $[X^{(l)}, X^{(l)}]_T$, $l = 1,2$. If we further choose $K_\varepsilon > 0$ such that $\mathbb{P}([X^{(l)}, X^{(l)}]_T > K_\varepsilon(1-\xi)) \leq \varepsilon$, $l = 1,2$, for some $\xi \in (0,1)$ we obtain

$$\limsup_{n \to \infty} \mathbb{P}(Y(n) > 2N_\varepsilon K_\varepsilon)$$
$$< \limsup_{n \to \infty} \mathbb{P}(\Omega(n,\varepsilon)) + \limsup_{n \to \infty} \sum_{l=1,2} \mathbb{P}\Big(\sum_{i:t_{i,n}^{(l)} \leq T} (\Delta_{i,n}^{(l)} X^{(l)})^2 > K_\varepsilon \Big)$$
$$\leq \varepsilon + \limsup_{n \to \infty} \sum_{l=1,2} \mathbb{P}\Big(\Big| \sum_{i:t_{i,n}^{(l)} \leq T} (\Delta_{i,n}^{(l)} X^{(l)})^2 - [X^{(l)}, X^{(l)}]_T \Big| > K_\varepsilon \xi \Big)$$
$$+ \limsup_{n \to \infty} \sum_{l=1,2} \mathbb{P}\big([X^{(l)}, X^{(l)}]_T \big)^2 > K_\varepsilon(1-\xi) \big)$$
$$\leq 3\varepsilon.$$

As $\varepsilon > 0$ can be chosen arbitrarily this yields the boundedness in probability of $Y(n)$, $n \in \mathbb{N}$. $\qquad\square$

2.2 Normalized Functionals

In Section 2.1 we have seen that the functional $V(f, \pi_n)_T$ converges to a limit which depends only on the jump part of X for functions f that decay sufficiently fast in a neighbourhood of zero. This is necessary because we need that for such functions f the contribution of the continuous part in $V(f, \pi_n)_T$ becomes asymptotically negligible. The jump part has the property that the magnitude of its increments remains constant as $|\pi_n|_T \to 0$ while for the continuous martingale part

$$C_t = \int_0^t \sigma_s dW_s, \quad t \geq 0,$$

the magnitude of the „*normalized*" increment $|\mathcal{I}_{i,n}^{(l)}|^{-1/2} \Delta_{i,n}^{(l)} C^{(l)}$ remains constant. Hence if we would like to learn something about the continuous part of X it is reasonable to look at functionals of the normalized increments $|\mathcal{I}_{i,n}^{(l)}|^{-1/2} \Delta_{i,n}^{(l)} X^{(l)}$.

As an illustration for the upcoming results consider the toy example

$$X_t^{toy} = \sigma W_t \tag{2.31}$$

where the volatility matrix

$$\sigma = \begin{pmatrix} \sigma^{(1)} & 0 \\ \rho\sigma^{(2)} & \sqrt{1-\rho^2}\sigma^{(2)} \end{pmatrix}$$

is constant in time with $\sigma^{(1)}, \sigma^{(2)} > 0$ and $\rho \in [-1, 1]$. Suppose the observation scheme is exogenous, compare Definition 1.2, and synchronous. Under Condition 1.3 we then obtain using the notation from (2.1)

$$\sum_{i:t_{i,n} \leq T} |\mathcal{I}_{i,n}| f(|\mathcal{I}_{i,n}|^{-1/2} \Delta_{i,n} X^{toy}) = \sum_{i:t_{i,n} \leq T} |\mathcal{I}_{i,n}| f(\sigma Z_i^n) \xrightarrow{\mathbb{P}} T\mathbb{E}[f(\sigma Z)]$$

$$\tag{2.32}$$

because of $|\pi_n|_T \xrightarrow{\mathbb{P}} 0$, where Z and $Z_i^n = |\mathcal{I}_{i,n}|^{-1/2} \Delta_{i,n} W$, $i \in \mathbb{N}_0$, are i.i.d. two-dimensional standard normal random variables for each $n \in \mathbb{N}$. Functionals of this form are discussed in Section 14.2 of [30]. Two straightforward generalizations of this approach to the setting of asynchronous observation times lead to functionals of the form

$$\sum_{i,j:t_{i,n}^{(1)} \vee t_{j,n}^{(2)} \leq T} (|\mathcal{I}_{i,n}^{(1)}||\mathcal{I}_{j,n}^{(2)}|)^{1/2} f(|\mathcal{I}_{i,n}^{(1)}|^{-1/2} \Delta_{i,n}^{(1)} X^{(1)}, |\mathcal{I}_{j,n}^{(2)}|^{-1/2} \Delta_{j,n}^{(2)} X^{(2)}) \mathbb{1}_{\{\mathcal{I}_{i,n}^{(1)} \cap \mathcal{I}_{j,n}^{(2)} \neq \emptyset\}}$$

$$\tag{2.33}$$

and

$$\sum_{i,j:t_{i,n}^{(1)} \vee t_{j,n}^{(2)} \leq T} |\mathcal{I}_{i,n}^{(1)} \cap \mathcal{I}_{j,n}^{(2)}| f(|\mathcal{I}_{i,n}^{(1)}|^{-1/2} \Delta_{i,n}^{(1)} X^{(1)}, |\mathcal{I}_{j,n}^{(2)}|^{-1/2} \Delta_{j,n}^{(2)} X^{(2)}) \mathbb{1}_{\{\mathcal{I}_{i,n}^{(1)} \cap \mathcal{I}_{j,n}^{(2)} \neq \emptyset\}}.$$

Here the main difference compared to the functional from (2.32) in the synchronous setting is that the law of $(|\mathcal{I}_{i,n}^{(1)}|^{-1/2}\Delta_{i,n}^{(1)}X^{(1)}, |\mathcal{I}_{j,n}^{(2)}|^{-1/2}\Delta_{j,n}^{(2)}X^{(2)})$ is in general not independent of the observation scheme π_n. This property is due to the fact that e.g. the correlation of $|\mathcal{I}_{i,n}^{(1)}|^{-1/2}\Delta_{i,n}^{(1)}X^{toy,(1)}$ and $|\mathcal{I}_{j,n}^{(2)}|^{-1/2}\Delta_{j,n}^{(2)}X^{toy,(2)}$ equals

$$\rho \frac{|\mathcal{I}_{i,n}^{(1)} \cap \mathcal{I}_{j,n}^{(2)}|}{|\mathcal{I}_{i,n}^{(1)}|^{1/2}|\mathcal{I}_{j,n}^{(2)}|^{1/2}} \tag{2.34}$$

as the increments of $X^{toy,(1)}$ and $X^{toy,(2)}$ are correlated only over the overlapping part $\mathcal{I}_{i,n}^{(1)} \cap \mathcal{I}_{j,n}^{(2)}$. This difference is also the main reason why it is more difficult to derive convergence results for normalized functionals as we will see later on. Further, regarding (2.33) the quantity

$$\sum_{i,j: t_{i,n}^{(1)} \vee t_{j,n}^{(2)} \leq T} (|\mathcal{I}_{i,n}^{(1)}||\mathcal{I}_{j,n}^{(2)}|)^{1/2} \mathbb{1}_{\{\mathcal{I}_{i,n}^{(1)} \cap \mathcal{I}_{j,n}^{(2)} \neq \emptyset\}}$$

might diverge for $|\pi_n|_T \to 0$ as has been shown in Example 2.8.

Due to these observations we will pick another approach where we use a global normalization instead of locally normalizing each $\Delta_{i,n}^{(l)}X^{(l)}$ with $|\mathcal{I}_{i,n}^{(l)}|^{1/2}$. Precisely, we have to assume that the „average" observation frequency increases with rate n. We then look at functionals of the form

$$\sum_{i,j: t_{i,n}^{(1)} \vee t_{j,n}^{(2)} \leq T} n^{-1} f(n^{1/2}\Delta_{i,n}^{(1)}X^{(1)}, n^{1/2}\Delta_{j,n}^{(2)}X^{(2)}) \mathbb{1}_{\{\mathcal{I}_{i,n}^{(1)} \cap \mathcal{I}_{j,n}^{(2)} \neq \emptyset\}}. \tag{2.35}$$

Such functionals also appear to occur more naturally in the applications discussed in Part II.

The most common functions for which the functionals (2.35) are studied are the power functions $g_p(x) = x^p$, $\overline{g}_p = |x|^p$ and $f_{(p_1,p_2)} = x_1^{p_1} x_2^{p_2}$, $\overline{f}_{(p_1,p_2)} = |x_1|^{p_1}|x_2|^{p_2}$ where $p, p_1, p_2 \geq 0$. Those functions are members of the following more general classes of functions; compare Section 3.4.1 in [30].

Definition 2.16. *A function $f\colon \mathbb{R}^d \to \mathbb{R}$ is called positively homogeneous of degree $p \geq 0$, if $f(\lambda x) = \lambda^p f(x)$ for all $x \in \mathbb{R}^d$ and $\lambda \geq 0$. Further f is called positively homogeneous with degree $p_i \geq 0$ in the i-th argument if the function $x \mapsto f(x_1, \ldots, x_{i-1}, x, x_{i+1}, \ldots, x_d)$ is positively homogeneous of degree p_i for any choice of $(x_1, \ldots, x_{i-1}, x_{i+1}, \ldots, x_d) \in \mathbb{R}^{d-1}$.* $\square$

If the function f is positively homogeneous with degree p_1 in the first argument and with degree p_2 in the second argument, (2.35) becomes

$$n^{(p_1+p_2)/2-1} \sum_{i,j: t_{i,n}^{(1)} \vee t_{j,n}^{(2)} \leq T} f(\Delta_{i,n}^{(1)}X^{(1)}, \Delta_{j,n}^{(2)}X^{(2)}) \mathbb{1}_{\{\mathcal{I}_{i,n}^{(1)} \cap \mathcal{I}_{j,n}^{(2)} \neq \emptyset\}}.$$

As we are going to derive results only for such functions we denote by

$$\overline{V}(p,f,\pi_n)_T = n^{p/2-1} \sum_{i,j:t_{i,n}^{(1)} \vee t_{j,n}^{(2)} \leq T} f(\Delta_{i,n}^{(1)} X^{(1)}, \Delta_{j,n}^{(2)} X^{(2)}) \mathbb{1}_{\{\mathcal{I}_{i,n}^{(1)} \cap \mathcal{I}_{j,n}^{(2)} \neq \emptyset\}},$$

(2.36)

$f \colon \mathbb{R}^2 \to \mathbb{R}$, the functional whose asymptotics we are going to study in this section. Further we set

$$\overline{V}^{(l)}(p,g,\pi_n)_T = n^{p/2-1} \sum_{i:t_{i,n}^{(l)} \leq T} g(\Delta_{i,n}^{(l)} X^{(l)}), \quad l = 1,2, \qquad (2.37)$$

for functions $g \colon \mathbb{R} \to \mathbb{R}$. As in Section 2.1 we will also derive an asymptotic result for $\overline{V}^{(l)}(p,g,\pi_n)_T$ to compare the results in the setting of asynchronously observed bivariate processes to those in simpler settings.

To describe the limits of the normalized functionals (2.36) and (2.37) in the upcoming results we need to introduce some notation. Denote by

$$m_\Sigma(h) = \mathbb{E}[h(Z)], \quad Z \sim \mathcal{N}(0,\Sigma), \qquad (2.38)$$

the expectation of a function $h \colon \mathbb{R}^d \to \mathbb{R}$ evaluated at a d–dimensional centered normal distributed random variable with covariance matrix Σ. Further we define the expressions

$$G_p^{(l),n}(t) = n^{p/2-1} \sum_{i:t_{i,n}^{(l)} \leq t} |\mathcal{I}_{i,n}^{(l)}|^{p/2},$$

$$G_{p_1,p_2}^n(t) = n^{(p_1+p_2)/2-1} \sum_{i,j:t_{i,n}^{(1)} \vee t_{j,n}^{(2)} \leq t} |\mathcal{I}_{i,n}^{(1)}|^{p_1/2} |\mathcal{I}_{j,n}^{(2)}|^{p_2/2} \mathbb{1}_{\{\mathcal{I}_{i,n}^{(1)} \cap \mathcal{I}_{j,n}^{(2)} \neq \emptyset\}},$$

(2.39)

$$H_{k,m,p}^n(t) = n^{p/2-1} \sum_{i,j:t_{i,n}^{(1)} \vee t_{j,n}^{(2)} \leq t} |\mathcal{I}_{i,n}^{(1)} \setminus \mathcal{I}_{j,n}^{(2)}|^{k/2} |\mathcal{I}_{j,n}^{(2)} \setminus \mathcal{I}_{i,n}^{(1)}|^{m/2}$$

$$\times |\mathcal{I}_{i,n}^{(1)} \cap \mathcal{I}_{j,n}^{(2)}|^{(p-(k+m))/2} \mathbb{1}_{\{\mathcal{I}_{i,n}^{(1)} \cap \mathcal{I}_{j,n}^{(2)} \neq \emptyset\}},$$

whose limits, if they exist, will occur in the limits of $\overline{V}^{(l)}(p,g,\pi_n)_T$ and $\overline{V}(p,f,\pi_n)_T$.

2.2.1 The Results

We start with a result for $\overline{V}(p,f,\pi_n)_T$ under the restriction that the observations are synchronous as in (2.1). If we consider a function $f \colon \mathbb{R}^2 \to \mathbb{R}$ which is positively homogeneous of degree p in our toy example (2.31) we get

$$\mathbb{E}[\overline{V}(p,f,\pi_n)_T | \mathcal{S}] = n^{p/2-1} \sum_{i:t_{i,n} \leq T} |\mathcal{I}_{i,n}|^{p/2} \mathbb{E}[f(|\mathcal{I}_{i,n}|^{-1/2} \Delta_{i,n} X^{toy}) | \mathcal{S}]$$

$$= m_{\sigma\sigma^*}(f)G_p^{(1),n}(T)$$

where $\mathcal{S}$ denotes the σ-algebra generated by $(\pi_n)_{n\in\mathbb{N}}$. Therefore it appears to be a necessary condition that $G_p^{(1),n}(T)$ converges in order for $\overline{V}(p,f,\pi_n)_T$ to converge as well. This reasoning also carries over to the case of non-constant σ_s via an approximation of σ_s by piecewise constant stochastic processes. We then obtain the following result which covers the whole class of positively homogeneous functions $f\colon \mathbb{R}^2 \to \mathbb{R}$.

Theorem 2.17. *Let $p \geq 0$ and suppose that Condition 1.3 is fulfilled, and that the observation scheme is exogenous and synchronous. Further assume that*

$$G_p^{(1),n}(t) = G_p^{(2),n}(t) \xrightarrow{\mathbb{P}} G_p(t), \quad t \in [0,T],$$

for a (possibly random) continuous function $G_p\colon [0,T] \to \mathbb{R}_{\geq 0}$ and that we have one of the following two conditions:

a) $p \in [0,2)$,

b) $p \geq 2$ and X is continuous.

Then for all continuous positively homogeneous functions $f\colon \mathbb{R}^2 \to \mathbb{R}$ of degree p it holds that

$$\overline{V}(p,f,\pi_n)_T \xrightarrow{\mathbb{P}} \int_0^T m_{c_s}(f)dG_p(s)$$

where $c_s = \sigma_s\sigma_s^*$.

Remark 2.18. *Note that the integral $\int_0^T m_{c_s}(f)dG_p(s)$ is well-defined as a Lebesgue-Stieltjes integral because $m_{c_s}(f)$ is càdlàg as σ_s is càdlàg and because the function $t \mapsto G_p(t)$ is by assumption continuous and it is also increasing as the functions $t \mapsto G_p^n(t)$, $n \in \mathbb{N}$, are all increasing.* $\square$

As a corollary of Theorem 2.17, we directly obtain the following convergence result for the functionals $V^{(l)}(p,g,\pi_n)_T$.

Corollary 2.19. *Let $l = 1$ or $l = 2$, $p \geq 0$, and suppose that Condition 1.3 is fulfilled and that the observation scheme is exogenous. Further assume that*

$$G_p^{(l),n}(t) \xrightarrow{\mathbb{P}} G_p^{(l)}(t), \quad t \in [0,T],$$

for a (possibly random) continuous function $G_p^{(l)}\colon [0,T] \to \mathbb{R}_{\geq 0}$ and that we have one of the following two conditions:

a) $p \in [0,2)$,

b) $p \geq 2$ and $X^{(l)}$ is continuous.

Then for all positively homogeneous functions $g \colon \mathbb{R} \to \mathbb{R}$ of degree p it holds that

$$\overline{V}^{(l)}(p, g, \pi_n)_T \xrightarrow{\mathbb{P}} m_1(g) \int_0^T (\sigma_s^{(l)})^p dG_p^{(l)}(s).$$

Remark 2.20. *In Chapter 14 of [30] synchronous observation schemes of the form*

$$t_{i,n} = t_{i-1,n} + \theta_{t_{i-1,n}}^n \varepsilon_{i,n}$$

are investigated where $\theta^n = (\theta_t^n)_{t \geq 0}$ is a strictly positive process which is adapted to the filtration $(\mathcal{F}_t^n)_{t \geq 0}$, where $\mathcal{F}_t^n$ denotes the smallest σ-algebra containing the filtration $(\mathcal{F}_t)_{t \geq 0}$ with respect to which X is defined and which has the property that all $t_{i,n}$ are $(\mathcal{F}_t^n)_{t \geq 0}$-stopping times; compare Definition 14.1.1 in [30]. $\varepsilon_{i,n}$ is supposed to be an i.i.d. sequence of positive random variables in i for fixed n. The $\varepsilon_{i,n}$ are independent of the process X and its components. If $n\theta^n$ converges in u.c.p. to some $(\mathcal{F}_t)_{t \geq 0}$-adapted process θ and the moments $\mathbb{E}[(\varepsilon_{i,n})^{p/2}] = \kappa_{p/2}^n$ converge to some $\kappa_{p/2} < \infty$ then Lemma 14.1.5 in [30] yields

$$G_p^n(t) \xrightarrow{\mathbb{P}} \kappa_{p/2} \int_0^t (\theta_s)^{p/2-1} ds =: G_p(t) \quad \forall t \in [0, T],$$

and using Theorem 14.2.1 in [30] we conclude

$$\overline{V}(p, f, \pi_n)_T \xrightarrow{\mathbb{P}} \kappa_{p/2} \int_0^T m_{c_s}(f)(\theta_s)^{p/2-1} ds = \int_0^T m_{c_s}(f) dG_p(s)$$

under the assumptions on f and X made in Theorem 2.17. The assumptions on the observation scheme in [30] are weaker than the assumptions made in this paper in the sense that the observation scheme does not have to be exogenous, but there are still strong restrictions on the law of $t_{i,n} - t_{i-1,n}$. $\theta_{t_{i-1,n}}^n$ is known in advance at time $t_{i-1,n}$ and $(t_{i,n} - t_{i-1,n})/\theta_{t_{i,n}}^n = \varepsilon_{i,n}$ is an exogenous random variable whose law is independent of the other observation times. On the other hand our assumptions allow for observation schemes which do not fulfil the assumptions made in Chapter 14 of [30]. This is due to the fact that we need no analogon to the i.i.d. property of the $\varepsilon_{i,n}$. G_p^n in general already converges to a linear function if the ratio $\mathbb{E}[n^{p/2}|\mathcal{I}_{i,n}|^{p/2}]/\mathbb{E}[n|\mathcal{I}_{i,n}|]$ remains constant; compare Lemma 5.3. Asynchronous observation schemes are not considered in [30]. $\qquad\square$

In the case of asynchronous observation times it is more difficult to derive results similar to Theorem 2.17. For functions f which are like $f_{(p_1,p_2)}$, $\overline{f}_{(p_1,p_2)}$ positively homogeneous with degree p_1 in the first argument and with degree p_2 in the second argument it holds that

$$\mathbb{E}[\overline{V}(p_1 + p_2, f, \pi_n)_T | \mathcal{S}] = n^{(p_1+p_2)/2-1} \sum_{i,j:t_{i,n}^{(1)} \vee t_{j,n}^{(2)} \leq T} |\mathcal{I}_{i,n}^{(1)}|^{p_1/2} |\mathcal{I}_{j,n}^{(2)}|^{p_2/2}$$

$$\times \mathbb{E}[f(|\mathcal{I}_{i,n}^{(1)}|^{-1/2}\Delta_{i,n}^{(1)}X^{(1)}, |\mathcal{I}_{j,n}^{(2)}|^{-1/2}\Delta_{j,n}^{(2)}X^{(2)}) | \mathcal{S}] \mathbb{1}_{\{\mathcal{I}_{i,n}^{(1)} \cap \mathcal{I}_{j,n}^{(2)} \neq \emptyset\}}.$$

However unlike in the case of synchronous observation times the law of

$$(|\mathcal{I}_{i,n}^{(1)}|^{-1/2}\Delta_{i,n}^{(1)}X^{(1)}, |\mathcal{I}_{j,n}^{(2)}|^{-1/2}\Delta_{j,n}^{(2)}X^{(2)})$$

is in general not independent of π_n as explained in (2.34). The process X^{toy} from (2.31) has the simplest form of all processes for which the functionals discussed in this section yield a non-trivial limit and hence it makes sense to first investigate conditions which grant convergence of $\overline{V}(p_1 + p_2, f, \pi_n)_T$ for X^{toy}. Also, the arguments used in the proof of Theorem 2.17 rely on an approximation of σ_s by piecewise constant processes in time, which then makes it possible to use results for processes like X^{toy} with a constant σ_s. In particular, we needed that $f(\Delta_{i,n}X^{toy})$ factorizes into a term that depends only on $\mathcal{S}$ and a term that is independent of $\mathcal{S}$. This technique of proof can be extended to the asynchronous setting whenever we can find a fixed natural number $N \in \mathbb{N}$ and functions g_k, h_k for $k = 1, \ldots, N$ such that we can write

$$\mathbb{E}[f(\Delta_{i,n}^{(1)}X^{toy,(1)}, \Delta_{j,n}^{(2)}X^{toy,(2)}) | \mathcal{S}]$$

$$= \sum_{k=1}^{N} g_k(|\mathcal{I}_{i,n}^{(1)}|, |\mathcal{I}_{j,n}^{(2)}|, |\mathcal{I}_{j,n}^{(2)} \cap \mathcal{I}_{j,n}^{(2)}|) h_k(\sigma^{(1)}, \sigma^{(2)}, \rho) \quad (2.40)$$

in our toy example, because in that case we have

$$\mathbb{E}[\overline{V}(p_1 + p_2, f, \pi_n)_T | \mathcal{S}] = \sum_{k=1}^{N} h_k(\sigma^{(1)}, \sigma^{(2)}, \rho)$$

$$\times n^{(p_1+p_2)/2-1} \sum_{i,j:t_{i,n}^{(1)} \vee t_{j,n}^{(2)} \leq T} g_k(|\mathcal{I}_{i,n}^{(1)}|, |\mathcal{I}_{j,n}^{(2)}|, |\mathcal{I}_{j,n}^{(2)} \cap \mathcal{I}_{j,n}^{(2)}|) \mathbb{1}_{\{\mathcal{I}_{i,n}^{(1)} \cap \mathcal{I}_{j,n}^{(2)} \neq \emptyset\}}$$

where the right-hand side converges if we assume that the expression in the last line converges as $n \to \infty$ for each $k = 1, \ldots, N$.

It is in general not possible to find a representation of the form (2.40) for an arbitrary function f which is positively homogeneous in both arguments. However, there are two interesting cases where such a representation is available. The first case is when $X^{toy,(1)}$ and $X^{toy,(2)}$ are uncorrelated, i.e. if $\rho \equiv 0$ on $[0,T]$, because then

$$\mathbb{E}[f(\Delta_{i,n}^{(1)} X^{toy,(1)}, \Delta_{j,n}^{(2)} X^{toy,(2)})|\mathcal{S}]$$
$$= (\sigma^{(1)})^{p_1} (\sigma^{(2)})^{p_2} \mathbb{E}[f(Z, Z')] |\mathcal{I}_{i,n}^{(1)}|^{p_1/2} |\mathcal{I}_{j,n}^{(2)}|^{p_2/2}$$

holds for independent standard normal random variables Z, Z'. The result obtained in this case is stated in Theorem 2.21. In the second case we consider the functions $f_{(p_1,p_2)}$ with $p_1, p_2 \in \mathbb{N}_0$. Indeed it holds

$$f_{(p_1,p_2)}(\Delta_{i,n}^{(1)} X^{toy,(1)}, \Delta_{j,n}^{(2)} X^{toy,(2)})$$
$$\stackrel{\mathcal{LS}}{=} (\sigma^{(1)})^{p_1} (\sigma^{(2)})^{p_2} (|\mathcal{I}_{i,n}^{(1)} \setminus \mathcal{I}_{j,n}^{(2)}|^{1/2} Z_1 + |\mathcal{I}_{i,n}^{(1)} \cap \mathcal{I}_{j,n}^{(2)}|^{1/2} Z_2)^{p_1}$$
$$\times (|\mathcal{I}_{i,n}^{(1)} \cap \mathcal{I}_{j,n}^{(2)}|^{1/2} (\rho Z_2 + \sqrt{1-\rho^2} Z_3) + |\mathcal{I}_{j,n}^{(2)} \setminus \mathcal{I}_{i,n}^{(1)}|^{1/2} Z_4)^{p_2}$$

where $\stackrel{\mathcal{LS}}{=}$ denotes equality of the $\mathcal{S}$-conditional law and $Z_1, \ldots, Z_4$ are i.i.d. standard normal random variables. Here the right-hand side can be brought into the form (2.40) using the multinomial theorem. The result obtained in this case is stated in Theorem 2.22. Unfortunately, for the functions $\overline{f}_{(p_1,p_2)}$ there exists no similar representation: however, for even p_1, p_2 we have $f_{(p_1,p_2)} = \overline{f}_{(p_1,p_2)}$ and Theorem 2.22 also applies there.

Theorem 2.21. *Let $p_1, p_2 \geq 0$ and suppose that Condition 1.3 is fulfilled, that the observation scheme is exogenous and that we have $\rho \equiv 0$ on $[0,T]$. Further assume that*

$$G_{p_1,p_2}^n(t) \xrightarrow{\mathbb{P}} G_{p_1,p_2}(t), \ t \in [0,T],$$

for a (possibly random) continuous function $G_{p_1,p_2} \colon [0,T] \to \mathbb{R}_{\geq 0}$ and that we have one of the following two conditions:

a) $p_1 + p_2 \in [0,2)$,

b) $p_1 + p_2 \geq 2$ and X is continuous.

Then for all continuous functions $f \colon \mathbb{R}^2 \to \mathbb{R}$ which are positively homogeneous of degree p_1 in the first argument and positively homogeneous of degree p_2 in the second argument it holds that

$$\overline{V}(p_1 + p_2, f, \pi_n)_T \xrightarrow{\mathbb{P}} m_{I_2}(f) \int_0^T (\sigma_s^{(1)})^{p_1} (\sigma_s^{(2)})^{p_2} dG_{p_1,p_2}(s). \qquad (2.41)$$

Here I_2 denotes the two-dimensional identity matrix.

Theorem 2.22. *Let $p_1, p_2 \in \mathbb{N}_0$ and suppose that Condition 1.3 is fulfilled and that the observation scheme is exogenous. Define*

$$L(p_1, p_2) = \{(k, l, m) \in (2\mathbb{N}_0)^3 : k \le p_1, l + m \le p_2, p_1 + p_2 - (k + l + m) \in 2\mathbb{N}_0\}.$$

Assume that for all $k, m \in \mathbb{N}_0$ for which an $l \in \mathbb{N}_0$ exists with $(k, l, m) \in L(p_1, p_2)$ there exist (possibly random) continuous functions $H_{k,m,p_1+p_2} : [0, T] \to \mathbb{R}_{\ge 0}$ which fulfill

$$H^n_{k,m,p_1+p_2}(t) \xrightarrow{\mathbb{P}} H_{k,m,p_1+p_2}(t), \quad t \in [0, T].$$

Further we assume that we have one of the following two conditions:

a) $p_1 + p_2 \in \{0, 1\}$,

b) $p_1 + p_2 \ge 2$ and X is continuous.

Then

$$\overline{V}(p_1 + p_2, f_{(p_1,p_2)}, \pi_n)_T$$
$$\xrightarrow{\mathbb{P}} \int_0^T (\sigma_s^{(1)})^{p_1} (\sigma_s^{(2)})^{p_2} \Big(\sum_{(k,l,m) \in L(p_1,p_2)} \binom{p_1}{k} \binom{p_2}{l,m} m_1(x^k) m_1(x^l) m_1(x^m)$$
$$\times m_1(x^{p_1+p_2-(k+l+m)})(1 - \rho_s^2)^{l/2} (\rho_s)^{p_2-(l+m)} dH_{k,m,p_1+p_2}(s) \Big) \quad (2.42)$$

holds. Here $\binom{p_2}{l,m} = p_2!/(l!m!(p_2 - l - m)!)$ denotes the multinomial coefficient.

Example 2.23. *In this example we will use Theorem 2.22 to find the limit of $\overline{V}(p_1 + p_2, f_{(p_1,p_2)}, \pi_n)_T$ for a few non-trivial cases with small p_1, p_2. For $p_1 = p_2 = 1$ the set $L(1, 1)$ contains only $(0, 0, 0)$ and we get*

$$\overline{V}(2, f_{(1,1)}, \pi_n) \xrightarrow{\mathbb{P}} \int_0^T \rho_s \sigma_s^{(1)} \sigma_s^{(2)} dH_{0,0,2}(s) = \int_0^T \rho_s \sigma_s^{(1)} \sigma_s^{(2)} ds$$

as $H_{0,0,2}(t) \xrightarrow{\mathbb{P}} \int_0^T t$. Hence $\overline{V}(2, f_{(1,1)}, \pi_n)$ converges to the covariation of $X^{(1)}, X^{(2)}$ for continuous processes X and we have retrieved (2.4) for continuous semimartingales X. For $p_1 = 1$, $p_2 = 2$ we get $L(1, 2) = \emptyset$ as k, l, m and $3 - (k + l + m)$ cannot be all divisible by 2. Hence $\overline{V}(3, f_{(1,2)}, \pi_n) \xrightarrow{\mathbb{P}} 0$. This holds for all (p_1, p_2) where $p_1 + p_2$ is odd.

Further we define

$$G^n_{k,m,p}(t) = n^{p/2-1} \sum_{i,j : t^{(1)}_{i,n} \vee t^{(2)}_{j,n} \le t} |\mathcal{I}^{(1)}_{i,n}|^{k/2} |\mathcal{I}^{(2)}_{j,n}|^{m/2}$$
$$\times |\mathcal{I}^{(1)}_{i,n} \cap \mathcal{I}^{(2)}_{j,n}|^{(p-k-m)/2} \mathbb{1}_{\{\mathcal{I}^{(1)}_{i,n} \cap \mathcal{I}^{(2)}_{j,n} \neq \emptyset\}}$$

and define $G_{k,m,p}(t)$ as the limit of $G_{k,m,p}^n(t)$ in probability as $n \to \infty$ if it exists. For $p_1 = p_2 = 2$ we have to consider the set $L(2,2) = \{0,2\} \times \{(0,0),(0,2),(2,0)\}$ and then obtain using the above notation

$$\overline{V}(4, f_{(2,2)}, \pi_n) \xrightarrow{\mathbb{P}} \int_0^T (\sigma_s^{(1)} \sigma_s^{(2)})^2 (3\rho_s^2 dH_{0,0,4}(s) + dH_{0,2,4}(s)$$

$$+ (1-\rho_s^2)dH_{0,0,4}(s) + \rho_s^2 dH_{2,0,4}(s) + dH_{2,2,4}(s) + (1-\rho_s^2)dH_{2,0,4}(s))$$

$$= \int_0^T (\sigma_s^{(1)} \sigma_s^{(2)})^2 (2\rho_s^2 dH_{0,0,4}(s) + dG_{2,2,4}(s)) \tag{2.43}$$

where we used $G_{2,2,p}(s) = H_{0,0,p}(s) + H_{0,2,p}(s) + H_{2,0,p}(s) + H_{2,2,p}(s)$, $p \geq 4$, which follows from the identity

$$|\mathcal{I}_{i,n}^{(1)}||\mathcal{I}_{j,n}^{(2)}| = (|\mathcal{I}_{i,n}^{(1)} \cap \mathcal{I}_{j,n}^{(2)}| + |\mathcal{I}_{i,n}^{(1)} \setminus \mathcal{I}_{j,n}^{(2)}|)(|\mathcal{I}_{i,n}^{(1)} \cap \mathcal{I}_{j,n}^{(2)}| + |\mathcal{I}_{j,n}^{(2)} \setminus \mathcal{I}_{i,n}^{(1)}|).$$

Without presenting detailed computations we state two more results to demonstrate that the limit in Theorem 2.22 after simplification sometimes has a much shorter representation compared to the general form in (2.42). For $p_1 = p_2 = 3$ we have $L(3,3) = L(2,2)$ and we get after simplification

$$\overline{V}(6, f_{(3,3)}, \pi_n) \xrightarrow{\mathbb{P}} \int_0^T (\sigma_s^{(1)} \sigma_s^{(2)})^3 (6\rho_s^3 dH_{0,0,6}(s) + 9\rho_s dG_{2,2,6}(s))$$

and for $p_1 = p_2 = 4$ we obtain

$$\overline{V}(8, f_{(4,4)}, \pi_n)$$

$$\xrightarrow{\mathbb{P}} \int_0^T (\sigma_s^{(1)} \sigma_s^{(2)})^4 (24\rho_s^4 dH_{0,0,8}(s) + 72\rho_s^2 dG_{2,2,8}(s) + 9dG_{4,4,8}(s)).$$

$\square$

Two not very difficult generalizations can be made for the statements in Theorems 2.17, 2.21 and 2.22. The previous results were only stated in a more specific form to keep the notation and the proofs clearer and to direct the reader's focus to the key aspects. First, throughout this section the rate n was chosen rather arbitrarily as the appropriate scaling factor by which the average interval lengths decrease and such that we obtain convergence for the functions $G_p^{(l),n}(t)$, $G_{p_1,p_2}^n(t)$ and $H_{k,m,p}^n(t)$.

Remark 2.24. *Let $r : \mathbb{N} \to [0,\infty)$ be a function with $r(n) \to \infty$ for $n \to \infty$. Then we obtain the same result as in Corollary 2.19 if we set*

$$\overline{V}^{(l)}(p, g, \pi_n)_T = (r(n))^{p/2-1} \sum_{i:t_{i,n}^{(l)} \leq T} g(\Delta_{i,n}^{(l)} X^{(l)}),$$

$$G_p^{(l),n}(t) = (r(n))^{p/2-1} \sum_{i:t_{i,n}^{(l)} \leq t} |\mathcal{I}_{i,n}^{(l)} X^{(l)}|^{p/2}.$$

Similarly the results from Theorems 2.17, 2.21 and 2.22 hold as well if we replace n by $r(n)$ in the definition of the functional $\overline{V}(p, f, \pi_n)_T$ and the functions $G_{p_1,p_2}^n(t)$, $H_{k,m,p}^n(t)$. The proofs for these claims are identical to the proofs in the more specific case $r(n) = n$. Hence we only need that the observation scheme scales with a deterministic rate $r(n)$ to obtain the results in this section. □

Further, only increments of the continuous martingale part of X contribute to the limits in Theorems 2.17, 2.21 and 2.22 and the increments of the continuous part of X tend to get very small as the observation intervals become shorter. Hence only function evaluations $f(x)$ at very small x and especially the behaviour of $f(x)$ for $x \to 0$ has an influence on the asymptotics. These arguments motivate that the convergences in the above theorems do not only hold for positively homogeneous functions but as the following corollary shows also for functions f which are very close to being positively homogeneous for $x \to 0$; compare Corollary 3.4.3 in [30].

Corollary 2.25. *Suppose that the convergence in one of the Theorems 2.17, 2.21 and 2.22 holds for the function $f \colon \mathbb{R}^2 \to \mathbb{R}$. Then the corresponding convergence also holds for all functions $\tilde{f} \colon \mathbb{R}^2 \to \mathbb{R}$ which can be written as*

$$\tilde{f}(x) = L(x)f(x)$$

for a locally bounded function $L(x)$ that fulfils $\lim_{x\to 0} L(x) = 1$.

Example 2.26. *In the setting of Poisson sampling the assumptions of Theorems 2.17, 2.21 and 2.22 are fulfilled. Indeed the functions $G_p^{(l),n}$, G_{p_1,p_2}^n, $H_{k,m,p}^n$ converge by Corollary 5.6 to deterministic linear and hence continuous functions. Although the functions $G_p^{(l)}, G_{p_1,p_2}, H_{k,m,p}$ are in general unknown they can easily be estimated by $G_p^{(l),n}, G_{p_1,p_2}^n, H_{k,m,p}^n$.* □

2.2.2 The Proofs

We will start this section by deriving some elementary inequalities for positively homogeneous functions. First, we need the following estimate for increments of the functions $x \mapsto \|x\|^p$.

Lemma 2.27. *Let $p \geq 0$ and $x, y \in \mathbb{R}^d$. Then it holds*

$$\left| \|x + y\|^p - \|x\|^p \right| \leq \begin{cases} \|y\|^p & p \in [0, 1], \\ K_p(\varepsilon^p \|x\|^p + K_{p,\varepsilon} \|y\|^p) & p > 1, \end{cases}$$

for all $\varepsilon > 0$ where K_p and $K_{p,\varepsilon}$ denote constants depending only on p respectively on p and ε.

Proof. The case $p = 0$ is trivial. For all $a, b \geq 0$ and $p \in (0, 1]$ it holds

$$a^p = \int_0^a p z^{p-1} dz \geq \int_b^{a+b} p z^{p-1} dz = (a+b)^p - b^p$$

where the inequality holds because of $p > 0$ and $z \mapsto z^{p-1}$ being monotonically decreasing. Using this inequality we obtain

$$\|x + y\|^p \leq (\|x\| + \|y\|)^p \leq \|x\|^p + \|y\|^p \qquad (2.44)$$

which yields the claim (together with the analogous inequality for $x = (x + y) - y$) in the case $p \in (0, 1]$.

In the case $p > 1$ the mean value theorem applied for the function $x \mapsto x^p$ states that there exists a $\xi \in [0, 1]$ with

$$\left| \|x + y\|^p - \|x\|^p \right| = \left| (\|x + y\| - \|x\|) p (\xi \|x + y\| + (1 - \xi)\|x\|)^{p-1} \right|$$
$$\leq p\|y\|(\|x + y\| + \|x\|)^{p-1}$$
$$\leq 2^{p-1} p\|y\|(\|y\| + \|x\|)^{p-1}$$

where the first inequality follows because $\|x\| \mapsto \|x\|^{p-1}$ is monotonically increasing on $[0, \infty)$ and hence the maximum of $(\xi\|x + y\| + (1 - \xi)\|y\|)^{p-1}$ is attained at $\xi = 0$ or $\xi = 1$. From Jensen's inequality we obtain

$$\left(\frac{a+b}{2} \right)^{p-1} \leq \frac{a^{p-1}}{2} + \frac{b^{p-1}}{2}$$

for $p > 2$. Using this relation and inequality (2.44) for $p \in (1, 2]$ yields

$$\left| \|x + y\|^p - \|x\|^p \right| \leq K_p (\|y\|^p + \|y\|\|x\|^{p-1}).$$

Further we obtain from Muirhead's inequality, compare Theorem 45 in [19], for $a_1 = \|y\|\varepsilon^{-(p-1)}$, $a_2 = \|x\|\varepsilon$ and the exponent vectors $(p - 1, 1) \prec (p, 0)$

$$\|y\|\|x\|^{p-1} \leq (\|y\|\varepsilon^{-(p-1)})(\varepsilon\|x\|)^{p-1} + (\|y\|\varepsilon^{-(p-1)})^{p-1}(\varepsilon\|x\|)$$
$$\leq \varepsilon^{-p(p-1)}\|y\|^p + \varepsilon^p\|x\|^p$$

which yields the claim. $\qquad\square$

Using the above lemma we derive the following estimates for continuous and positively homogeneous functions.

Lemma 2.28. *Let* $f \colon \mathbb{R}^{d_1} \times \mathbb{R}^{d_2} \to \mathbb{R}$, $d_1, d_2 \in \mathbb{N}_0$, *be a continuous function which is positively homogeneous with degree* $p_1 \geq 0$ *in the first argument and with degree* $p_2 \geq 0$ *in the second argument. Then there exists a constant* K *with*

$$|f(x_1, x_2)| \leq K \|x_1\|^{p_1} \|x_2\|^{p_2} \quad \forall x_1 \in \mathbb{R}^{d_1}, x_2 \in \mathbb{R}^{d_2}. \tag{2.45}$$

Further there exists a function $\theta \colon [0, \infty) \to [0, \infty)$ *depending on* f *and* p *with* $\theta(\varepsilon) \to 0$ *for* $\varepsilon \to 0$ *and a constant* $K_{p_1, p_2, \varepsilon}$ *which may depend on* f *such that*

$$|f(x_1 + y_1, x_2 + y_2) - f(x_1, x_2)| \leq \theta(\varepsilon) \|x_1\|^{p_1} \|x_2\|^{p_2}$$
$$+ K_{p_1, p_2, \varepsilon}(\|y_1\|^{p_1}(\|x_2\|^{p_2} + \|y_2\|^{p_2}) + \|y_2\|^{p_2}(\|x_1\|^{p_1} + \|y_1\|^{p_1})) \tag{2.46}$$

holds for all $x_1, y_1 \in \mathbb{R}^{d_1}$ *and* $x_2, y_2 \in \mathbb{R}^{d_2}$. *In the case* $p_2 = 0$ *the inequality* (2.46) *can be replaced by*

$$|f(x_1 + y_1, x_2 + y_2) - f(x_1, x_2)| \leq \theta(\varepsilon) \|x_1\|^{p_1} + K_{p_1, \varepsilon} \|y_1\|^{p_1} \tag{2.47}$$

for all $x_1, y_1 \in \mathbb{R}^{d_1}$ *and* $x_2, y_2 \in \mathbb{R}^{d_2}$. *The analogous result holds if* $p_1 = 0$.

Proof. From the defining property of a positively homogeneous function we obtain

$$f(x_1, x_2) = \|x_1\|^{p_1} \|x_2\|^{p_2} f(x_1 / \|x_1\|, x_2 / \|x_2\|)$$
$$\leq \|x_1\|^{p_1} \|x_2\|^{p_2} \sup_{(z_1, z_2) \in \mathbb{R}^d : \|z_1\| = \|z_2\| = 1} |f(z)|.$$

(2.45) then follows because the function f is continuous which yields that f is bounded on any compact set.

Next we prove (2.47). To this end note that for $p_2 = 0$ it holds

$$f(x, y) = \lim_{\lambda \downarrow 0} \lambda^0 f(x, y) = \lim_{\lambda \downarrow 0} f(x, \lambda y) = f(x, 0) \tag{2.48}$$

by the continuity of f for any $x \in \mathbb{R}^{d_1}$ and $y \in \mathbb{R}^{d_2}$. Hence for $p_2 = 0$ the function f does not depend on the second argument at all and for showing (2.47) it suffices to prove that

$$|\tilde{f}(x + y) - \tilde{f}(x)| \leq \theta(\varepsilon) \|x\|^{p_1} + K_{p_1, \varepsilon} \|y\|^{p_1} \tag{2.49}$$

holds for any $x, y \in \mathbb{R}^{d_1}$ where $\tilde{f} \colon \mathbb{R}^{d_1} \to \mathbb{R}$ denotes a continuous and positively homogeneous function with degree $p_1 \geq 0$ and θ denotes a function as described above.

For proving (2.49) we observe

$$|\tilde{f}(x + y) - \tilde{f}(x)| = \left| \|x + y\|^{p_1} \tilde{f}\left(\frac{x + y}{\|x + y\|}\right) - \|x\|^{p_1} \tilde{f}\left(\frac{x}{\|x\|}\right) \right|$$
$$\leq \left| \|x + y\|^{p_1} - \|x\|^{p_1} \right| \sup_{z \in \mathbb{R}^{d_1} : \|z\| = 1} |\tilde{f}(z)| + \|x\|^{p_1} \left| \tilde{f}\left(\frac{x + y}{\|x + y\|}\right) - \tilde{f}\left(\frac{x}{\|x\|}\right) \right|. \tag{2.50}$$

Defining

$$\tilde{\theta}(\varepsilon) = \sup_{x,y \in \mathbb{R}^{d_1} : \|x\| = \|y\| = 1 \wedge \|x-y\| \le \varepsilon} |\tilde{f}(x) - \tilde{f}(y)|$$

we obtain

$$\left| \tilde{f}\left(\frac{x+y}{\|x+y\|}\right) - \tilde{f}\left(\frac{x}{\|x\|}\right) \right| \le \tilde{\theta}(\varepsilon) + \left(2 \sup_{z \in \mathbb{R}^{d_1} : \|z\| = 1} |\tilde{f}(z)|\right) \varepsilon^{-p_1} \left\| \frac{x+y}{\|x+y\|} - \frac{x}{\|x\|} \right\|^{p_1}$$

$$\le \tilde{\theta}(\varepsilon) + K_{p_1,\varepsilon} \|x+y\|^{p_1} \left| \frac{1}{\|x+y\|} - \frac{1}{\|x\|} \right|^{p_1} + K_{p_1,\varepsilon} \frac{\|y\|^{p_1}}{\|x\|^{p_1}} \le \tilde{\theta}(\varepsilon) + K_{p_1,\varepsilon} \frac{\|y\|^{p_1}}{\|x\|^{p_1}}$$

where we used $(a+b)^{p_1} \le K_{p_1}(a^{p_1} + b^{p_1})$ for all $p_1 \ge 0$ and $a, b \ge 0$ which we showed in the proof of Lemma 2.27. Using this estimate in (2.50) we get

$$|\tilde{f}(x+y) - \tilde{f}(x)| \le K |\|x+y\|^{p_1} - \|x\|^{p_1}| + \tilde{\theta}(\varepsilon)\|x\|^{p_1} + K_\varepsilon \|y\|^{p_1}$$

which together with Lemma 2.27 yields (2.49). Note therefore that $\tilde{\theta}(\varepsilon) \to 0$ as $\varepsilon \to 0$ because f is uniformly continuous on the compact unit sphere.

The proof of (2.46) is as the proof of (2.47) mainly based on Lemma 2.27 and uses very similar estimates. It is therefore skipped here.   □

To discuss the synchronous setting and the asynchronous setting simultaneously we consider a $(d_1 + d_2)$-dimensional Itô semimartingale $\widetilde{X}$ of the form (1.1). $\widetilde{X}^{(1)}$ denotes the vector-valued process containing the first d_1 components of $\widetilde{X}$ and is observed at the observation times $t_{i,n}^{(1)}$, $\widetilde{X}^{(2)}$ contains the remaining d_2 components of $\widetilde{X}$ and is observed at the observation times $t_{i,n}^{(2)}$. Then the synchronous setting and the asynchronous setting discussed in Section 2.2 correspond to $d_1 = 2$, $d_2 = 0$, $p_1 = p$, $p_2 = 0$, and $d_1 = d_2 = 1$ respectively. Here the notion of a zero-dimensional semimartingale remains ambiguous. However, we will only plug in increments of this zero-dimensional process into the argument of f in which f is positively homogeneous of degree $p_2 = 0$ and then f constant in this argument; compare (2.48). Hence it is not necessary to specify the notion of a zero-dimensional semimartingale as we are going to use it only to indicate the case where the function f solely depends on the first argument. Further we define

$$\overline{V}^*(p, f, \pi_n)_T = n^{p/2-1} \sum_{i,j \ge 0 : t_{i,n}^{(1)} \vee t_{j,n}^{(2)} \le T} f(\Delta_{i,n}^{(1)} \widetilde{X}^{(1)}, \Delta_{j,n}^{(2)} \widetilde{X}^{(2)}) \mathbf{1}_{d_1,d_2}^{*,n}(i,j)$$

$$(2.51)$$

for all functions $f \colon \mathbb{R}^{d_1} \times \mathbb{R}^{d_2} \to \mathbb{R}$ where we set $\mathcal{I}_{0,n}^{(l)} = \emptyset$, $\Delta_{0,n}^{(l)} X^{(l)} = 0$, $l = 1, 2$, and

$$\mathbf{1}_{d_1,d_2}^{*,n}(i,j) = \begin{cases} \mathbf{1}_{\{\mathcal{I}_{i,n}^{(1)} \cap \mathcal{I}_{j,n}^{(2)} \ne \emptyset\}}, & d_1 > 0, d_2 > 0, \\ \mathbf{1}_{\{i>0, j=0\}}, & d_1 > 0, d_2 = 0, \\ \mathbf{1}_{\{i=0, j>0\}}, & d_1 = 0, d_2 > 0. \end{cases} \quad (2.52)$$

That means, we start the sum in (2.51) at zero and define the indicator $\mathbf{1}^{*,n}_{d_1,d_2}(i,j)$ in such a way that, whenever $d_{3-l} = 0$, for any $i \in \mathbb{N}$ with $t^{(l)}_{i,n} \leq T$ exactly one summand depending on $\Delta^{(l)}_{i,n} \widetilde{X}^{(l)}$ occurs in the sum. Hence $\overline{V}^*(p, f, \pi_n)_T$ corresponds to $\overline{V}(p, f, \pi_n)_T$ in the asynchronous setting for $d_1 = d_2 = 1$ and $\overline{V}^*(p, f, \pi_n)_T$ corresponds to $\overline{V}(p, f, \pi_n)_T$ in the synchronous setting for $d_l = 2$, $d_{3-l} = 0$.

Definition 2.29. *We denote by*

$$\overline{C}^*(p, f, \pi_n)_T = n^{p/2-1} \sum_{i,j:t^{(1)}_{i,n} \vee t^{(2)}_{j,n} \leq T} f\big(\Delta^{(1)}_{i,n} \widetilde{C}^{(1)}, \Delta^{(2)}_{j,n} \widetilde{C}^{(2)}\big) \mathbf{1}^{*,n}_{d_1,d_2}(i,j)$$

the functional $\overline{V}^*(p, f, \pi_n)_T$ *evaluated at the continuous martingale part* $\widetilde{C}$ *instead of at* $\widetilde{X}$ *itself.* $\overline{C}(p, f, \pi_n)_T$ *is defined as* $\overline{C}^*(p, f, \pi_n)_T$ *above only with* $\widetilde{C}$ *replaced by* C. $\qquad\square$

As for the indicator in (2.52) we also define a unifying notation for the functions $G^n_{p_1,p_2}(t)$ and $G^{(l),n}_p$ via

$$\widetilde{G}^{(d_1,d_2),n}_{p_1,p_2}(t) = \begin{cases} G^n_{p_1,p_2}(t), & d_1 > 0, d_2 > 0, \\ G^{(1),n}_{p_1}(t), & d_1 > 0, d_2 = 0, \\ G^{(2),n}_{p_2}(t), & d_1 = 0, d_2 > 0. \end{cases} \tag{2.53}$$

The following proposition yields that by specifying d_1, d_2 appropriately as discussed above, it suffices to prove the convergences in Theorems 2.17, 2.21 and 2.22 for $\overline{C}(p, f, \pi_n)_T$, $\overline{C}(p_1+p_2, f, \pi_n)_T$ instead of $\overline{V}(p, f, \pi_n)_T$, $\overline{V}(p_1+p_2, f, \pi_n)_T$.

Proposition 2.30. *Let* $f : \mathbb{R}^{d_1} \times \mathbb{R}^{d_2} \to \mathbb{R}$ *be a function as in Lemma 2.28. Suppose that* $\widetilde{G}^{(d_1,d_2),n}_{p_1,p_2}(T) = O_{\mathbb{P}}(1)$ *and let either* $p_1 + p_2 \in [0,2)$ *or* $p_1 + p_2 \geq 2$ *and assume that* $\widetilde{X}$ *is continuous. Further we assume* $d_l = 0 \Rightarrow p_l = 0$, $l = 1, 2$. *Then it holds that*

$$\overline{V}^*(p_1 + p_2, f, \pi_n)_T - \overline{C}^*(p_1 + p_2, f, \pi_n)_T \xrightarrow{\mathbb{P}} 0 \tag{2.54}$$

as $n \to \infty$.

Proof. In the following, we denote by $\widetilde{B}(q), \widetilde{C}, \widetilde{M}(q), \widetilde{N}(q)$ a decomposition of $\widetilde{X}$ similar to (1.7). Using (2.46) we obtain

$$\big|\overline{V}^*(p_1 + p_2, f, \pi_n)_T - \overline{C}^*(p_1 + p_2, f, \pi_n)_T\big|$$

$$\leq n^{(p_1+p_2)/2-1} \sum_{i,j:t^{(1)}_{i,n} \vee t^{(2)}_{j,n} \leq T} \mathbf{1}^{*,n}_{d_1,d_2}(i,j)\big[\theta(\varepsilon)\|\Delta^{(1)}_{i,n}\widetilde{C}^{(1)}\|^{p_1}\|\Delta^{(2)}_{j,n}\widetilde{C}^{(2)}\|^{p_2} \tag{2.55}$$

$$+ K_{p_1,p_2,\varepsilon}\|\Delta^{(1)}_{i,n}(\widetilde{X} - \widetilde{C})^{(1)}\|^{p_1}(\|\Delta^{(?)}_{j,n}\widetilde{C}^{(2)}\|^{p_2} + \|\Delta^{(2)}_{j,n}(\widetilde{X} - \widetilde{C})^{(2)}\|^{p_2}) \tag{2.56}$$

$$+ K_{p_1,p_2,\varepsilon}\|\Delta^{(2)}_{j,n}(\widetilde{X} - \widetilde{C})^{(2)}\|^{p_2}(\|\Delta^{(1)}_{i,n}\widetilde{C}^{(1)}\|^{p_1} + \|\Delta^{(1)}_{i,n}(\widetilde{X} - \widetilde{C})^{(1)}\|^{p_1})\big]. \tag{2.57}$$

For (2.55) we get using the Cauchy-Schwarz inequality and inequality (1.9), as Lemma 1.4 holds for Itô semimartingales of arbitrary dimension,

$$\mathbb{E}\Big[n^{(p_1+p_2)/2-1} \sum_{i,j:t_{i,n}^{(1)}\vee t_{j,n}^{(2)}\leq T} \theta(\varepsilon)\|\Delta_{i,n}^{(1)}\widetilde{C}^{(1)}\|^{p_1}\|\Delta_{j,n}^{(2)}\widetilde{C}^{(2)}\|^{p_2}\mathbf{1}_{d_1,d_2}^{*,n}(i,j)\Big|\mathcal{S}\Big]$$

$$\leq \theta(\varepsilon)K\widetilde{G}_{p_1,p_2}^{(d_1,d_2),n}(T).$$

Hence by the assumption $\widetilde{G}_{p_1,p_2}^{(d_1,d_2),n}(T) = O_\mathbb{P}(1)$ and Lemma 2.15 we obtain

$$\lim_{\varepsilon\to 0}\limsup_{n\to 0}\mathbb{P}\Big(\theta(\varepsilon)n^{(p_1+p_2)/2-1}\sum_{i,j:t_{i,n}^{(1)}\vee t_{j,n}^{(2)}\leq T}\|\Delta_{i,n}^{(1)}\widetilde{C}^{(1)}\|^{p_1}\|\Delta_{j,n}^{(2)}\widetilde{C}^{(2)}\|^{p_2}\mathbf{1}_{d_1,d_2}^{*,n}(i,j) > \delta\Big) = 0$$

for any $\delta > 0$. To prove (2.54) it now remains to show that (2.56) vanishes as $n\to\infty$ for any $\varepsilon > 0$ because then (2.57) can be dealt with analogously by symmetry. We will separately discuss the cases $p_1 > 0$ and $p_1 = 0$.

Case 1. We first consider the case where $p_1 > 0$. In the situation $p_1 + p_2 \geq 2$ we have $\widetilde{X} = \widetilde{B} + \widetilde{C}$ with $\widetilde{B}_t = \int_0^t \widetilde{b}_s ds$ for some bounded process $\widetilde{b}$. Hence we get that the $\mathcal{S}$-conditional expectation of (2.56) is bounded by

$$K_{p_1,p_2,\varepsilon}n^{(p_1+p_2)/2-1}\sum_{i,j:t_{i,n}^{(1)}\vee t_{j,n}^{(2)}\leq T}|\mathcal{I}_{i,n}^{(1)}|^{p_1}(|\mathcal{I}_{j,n}^{(2)}|^{p_2}+|\mathcal{I}_{j,n}^{(2)}|^{p_2/2})\mathbf{1}_{d_1,d_2}^{*,n}(i,j)$$

$$\leq K_{p_1,p_2,\varepsilon}(|\pi_n|_T)^{p_1/2}\widetilde{G}_{p_1,p_2}^{(d_1,d_2),n}(T)$$

which vanishes as $n\to\infty$ for $p_1 > 0$.

Next we consider (2.56) in the case $p_1 + p_2 < 2$. Using

$$\|\Delta_{i,n}^{(1)}(\widetilde{X}-\widetilde{C})^{(1)}\|^{p_1} \leq K_{p_1}(\|\Delta_{i,n}^{(1)}(\widetilde{B}(q)+\widetilde{M}(q))^{(1)}\|^{p_1}+\|\Delta_{i,n}^{(1)}\widetilde{N}^{(1)}(q)\|^{p_1})$$

allows to treat the different components of $\Delta_{i,n}^{(1)}(\widetilde{X}-\widetilde{C})$ separately. Applying Hölder's inequality for $p' = 2/(2-p_2)$, $q' = 2/p_2$ and using the inequalities from Lemma 1.4 (note that $p_1 < 2 - p_2$) yields

$$\mathbb{E}\Big[n^{(p_1+p_2)/2-1}\sum_{i,j:t_{i,n}^{(1)}\vee t_{j,n}^{(2)}\leq T}K_{p_1,p_2,\varepsilon}\|\Delta_{i,n}^{(1)}(\widetilde{B}(q)+\widetilde{M}(q))^{(1)}\|^{p_1}$$

$$\times (\|\Delta_{j,n}^{(2)}\widetilde{C}^{(2)}\|^{p_2}+\|\Delta_{j,n}^{(2)}(\widetilde{X}-\widetilde{C})^{(2)}\|^{p_2})\mathbf{1}_{d_1,d_2}^{*,n}(i,j)\Big|\mathcal{S}\Big]$$

$$\leq K_{p_1,p_2,\varepsilon}n^{(p_1+p_2)/2-1}\sum_{i,j:t_{i,n}^{(1)}\vee t_{j,n}^{(2)}\leq T}\mathbb{E}[\|\Delta_{i,n}^{(1)}(\widetilde{B}(q)+\widetilde{M}(q))^{(1)}\|^{\frac{2p_1}{2-p_2}}|\mathcal{S}]^{\frac{2-p_2}{2}}$$

$$\times (\mathbb{E}[\|\Delta_{j,n}^{(2)}\widetilde{C}^{(2)}\|^2|\mathcal{S}]^{p_2/2}+\mathbb{E}[\|\Delta_{j,n}^{(2)}(\widetilde{X}-\widetilde{C})^{(2)}\|^2|\mathcal{S}]^{p_2/2})\mathbf{1}_{d_1,d_2}^{*,n}(i,j)$$

$$\leq K_{p_1,p_2,\varepsilon}(K_q(|\pi_n|_T)^{p_1/2}+(e_q)^{p_1/2})\widetilde{G}_{p_1,p_2}^{(d_1,d_2),n}(T)$$

which vanishes as $n, q \to \infty$ for any $\varepsilon > 0$ if $p_1 > 0$.

Finally consider

$$n^{(p_1+p_2)/2-1} K_{p_1,p_2,\varepsilon} \sum_{i,j:t_{i,n}^{(1)} \vee t_{j,n}^{(2)} \leq T} \|\Delta_{i,n}^{(1)} \widetilde{N}^{(1)}(q)\|^{p_1}$$

$$\times (\|\Delta_{j,n}^{(2)} \widetilde{C}^{(2)}\|^{p_2} + \|\Delta_{j,n}^{(2)} (\widetilde{X} - \widetilde{C})^{(2)}\|^{p_2}) \mathbf{1}_{d_1,d_2}^{*,n}(i,j)$$

$$\leq n^{(p_1+p_2)/2-1} K_{p_1,p_2,\varepsilon} \sum_{i,j:t_{i,n}^{(1)} \vee t_{j,n}^{(2)} \leq T} \|\Delta_{i,n}^{(1)} \widetilde{N}^{(1)}(q)\|^{p_1}$$

$$\times (\|\Delta_{j,n}^{(2)} \widetilde{B}^{(2)}(q)\|^{p_2} + \|\Delta_{j,n}^{(2)} \widetilde{C}^{(2)}\|^{p_2} + \|\Delta_{j,n}^{(2)} \widetilde{M}^{(2)}(q)\|^{p_2}) \mathbf{1}_{d_1,d_2}^{*,n}(i,j) \quad (2.58)$$

$$+ n^{(p_1+p_2)/2-1} K_{p_1,p_2,\varepsilon} \sum_{i,j:t_{i,n}^{(1)} \vee t_{j,n}^{(2)} \leq T} \|\Delta_{i,n}^{(1)} \widetilde{N}^{(1)}(q)\|^{p_1} \|\Delta_{j,n}^{(2)} \widetilde{N}^{(2)}(q)\|^{p_2}$$

$$\times \mathbf{1}_{d_1,d_2}^{*,n}(i,j). \quad (2.59)$$

(2.59) vanishes as $n \to \infty$ due to $p_1 + p_2 < 2$ and because the finitely many jumps of $N(q)$ are asymptotically separated by the observation scheme. Further choose $\delta > 0$ such that $p_1 \vee 1 < 2 - \delta$, $2(p_1 + p_2) + (2 - p_2)\delta < 4$. Then the $\mathcal{S}$-conditional expectation of (2.58) is by Hölder's inequality for $p' = (2 - \delta)/p_1$ and $q' = p'/(p' - 1)$ and inequalities (1.8)–(1.11) bounded by

$$n^{(p_1+p_2)/2-1} K_{p_1,p_2,\varepsilon} \sum_{i,j:t_{i,n}^{(1)} \vee t_{j,n}^{(2)} \leq T} \left(\mathbb{E}[\|\Delta_{i,n}^{(1)} \widetilde{N}^{(1)}(q)\|^{2-\delta}|\mathcal{S}]\right)^{\frac{p_1}{2-\delta}} \mathbf{1}_{d_1,d_2}^{*,n}(i,j)$$

$$\times \left(\mathbb{E}[\|\Delta_{j,n}^{(2)} \widetilde{B}^{(2)}(q)\|^{p_2 \frac{2-\delta}{2-\delta-p_1}} + \|\Delta_{j,n}^{(2)} \widetilde{C}^{(2)}\|^{p_2 \frac{2-\delta}{2-\delta-p_1}}\right.$$

$$\left. + \|\Delta_{j,n}^{(2)} \widetilde{M}^{(2)}(q)\|^{p_2 \frac{2-\delta}{2-\delta-p_1}}|\mathcal{S}]\right)^{\frac{2-\delta-p_1}{2-\delta}}$$

$$\leq n^{(p_1+p_2)/2-1} K_{p_1,p_2,\varepsilon} \sum_{i,j:t_{i,n}^{(1)} \vee t_{j,n}^{(2)} \leq T} K_q\left(|\mathcal{I}_{i,n}^{(1)}|^{\frac{p_1}{2-\delta}} + |\mathcal{I}_{i,n}^{(1)}|^{p_1}\right) K_q |\mathcal{I}_{j,n}^{(2)}|^{\frac{p_2}{2}}$$

$$\times \mathbf{1}_{d_1,d_2}^{*,n}(i,j)$$

$$\leq K_{p_1,p_2,\varepsilon,q} (|\pi_n|_T)^{\frac{p_1}{2-\delta} - \frac{p_1}{2}} \widetilde{G}_{p_1,p_2}^{(d_1,d_2),n}(T)$$

where we used

$$p_2 q' = p_2 \frac{p'}{p' - 1} = p_2 \frac{2 - \delta}{2 - \delta - p_1} < 2 \Leftrightarrow 2(p_1 + p_2) + (2 - p_2)\delta < 4.$$

Hence (2.58) and then also (2.56) vanish by Lemma 2.15 as $n \to \infty$ for any ε and any $q > 0$ if $p_1 > 0$.

Case 2. Now we consider the case $p_1 = 0$. As (2.54) is trivial for $p_1 = p_2 = 0$ it remains to discuss $p_1 = 0$, $p_2 > 0$. In that case

$$|\overline{V}^*(p_1 + p_2, f, \pi_n)_T - \overline{C}^*(p_1 + p_2, f, \pi_n)_T|$$

is by (2.47) (or rather the analogous result with $p_1 = 0$ and $p_2 > 0$) bounded by

$$n^{(p_1+p_2)/2-1} \sum_{i,j:t_{i,n}^{(1)} \vee t_{j,n}^{(2)} \leq T} [\theta(\varepsilon)\|\Delta_{j,n}^{(2)}\widetilde{C}^{(2)}\|^{p_2} + K_{p_1,p_2,\varepsilon}\|\Delta_{j,n}^{(2)}(\widetilde{X} - \widetilde{C})^{(2)}\|^{p_2}]\mathbf{1}_{d_1,d_2}^{*,n}(i,j).$$

Here the first term in the sum corresponds to (2.55) and the second term to (2.57). Hence the term (2.56) does not have to be dealt with if $p_1 = 0$ because such a term simply does not occur in the upper bound in this situation.

By symmetry (2.57) can be discussed in the same way as (2.56) with the difference that for (2.57) we have to discuss the cases $p_2 > 0$ and $p_2 = 0$ separately. Hence (2.54) follows because we have shown that (2.55)–(2.57) vanish.   □

Next, we define discretizations of σ and C by

$$\sigma(r)_s = \sigma_{(k-1)T/2^r}, \quad s \in [(k-1)T/2^r, kT/2^r),$$
$$C(r)_t = \int_0^t \sigma_s(r)ds. \tag{2.60}$$

Similarly we define discretizations of $\widetilde{\sigma}$ and $\widetilde{C}$ for the $(d_1 + d_2)$-dimensional process $\widetilde{X}$ and denote

$$\overline{C}^{*,r}(p, f, \pi_n)_T = n^{p/2-1} \sum_{i,j:t_{i,n}^{(1)} \vee t_{j,n}^{(2)} \leq T} f(\Delta_{i,n}^{(1)}\widetilde{C}^{(1)}(r), \Delta_{j,n}^{(2)}\widetilde{C}^{(2)}(r))\mathbf{1}_{d_1,d_2}^{*,n}(i,j).$$

Proposition 2.31. *Let* $f: \mathbb{R}^{d_1} \times \mathbb{R}^{d_2} \to \mathbb{R}$ *be a function as in Lemma 2.28 and assume that* $\widetilde{G}_{p_1,p_2}^{(d_1,d_2),n}(T) = O_\mathbb{P}(1)$. *Further we assume* $d_l = 0 \Rightarrow p_l = 0$, $l = 1, 2$. *Then*

$$\lim_{r\to\infty} \limsup_{n\to\infty} \mathbb{P}(|\overline{C}^*(p_1 + p_2, f, \pi_n)_T - \overline{C}^{*,r}(p_1 + p_2, f, \pi_n)_T| > \delta) = 0$$

holds for any $\delta > 0$.

Proof. We obtain using inequality (2.46)

$$\mathbb{E}\big[|\overline{C}^*(p_1+p_2,f,\pi_n)_T - \overline{C}^{*,r}(p_1+p_2,f,\pi_n)_T|\,\big|\mathcal{S}\big]$$

$$\leq n^{(p_1+p_2)/2-1} \sum_{i,j:t_{i,n}^{(1)}\vee t_{j,n}^{(2)}\leq T} \mathbb{E}\big[|f(\Delta_{i,n}^{(1)}\widetilde{C}^{(1)},\Delta_{j,n}^{(2)}\widetilde{C}^{(2)})$$

$$- f(\Delta_{i,n}^{(1)}\widetilde{C}^{(1)}(r),\Delta_{j,n}^{(2)}\widetilde{C}^{(2)}(r))|\,\big|\mathcal{S}\big]\mathbf{1}_{d_1,d_2}^{*,n}(i,j)$$

$$\leq n^{(p_1+p_2)/2-1} \sum_{i,j:t_{i,n}^{(1)}\vee t_{j,n}^{(2)}\leq T} \mathbf{1}_{d_1,d_2}^{*,n}(i,j)\Big(\mathbb{E}\big[\theta(\varepsilon)\|\Delta_{i,n}^{(1)}\widetilde{C}^{(1)}\|^{p_1}|\Delta_{j,n}^{(2)}\widetilde{C}^{(2)}\|^{p_2}\big|\mathcal{S}\big]$$

$$+ K_\varepsilon\mathbb{E}\big[\|\Delta_{i,n}^{(1)}(\widetilde{C}-\widetilde{C}(r))^{(1)}\|^{p_1}(\|\Delta_{j,n}^{(2)}\widetilde{C}^{(2)}\|^{p_2}+\|\Delta_{j,n}^{(2)}\widetilde{C}^{(2)}(r)\|^{p_2})\big|\mathcal{S}\big]$$

$$+ K_\varepsilon\mathbb{E}\big[\|\Delta_{j,n}^{(2)}(\widetilde{C}-\widetilde{C}(r))^{(2)}\|^{p_2}(\|\Delta_{i,n}^{(1)}\widetilde{C}^{(1)}\|^{p_1}+\|\Delta_{i,n}^{(1)}\widetilde{C}^{(1)}(r)\|^{p_1})\big|\mathcal{S}\big]\Big)$$

$$\leq \theta(\varepsilon)\widetilde{G}_{p_1,p_2}^{(d_1,d_2),n}(T) + K_\varepsilon n^{(p_1+p_2)/2-1} \sum_{l=1,2}\sum_{i,j:t_{i,n}^{(l)}\vee t_{j,n}^{(3-l)}\leq T}$$

$$\mathbb{E}\Big[\Big(\int_{t_{i-1,n}^{(l)}}^{t_{i,n}^{(l)}}\|\tilde\sigma_s-\tilde\sigma_s(r)\|^2 ds\Big)^{p_l\vee\frac{1}{2}}\big|\mathcal{S}\Big]^{\frac{1}{2}\wedge p_l}K|\mathcal{I}_{j,n}^{(3-l)}|^{\frac{p_{3-l}}{2}}\mathbf{1}_{d_1,d_2}^{*,n}(i,j) \qquad (2.61)$$

where we applied the Cauchy-Schwarz inequality, inequality (1.8) and (2.1.34) from [30] together with Jensen's inequality for $p_l < 1/2$. Using the trivial inequality $a^x \leq \tilde\eta^x + \tilde\eta^{x-1}a$ which holds for any $\tilde\eta, a > 0$ and $x \in [0,1]$ for $x = (1/2)\wedge p_l \leq 1$,

$$\tilde\eta = (\eta|\mathcal{I}_{i,n}^{(l)}|)^{p_l/(1\wedge 2p_l)}, \quad a = \mathbb{E}\Big[\Big(\int_{t_{i-1,n}^{(l)}}^{t_{i,n}^{(l)}}\|\tilde\sigma_s-\tilde\sigma_s(r)\|^2 ds\Big)^{p_l\vee\frac{1}{2}}\big|\mathcal{S}\Big]$$

yields that (2.61) is bounded by (note $\frac{((1/2)\wedge p_l-1)p_l}{1\wedge 2p_l} + (p_l\vee\frac{1}{2}) = \frac{p_l}{2}$)

$$(\theta(\varepsilon) + K_\varepsilon\eta^{p_l/2})\widetilde{G}_{p_1,p_2}^{(d_1,d_2),n}(T)$$

$$+ K_{\varepsilon,\eta}n^{(p_1+p_2)/2-1} \sum_{l=1,2}\sum_{i,j:t_{i,n}^{(l)}\vee t_{j,n}^{(3-l)}\leq T}|\mathcal{I}_{i,n}^{(l)}|^{\frac{((1/2)\wedge p_l-1)p_l}{1\wedge 2p_l}}$$

$$\times \mathbb{E}\Big[\Big(\int_{t_{i-1,n}^{(l)}}^{t_{i,n}^{(l)}}\|\tilde\sigma_s-\tilde\sigma_s(r)\|^2 ds\Big)^{p_l\vee\frac{1}{2}}\big|\mathcal{S}\Big]|\mathcal{I}_{j,n}^{(3-l)}|^{\frac{p_{3-l}}{2}}\mathbf{1}_{d_1,d_2}^{*,n}(i,j)$$

$$\leq (\theta(\varepsilon) + K_\varepsilon\eta^{p_l/2} + K_{\varepsilon,\eta}(\delta^{2p_1\vee 1}+\delta^{2p_2\vee 1}))\widetilde{G}_{p_1,p_2}^{(d_1,d_2),n}(T) + K_{\varepsilon,\eta}n^{(p_1+p_2)/2-1}$$

$$\times \sum_{l=1,2}\mathbb{E}\Big[\sum_{i,j:t_{i,n}^{(l)}\vee t_{j,n}^{(3-l)}\leq T}|\mathcal{I}_{i,n}^{(l)}|^{\frac{((1/2)\wedge p_l-1)\eta_l}{1\wedge 2p_l}}|\mathcal{I}_{j,n}^{(3-l)}|^{\frac{p_{3-l}}{2}}\mathbf{1}_{d_1,d_2}^{*,n}(i,j)$$

$$\times \Big(\int_{t_{i-1,n}^{(l)}}^{t_{i,n}^{(l)}} \|\tilde{\sigma}_s - \tilde{\sigma}_s(r)\|^2 ds \Big)^{p_l \vee \frac{1}{2}} \mathbb{1}_{\{\sup_{s \in (t_{i-1,n}^{(l)}, t_{i,n}^{(l)}]} \|\tilde{\sigma}_s - \tilde{\sigma}_s(r)\| > \delta\}} \Big| \mathcal{S} \Big]. \quad (2.62)$$

Denote by $\Omega(N, n, r, \delta)$ the set where $\tilde{\sigma}$ has at most N jump times S_p in $[0, T]$ with $\|\Delta \tilde{\sigma}_{S_p}\| > \delta/2$, two different such jumps are further apart than $|\pi_n|_T$ and $\|\tilde{\sigma}_t - \tilde{\sigma}_s\| \leq \delta$ for all $s, t \in [0, T]$ with $s < t$, $|t-s| < 2^{-r} + |\pi_n|_T$ and $\nexists p : S_p \in [t, s]$. Then $\mathbb{P}(\Omega(N, n, r, \delta)) \to 1$ as $N, n, r \to \infty$ for any $\delta > 0$ because σ is càdlàg. Using the assumption that $\tilde{\sigma}$ is bounded we get that (2.62) is less or equal than

$$\big(\theta(\varepsilon) + K_\varepsilon \eta^{p_l/2} + K_{\varepsilon,\eta}[\delta^{2p_1 \vee 1} + \delta^{2p_2 \vee 1} + \mathbb{P}((\Omega(N, n, r, \delta))^c | \mathcal{S})]\big) \widetilde{G}_{p_1, p_2}^{(d_1, d_2), n}(T)$$

$$+ K_{\varepsilon,\eta} N \sup_{0 \leq s < t, |t-s| \leq 2^{-r} + |\pi_n|_T} \big(\widetilde{G}_{p_1, p_2}^{(d_1, d_2), n}(t) - \widetilde{G}_{p_1, p_2}^{(d_1, d_2), n}(s) \big)$$

which yields

$$\lim_{\varepsilon \to 0} \limsup_{\eta \to 0} \limsup_{\delta \to 0} \limsup_{N \to \infty}$$

$$\limsup_{r \to \infty} \limsup_{n \to \infty} \mathbb{P}(|\overline{C}^*(p_1 + p_2, f, \pi_n)_T - \overline{C}^{*,r}(p_1 + p_2, f, \pi_n)_T| > \varepsilon') = 0$$

for all $\varepsilon' > 0$. $\square$

Proof of Theorem 2.17. By Proposition 2.30 it suffices to prove Theorem 2.17 only in the case $X_t = C_t$.

We consider the discretization (2.60) and denote $c_s(r) = \sigma_s(r)\sigma_s(r)^*$. Setting

$$R_n = \overline{C}(p, f, \pi_n)_T,$$

$$R = \int_0^T m_{c_s}(f) dG_p(s),$$

$$R_n(r) = n^{p/2-1} \sum_{i:t_{i,n} \leq T} f\big(\Delta_{i,n} C(r)\big),$$

$$R(r) = \int_0^T m_{c_s(r)}(f) dG_p(s),$$

we will prove

$$\lim_{r \to \infty} \limsup_{n \to \infty} \mathbb{P}\big(|R - R(r)| + |R(r) - R_n(r)| + |R_n(r) - R_n| > \delta\big) = 0 \quad \forall \delta > 0.$$

Step 1. As c_s is càdlàg and $G_p(s)$ is continuous it holds

$$R = \int_0^T m_{c_{s-}}(f) dG_p(s).$$

Deonte by Φ_{0,I_2} the distribution function of a two-dimensional standard normal random variable. Further consider a function $\psi\colon [0,\infty) \to [0,1]$ as in the proof of Theorem 2.3 with $\mathbb{1}_{[1,\infty)}(x) \le \psi(x) \le \mathbb{1}_{[1/2,\infty)}(x)$ and define $\psi_A(x) = \psi(x/A)$ and $\psi'_A = 1 - \psi_A$ for $A > 0$. Note that

$$
\begin{aligned}
|R - R(r)| &= \left| \int_0^T \int_{\mathbb{R}^2} (f(\sigma_s-x) - f(\sigma_s(r)x))\Phi_{0,I_2}(dx)dG_p(s) \right| \\
&\le \int_0^T \int_{\mathbb{R}^2} |(f\psi_A)(\sigma_s-x) - (f\psi_A)(\sigma_s(r)x)|\Phi_{0,I_2}(dx)dG_p(s) \\
&\quad + \int_0^T \int_{\mathbb{R}^2} |(f\psi'_A)(\sigma_s-x) - (f\psi'_A)(\sigma_s(r)x)|\Phi_{0,I_2}(dx)dG_p(s)
\end{aligned}
$$
(2.63)

By (2.45) for $p_1 = p$, $p_2 = 0$ we obtain $|(f\psi'_A)(x)| \le KA^p$ and hence $(f\psi'_A)$ is bounded. Then the fact that $(f\psi'_A)$ is continuous together with the pointwise convergence $\sigma_s(r) \to \sigma_{s-}$ yields by dominated convergence that the second summand in (2.63) vanishes for any $A > 0$ as $r \to \infty$.

The first summand in (2.63) is bounded by

$$
K \int_0^T \int_{\mathbb{R}^2} \left(\|\sigma_s-x\|^p \mathbb{1}_{\{\|\sigma_s-x\| \ge A/2\}} + \|\sigma_s(r)x\|^p \mathbb{1}_{\{\|\sigma_o(r)x\| \ge A/2\}} \right) \Phi_{0,I_2}(dx)dG_p(s)
$$
(2.64)

where the inner integral is increasing in σ_s. As we assume that σ is bounded on $[0,T]$ this yields that there exists a constant

$$
K' = \operatorname*{ess\,sup}_{s\in[0,T],\omega\in\Omega} (|\sigma_s^{(1)}(\omega)| + |\sigma_s^{(2)}(\omega)|)
$$

such that (2.64) is bounded by

$$
K \int_{\mathbb{R}^2} \left(\|K'x\|^p \mathbb{1}_{\{\|K'x\| \ge A/2\}} + \|K'x\|^p \mathbb{1}_{\{\|K'x\| \ge A/2\}} \right) \Phi_{0,I_2}(dx) \int_0^T dG_p(s)
$$

which vanishes as $A \to \infty$. Hence we have shown $|R - R(r)| \to 0$ almost surely as $r \to \infty$.

Step 2. In order to prove $|R(r) - R_n(r)| \xrightarrow{\mathbb{P}} 0$ as $n \to \infty$ we apply Lemma C.2 with

$$
\xi_k^n = n^{p/2-1} \sum_{i\in L(n,k,T)} f(\Delta_{i,n}C(r)),
$$

$L(n, k, T) = \{i : t_{i-1,n} \in [(k-1)T/2^{r_n}, kT/2^{r_n})\}$, $k = 1, 2, \ldots, 2^{r_n}$, and set
$\mathcal{G}_k^n = \sigma(\mathcal{F}_{(k-1)T/2^{r_n}} \cup \mathcal{S})$. Here, r_n is a sequence of real numbers with $r_n \geq r$,
$r_n \to \infty$ and

$$2^{r_n} \sup_{s \in [0,T]} |G_p(s) - G_p^{(l),n}(s)| = o_{\mathbb{P}}(1),$$

$$2^{r_n} \sup_{s,t \in [0,T], |t-s| \leq |\pi_n|_T} |G_p^{(l),n}(t) - G_p^{(l),n}(s)| = o_{\mathbb{P}}(1).$$

(2.65)

Such a sequence exists, because $G_p^{(l),n}$ and hence G_p are nondecreasing functions
such that pointwise convergence implies uniform convergence on $[0, T]$ to the
continuous function G_p. We then get

$$\mathbb{E}[\xi_k^n | \mathcal{G}_{k-1}^n] = n^{p/2-1} \sum_{i \in L(n,k,T)} \mathbb{E}[f(\Delta_{i,n}C(r)) | \mathcal{G}_{k-1}^n]$$

$$= n^{p/2-1} \sum_{i \in L(n,k,T)} |\mathcal{I}_{i,n}|^{p/2} \mathbb{E}[f(|\mathcal{I}_{i,n}|^{-1/2} \Delta_{i,n}C(r)) | \mathcal{G}_{k-1}^n]$$

$$= n^{p/2-1} m_{c_{(k-1)T/2^{r_n}}}(r)(f) \sum_{i : \mathcal{I}_{i,n} \subset ((k-1)T/2^{r_n}, kT/2^{r_n}]} |\mathcal{I}_{i,n}|^{p/2}$$

$$+ K \sup_{s,t \in [0,T], |t-s| \leq |\pi_n|_T} |G_p^{(l),n}(t) - G_p^{(l),n}(s)|$$

$$= m_{c_{(k-1)T/2^{r_n}}}(r)(f) \left(G_p^{(l),n}(kT/2^{r_n}) - G_p^{(l),n}((k-1)T/2^{r_n}) \right)$$

$$+ K \sup_{s,t \in [0,T], |t-s| \leq |\pi_n|_T} |G_p^{(l),n}(t) - G_p^{(l),n}(s)|$$

because $|\mathcal{I}_{i,n}|^{-1/2} \Delta_{i,n} C(r)$ is conditional on $\mathcal{G}_{k-1}^n$ centered normal distributed
with covariance matrix $c_{(k-1)T/2^{r_n}}(r)$. The term

$$K \sup_{s,t \in [0,T], |t-s| \leq |\pi_n|_T} |G_p^{(l),n}(t) - G_p^{(l),n}(s)|$$

is due to the summand with $kT/2^{r_n} \in \mathcal{I}_{i,n}$ which has to be treated separately as
in the corresponding interval the process $\sigma(r)$ might jump. Further as in Step 1
the boundedness of σ_s together with $|f(x)| \leq K\|x\|^p$ yields that $m_{c_s}(f)$ is also
bounded which together with the previous computations yields

$$\left| R(r) - \sum_{k=1}^{2^{r_n}} \mathbb{E}[\xi_k^n | \mathcal{G}_{k-1}^n] \right|$$

$$\leq K2^{r_n} \sup_{s \in [0,T]} |G_p(s) - G_p^n(s)| + K2^{r_n} \sup_{s,t \in [0,T], |t-s| \leq |\pi_n|_T} |G_p^{(l),n}(t) - G_p^{(l),n}(s)|$$

where the right-hand side is $o_{\mathbb{P}}(1)$ by (2.65). Hence the sum over the $\mathbb{E}[\xi_k^n|\mathcal{G}_{k-1}^n]$ converges in probability to $R(r)$.

Using the Cauchy-Schwarz inequality, inequality (1.9), the definition of G_p^n and telescoping sums we also get

$$\sum_{k=1}^{2^{rn}} \mathbb{E}[|\xi_k^n|^2|\mathcal{G}_{k-1}^n] \leq \sum_{k=1}^{2^{rn}} \mathbb{E}\Big[\big(n^{p/2-1} \sum_{i\in L(n,k,T)} K\|\Delta_{i,n}C(r)\|^p\big)^2\big|\mathcal{G}_{k-1}^n\Big]$$

$$= K \sum_{k=1}^{2^{rn}} n^{p-2} \sum_{i\in L(n,k,T)} \sum_{j\in L(n,k,T)} \mathbb{E}\big[\|\Delta_{i,n}C(r)\|^p\|\Delta_{j,n}C(r)\|^p\big|\mathcal{G}_{k-1}^n\big]$$

$$\leq K \sum_{k=1}^{2^{rn}} n^{p-2} \sum_{i\in L(n,k,T)} \sum_{j\in L(n,k,T)} \big(\prod_{m=i,j} \mathbb{E}\big[\|\Delta_{m,n}C(r)\|^{2p}\big|\mathcal{G}_{k-1}^n\big]\big)^{1/2}$$

$$\leq K \sum_{k=1}^{2^{rn}} n^{p-2} \sum_{i\in L(n,k,T)} \sum_{j\in L(n,k,T)} |\mathcal{I}_{i,n}|^{p/2}|\mathcal{I}_{j,n}|^{p/2}$$

$$\leq K \sum_{k=1}^{2^{rn}} \big(n^{p/2-1} \sum_{i\in L(n,k,T)} |\mathcal{I}_{i,n}|^{p/2}\big)^2$$

$$\leq K G_p^n(T) \sup_{u,s\in[0,T],|u-s|\leq T2^{-rn}+|\pi_n|_T} |G_p^{(l),n}(u) - G_p^{(l),n}(s)|$$

where the right-hand side converges to zero in probability, since G_p^n converges uniformly to a continuous function G_p. Hence we have shown

$$\sum_{k=1}^{2^{rn}} \mathbb{E}[\xi_k^n|\mathcal{G}_{k-1}^n] \xrightarrow{\mathbb{P}} R(r), \qquad \sum_{k=1}^{2^{rn}} \mathbb{E}[|\xi_k^n|^2|\mathcal{G}_{k-1}^n] \xrightarrow{\mathbb{P}} 0.$$

Further the ξ_k^n are $\mathcal{G}_k^n$-measurable and hence $\xi_k^n - \mathbb{E}[\xi_k^n|\mathcal{G}_{k-1}^n]$ are martingale differences. Lemma C.2 then yields

$$R_n(r) = \sum_{k=1}^{2^{rn}} \xi_k^n \xrightarrow{\mathbb{P}} R(r)$$

for any $r \in \mathbb{N}$.

Step 3. Finally we obtain

$$\lim_{r\to\infty} \limsup_{n\to\infty} \mathbb{P}(|R_n - R_n(r)| > \delta) = 0$$

for any $\delta > 0$ from Proposition 2.31 with $d_1 = 2$, $d_2 = 0$, $p_1 = p$ and $p_2 = 0$. $\qquad\square$

Proof of Corollary 2.19. This is Theorem 2.17 with the function

$$f(x_1, x_2) = \operatorname{sgn}(g(x_1))|g(x_1)g(x_2)|^{1/2}$$

applied to the process $\widetilde{X}_t = (X_t^{(l)}, X_t^{(l)})^*$. Here, $\operatorname{sgn}(x)$ denotes the signum function of a real number $x \in \mathbb{R}$. Further note that any positively homogeneous function g in dimension 1 is continuous because it holds

$$g(x) = |x|g(1)\mathbb{1}_{\{x>0\}} + |x|g(-1)\mathbb{1}_{\{x<0\}}.$$

$\square$

Proof of Theorem 2.21. By Proposition 2.30 it suffices to prove Theorem 2.21 if $X_t = C_t$. We will proceed as in the proof of Theorem 2.17 and define

$$R_n = \overline{C}(p_1 + p_2, f, \pi_n)_T,$$

$$R = m_{I_2}(f) \int_0^T (\sigma_s^{(1)})^{p_1}(\sigma_s^{(2)})^{p_2} dG_{p_1,p_2}(s),$$

$$R_n(r) = n^{(p_1+p_2)/2-1} \sum_{i,j:t_{i,n}^{(1)} \vee t_{j,n}^{(2)} \leq T} f\big(\Delta_{i,n}^{(1)} C^{(1)}(r), \Delta_{j,n}^{(2)} C^{(2)}(r)\big)\mathbb{1}_{\{\mathcal{I}_{j,n}^{(2)} \cap \mathcal{I}_{j,n}^{(2)} \neq \emptyset\}},$$

$$R(r) = m_{I_2}(f) \int_0^T (\sigma_s^{(1)}(r))^{p_1}(\sigma_s^{(2)}(r))^{p_2} dG_{p_1,p_2}(s).$$

Step 1. As σ is càdlàg we obtain $\sigma_s(r) \to \sigma_{s-}$ pointwise for $r \to \infty$. As σ is further bounded we derive by dominated convergence

$$R(r) \to m_{I_2}(f) \int_0^T (\sigma_{s-}^{(1)})^{p_1}(\sigma_{s-}^{(2)})^{p_2} dG_{p_1,p_2}(s)$$

where the right-hand side equals R because G_{p_1,p_2} is continuous.

Step 2. Comparing Step 2 in the proof of Theorem 2.17 we define

$$\xi_k^n = n^{(p_1+p_2)/2-1} \sum_{(i,j) \in L(n,k,T)} f\big(\Delta_{i,n}^{(1)} C^{(1)}(r), \Delta_{j,n}^{(2)} C^{(2)}(r)\big)\mathbb{1}_{\{\mathcal{I}_{i,n}^{(1)} \cap \mathcal{I}_{j,n}^{(2)} \neq \emptyset\}}$$

where the sequence r_n fulfils similar properties as in the proof of Theorem 2.21 and $L(n,k,T) = \{(i,j) : t_{i-1,n}^{(1)} \wedge t_{j-1,n}^{(2)} \in [(k-1)T/2^{r_n}, kT/2^{r_n})\}$ for $k = 1, \ldots, 2^{r_n}$. Then we obtain the identity

$$\mathbb{E}[\xi_k^n | \mathcal{G}_{k-1}^n] = n^{(p_1+p_2)/2-1} \sum_{(i,j) \in L(n,k,T)} (\sigma_{(k-1)T/2^{r_n}}^{(1)}(r))^{p_1}(\sigma_{(k-1)T/2^{r_n}}^{(2)}(r))^{p_2}$$

$$\times |\mathcal{I}_{i,n}^{(1)}|^{p_1/2}|\mathcal{I}_{j,n}^{(2)}|^{p_2/2} m_{I_2}(f)\mathbb{1}_{\{\mathcal{I}_{i,n}^{(1)} \cap \mathcal{I}_{j,n}^{(2)} \neq \emptyset\}}$$

$$+ K \sup_{s,t \in [0,T], |t-s| \leq 3|\pi_n|_T} |G_{p_1,p_2}^n(t) - G_{p_1,p_2}^n(s)|$$

were we used that f is positively homogeneous in both arguments and that

$$(\Delta_{i,n}^{(1)}C^{(1)}(r)/(\sigma_{(k-1)T/2^{rn}}^{(1)}(r)|\mathcal{I}_{i,n}^{(1)}|^{1/2}),\Delta_{j,n}^{(2)}C^{(2)}(r)/(\sigma_{(k-1)T/2^{rn}}^{(2)}(r)|\mathcal{I}_{j,n}^{(2)}|^{1/2}))$$

is for $\mathcal{I}_{i,n}^{(1)}\cup\mathcal{I}_{j,n}^{(2)} \subset [(k-1)T/2^{rn}, kT/2^{rn})$ conditionally on $\mathcal{G}_{k-1}^n$ standard normally distributed due to $\rho \equiv 0$. The term

$$K \sup_{s,t\in[0,T],|t-s|\leq 3|\pi_n|_T} |G_{p_1,p_2}^n(t) - G_{p_1,p_2}^n(s)|$$

originates similarly as before from summands with $kT/2^{rn} \in \mathcal{I}_{i,n}^{(1)} \cup \mathcal{I}_{j,n}^{(2)}$ which have to be treated separately as in the corresponding intervals the process $\sigma(r)$ might jump.

This yields

$$\left|R(r) - \sum_{k=1}^{2^{rn}} \mathbb{E}[\xi_k^n|\mathcal{G}_{k-1}^n]\right| \leq K2^{rn} \sup_{s\in[0,T]} |G_{p_1,p_2}(s) - G_{p_1,p_2}^n(s)|$$

$$+ K2^{rn} \sup_{s,t\in[0,T],|t-s|\leq 3|\pi_n|_T} |G_{p_1,p_2}^n(t) - G_{p_1,p_2}^n(s)|$$

and we get similarly as in Step 2 in the proof of Theorem 2.17

$$\sum_{k=1}^{2^{rn}} \mathbb{E}[|\xi_k^n|^2|\mathcal{G}_{k-1}^n]$$

$$\leq \sum_{k=1}^{2^{rn}} \mathbb{E}\Big[\Big(n^{(p_1+p_2)/2-1} \sum_{(i,j)\in L(n,k,T)} K|\Delta_{i,n}^{(1)}C^{(1)}(r)|^{p_1}|\Delta_{j,n}^{(2)}C^{(2)}(r)|^{p_2}$$

$$\times \mathbb{1}_{\{\mathcal{I}_{i,n}^{(1)}\cap\mathcal{I}_{j,n}^{(2)}\neq\emptyset\}}\Big)^2\Big|\mathcal{G}_{k-1}^n\Big]$$

$$\leq K \sum_{k=1}^{2^{rn}} n^{p_1+p_2-2} \sum_{(i,j),(i',j')\in L(n,k,T)} (\mathbb{E}[|\Delta_{i,n}^{(1)}C^{(1)}(r)|^{4p_1}|\mathcal{G}_{k-1}^n]$$

$$\times \mathbb{E}[|\Delta_{j,n}^{(2)}C^{(2)}(r)|^{4p_2}|\mathcal{G}_{k-1}^n])^{1/4}$$

$$\times (\mathbb{E}[|\Delta_{i',n}^{(1)}C^{(1)}(r)|^{4p_1}|\mathcal{G}_{k-1}^n]\mathbb{E}[|\Delta_{j',n}^{(2)}C^{(2)}(r)|^{4p_2}|\mathcal{G}_{k-1}^n])^{1/4}\mathbb{1}_{\{\mathcal{I}_{i,n}^{(1)}\cap\mathcal{I}_{j,n}^{(2)}\neq\emptyset\}}$$

$$\times \mathbb{1}_{\{\mathcal{I}_{i',n}^{(1)}\cap\mathcal{I}_{j',n}^{(2)}\neq\emptyset\}}$$

$$\leq K \sum_{k=1}^{2^{rn}} n^{p_1+p_2-2} \sum_{(i,j),(i',j')\in L(n,k,T)} |\mathcal{I}_{i,n}^{(1)}|^{p_1/2}|\mathcal{I}_{j,n}^{(2)}|^{p_2/2}|\mathcal{I}_{i',n}^{(1)}|^{p_1/2}|\mathcal{I}_{j',n}^{(2)}|^{p_2/2}$$

$$\times \mathbb{1}_{\{\mathcal{I}_{i,n}^{(1)}\cap\mathcal{I}_{j,n}^{(2)}\neq\emptyset\}}\mathbb{1}_{\{\mathcal{I}_{i',n}^{(1)}\cap\mathcal{I}_{j',n}^{(2)}\neq\emptyset\}}$$

$$= K \sum_{k=1}^{2^{rn}} \left(n^{(p_1+p_2)/2-1} \sum_{(i,j)\in L(n,k,T)} |\mathcal{I}_{i,n}^{(1)}|^{p_1/2} |\mathcal{I}_{j,n}^{(2)}|^{p_2/2} \mathbb{1}_{\{\mathcal{I}_{i,n}^{(1)}\cap\mathcal{I}_{j,n}^{(2)}\neq\emptyset\}} \right)^2$$

$$\leq K G_{p_1,p_2}^n(T) \sup_{u,s\in[0,T],|u-s|\leq T2^{-rn}+|\pi_n|_T} \left| G_{p_1,p_2}^n(u) - G_{p_1,p_2}^n(s) \right|$$

which then yields $R_n(r) \xrightarrow{\mathbb{P}} R(r)$ as $n \to \infty$ by Lemma C.2.

Step 3. Finally we obtain

$$\lim_{r\to\infty} \limsup_{n\to\infty} \mathbb{P}(|R_n - R_n(r)| > \delta) = 0$$

for any $\delta > 0$ from Proposition 2.31 with $d_1 = d_2 = 1$. $\square$

Proof of Theorem 2.22. Proposition 2.30 yields that it suffices to prove Theorem 2.22 in the case $X_t = C_t$. Further, following the proof of Theorem 2.21 we observe that Step 1 and Step 3 do not make use of the assumption $\rho = 0$. Hence the arguments therein also apply here. The only difference occurs if we want to adapt Step 2 in the proof of Theorem 2.21 for the proof of Theorem 2.22. In fact, if we look at

$$\xi_k^n = n^{(p_1+p_2)/2-1} \sum_{(i,j)\in L(n,k,T)} (\Delta_{i,n}^{(1)} C^{(1)}(r))^{p_1} (\Delta_{j,n}^{(2)} C^{(2)}(r))^{p_2} \mathbb{1}_{\{\mathcal{I}_{i,n}^{(1)}\cap\mathcal{I}_{j,n}^{(2)}\neq\emptyset\}}$$

and denote by Φ_{0,I_4} the distribution function of a four-dimensional standard normal random variable we get

$$\mathbb{E}[\xi_k^n | \mathcal{G}_{k-1}^n] = n^{(p_1+p_2)/2-1} \sum_{(i,j)\in L(n,k,T)} \mathbb{E}[(\Delta_{i,n}^{(1)} C^{(1)}(r))^{p_1} (\Delta_{j,n}^{(2)} C^{(2)}(r))^{p_2} | \mathcal{G}_k^n]$$

$$\times \mathbb{1}_{\{\mathcal{I}_{i,n}^{(1)}\cap\mathcal{I}_{j,n}^{(2)}\neq\emptyset\}}$$

$$= \prod_{l=1,2} (\sigma_{(k-1)T/2^{rn}}^{(l)})^{p_l} \sum_{(i,j)\in L(n,k,T)} \int_{\mathbb{R}^4} \left(|\mathcal{I}_{i,n}^{(1)} \setminus \mathcal{I}_{j,n}^{(2)}|^{1/2} x_1 + |\mathcal{I}_{i,n}^{(1)} \cap \mathcal{I}_{j,n}^{(2)}|^{1/2} x_2 \right)^{p_1}$$

$$\times \left(\rho_{(k-1)T/2^{rn}} |\mathcal{I}_{i,n}^{(1)} \cap \mathcal{I}_{j,n}^{(2)}|^{1/2} x_2 + (1 - (\rho_{(k-1)T/2^{rn}})^2)^{1/2} |\mathcal{I}_{i,n}^{(1)} \cap \mathcal{I}_{j,n}^{(2)}|^{1/2} x_3 \right.$$

$$\left. + |\mathcal{I}_{j,n}^{(2)} \setminus \mathcal{I}_{i,n}^{(1)}|^{1/2} x_4 \right)^{p_2} \Phi_{0,I_4}(dx) \mathbb{1}_{\{\mathcal{I}_{i,n}^{(1)}\cap\mathcal{I}_{j,n}^{(2)}\neq\emptyset\}}$$

$$+ K \sup_{s,t\in[0,T],|t-s|\leq 3|\pi_n|_T} |G_{p_1,p_2}^n(t) - G_{p_1,p_2}^n(s)|$$

$$= (\sigma_{(k-1)T/2^{rn}}^{(1)})^{p_1} (\sigma_{(k-1)T/2^{rn}}^{(2)})^{p_2} \sum_{(i,j)\in L(n,k,T)} \sum_{k=0}^{p_1} \sum_{l,m=0}^{p_2} \binom{p_1}{k} \binom{p_2}{l,m}$$

$$\times \int_{\mathbb{R}^4} |\mathcal{I}_{i,n}^{(1)} \setminus \mathcal{I}_{j,n}^{(2)}|^{k/2} (x_1)^k (\rho_{(k-1)T/2^{rn}})^{p_2-(l+m)} |\mathcal{I}_{i,n}^{(1)} \cap \mathcal{I}_{j,n}^{(2)}|^{(p_1+p_2-(k+l+m))/2}$$

$$\times (x_2)^{p_1+p_2-(k+l+m)} (1 - (\rho_{(k-1)T/2^{rn}})^2)^{l/2} |\mathcal{I}_{i,n}^{(1)} \cap \mathcal{I}_{j,n}^{(2)}|^{l/2} (x_3)^l$$

$$\times |\mathcal{I}_{j,n}^{(2)} \setminus \mathcal{I}_{i,n}^{(1)}|^{m/2}(x_4)^m \Phi_{0,I_4}(dx) \mathbb{1}_{\{\mathcal{I}_{i,n}^{(1)} \cap \mathcal{I}_{j,n}^{(2)} \neq \emptyset\}}$$
$$+ K \sup_{s,t \in [0,T], |t-s| \leq 3|\pi_n|_T} |G_{p_1,p_2}^n(t) - G_{p_1,p_2}^n(s)|$$
$$= (\sigma_{(k-1)T/2^{r_n}}^{(1)})^{p_1}(\sigma_{(k-1)T/2^{r_n}}^{(2)})^{p_2}\Bigg[\sum_{(k,l,m) \in L(p_1,p_2)} \binom{p_1}{k}\binom{p_2}{l,m} m_1(x^k)m_1(x^l)$$
$$\times m_1(x^m) m_1(x^{p_1+p_2-(k+l+m)})(\rho_{(k-1)T/2^{r_n}})^{p_2-(l+m)}(1 - \rho_{(k-1)T/2^{r_n}}^2)^{l/2}$$
$$\times \left(H_{k,m,p_1+p_2}^n(kT/2^{r_n}) - H_{k,m,p_1+p_2}^n((k-1)T/2^{r_n}) \right)\Bigg]$$
$$+ K \sup_{s,t \in [0,T], |t-s| \leq 3|\pi_n|_T} |G_{p_1,p_2}^n(t) - G_{p_1,p_2}^n(s)|$$

where we used the multinomial theorem, $\mathbb{E}[X^k] = 0$ for $X \sim \mathcal{N}(0,1)$, k odd, and

$$\begin{pmatrix} \Delta_{i,n}^{(1)} C^{(1)}(r) \\ \Delta_{j,n}^{(2)} C^{(2)}(r) \end{pmatrix} \stackrel{\mathcal{L}_{\mathcal{G}_{k-1}^n}}{=} \begin{pmatrix} v_1 \\ v_2 \end{pmatrix} U$$

with

$$v_1 = \begin{pmatrix} \sigma_{(k-1)T/2^{r_n}}^{(1)} |\mathcal{I}_{i,n}^{(1)} \setminus \mathcal{I}_{j,n}^{(2)}|^{1/2} \\ \sigma_{(k-1)T/2^{r_n}}^{(1)} |\mathcal{I}_{i,n}^{(1)} \cap \mathcal{I}_{j,n}^{(2)}|^{1/2} \\ 0 \\ 0 \end{pmatrix}^*$$

$$v_2 = \begin{pmatrix} 0 \\ \rho_{(k-1)T/2^{r_n}}\sigma_{(k-1)T/2^{r_n}}^{(2)} |\mathcal{I}_{i,n}^{(1)} \cap \mathcal{I}_{j,n}^{(2)}|^{1/2} \\ (1 - (\rho_{(k-1)T/2^{r_n}})^2)^{1/2}\sigma_{(k-1)T/2^{r_n}}^{(2)} |\mathcal{I}_{i,n}^{(1)} \cap \mathcal{I}_{j,n}^{(2)}|^{1/2} \\ (\sigma_{(k-1)T/2^{r_n}}^{(2)})|\mathcal{I}_{j,n}^{(2)} \setminus \mathcal{I}_{i,n}^{(1)}|^{1/2} \end{pmatrix}^*$$

for $\mathcal{I}_{i,n}^{(1)} \cup \mathcal{I}_{j,n}^{(2)} \subset [(k-1)T/2^{r_n}, kT/2^{r_n})$ where $U = (U_1, U_2, U_3, U_4)^*$ is $\mathcal{N}(0,I_4)$-distributed and independent of $\mathcal{F}$. Here, $\mathcal{L}_{\mathcal{G}_{k-1}^n}$ denotes identity of the $\mathcal{G}_{k-1}^n$-conditional distributions. The rest of Step 2 from the proof of Theorem 2.21 also applies here without modification. $\qquad\square$

Remark 2.32. *Propositions 2.30 and 2.31 suggest that similar results as in Theorems 2.17, 2.21 and 2.22 could also be derived for $(d_1 + d_2)$-dimensional Itô semimartingales $X = (X^{(1)}, X^{(2)})$ where all components of the d_l-dimensional process $X^{(l)}$, $l = 1, 2$, are observed snychronously and functions $f: \mathbb{R}^{d_1+d_2} \to \mathbb{R}$. However, as interesting properties and challenges due to the asynchronicity of the observation scheme already arise in the bivariate setting, I decided to restrict myself to the bivariate setting also to keep the notation and statements clearer. Further, in Part II we are going to discuss applications only for bivariate Itô semimartingales as well.* $\qquad\square$

Proof of Corollary 2.25. It suffices to prove

$$\overline{V}^*(p_1 + p_2, f, \pi_n)_T - \overline{V}^*(p_1 + p_2, \tilde{f}, \pi_n)_T \xrightarrow{\mathbb{P}} 0.$$

For $\varepsilon > 0$ pick $\delta > 0$ such that $|L(x) - 1| < \varepsilon$ for all x with $\|x\| \in [0, \delta]$. Then it holds

$$\left| \overline{V}^*(p_1 + p_2, f, \pi_n)_T - \overline{V}^*(p_1 + p_2, \tilde{f}, \pi_n)_T \right|$$

$$= \left| n^{(p_1+p_2)/2-1} \sum_{i,j:t_{i,n}^{(1)} \vee t_{j,n}^{(2)} \leq T} (1 - L(\Delta_{i,n}^{(1)} \widetilde{X}^{(1)}, \Delta_{j,n}^{(2)} \widetilde{X}^{(2)})) \right.$$

$$\left. \times f(\Delta_{i,n}^{(1)} \widetilde{X}^{(1)}, \Delta_{j,n}^{(2)} \widetilde{X}^{(2)}) \mathbf{1}_{d_1,d_2}^{*,n}(i,j) \right|$$

$$\leq \varepsilon \overline{V}^*(p_1 + p_2, |f|, \pi_n)_T$$

$$+ n^{(p_1+p_2)/2-1} \sum_{i,j:t_{i,n}^{(1)} \vee t_{j,n}^{(2)} \leq T} |1 - L(\Delta_{i,n}^{(1)} \widetilde{X}^{(1)}, \Delta_{j,n}^{(2)} \widetilde{X}^{(2)})|$$

$$\times |f(\Delta_{i,n}^{(1)} \widetilde{X}^{(1)}, \Delta_{j,n}^{(2)} \widetilde{X}^{(2)})| \mathbf{1}_{\{\|(\Delta_{i,n}^{(1)} \widetilde{X}^{(1)}, \Delta_{j,n}^{(2)} \widetilde{X}^{(2)})\| > \delta\}} \mathbf{1}_{d_1,d_2}^{*,n}(i,j). \quad (2.66)$$

The function $|f| : x \mapsto |f(x)|$ is positively homogeneous of degree p_1 in the first argument and positively homogeneous of degree p_2 in the second argument. Hence using Proposition 2.30 and

$$\mathbb{E}[\overline{C}^*(p_1 + p_2, |f|, \pi_n)_T | \mathcal{S}]$$

$$\leq \sum_{i,j:t_{i,n}^{(1)} \vee t_{j,n}^{(2)} \leq T} K \mathbb{E}[|\Delta_{i,n}^{(1)} \widetilde{C}^{(1)}|^{p_1} |\Delta_{j,n}^{(2)} \widetilde{C}^{(2)}|^{p_2} | \mathcal{S}] \mathbf{1}_{d_1,d_2}^{*,n}(i,j)$$

$$\leq K \widetilde{G}_{p_1,p_2}^{(d_1,d_2),n}(T),$$

which follows by (2.45), we obtain $\overline{V}^*(p_1 + p_2, |f|, \pi_n)_T = O_{\mathbb{P}}(1)$ as $n \to \infty$ and hence the first term in (2.66) vanishes as $n \to \infty$ and then $\varepsilon \to 0$. If $\widetilde{X}$ is continuous, then the second term in (2.66) converges almost surely to 0 as $n \to \infty$. If $\widetilde{X}$ may be discontinuous we have $p_1 + p_2 < 2$. We then denote by $\Omega(n, q, N, \delta)$ the set where $\|(\Delta_{i,n}^{(1)} \widetilde{X}^{(1)}, \Delta_{j,n}^{(2)} \widetilde{X}^{(2)})\| > \delta$ implies $\Delta_{i,n}^{(1)} N^{(1)}(q) \neq 0$ or $\Delta_{j,n}^{(2)} N^{(2)}(q) \neq 0$, it holds $\|\Delta \widetilde{X}_s\| \leq N$ for all $s \in [0, T]$ and $\|(\Delta_{i,n}^{(1)} \widetilde{X}^{(1)}, \Delta_{j,n}^{(2)} \widetilde{X}^{(2)})\| \leq 2N$ for all i, j with $t_{i,n}^{(1)} \vee t_{j,n}^{(2)} \leq T$. On this set the second term in (2.66) is by the local boundedness of L bounded by

$$n^{(p_1+p_2)/2-1} \sum_{i,j:t_{i,n}^{(1)} \vee t_{j,n}^{(2)} \leq T} K |f(\Delta_{i,n}^{(1)} \widetilde{X}^{(1)}, \Delta_{j,n}^{(2)} \widetilde{X}^{(2)})|$$

$$\times \mathbf{1}_{\{\|(\Delta_{i,n}^{(1)} \widetilde{N}^{(1)}(q), \Delta_{j,n}^{(2)} \widetilde{N}^{(2)}(q))\| \neq 0\}} \mathbf{1}_{d_1,d_2}^{*,n}(i,j)$$

which vanishes due to $p_1 + p_2 < 2$ using (2.45) and the arguments used for the discussion of (2.58) and (2.59) in the proof of Proposition 2.30. The proof is then finished by observing $\mathbb{P}(\Omega(n, q, N, \delta)) \to 1$ as $n, q, N \to \infty$ for any $\delta > 0$. $\qquad\square$

2.3 Functionals of Truncated Increments

In Section 2.1 only the jump part of X contributes to the limit while in Section 2.2 only the continuous part of X contributes to the limit. Hence for the non-normalized functionals only large increments $\Delta_{i,n}^{(l)} X^{(l)}$ contribute in the limit while for the normalized functionals only small increments $\Delta_{i,n}^{(l)} X^{(l)}$ contribute in the limit. Following up on these observations one might expect that expressions of the form

$$\sum_{i,j:t_{i,n}^{(1)} \vee t_{j,n}^{(2)} \leq T} f(\Delta_{i,n}^{(1)} X^{(1)}, \Delta_{j,n}^{(2)} X^{(2)}) \mathbb{1}_{\{\mathcal{I}_{i,n}^{(1)} \cap \mathcal{I}_{j,n}^{(2)} \neq \emptyset\}} \mathbb{1}_{\{|\Delta_{i,n}^{(1)} X^{(1)}|, |\Delta_{j,n}^{(2)} X^{(2)}| \text{ „large“}\}},$$

$$n^{p/2-1} \sum_{i,j:t_{i,n}^{(1)} \vee t_{j,n}^{(2)} \leq T} f(\Delta_{i,n}^{(1)} X^{(1)}, \Delta_{j,n}^{(2)} X^{(2)}) \mathbb{1}_{\{\mathcal{I}_{i,n}^{(1)} \cap \mathcal{I}_{j,n}^{(2)} \neq \emptyset\}} \mathbb{1}_{\{|\Delta_{i,n}^{(1)} X^{(1)}|, |\Delta_{j,n}^{(2)} X^{(2)}| \text{ „small“}\}}$$

$$(2.67)$$

converge to the same limits as $V(f, \pi_n)_T$ and $\overline{V}(p, f, \pi_n)_T$ and that the convergence of sums based on selected summands might be faster than for the sums which include all summands. Furthermore we may hope that we obtain convergence for functionals of the form (2.67) for a wider class of functions f than we did in Sections 2.1 and 2.2 for the functionals $V(f, \pi_n)_T$ and $\overline{V}(p, f, \pi_n)_T$. This is due to the fact that most of the conditions on the functions f made in Sections 2.1 and 2.2 were needed to show that the contribution of the continuous respectively of the jump part vanishes in the limit. But these contributions mainly stem from „small“ respectively „large“ increments which we here explicitly exclude from the sums.

It remains to specify which increments we classify to be small and which to be large. We therefore recall the discussions from the beginning of Section 2.2 were we observed that the increments of the jump part over an interval $\mathcal{I}_{i,n}^{(l)}$ are constant in magnitude as $|\pi_n|_T \to 0$ while the increments of the continuous part are of magnitude $|\mathcal{I}_{i,n}^{(l)}|^{1/2}$ as $|\pi_n|_T \to 0$. Hence by checking whether

$$|\Delta_{i,n}^{(l)} X| \leq \beta |\mathcal{I}_{i,n}^{(l)}|^{\varpi}$$

for some $\beta > 0$ and $\varpi \in (0, 1/2)$ is fulfilled or not we are asymptotically able to distinguish whether $\Delta_{i,n}^{(l)} X^{(l)}$ is dominated by the jump part or the continuous part of $X^{(l)}$; compare also Chapter 9 of [30]. Here the rate $\varpi \in (0, 1/2)$ should lie between the rate 0 by which increments of the jump parts decrease (or rather remain constant) and the rate $1/2$ by which increments of the continuous part decrease. Building on this motivation we define the functionals

$$V_+(f, \pi_n, (\beta, \varpi))_T = \sum_{i,j: t_{i,n}^{(1)} \vee t_{j,n}^{(2)} \leq T} f\big(\Delta_{i,n}^{(1)} X^{(1)}, \Delta_{j,n}^{(2)} X^{(2)}\big) \mathbb{1}_{\{\mathcal{I}_{i,n}^{(1)} \cap \mathcal{I}_{j,n}^{(2)} \neq \emptyset\}}$$

$$\times \mathbb{1}_{\{|\Delta_{i,n}^{(1)} X^{(1)}| > \beta |\mathcal{I}_{i,n}^{(1)}|^\varpi, |\Delta_{j,n}^{(2)} X^{(2)}| > \beta |\mathcal{I}_{j,n}^{(2)}|^\varpi\}},$$

$$\overline{V}_-(p, f, \pi_n, (\beta, \varpi))_T = n^{p/2-1} \sum_{i,j: t_{i,n}^{(1)} \vee t_{j,n}^{(2)} \leq T} f\big(\Delta_{i,n}^{(1)} X^{(1)}, \Delta_{j,n}^{(2)} X^{(2)}\big)$$

$$\times \mathbb{1}_{\{\mathcal{I}_{i,n}^{(1)} \cap \mathcal{I}_{j,n}^{(2)} \neq \emptyset\}} \mathbb{1}_{\{|\Delta_{i,n}^{(1)} X^{(1)}| \leq \beta |\mathcal{I}_{i,n}^{(1)}|^\varpi, |\Delta_{j,n}^{(2)} X^{(2)}| \leq \beta |\mathcal{I}_{j,n}^{(2)}|^\varpi\}}$$

for functions $f: \mathbb{R}^2 \to \mathbb{R}$ and

$$V_+^{(l)}(g, \pi_n, (\beta, \varpi))_T = \sum_{i: t_{i,n}^{(l)} \leq T} g\big(\Delta_{i,n}^{(l)} X^{(l)}\big) \mathbb{1}_{\{|\Delta_{i,n}^{(l)} X^{(l)}| > \beta |\mathcal{I}_{i,n}^{(l)}|^\varpi\}},$$

$$\overline{V}_-^{(l)}(p, g, \pi_n, (\beta, \varpi))_T = n^{p/2-1} \sum_{i: t_{i,n}^{(l)} \leq T} g\big(\Delta_{i,n}^{(l)} X^{(l)}\big) \mathbb{1}_{\{|\Delta_{i,n}^{(l)} X^{(l)}| \geq \beta |\mathcal{I}_{i,n}^{(l)}|^\varpi\}},$$

for functions $g: \mathbb{R} \to \mathbb{R}$ and $l = 1, 2$.

2.3.1 The Results

As discussed before Remark 2.7 we cannot get convergence of the non-normalized functionals for functions f that do not fulfil $f(x, y) = O(\|(x, y)\|^2)$ as $(x, y) \to 0$ because for such f the quantities $B(f)$, $B^*(f)$ are in general not well-defined. Hence in Theorem 2.1 and Corollary 2.2 we only get an improvement from functions f with $f(x, y) = o(\|x\|^2)$ as $x \to 0$ to functions with $f(x) = O(\|x\|^2)$ as $x \to 0$. Theorem 9.1.1 in [30] states this result only for equidistant observation times $t_{i,n}^{(1)} = t_{i,n}^{(2)} = i/n$. In the following theorem we extend their result to the setting of irregular, exogenous and synchronous observations.

Theorem 2.33. *Suppose the observation scheme is exogenous, synchronous and Condition 1.3 holds. Further let $\beta > 0$ and $\varpi \in (0, 1/2)$. Then we have with the notation from (2.1)*

$$\sum_{i: t_{i,n} \leq T} f(\Delta_{i,n} X) \mathbb{1}_{\{\|\Delta_{i,n} X\| > \beta |\mathcal{I}_{i,n}|^\varpi\}} \xrightarrow{\mathbb{P}} B(f)_T$$

for all continuous functions $f: \mathbb{R}^2 \to \mathbb{R}$ with $f(x) = O(\|x\|^2)$ as $x \to 0$.

Again we obtain the convergence result for $V_+^{(l)}(g, \pi_n, (\beta, \varpi))_T$ as a corollary.

Corollary 2.34. *Let* $\beta > 0$, $\varpi \in (0, 1/2)$ *and assume that Condition 1.3 is fulfilled and that the observation scheme is exogenous. Then it holds*

$$V_+^{(l)}(g, \pi_n, (\beta, \varpi))_T \xrightarrow{\mathbb{P}} B^{(l)}(g)_T$$

for all continuous functions $g \colon \mathbb{R} \to \mathbb{R}$ *with* $g(x) = O(x^2)$ *as* $x \to 0$.

In the asynchronous setting in Theorems 2.3 and 2.8 the order condition on f is way stronger compared to the theoretical bound of $f(x, y) = O(\|(x, y)\|^2)$ as $(x, y) \to 0$ than in the synchronous setting. Hence we might expect a larger improvement for the condition on f in the asynchronous setting than in the synchronous setting when using the functionals $V_+(f, \pi_n, (\beta, \varpi))_T$ instead of $V(f, \pi_n,)_T$. However in the setting of Theorem 2.3 we will get no improvement at all which will be illustrated in Example 2.36. In Theorem 2.8 on the other hand we can relax the assumption (2.8). In the general case we get an improvement if $p_1 \vee p_2 < 2$, while we always get an improvement in the case where X has only finite jump activity. The improvement is larger if ϖ is smaller i.e. if increments are classified to be large via a higher threshold. The relaxation of (2.8) leads to the fact that the class of functions f for which we obtain convergence under an identical condition is larger for $V_+(f, \pi_n, (\beta, \varpi))_T$ than for $V(f, \pi_n)_T$; compare Remark 2.37.

Theorem 2.35. *Suppose Condition 1.3 is fulfilled and let* $\beta > 0$ *and* $\varpi \in (0, 1/2)$.

a) It holds

$$V_+(f, \pi_n, (\beta, \varpi))_T \xrightarrow{\mathbb{P}} B^*(f)_T$$

for all continuous functions $f \colon \mathbb{R}^2 \to \mathbb{R}$ *with* $f(x, y) = O(x^2 y^2)$ *as* $|xy| \to 0$.

b) Let $p_1, p_2 > 0$ *with* $p_1 + p_2 \geq 2$. *Further assume that the observation scheme is exogenous and that we have one of the following:*

(i) We have

$$\sum_{i,j : t_{i,n}^{(1)} \vee t_{j,n}^{(2)} \leq T} |\mathcal{I}_{i,n}^{(1)}|^{\frac{p_1 + (1/2 - \varpi)(2 - p_1 \vee p_2)^+}{2} \wedge 1}$$

$$\times |\mathcal{I}_{j,n}^{(2)}|^{\frac{p_1 + (1/2 - \varpi)(2 - p_1 \vee p_2)^+}{2} \wedge 1} \mathbb{1}_{\{\mathcal{I}_{i,n}^{(1)} \cap \mathcal{I}_{j,n}^{(2)} \neq \emptyset\}} = O_{\mathbb{P}}(1) \quad (2.68)$$

as $n \to \infty$ *where* $(2 - p_1 \vee p_2)^+ = \max\{2 - p_1 \vee p_2, 0\}$ *denotes the positive part of* $2 - p_1 \vee p_2$.

(ii) We have

$$\sum_{i,j:t_{i,n}^{(1)}\vee t_{j,n}^{(2)}\leq T} |\mathcal{I}_{i,n}^{(1)}|^{\frac{p_1+(1/2-\varpi)}{2}} |\mathcal{I}_{j,n}^{(2)}|^{\frac{p_2+(1/2-\varpi)}{2}} \mathbb{1}_{\{\mathcal{I}_{i,n}^{(1)}\cap \mathcal{I}_{j,n}^{(2)}\neq \emptyset\}} = O_{\mathbb{P}}(1)$$

$$(2.69)$$

and X has almost surely only finitely many jumps in $[0,T]$.

Then it holds

$$V_+(f,\pi_n,(\beta,\varpi))_T \xrightarrow{\mathbb{P}} B^*(f)_T$$

for all continuous functions $f\colon \mathbb{R}^2 \to \mathbb{R}$ with $f(x,y) = o(|x|^{p_1}|y|^{p_2})$ as $|xy| \to 0$.

The following example shows that if $V_+(f,\pi_n,(\beta,\varpi))_T$ is supposed to converge for any Itô semimartingale X and any observation scheme fulfilling Condition 1.3 then the function f has to fulfil the same order condition $f(x,y) = O(x^2y^2)$ for $|xy| \to 0$ as was required in Theorem 2.3.

Example 2.36. *Let $X^{(1)} = \mathbb{1}_{\{t\geq U\}}$, $U \sim \mathcal{U}[0,1]$, and the observation scheme be given by $t_{i,n}^{(1)} = i/n$ and $t_{i,n}^{(2)} = i/n^{1+\gamma}$, $\gamma > 0$, as in Example 2.5 and $f(x,y) = |x|^{p_1}|y|^{p_2}$, $p_1,p_2 \geq 0$. Further let $X^{(2)}$ be an α-stable Lévy motion, $\alpha \in (0,2)$, as defined in Example 3.1.3 of [43] i.e. $X^{(2)}$ is a stationary process with independent increments and*

$$X_t^{(2)} - X_s^{(2)} \overset{\mathcal{L}}{=} (t-s)^{1/\alpha} X_1^{(2)} \quad \forall t \geq s \geq 0.$$

Then for n large enough and i.i.d. random variables $Z_i \sim X_1^{(2)}$ it holds

$$V_+(f,\pi_n,(\beta,\varpi))_1 = \sum_{i=\lfloor n^{1+\gamma}(\lceil nU\rceil -1)/n\rfloor /n^{1+\gamma}+1}^{\lceil n^{1+\gamma}\lceil nU\rceil /n\rceil /n^{1+\gamma}} \left| X_{i/n^{1+\gamma}}^{(2)} - X_{(i-1)/n^{1+\gamma}}^{(2)} \right|^{p_2}$$

$$\times \mathbb{1}_{\{|\Delta_{i,n}^{(2)}X^{(2)}|>\beta n^{-(1+\gamma)\varpi}\}}$$

$$\geq \sum_{i=\lfloor n^{1+\gamma}(\lceil nU\rceil -1)/n\rfloor +1}^{\lfloor n^{1+\gamma}(\lceil nU\rceil -1)/n\rfloor +\lceil n^\gamma\rceil} \beta^{p_2} n^{-(1+\gamma)\varpi p_2} \mathbb{1}_{\{|\Delta_{i,n}^{(2)}X^{(2)}|>\beta n^{-(1+\gamma)\varpi}\}}$$

$$\overset{\mathcal{L}}{=} \beta^{p_2} n^{-(1+\gamma)\varpi p_2} \sum_{i=1}^{\lfloor n^\gamma\rfloor} \mathbb{1}_{\{|Z_i|>\beta(n^{(1+\gamma)(1/\alpha-\varpi)})\}}.$$

Using the estimate (1.2.8) from [43] then yields that the expectation of the above quantity is as $n \to \infty$ approximately equal to

$$\beta^{p_2} n^{-(1+\gamma)\varpi p_2} n^\gamma K(\beta(n^{(1+\gamma)(1/\alpha - \varpi)})^{-\alpha}$$

for some constant K which depends on $X^{(2)}$. Hence $V_+(f, \pi_n, (\beta, \varpi))_1$ diverges as $n \to \infty$ if

$$\gamma \varpi(\alpha - p_2) + (\varpi(\alpha - p_2) - 1) > 0.$$

If $\alpha > p_2$ we can find some $\gamma > 0$ large enough such that this is fulfilled and whenever $p_2 < 2$ we are able to find an α with $\alpha > p_2$ such that there exists an α-stable process.   □

For any $\varpi \in (0, 1/2)$ the condition (2.68) is weaker than (2.9) as shown in the following remark.

Remark 2.37. *In the case $p_1 = p_2 = 1$ both conditions (2.68) and (2.69) become*

$$\sum_{i,j:t_{i,n}^{(1)} \vee t_{j,n}^{(2)} \leq T} |\mathcal{I}_{i,n}^{(1)}|^{\frac{1+(1/2-\varpi)}{2}} |\mathcal{I}_{j,n}^{(2)}|^{\frac{1+(1/2-\varpi)}{2}} \mathbb{1}_{\{\mathcal{I}_{i,n}^{(1)} \cap \mathcal{I}_{j,n}^{(2)} \neq \emptyset\}} = O_{\mathbb{P}}(1).$$

Hence the exponents of the observation interval lengths in the sum above both increase by $1/2 - \varpi > 0$ compared to the similar condition in Theorem 2.6. The improvement here is larger if ϖ is smaller.

The improvement in Theorem 2.35 yields that we get convergence of the functional $V_+(f, \pi_n, (\beta, \varpi))_T$ for the same class of functions as for $V(f, \pi_n)_T$ but under a weaker condition. However, it might also be interpreted such that we get convergence for a wider class of functions under an identical condition. To illustrate this interpretation consider the case $p = p_1 = p_2 < 2$. Then the condition from Theorem 2.6 reads

$$O_{\mathbb{P}}(1) = \sum_{i,j:t_{i,n}^{(1)} \vee t_{j,n}^{(2)} \leq T} |\mathcal{I}_{i,n}^{(1)}|^{\frac{p}{2}} |\mathcal{I}_{j,n}^{(2)}|^{\frac{p}{2}} \mathbb{1}_{\{\mathcal{I}_{i,n}^{(1)} \cap \mathcal{I}_{j,n}^{(2)} \neq \emptyset\}}$$

$$= \sum_{i,j:t_{i,n}^{(1)} \vee t_{j,n}^{(2)} \leq T} |\mathcal{I}_{i,n}^{(1)}|^{\frac{p'+(1/2-\varpi)(2-p')}{2}} |\mathcal{I}_{j,n}^{(2)}|^{\frac{p+(1/2-\varpi)(2-p')}{2}} \mathbb{1}_{\{\mathcal{I}_{i,n}^{(1)} \cap \mathcal{I}_{j,n}^{(2)} \neq \emptyset\}}$$

for $p' = (p - 2(1/2 - \varpi))/(1/2 + \varpi) < p$. Hence under the above condition we get the convergence of $V_+(f, \pi_n, (\beta, \varpi))_T$ for all functions f with $f(x, y) = o(|xy|^{p'})$ as $|xy| \to 0$ while we obtain convergence of $V(f, \pi_n)_T$ only for functions f with $f(x, y) = o(|xy|^p)$ as $|xy| \to 0$.   □

For the normalized functionals the only improvement from using the truncated functionals compared to the ordinary ones is that X does not have to be continuous in the cases $p = 2$ respectively $p_1 + p_2 = 2$ such that we get convergence of $\overline{V}_-(p, f, \pi_n, (\beta, \varpi))_T$, $\overline{V}_-^{(l)}(p, g, \pi_n, (\beta, \varpi))_T$ and $\overline{V}_-(p_1 + p_2, f, \pi_n, (\beta, \varpi))_T$ in the settings of Theorem 2.17, Corollary 2.19 and Theorems 2.21, 2.22. This is the same improvement obtained in the setting of equidistant and synchronous observation times $t_{i,n}^{(1)} = t_{i,n}^{(2)} = i/n$; compare Theorem 9.2.1 in [30].

Theorem 2.38. *Let $\beta > 0$ and $\varpi \in (0, 1/2)$.*

  a) *Suppose the assumptions made for Theorem 2.17 respectively Corollary 2.19 are fulfilled with the only difference that we have either $p \in [0, 2]$ or $p > 2$ and X is continuous. Then it holds*

$$\overline{V}_-(p, f, \pi_n, (\beta, \varpi))_T \xrightarrow{\mathbb{P}} \int_0^T \rho_{c_s}(f) dG_p(s),$$

  respectively

$$\overline{V}_-^{(l)}(p, g, \pi_n, (\beta, \varpi))_T \xrightarrow{\mathbb{P}} \rho_1(g) \int_0^T (\sigma_s^{(l)})^p dG_p^{(l)}(s), \quad l = 1, 2.$$

  b) *Suppose the assumptions made for Theorems 2.21 respectively Theorem 2.22 are fulfilled with the only difference that we have either $p_1 + p_2 \in [0, 2]$ or $p_1 + p_2 > 2$ and X is continuous. Then $\overline{V}_-(p_1 + p_2, f, \pi_n, (\beta, \varpi))_T$ converges to the limits in (2.41) respectively (2.42).*

Example 2.39. *A special case in Theorem 2.22 was $p_1 = p_2 = 1$ where we obtained*

$$\overline{V}(2, f_{(1,1)}, \pi_n)_T = \sum_{i,j: t_{i,n}^{(1)} \vee t_{j,n}^{(2)}} \Delta_{i,n}^{(1)} X^{(1)} \Delta_{j,n}^{(2)} X^{(2)} \mathbb{1}_{\{\mathcal{I}_{i,n}^{(1)} \cap \mathcal{I}_{j,n}^{(2)} \neq \emptyset\}}$$

$$\xrightarrow{\mathbb{P}} \int_0^T \rho_s \sigma_s^{(1)} \sigma_s^{(2)} ds$$

for continuous processes X. Using truncated increments we then obtain by Theorem 2.38 via

$$\overline{V}_-(2, f_{(1,1)}, \pi_n, (\beta, \varpi))_T \xrightarrow{\mathbb{P}} \int_0^T \rho_s \sigma_s^{(1)} \sigma_s^{(2)} ds$$

a consistent estimator for the integrated co-volatility of $X^{(1)}$ and $X^{(2)}$, which we will discuss in Chapter 7 in more detail, also for processes X with jumps. □

Notes. *In the textbooks [30] and [3] functionals of truncated increments are only considered in the setting of equidistant and synchronous observation times.*

[35] consider an estimator for the continuous part of the quadratic covariation of two processes $X^{(1)}$ and $X^{(2)}$ with finite jump activity based on asynchronous discrete observations which is of the form

$$\sum_{i,j:t_{i,n}^{(1)} \vee t_{j,n}^{(2)} \leq T} \Delta_{i,n}^{(1)} X^{(1)} \Delta_{j,n}^{(2)} X^{(2)} \mathbb{1}_{\{|\Delta_{i,n}^{(1)} X^{(1)}| \leq r(|\pi_n|_T), |\Delta_{j,n}^{(2)} X^{(1)}| \leq r(|\pi_n|_T)\}}$$

$$\times \mathbb{1}_{\{\mathcal{I}_{i,n}^{(1)} \cap \mathcal{I}_{j,n}^{(2)} \neq \emptyset\}}. \tag{2.70}$$

Contrary to our approach they use a uniform threshold $r(|\pi_n|_T)$ which depends not on the length of the specific observation interval but on the mesh $|\pi_n|_T$ of the observation scheme restricted to $[0, T]$. Here, $r: [0, \infty) \to [0, \infty)$ denotes a function with $r(x) \to 0$ sufficiently fast as $x \to 0$. They find that the estimator (2.70) is consistent for $\int_0^T \rho_s \sigma_s^{(1)} \sigma_s^{(2)} ds$ even for endogenous observation times provided that X has finite jump activity; compare Theorem 3.13 in [35]. Their result has been extended for processes of infinite jump activity and more general thresholds in [34].

2.3.2 The Proofs

Proof of Theorem 2.33. This proof is inspired by the proof of Theorem 9.1.1 in [30]. Define $f_\rho(x) = f(x)\psi(\|x\|/\rho)$ using a continuous function $\psi: [0, \infty) \to [0, 1]$ with $\psi(u) = 0$ for $u \leq 1/2$ and $\psi(u) = 1$ for $u \geq 1$ as in the proof of Theorem 2.3. Theorem 2.1 then yields

$$V(f_\rho, \pi_n)_T \xrightarrow{\mathbb{P}} B(f_\rho)_T$$

as $n \to \infty$ and similar arguments as in the proof of Lemma 2.14 yield the convergence $B(f_\rho)_T \to B(f)_T$ as $\rho \to 0$. Hence it remains to prove

$$\lim_{\rho \to 0} \limsup_{n \to \infty} \mathbb{P}\Big(\Big| \sum_{i:t_{i,n} \leq T} f(\Delta_{i,n} X) \mathbb{1}_{\{\|\Delta_{i,n} X\| > \beta |\mathcal{I}_{i,n}|^\varpi\}} - V(f_\rho, \pi_n)_T \Big| > \varepsilon \Big) = 0$$

$$\tag{2.71}$$

for all $\varepsilon > 0$. Therefore we compute

$$\Big| \sum_{i:t_{i,n} \leq T} f(\Delta_{i,n} X) \mathbb{1}_{\{\|\Delta_{i,n} X\| > \beta |\mathcal{I}_{i,n}|^\varpi\}} - V(f_\rho, \pi_n)_T \Big|$$

$$\leq K_\rho \sum_{i:t_{i,n} \leq T} \|\Delta_{i,n} X\|^2 \mathbb{1}_{\{\|\Delta_{i,n} X\| > \beta |\mathcal{I}_{i,n}|^\varpi\}} \mathbb{1}_{\{\|\Delta_{i,n} X\| < \rho/2\}} \tag{2.72}$$

$$+ 2K_\rho \sum_{i:t_{i,n} \leq T} \|\Delta_{i,n} X\|^2 \mathbb{1}_{\{\rho/2 \leq \|\Delta_{i,n} X\| < \rho\}} \tag{2.73}$$

$$+ \sum_{i:t_{i,n}\leq T} f(\Delta_{i,n}X)\mathbb{1}_{\{\|\Delta_{i,n}X\|\leq\beta|\mathcal{I}_{i,n}|^{\varpi}\}}\mathbb{1}_{\{\|\Delta_{i,n}X\|\geq\rho\}} \qquad (2.74)$$

where (2.74) vanishes as $n \to \infty$ for any $\rho > 0$ because of $|\pi_n|_T \xrightarrow{\mathbb{P}} 0$. Further the sum of (2.72) and (2.73) is bounded by

$$2K_\rho \sum_{i:t_{i,n}\leq T} \|\Delta_{i,n}B(q)\|^2 + 2K_\rho \sum_{i:t_{i,n}\leq T} \|\Delta_{i,n}C\|^2\big(\mathbb{1}_{\{\|\Delta_{i,n}X\|>\beta|\mathcal{I}_{i,n}|^{\varpi}\}} + \mathbb{1}_{\{\rho/2\leq\|\Delta_{i,n}X\|<\rho\}}\big)$$

$$+2K_\rho \sum_{i:t_{i,n}\leq T} \|\Delta_{i,n}M(q)\|^2 + 2K_\rho \sum_{i:t_{i,n}\leq T} \|\Delta_{i,n}N(q)\|^2\mathbb{1}_{\{\|\Delta_{i,n}X\|<\rho\}}.$$

$$(2.75)$$

The sum over $\|\Delta_{i,n}C\|^2\mathbb{1}_{\{\rho/2\leq\|\Delta_{i,n}X\|<\rho\}}$ vanishes as $n \to \infty$ for any $\rho > 0$ because C is continuous and because the number of indices i with $\rho/2 < \|\Delta_{i,n}X\| \leq \rho$ is bounded as $n \to \infty$ because these indices eventually correspond to intervals which contain jumps of X with $\|\Delta X_s\| \geq \rho/4$. The $\mathcal{S}$-conditional expectation of the remaining terms in (2.75) containing $B(q), C, M(q)$ is bounded by

$$K_\rho K_q|\pi_n|_T T + K_\rho(|\pi_n|_T)^{1/2-\varpi}T + K_\rho e_q T \qquad (2.76)$$

where we used inequalities (1.8)–(1.10) and

$$\mathbb{E}[\|\Delta_{i,n}C\|^2\mathbb{1}_{\{\|\Delta_{i,n}X\|>\beta|\mathcal{I}_{i,n}|^{\varpi}\}}|\mathcal{S}]$$
$$\leq (\mathbb{E}[\|\Delta_{i,n}C\|^4|\mathcal{S}])^{1/2}(\mathbb{P}(\|\Delta_{i,n}X\|^2 > \beta^2|\mathcal{I}_{i,n}|^{2\varpi}|\mathcal{S}))^{1/2}$$
$$\leq K|\mathcal{I}_{i,n}|\Big(\frac{\mathbb{E}[\|\Delta_{i,n}X|\mathcal{S}\|^2]}{\beta^2|\mathcal{I}_{i,n}|^{2\varpi}}\Big)^{1/2}$$
$$\leq K|\mathcal{I}_{i,n}|^{1+(1/2-\varpi)} \qquad (2.77)$$

which is obtained using the Cauchy-Schwarz inequality, Markov's inequality and (1.14), (1.16). The bound (2.76) vanishes as first $n \to \infty$ and then $q \to \infty$. Further the sum involving $N(q)$ in (2.75) converges to

$$\sum_{s:s\leq T} \|\Delta N(q)_s\|^2\mathbb{1}_{\{\|\Delta N(q)_s\|<\rho\}}$$

as $n \to \infty$ which vanishes for $\rho \to 0$. Hence using Lemma 2.15 we have proven (2.71). $\qquad\square$

Proof of Theorem 2.35. We will prove

$$V_+(f_\rho, \pi_n, (\beta, \varpi))_T \xrightarrow{\mathbb{P}} B^*(f_\rho), \quad n \to \infty, \qquad (2.78)$$

$$B^*(f_\rho) \to B^*(f), \quad \rho \to 0, \qquad (2.79)$$

$$\lim_{\rho\to0} \limsup_{n\to\infty} \mathbb{P}(|V_+(f, \pi_n, (\beta, \varpi))_T - V_+(f_\rho, \pi_n, (\beta, \varpi))_T| > \varepsilon) = 0, \quad \varepsilon > 0,$$

$$(2.80)$$

where f_ρ is defined as in the proof of Theorem 2.33. The convergence (2.79) directly follows from (2.21) because the limits are independent of the estimator.

For showing (2.78) note that it holds

$$
\begin{aligned}
& \left(V_+(f_\rho, \pi_n, (\beta, \varpi))_T - V(f_\rho, \pi_n)_T\right) \mathbb{1}_{\Omega(n, N, \rho)} \\
= & \sum_{i,j: t_{i,n}^{(1)} \vee t_{j,n}^{(2)} \leq T} f_\rho\left(\Delta_{i,n}^{(1)} X^{(1)}, \Delta_{j,n}^{(2)} X^{(2)}\right) \mathbb{1}_{\{\mathcal{I}_{i,n}^{(1)} \cap \mathcal{I}_{j,n}^{(2)} \neq \emptyset\}} \\
& \times \mathbb{1}_{\{|\Delta_{i,n}^{(1)} X^{(1)}| \leq \beta |\mathcal{I}_{i,n}^{(1)}|^\varpi \vee |\Delta_{j,n}^{(2)} X^{(2)}| \leq \beta |\mathcal{I}_{j,n}^{(2)}|^\varpi\}} \mathbb{1}_{\{|\Delta_{i,n}^{(1)} X^{(1)} \Delta_{j,n}^{(2)} X^{(2)}| > \rho/2\}} \\
& \times \mathbb{1}_{\Omega(n, N, \rho)} \\
= & \, 0
\end{aligned}
$$

where $\Omega(n, N, \rho)$ denotes the set where $|\Delta_{i,n}^{(l)} X^{(l)}| \leq N$ holds whenever $t_{i,n}^{(l)} \leq T$, $l = 1, 2$, and where it holds $\beta(|\pi_n|T)^\varpi N \leq \rho/2$. Then (2.78) follows from

$$
\lim_{N \to 0} \limsup_{n \to \infty} \mathbb{P}(\Omega(n, N, \rho)) = 1
$$

for any $\rho > 0$ and Lemma 2.13.

Finally we consider (2.80). In the setting from part a) the proof of (2.80) is identical to the proof of (2.22) because the estimate (2.23) also holds for

$$
|V_+(f, \pi_n, (\beta, \varpi))_T - V_+(f_\rho, \pi_n, (\beta, \varpi))_T|.
$$

For part b) note that we obtain using the same argument as in (2.27)

$$
\begin{aligned}
& |V_+(f, \pi_n, (\beta, \varpi))_T - V_+(f_\rho, \pi_n, (\beta, \varpi))_T| \\
\leq & \, 2K_\rho \sum_{i,j: t_{i,n}^{(1)} \vee t_{j,n}^{(2)} \leq T} |\Delta_{i,n}^{(1)} X^{(1)}|^{p_1} |\Delta_{j,n}^{(2)} X^{(2)}|^{p_2} \mathbb{1}_{\{\mathcal{I}_{i,n}^{(1)} \cap \mathcal{I}_{j,n}^{(2)} \neq \emptyset\}} \\
& \times \mathbb{1}_{\{|\Delta_{i,n}^{(1)} X^{(1)}| > \beta |\mathcal{I}_{i,n}^{(1)}|^\varpi, |\Delta_{j,n}^{(2)} X^{(2)}| > \beta |\mathcal{I}_{j,n}^{(2)}|^\varpi\}}. \quad (2.81)
\end{aligned}
$$

Using the notation from (2.29) and arguments from the proof of Theorem 2.6 we obtain that the $\mathcal{S}$-conditional expectation of the sum in (2.81) is bounded by

$$
\begin{aligned}
& K_{p_1, p_2} T \\
& + K_{p_1, p_2} \sum_{l=1,2} \sum_{k: T_k^n \leq T} \mathbb{E}\left[|\Delta_k^{n, l, -} X^{(l)}|^{p_l} (|\Delta_k^n X^{(3-l)}|^{p_{3-l}} + |\Delta_k^{n, 3-l, +} X^{(3-l)}|^{p_{3-l}})\right. \\
& \times \left. \mathbb{1}_{\{|\Delta_{i_n^{(l)}(T_k^n), n}^{(1)} X^{(1)}| > \beta |\mathcal{I}_{i_n^{(l)}(T_k^n), n}^{(1)}|^\varpi, |\Delta_{i_n^{(3-l)}(T_k^n), n}^{(3-l)} X^{(3-l)}| > \beta |\mathcal{I}_{i_n^{(3-l)}(T_k^n), n}^{(3-l)}|^\varpi\}} \Big| \mathcal{S} \right] \\
& + K_{p_1, p_2} \sum_{l=1,2} \sum_{k: T_k^n \leq T} \mathbb{E}\left[|\Delta_k^n X^{(l)}|^{p_l} |\Delta_k^{n, 3-l, +} X^{(3-l)}|^{p_{3-l}}\right. \\
& \times \left. \mathbb{1}_{\{|\Delta_{i_n^{(l)}(T_k^n), n}^{(1)} X^{(1)}| > \beta |\mathcal{I}_{i_n^{(l)}(T_k^n), n}^{(1)}|^\varpi, |\Delta_{i_n^{(3-l)}(T_k^n), n}^{(3-l)} X^{(3-l)}| > \beta |\mathcal{I}_{i_n^{(l)}(T_k^n), n}^{(3-l)}|^\varpi\}} \Big| \mathcal{S} \right]
\end{aligned}
$$

which in the case of $p_1 \vee p_2 < 2$ is using Hölder's inequality for $p = 2/(p_1 \vee p_2)$, $p' = 2/(2 - p_1 \vee p_2)$ bounded by

$$K_{p_1,p_2} T + K_{p_1,p_2} \sum_{l=1,2} \sum_{k:T_k^n \leq T} \mathbb{E}\big[|\Delta_k^{n,l,-} X^{(l)}|^{\frac{2p_l}{p_1 \vee p_2}}$$

$$\times \big(|\Delta_k^n X^{(3-l)}|^{\frac{2p_{3-l}}{p_1 \vee p_2}} + |\Delta_k^{n,3-l,+} X^{(3-l)}|^{\frac{2p_{3-l}}{p_1 \vee p_2}} \big) |\mathcal{S} \big]^{(p_1 \vee p_2)/2}$$

$$\times \mathbb{E}\big[\mathbb{1}_{\{ |\Delta^{(1)}_{i_n^{(l)}(T_k^n),n} X^{(1)}| > \beta |\mathcal{I}^{(1)}_{i_n^{(l)}(T_k^n),n}|^\varpi, |\Delta^{(3-l)}_{i_n^{(3-l)}(T_k^n),n} X^{(3-l)}| > \beta |\mathcal{I}^{(3-l)}_{i_n^{(l)}(T_k^n),n}|^\varpi \}} \cdots$$

$$\cdots |\mathcal{S}\big]^{(2 - p_1 \vee p_2)/2}$$

$$+ K_{p_1,p_2} \sum_{l=1,2} \sum_{k:T_k^n \leq T} \mathbb{E}\big[|\Delta_k^n X^{(l)}|^{\frac{2p_l}{p_1 \vee p_2}} |\Delta_k^{n,3-l,+} X^{(3-l)}|^{\frac{2p_{3-l}}{p_1 \vee p_2}} |\mathcal{S}\big]^{(p_1 \vee p_2)/2}$$

$$\times \mathbb{E}\big[\mathbb{1}_{\{ |\Delta^{(1)}_{i_n^{(l)}(T_k^n),n} X^{(1)}| > \beta |\mathcal{I}^{(1)}_{i_n^{(l)}(T_k^n),n}|^\varpi, |\Delta^{(3-l)}_{i_n^{(3-l)}(T_k^n),n} X^{(3-l)}| > \beta |\mathcal{I}^{(3-l)}_{i_n^{(l)}(T_k^n),n}|^\varpi \}} \cdots$$

$$\cdots |\mathcal{S}\big]^{(2 - p_1 \vee p_2)/2}$$

$$\leq K_{p_1,p_2} T +$$

$$+ K_{p_1,p_2} \sum_{i,j:t_{i,n}^{(1)} \vee t_{j,n}^{(2)} \leq T} |\mathcal{I}_{i,n}^{(1)}|^{\frac{p_1 + (1/2 - \varpi)(2 - p_1 \vee p_2)}{2}} |\mathcal{I}_{j,n}^{(2)}|^{\frac{p_2 + (1/2 - \varpi)(2 - p_1 \vee p_2)}{2}}$$

$$\times \mathbb{1}_{\{ \mathcal{I}_{i,n}^{(1)} \cap \mathcal{I}_{j,n}^{(2)} \neq \emptyset \}} \tag{2.82}$$

where we used iterated expectations, (1.12), the Cauchy-Schwarz inequality and the estimate for the indicator from (2.77) for the last inequality. In the case $p_1 \vee p_2 \geq 2$ we bound (2.81) by the same sum without the indicators and argue as in the proof of Theorem 2.6. Hence, (2.80) in the case (i) follows from (2.68) and $K_\rho \to 0$ for $\rho \to 0$.

Next we consider (2.80) in the case (ii). If X is of finite jump activity we have

$$|V_+(f, \pi_n(\beta, \varpi))_T - V_+(f_\rho, \pi_n(\beta, \varpi))_T|$$

$$\leq 2K_\rho \sum_{i,j:t_{i,n}^{(1)} \vee t_{j,n}^{(2)} \leq T} |\Delta_{i,n}^{(1)} X^{(1)}|^{p_1} |\Delta_{j,n}^{(2)} X^{(2)}|^{p_2} \mathbb{1}_{\{\mathcal{I}_{i,n}^{(1)} \cap \mathcal{I}_{j,n}^{(2)} \neq \emptyset\}}$$

$$\times \mathbb{1}_{\{ |\Delta_{i,n}^{(1)} X^{(1)}| > \beta |\mathcal{I}_{i,n}^{(1)}|^\varpi, |\Delta_{j,n}^{(2)} X^{(2)}| > \beta |\mathcal{I}_{j,n}^{(2)}|^\varpi \}}$$

$$= 2K_\rho \sum_{i,j:t_{i,n}^{(1)} \vee t_{j,n}^{(2)} \leq T} |\Delta_{i,n}^{(1)} (B + C)^{(1)}|^{p_1} |\Delta_{j,n}^{(2)} (B + C)^{(2)}|^{p_2} \mathbb{1}_{\{\mathcal{I}_{i,n}^{(1)} \cap \mathcal{I}_{j,n}^{(2)} \neq \emptyset\}}$$

$$\times \mathbb{1}_{\{ |\Delta_{i,n}^{(1)} X^{(1)}| > \beta |\mathcal{I}_{i,n}^{(1)}|^\varpi, |\Delta_{j,n}^{(2)} X^{(2)}| > \beta |\mathcal{I}_{j,n}^{(2)}|^\varpi \}} + K_\rho O_{\mathbb{P}}(1) \tag{2.83}$$

where

$$B_t = \int_0^t \left(b_s - \int_{\mathbb{R}^2} \delta(s,z) \mathbb{1}_{\{\|\delta(s,z)\| \le 1\}} \lambda(dz) \right) ds,$$

and the last identity in (2.83) holds because of

$$\Delta^{(l)}_{i_n^{(l)}(S_p),n} X^{(l)} \to \Delta^{(l)}_{i_n^{(l)}(S_p),n} X^{(l)}, \quad l = 1,2,$$

for any jump time S_p of X. The $\mathcal{S}$-conditional expectation of (2.83) is using the Cauchy-Schwarz inequality and (2.77) bounded by

$$K_\rho((|\pi_n|_T)^{(p_1+p_2)/2} + 1) \sum_{i,j:t^{(1)}_{i,n} \vee t^{(2)}_{j,n} \le T} |\mathcal{I}^{(1)}_{i,n}|^{\frac{p_1+(1/2-\varpi)}{2}} |\mathcal{I}^{(2)}_{j,n}|^{\frac{p_2+(1/2-\varpi)}{2}}$$

$$\times \mathbb{1}_{\{\mathcal{I}^{(1)}_{i,n} \cap \mathcal{I}^{(2)}_{j,n} \ne \emptyset\}}.$$

(2.80) then follows from (2.69) as in case (i). $\qquad\square$

Proof of Theorem 2.38. Comparing Proposition 2.30 and the proofs of Theorems 2.17, 2.21, 2.22 and Corollary 2.19 it suffices to prove

$$\overline{C}^*_-(p_1+p_2, f, \pi_n, (\beta, \varpi))_T - \overline{C}^*(p_1+p_2, f, \pi_n)_T \xrightarrow{\mathbb{P}} 0, \quad p_1+p_2 \ne 2, \quad (2.84)$$

$$\overline{V}^*_-(p_1+p_2, f, \pi_n, (\beta, \varpi))_T - \overline{C}^*(p_1+p_2, f, \pi_n)_T \xrightarrow{\mathbb{P}} 0, \quad p_1+p_2 = 2, \quad (2.85)$$

based on the assumptions made in Proposition 2.30 where we set

$$\overline{C}^*_-(p, f, \pi_n, (\beta, \varpi))_T = n^{p/2-1} \sum_{i,j \ge 0: t^{(1)}_{i,n} \vee t^{(2)}_{j,n} \le T} f(\Delta^{(1)}_{i,n} \widetilde{C}^{(1)}, \Delta^{(2)}_{j,n} \widetilde{C}^{(2)}) \mathbb{1}^{*,n}_{d_1,d_2}(i,j)$$

$$\times \mathbb{1}_{\{\|\Delta^{(1)}_{i,n} \widetilde{X}^{(1)}\| \le \beta |\mathcal{I}^{(1)}_{i,n}|^\varpi, \|\Delta^{(2)}_{j,n} \widetilde{X}^{(2)}\| \le \beta |\mathcal{I}^{(2)}_{j,n}|^\varpi\}}$$

$$\overline{V}^*_-(p, f, \pi_n, (\beta, \varpi))_T = n^{p/2-1} \sum_{i,j \ge 0: t^{(1)}_{i,n} \vee t^{(2)}_{j,n} \le T} f(\Delta^{(1)}_{i,n} \widetilde{X}^{(1)}, \Delta^{(2)}_{j,n} \widetilde{X}^{(2)}) \mathbb{1}^{*,n}_{d_1,d_2}(i,j)$$

$$\times \mathbb{1}_{\{\|\Delta^{(1)}_{i,n} \widetilde{X}^{(1)}\| \le \beta |\mathcal{I}^{(1)}_{i,n}|^\varpi, \|\Delta^{(2)}_{j,n} \widetilde{X}^{(2)}\| \le \beta |\mathcal{I}^{(2)}_{j,n}|^\varpi\}}$$

using the same notation as in (2.51). In the case that $\widetilde{X}^{(l)}$ is a zero-dimensional process we define $\|\Delta^{(l)}_{i,n} \widetilde{X}^{(l)}\| \le \beta |\mathcal{I}^{(l)}_{i,n}|^\varpi$ to be always true. Note that (2.84) is sufficient in the case $p_1 + p_2 \ne 2$ because

$$\overline{V}^*_-(p_1+p_2, f, \pi_n, (\beta, \varpi))_T - \overline{C}^*_-(p_1+p_2, f, \pi_n, (\beta, \varpi))_T \xrightarrow{\mathbb{P}} 0$$

can be proven just like Proposition 2.30 as we may simply drop the indicators.

For proving (2.84) we obtain using (2.45) and the estimate $\mathbb{1}_{A \vee B} \leq \mathbb{1}_A + \mathbb{1}_B$ for events A, B

$$
\mathbb{E}[\overline{C}^*_-(p, f, \pi_n, (\beta, \varpi))_T - \overline{C}^*(p, f, \pi_n)_T | \mathcal{S}]
$$

$$
\leq n^{p/2-1} \sum_{l=1,2} \sum_{i,j:t_{i,n}^{(l)} \vee t_{j,n}^{(3-l)} \leq T} K \mathbb{E}[\|\Delta_{i,n}^{(l)} \widetilde{C}^{(l)}\|^{p_l}
$$

$$
\times \|\Delta_{j,n}^{(3-l)} \widetilde{C}^{(3-l)}\|^{p_{3-l}} \mathbb{1}_{\{\|\Delta_{i,n}^{(l)} \widetilde{C}^{(l)}\| > \beta |\mathcal{I}_{i,n}^{(l)}|^\varpi\}} | \mathcal{S}] \mathbb{1}_{d_1,d_2}^{*,n}(i,j).
$$

Using the Cauchy-Schwarz inequality, inequality (1.9), arguments as in (2.77) and the notation from (2.53) this quantity is bounded by

$$
n^{p/2-1} \sum_{l=1,2} \sum_{i,j:t_{i,n}^{(l)} \vee t_{j,n}^{(3-l)} \leq T} K (\mathbb{E}[\|\Delta_{i,n}^{(l)} \widetilde{C}^{(l)}\|^{4p_l} | \mathcal{S}])^{1/4}
$$

$$
\times (\mathbb{P}(\|\Delta_{i,n}^{(l)} \widetilde{C}^{(l)}\| > \beta |\mathcal{I}_{i,n}^{(l)}|^\varpi | \mathcal{S}))^{1/4} (\mathbb{E}[\|\Delta_{j,n}^{(3-l)} \widetilde{C}^{(3-l)}\|^{2p_{3-l}} | \mathcal{S}])^{1/2} \mathbb{1}_{d_1,d_2}^{*,n}(i,j)
$$

$$
\leq 2K (|\pi_n|_T)^{(1-2\varpi)/4} \widetilde{G}_{p_1,p_2}^{(d_1,d_2),n}(T).
$$

Here the last bound vanishes as $n \to \infty$ which proves (2.84) using Lemma 2.15.

It remains to prove (2.85). To this end note that using inequality (2.46) we obtain the following estimate

$$
\mathbb{E}[\overline{V}^*_-(2, f, \pi_n, (\beta, \varpi))_T - \overline{C}^*(2, f, \pi_n)_T | \mathcal{S}]
$$

$$
\leq \sum_{i,j:t_{i,n}^{(1)} \vee t_{j,n}^{(2)} \leq T} \mathbb{E}[\theta(\varepsilon)\|\Delta_{i,n}^{(1)} \widetilde{C}^{(1)}\|^{p_1} \|\Delta_{j,n}^{(2)} \widetilde{C}^{(2)}\|^{p_2}
$$

$$
\times \mathbb{1}_{\{\|\Delta_{i,n}^{(1)} \widetilde{X}^{(1)}\| \leq \beta |\mathcal{I}_{i,n}^{(1)}|^\varpi, \|\Delta_{j,n}^{(2)} \widetilde{X}^{(2)}\| \leq \beta |\mathcal{I}_{j,n}^{(2)}|^\varpi\}} | \mathcal{S}] \mathbb{1}_{d_1,d_2}^{*,n}(i,j)
$$

$$
+ \sum_{l=1,2} \sum_{i,j:t_{i,n}^{(l)} \vee t_{j,n}^{(3-l)} \leq T} \mathbb{E}[K_\varepsilon \|\Delta_{i,n}^{(l)} (\widetilde{X} - \widetilde{C})^{(l)}\|^{p_l}
$$

$$
\times (\|\Delta_{j,n}^{(3-l)} \widetilde{C}^{(3-l)}\|^{p_{3-l}} + \|\Delta_{j,n}^{(3-l)} (\widetilde{X} - \widetilde{C})^{(3-l)}\|^{p_{3-l}})
$$

$$
\times \mathbb{1}_{\{\|\Delta_{i,n}^{(l)} \widetilde{X}^{(l)}\| \leq \beta |\mathcal{I}_{i,n}^{(l)}|^\varpi, \|\Delta_{j,n}^{(3-l)} \widetilde{X}^{(3-l)}\| \leq \beta |\mathcal{I}_{i,n}^{(3-l)}|^\varpi\}} | \mathcal{S}] \mathbb{1}_{d_1,d_2}^{*,n}(i,j)
$$

$$
+ \sum_{l=1,2} \sum_{i,j:t_{i,n}^{(l)} \vee t_{j,n}^{(3-l)} \leq T} \mathbb{E}[\|\Delta_{i,n}^{(1)} \widetilde{C}^{(1)}\|^{p_1} \|\Delta_{j,n}^{(2)} \widetilde{C}^{(2)}\|^{p_2} \mathbb{1}_{\{\|\Delta_{i,n}^{(l)} \widetilde{X}^{(l)}\| > \beta |\mathcal{I}_{i,n}^{(l)}|^\varpi\}} | \mathcal{S}]
$$

$$
\times \mathbb{1}_{d_1,d_2}^{*,n}(i,j).
$$

Let $p_1, p_2 > 0$. Then the quantity above is using Hölder's inequality with exponents $2/p_1$, $2/p_2$, Lemma 1.4 and an argument similar to (2.77) bounded by

$$\theta(\varepsilon) K \widetilde{G}_{p_1,p_2}^{(d_1,d_2),n}(T) + K_\varepsilon \sum_{l=1,2} (K_q |\pi_n|_T)^{p_l/2} + K e_q) K \widetilde{G}_{p_1,p_2}^{(d_1,d_2),n}(T)$$

$$+ \sum_{l=1,2} \sum_{i,j:t_{i,n}^{(l)} \vee t_{j,n}^{(3-l)} \leq T} \mathbb{E}[\|\Delta_{i,n}^{(l)} \widetilde{N}(q)^{(l)}\|^{p_l}$$

$$\times (\|\Delta_{j,n}^{(3-l)} \widetilde{C}^{(3-l)}\|^{p_{3-l}} + \|\Delta_{j,n}^{(3-l)} (\widetilde{X} - \widetilde{C})^{(3-l)}\|^{p_{3-l}}) | \mathcal{S}]$$

$$\times \mathbf{1}_{\{\|\Delta_{i,n}^{(l)} \widetilde{X}^{(l)}\| \leq \beta |\mathcal{I}_{i,n}^{(l)}|^\varpi, \|\Delta_{j,n}^{(3-l)} \widetilde{X}^{(3-l)}\| \leq \beta |\mathcal{I}_{j,n}^{(3-l)}|^\varpi\}} | \mathcal{S}] \mathbf{1}_{d_1,d_2}^{*,n}(i,j)$$

$$+ (|\pi_n|_T)^{1/2-\varpi} K \widetilde{G}_{p_1,p_2}^{(d_1,d_2),n}(T)$$

where the sums in the first and last line vanish as first $n \to \infty$ and then $q \to \infty$ because of $\widetilde{G}_{p_1,p_2}^{(d_1,d_2),n}(T) = O_\mathbb{P}(1)$. Further the sum in the second line vanishes as $n \to \infty$ for any $q > 0$ using dominated convergence because we have

$$\|\Delta_{i,n}^{(l)} \widetilde{N}(q)^{(l)}\| \neq 0 \Rightarrow \|\Delta_{i,n}^{(l)} \widetilde{X}^{(l)}\| > \beta |\mathcal{I}_{i,n}^{(l)}|^\varpi$$

with probability approaching one due to $|\pi_n|_T \xrightarrow{\mathbb{P}} 0$. Hence (2.85) follows using Lemma 2.15.

For $p_1 = 0$ or $p_2 = 0$ the proof of (2.85) is based on (2.47). The arguments used in the case $p_1 = 0$ and $p_2 = 0$ are very similar to those in the case $p_1, p_2 > 0$ and therefore not presented here. $\qquad \square$

2.4 Functionals of Increments over Multiple Observation Intervals

In Chapter 8 and Section 9.2 we will use statistics where we compare functionals based on data sampled at different frequencies. Therefore for some $k \geq 2$ we are interested in functionals which are of similar form as those discussed in Sections 2.1 and 2.2, but which are based on increments of $X^{(1)}$ and $X^{(2)}$ over the observation intervals

$$\mathcal{I}_{i,k,n}^{(l)} = (t_{i-k,n}^{(l)}, t_{i,n}^{(l)}], \quad l = 1, 2, \tag{2.86}$$

instead of increments over $\mathcal{I}_{i,n}^{(l)} = \mathcal{I}_{i,1,n}^{(l)}$. We introduce the notation

$$\Delta_{i,k,n}^{(l)} X^{(l)} = X_{t_{i,n}^{(l)}}^{(l)} - X_{t_{i-k,n}^{(l)}}^{(l)} \tag{2.87}$$

and for convenience we again set $\mathcal{I}_{i,k,n}^{(l)} = \emptyset$ and $\Delta_{i,k,n}^{(l)} X^{(l)} = 0$ if $i < k$. Then we define for functions $f \colon \mathbb{R}^2 \to \mathbb{R}$, $g \colon \mathbb{R} \to \mathbb{R}$ the functionals

$$
\begin{aligned}
V(f, [k], \pi_n)_T &= \sum_{i,j : t_{i,n}^{(1)} \vee t_{j,n}^{(2)} \leq T} f\big(\Delta_{i,k,n}^{(1)} X^{(1)}, \Delta_{j,k,n}^{(2)} X^{(2)}\big) \mathbb{1}_{\{\mathcal{I}_{i,k,n}^{(1)} \cap \mathcal{I}_{j,k,n}^{(2)} \neq \emptyset\}}, \\
V^{(l)}(g, [k], \pi_n)_T &= \sum_{i : t_{i,n}^{(l)} \leq T} g\big(\Delta_{i,k,n}^{(l)} X^{(l)}\big), \quad l = 1, 2,
\end{aligned}
\tag{2.88}
$$

and

$$
\overline{V}(p, f, [k], \pi_n)_T = \frac{n^{p/2-1}}{k^{p/2+1}} \sum_{i,j : t_{i,n}^{(1)} \vee t_{j,n}^{(2)} \leq T} f\big(\Delta_{i,k,n}^{(1)} X^{(1)}, \Delta_{j,k,n}^{(2)} X^{(2)}\big) \mathbb{1}_{\{\mathcal{I}_{i,k,n}^{(1)} \cap \mathcal{I}_{j,k,n}^{(2)} \neq \emptyset\}},
$$

$$
\overline{V}^{(l)}(p, g, [k], \pi_n)_T = \frac{n^{p/2-1}}{k^{p/2}} \sum_{i : t_{i,n}^{(l)} \leq T} g\big(\Delta_{i,n}^{(l)} X^{(l)}\big), \; l = 1, 2.
$$

To describe the limits of $\overline{V}(p, f, [k], \pi_n)_T$, $\overline{V}^{(l)}(p, g, [k], \pi_n)_T$ whenever they exist we define the functions

$$
G_p^{[k],(l),n}(t) = \frac{n^{p/2-1}}{k^{p/2}} \sum_{i : t_{i,n}^{(l)} \leq t} \big|\mathcal{I}_{i,k,n}^{(l)}\big|^{p/2},
$$

$$
G_{p_1,p_2}^{[k],n}(t) = \frac{n^{(p_1+p_2)/2-1}}{k^{(p_1+p_2)/2+1}} \sum_{i,j : t_{i,n}^{(1)} \vee t_{j,n}^{(2)} \leq t} \big|\mathcal{I}_{i,k,n}^{(1)}\big|^{p_1/2} \big|\mathcal{I}_{j,k,n}^{(2)}\big|^{p_2/2} \mathbb{1}_{\{\mathcal{I}_{i,k,n}^{(1)} \cap \mathcal{I}_{j,k,n}^{(2)} \neq \emptyset\}},
$$

$$
H_{\iota,m,p}^{[k],n}(t) = \frac{n^{p/2-1}}{k^{p/2+1}} \sum_{i,j : t_{i,n}^{(1)} \vee t_{j,n}^{(2)} \leq t} \big|\mathcal{I}_{i,k,n}^{(1)} \setminus \mathcal{I}_{j,k,n}^{(2)}\big|^{\iota/2} \big|\mathcal{I}_{j,k,n}^{(2)} \setminus \mathcal{I}_{i,k,n}^{(1)}\big|^{m/2}
$$
$$
\times \big|\mathcal{I}_{i,k,n}^{(1)} \cap \mathcal{I}_{j,k,n}^{(2)}\big|^{(p-(\iota+m))/2} \mathbb{1}_{\{\mathcal{I}_{i,k,n}^{(1)} \cap \mathcal{I}_{j,k,n}^{(2)} \neq \emptyset\}}.
\tag{2.89}
$$

Here, we rescale interval lengths with the factor n/k such that the magnitude of the rescaled interval lengths is independent of k and n. Further we have an extra factor k^{-1} in $\overline{V}(p, f, [k], \pi_n)_T$ because we have approximately k times as many summands in $\overline{V}(p, f, [k], \pi_n)_T$ as in $\overline{V}(p, f, [1], \pi_n)_T$. The fractions $n^{p/2-1}/k^{p/2+1}$ and $n^{p/2-1}/k^{p/2}$ in the definitions of $\overline{V}(p, f, [k], \pi_n)_T$ and $\overline{V}^{(l)}(p, g, [k], \pi_n)_T$ are chosen such that the magnitude of $\overline{V}(p, f, [k], \pi_n)_T$ and $\overline{V}^{(l)}(p, g, [k], \pi_n)_T$ does not depend on k at least for equidistant and synchronous observation times $t_{i,n}^{(1)} = t_{i,n}^{(2)} = i/n$.

2.4.1 The Results

The results we obtain for the asymptotic behaviour of the functionals based on increments over the intervals $\mathcal{I}_{i,k,n}^{(l)}$ are very similar to the results obtained in Sections 2.1 and 2.2. As a generalization of Corollary 2.2, Theorem 2.3 and Theorem 2.6 we obtain the following result for the non-normalized functionals.

Theorem 2.40. *Suppose Condition 1.3 is fulfilled.*

a) *It holds*

$$V^{(l)}(g,[k],\pi_n)_T \xrightarrow{\text{P}} kB^{(l)}(g)_T.$$

for all continuous functions $g\colon \mathbb{R} \to \mathbb{R}$ with $g(x) = o(x^2)$ as $x \to 0$.

b) *It holds*

$$V(f,[k],\pi_n)_T \xrightarrow{\text{P}} k^2 B(f)_T \qquad (2.90)$$

for all continuous functions $f\colon \mathbb{R} \to \mathbb{R}$ with $f(x,y) = O(x^2 y^2)$ as $|xy| \to 0$.

c) *Let $p_1, p_2 > 0$ with $p_1 + p_2 \geq 0$. We have (2.90) for all continuous functions $f\colon \mathbb{R} \to \mathbb{R}$ with $f(x,y) = o(|x|^{p_1}|y|^{p_2})$ as $|xy| \to 0$ if the observation scheme is exogenous and*

$$\sum_{i,j:t_{i,n}^{(1)} \vee t_{j,n}^{(2)} \leq t} \left|\mathcal{I}_{i,k,n}^{(1)}\right|^{\frac{p_1}{2}\wedge 1}\left|\mathcal{I}_{j,k,n}^{(2)}\right|^{\frac{p_2}{2}\wedge 1}\mathbb{1}_{\{\mathcal{I}_{i,k,n}^{(1)} \cap \mathcal{I}_{j,k,n}^{(2)} \neq \emptyset\}} = O_{\text{P}}(1). \quad (2.91)$$

For the normalized functionals we get the following results where Theorem 2.41 contains the results in the setting of synchronous observation times and Theorem 2.42 the results in the setting of asynchronous observation times.

Theorem 2.41. *Suppose Condition 1.3 is fulfilled and the observation scheme is exogenous. Further assume that it either holds $p \in [0,2)$ or we have $p \geq 2$ and X is continuous.*

a) *For all positively homogeneous functions $g\colon \mathbb{R} \to \mathbb{R}$ of degree p it holds*

$$\overline{V}^{(l)}(p,g,[k],\pi_n)_T \xrightarrow{\text{P}} \rho_1(g)\int_0^T (\sigma_s^{(l)})^p dG_p^{[k],(l)}(s)$$

if there exists some (possibly random) continuous function $G_p^{[k],(l)}\colon [0,T] \to [0,\infty)$ with

$$G_p^{[k],(l),n}(t) \xrightarrow{\text{P}} G_p^{[k],(l)}(t), \quad t \in [0,T].$$

b) *Further if the observation scheme is synchronous and if there exists some (possibly random) continuous function $G_p^{[k]} : [0, T] \to [0, \infty)$ with*

$$G_p^{[k],(1),n}(t) \xrightarrow{\mathbb{P}} G_p^{[k]}(t), \quad t \in [0, T],$$

then it holds

$$\overline{V}(p, f, [k], \pi_n)_T \xrightarrow{\mathbb{P}} \int_0^T \rho_{c_s}(f) dG_p^{[k]}(s)$$

for all positively homogeneous functions $f : \mathbb{R}^2 \to \mathbb{R}$ of degree p.

Theorem 2.42. *Suppose Condition 1.3 is fulfilled and the observation scheme is exogenous. Further assume that either $p_1, p_2 \geq 0$ with $p_1 + p_2 \in [0, 2)$ or we have $p_1 + p_2 \geq 2$ and X is continuous.*

a) *Suppose $\rho \equiv 0$ on $[0, T]$ and assume that*

$$G_{p_1,p_2}^{[k],n}(t) \xrightarrow{\mathbb{P}} G_{p_1,p_2}^{[k]}(t), \quad t \in [0, T],$$

holds for some (possibly random) continuous function $G_{p_1,p_2}^{[k]} : [0, T] \to [0, \infty)$. Then for all continuous functions $f : \mathbb{R}^2 \to \mathbb{R}$ which are positively homogeneous of degree p_1 in the first component and positively homogeneous of degree p_2 in the second component it holds

$$\overline{V}(p_1 + p_2, f, [k], \pi_n)_T \xrightarrow{\mathbb{P}} \rho_{I_2}(f) \int_0^T (\sigma_s^{(1)})^{p_1} (\sigma_s^{(2)})^{p_2} dG_{p_1,p_2}^{[k]}(s).$$

Here I_2 denotes the two-dimensional identity matrix.

b) *Let p_1, p_2 be non-negative integers and assume that $G_{p_1,p_2}^{[k],n}(T) = O_{\mathbb{P}}(1)$ holds. Further assume that for all $\iota, m \in \mathbb{N}_0$ for which an $l \in \mathbb{N}_0$ exists with $(\iota, l, m) \in L(p_1, p_2)$, for the notation compare Theorem 2.22, there exist (possibly random) continuous functions $H_{\iota,m,p_1+p_2}^{[k]} : [0, T] \to [0, \infty)$ which fulfill*

$$H_{\iota,m,p_1+p_2}^{[k],n}(t) \xrightarrow{\mathbb{P}} H_{\iota,m,p_1+p_2}^{[k]}(t), \quad t \in [0, T].$$

Then it holds

$$\overline{V}(p_1 + p_2, f_{(p_1,p_2)}, [k], \pi_n)_T$$

$$\xrightarrow{\mathbb{P}} \int_0^T (\sigma_s^{(1)})^{p_1} (\sigma_s^{(2)})^{p_2} \Big(\sum_{(\iota,l,m) \in L(p_1,p_2)} \binom{p_1}{\iota,l} \binom{p_2}{m} \rho_1(x^{\iota}) \rho_1(x^{l}) \rho_1(x^{m})$$

$$\times \rho_1(x^{p_1+p_2-(\iota+l+m)}) (1 - \rho_s^2)^{l/2} (\rho_s)^{p_1-(\iota+l)} dH_{\iota,m,p_1+p_2}^{[k]}(s) \Big).$$

where $f_{(p_1,p_2)}(x, y) = x^{p_1} y^{p_2}$.

Example 2.43. *The assumptions from Theorems 2.40–2.42 are fulfilled in the setting of Poisson sampling; compare Definition 5.1. Indeed that the assumption (2.91) is fullfilled and that the functions $G_p^{[k],(l),n}$, $G_{p_1,p_2}^{[k],n}$, $H_{l,m,p}^{[k],n}$ converge in probability to linear deterministic functions follows directly from Corollary 5.8.* $\square$

2.4.2 The Proofs

Proof of Theorem 2.40. It holds

$$V^{(l)}(g,[k],\pi_n)_T = \sum_{\iota=1}^{k} \sum_{i:t_{ik+\iota,n}^{(l)} \leq T} g(\Delta_{ik+\iota,k,n}^{(l)} X^{(l)})$$

where each of the sums

$$\sum_{i:t_{ik+\iota,n}^{(l)} \leq T} g(\Delta_{ik+\iota,k,n}^{(l)} X^{(l)}), \quad \iota = 0,\dots,k-1,$$

converges to $B^{(l)}(g)_T$ by Corollary 2.2. Therefore note that

$$(X_{t_{\iota,n}^{(l)}}^{(l)} - X_{t_{0,n}^{(l)}}^{(l)}) + \sum_{i:t_{ik+\iota,n}^{(l)} \leq T} g(\Delta_{ik+\iota,k,n}^{(l)} X^{(l)}) = V^{(l)}(g,\pi_n^\iota)_T$$

with $\pi_n^\iota = \{(0, t_{k+\iota,n}^{(l)}, t_{2k+\iota,n}^{(l)}, \dots) | l = 1,2\}$ and

$$(X_{t_{\iota,n}^{(l)}}^{(l)} - X_{t_{0,n}^{(l)}}^{(l)}) = o_{\mathbb{P}}(1), \quad \iota = 0,\dots k-1,$$

as $n \to \infty$.

Similarly we proof parts b) and c) using

$$V(f,[k],\pi_n)_T = \sum_{\iota,\iota'=1}^{k} \sum_{i,j:t_{ik+\iota,n}^{(1)} \vee t_{jk+\iota',n}^{(2)} \leq T} f(\Delta_{ik+\iota,k,n}^{(1)} X^{(1)}, \Delta_{(jk+\iota',k,n}^{(2)} X^{(2)})$$

$$\times \mathbb{1}_{\{\mathcal{I}_{ik+\iota,k,n}^{(1)} \cap \mathcal{I}_{jk+\iota,k,n}^{(2)} \neq \emptyset\}}$$

and Theorems 2.3, 2.6. For part c) note that (2.8) is fulfilled for each (ι, ι') because of (2.91). $\square$

Proof of Theorems 2.41 and 2.42. The proofs are identical to the proofs in Section 2.2. $\square$

3 Central Limit Theorems

In this chapter, we will discuss central limit theorems for some of the functionals introduced in Chapter 2. Further we develop general techniques that will later in Chapters 7–9 be applied to find central limit theorems also for other more specific statistics which are based on functionals from Chapter 2.

Throughout this chapter we will assume that the observation scheme is exogenous. In practice a theory for endogenous observation times might be desirable as well. However, previous research shows that for such observation schemes it is difficult to derive central limit theorems even in simple situations; compare [18] or [47]. For this reason we restrict ourselves to exogenous observation times. This setting is already far more general than the setting of equidistant and synchronous observation times $t_{i,n}^{(1)} = t_{i,n}^{(2)} = i/n$ which is mainly discussed in the literature. It covers a lot of interesting random and irregular sampling schemes where the most prominent one is Poisson sampling; see Definition 5.1. Already in the setting of exogenous observation times the resulting central limit theorems have a more complicated structure as the limit not only depends on the process X but also on the asymptotics of the observation scheme. Hence we will need to impose further assumptions that guarantee an appropriate behaviour of the observation schemes π_n for $n \to \infty$.

As common in this field of high-frequency statistics the asymptotic variances in the upcoming central limit theorems are themselves random. They e.g. depend via the volatility process σ and jump heights ΔX_s, $s \in [0, T]$, on the observed path of the process X. For this reason we derive *stable* central limit theorems as is done frequently in related literature. This means that we do not look for convergence in law but *stable convergence in law* of the properly rescaled error term to a certain random limit. Stable convergence has the advantage that a consistent estimator for the random asymptotic variance can be used to construct asymptotically exact confidence intervals. To illustrate this property let us consider the stable convergence

$$\sqrt{n}(Y_n - Y) \overset{\mathcal{L}-s}{\longrightarrow} VU$$

for random variables Y, $(Y_n)_{n \in \mathbb{N}}$, V and $U \sim \mathcal{N}(0,1)$ where V is not independent of $Y, (Y_n)_{n \in \mathbb{N}}$. If we now can find a consistent sequence of estimators $(V_n)_{n \in \mathbb{N}}$ with $V_n \overset{\mathbb{P}}{\longrightarrow} V$ we also obtain

$$\frac{\sqrt{n}(Y_n - Y)}{V_n} \overset{\mathcal{L}-s}{\longrightarrow} U.$$

© Springer Fachmedien Wiesbaden GmbH, part of Springer Nature 2019
O. Martin, *High-Frequency Statistics with Asynchronous and Irregular Data*,
Mathematische Optimierung und Wirtschaftsmathematik | Mathematical Optimization
and Econömathematics, https://doi.org/10.1007/978-3-658-28418-3_3

This implication in general does not hold if we consider ordinary convergence in law instead of stable convergence in law. In particular the above implication holds in general for ordinary convergence in law only if V is deterministic. A more detailed introduction of the concept of stable convergence in law and the discussion of some of its properties which are used in the upcoming proofs can be found in Appendix B.

3.1 Central Limit Theorem for Non-Normalized Functionals

In this section we will derive central limit theorems for the non-normalized functionals $V^{(l)}(g, \pi_n)_T$ and $V(f, \pi_n)_T$. We will first sketch the results using so-called toy examples. These are models for the process $X^{(l)}$ respectively X which have the simplest form among all processes which yield a central limit theorem of the general form i.e. they contain all relevant terms that contribute in the limit but were freed of terms that do not.

We start with motivating the result in the univariate setting because in that situation the structure of the result is simpler. To this end we consider

$$X_t^{toy,(l)} = \int_0^t \sigma_s^{(l)} dW_s^{(l)} + \sum_{s \leq t} \Delta N^{(l)}(q)_s, \quad t \geq 0,$$

for some $q > 0$ where the process $X^{toy,(l)}$ consists of a Brownian part with volatility process $\sigma^{(l)}$ and a finite activity jump part. Finite jump activity means that $X_t^{toy,(l)}$ almost surely has only finitely many jumps in any interval $[0, t]$, $t \geq 0$. Further let $g \colon \mathbb{R} \to \mathbb{R}$ denote a twice continuously differentiable function with $g(0) = g'(0) = 0$ and $g''(x) = o(|x|)$ as $x \to 0$. We denote $C_t^{toy,(l)} = \int_0^t \sigma_s^{(l)} dW_s^{(l)}$. By means of a Taylor expansion we then obtain the following identity

$$\sqrt{n}\big(V^{(l)}(g, \pi_n)_T - B^{(l)}(g)_T\big)$$
$$= \sqrt{n}\Big(\sum_{i : t_{i,n}^{(l)} \leq T} \big(g(\Delta_{i,n}^{(l)} N^{(l)}(q)) + g'(\Delta_{i,n}^{(l)} N^{(l)}(q))\Delta_{i,n}^{(l)} C^{toy,(l)}$$
$$+ \frac{1}{2} g''(\Delta_{i,n}^{(l)} N^{(l)}(q) + \theta_{i,n}\Delta_{i,n}^{(l)} C^{toy,(l)})(\Delta_{i,n}^{(l)} C^{toy,(l)})^2\big) - \sum_{s \leq T} g(\Delta N^{(l)}(q)_s)\Big)$$

$$(3.1)$$

for some (random) $\theta_{i,n} \in [0, 1]$. Denote by $S_p^{(l)}$, $p \in \mathbb{N}$, an enumeration of the (countably many) jump times of $N^{(l)}(q)$ on $[0, \infty)$. Conditional on the set where two jumps of $N^{(l)}(q)$ in $[0,T]$ are further apart than $|\pi_n|_T$ the expression (3.1) is due to $g(0) = g'(0) = 0$ equal to

$$\sqrt{n} \sum_{p:S_p^{(l)} \leq T} g'(\Delta N^{(l)}(q)_{S_p^{(l)}}) \Delta_{i_n^{(l)}(S_p^{(l)}),n}^{(l)} C^{toy,(l)} \tag{3.2}$$

$$+ \frac{\sqrt{n}}{2} \sum_{i:t_{i,n}^{(l)} \leq T} g''(\Delta_{i,n}^{(l)} N^{(l)}(q) + \theta_{i,n} \Delta_{i,n}^{(l)} C^{toy,(l)})(\Delta_{i,n}^{(l)} C^{toy,(l)})^2. \tag{3.3}$$

For the sum (3.3) we obtain by the assumption $g''(x) = o(|x|)$

$$\frac{\sqrt{n}}{2} \sum_{i:t_{i,n}^{(l)} \leq T} g''(\Delta_{i,n}^{(l)} N^{(l)}(q) + \theta_{i,n} \Delta_{i,n}^{(l)} C^{toy,(l)})(\Delta_{i,n}^{(l)} C^{toy,(l)})^2$$

$$= \frac{\sqrt{n}}{2} \sum_{p:S_p^{(l)} \leq T} g''(\Delta N^{(l)}(q)_{S_p^{(l)}} + \theta_{i_n^{(l)}(S_p^{(l)}),n} \Delta_{i_n^{(l)}(S_p^{(l)}),n}^{(l)} C^{toy,(l)})$$

$$\times (\Delta_{i_n^{(l)}(S_p^{(l)}),n}^{(l)} C^{toy,(l)})^2 \tag{3.4}$$

$$+ o_{\mathbb{P}}\left(\sqrt{n} \sum_{i:t_{i,n}^{(l)} \leq T} |\Delta_{i_n^{(l)}(S_p^{(l)}),n}^{(l)} C^{toy,(l)}|^3\right). \tag{3.5}$$

Here, the sum (3.4) contains almost surely only finitely many summands of which each is of order $\sqrt{n}|\mathcal{I}_{i_n^{(l)}(S_p^{(l)}),n}^{(l)}|$. Hence (3.4) vanishes for $n \to \infty$ because we will assume $|\mathcal{I}_{i_n^{(l)}(S_p^{(l)}),n}^{(l)}| = O_{\mathbb{P}}(1/n)$. Further it holds by inequality (1.9)

$$\mathbb{E}\left[\sqrt{n} \sum_{i:t_{i,n}^{(l)} \leq T} |\Delta_{i_n^{(l)}(S_p^{(l)}),n}^{(l)} C^{toy,(l)}|^3 \Big| S\right] \leq K G_3^{(l),n}(T).$$

Hence (3.5) also vanishes as $n \to \infty$ if we assume $G_3^{(l),n}(T) = O_{\mathbb{P}}(1)$ as $n \to \infty$. Combining these observations we obtain that (3.3) is asymptotically negligible by the assumption on g.

For the sum (3.2) we obtain the approximation

$$\sqrt{n} \sum_{p:S_p^{(l)} \leq T} g'(\Delta N^{(l)}(q)_{S_p^{(l)}}) \Delta_{i_n^{(l)}(S_p^{(l)}),n}^{(l)} C^{toy,(l)} \overset{\mathcal{L}}{\approx} \sum_{p:S_p^{(l)} \leq T} g'(\Delta N^{(l)}(q)_{S_p^{(l)}})$$

$$\times (\sigma_{S_p^{(l)}-}^{(l)} (n\delta_{n,-}^{(l)}(S_p^{(l)}))^{1/2} U_-^{(l)}(S_p^{(l)}) + \sigma_{S_p^{(l)}}^{(l)} (n\delta_{n,+}^{(l)}(S_p^{(l)}))^{1/2} U_+^{(l)}(S_p^{(l)}))$$

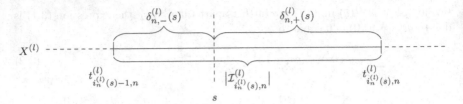

Figure 3.1: Illustration of the terms $\delta_{n,-}^{(l)}(s)$ and $\delta_{n,+}^{(l)}(s)$.

where $U_-^{(l)}(s), U_+^{(l)}(s)$, $s \in [0,T]$, are i.i.d. standard normal random variables independent of $N^{(l)}(q)$ and $\sigma^{(l)}$ and the random variables

$$(\delta_{n,-}^{(l)}(s), \delta_{n,+}^{(l)}(s)) = \big(s - t_{i_n^{(l)}(s)-1,n}^{(l)}, t_{i_n^{(l)}(s),n}^{(l)} - s\big), \quad s \geq 0, \qquad (3.6)$$

describe the distances of s to previous and upcoming observation times; compare Figure 3.1.

If we assume that the random variables $(n\delta_{n,-}^{(l)}(s), n\delta_{n,+}^{(l)}(s))$, $s \in [0,T]$, converge in a suitable sense to random variables $(\delta_-^{(l)}(s), \delta_+^{(l)}(s))$, $s \in [0,T]$, we obtain the stable convergence

$$\sqrt{n}\big(V^{(l)}(g,\pi_n)_T - B^{(l)}(g)_T\big) \overset{\mathcal{L}-s}{\longrightarrow} \sum_{p:S_p^{(l)}\leq T} g'(\Delta N^{(l)}(q)_{S_p^{(l)}})$$

$$\times (\sigma_{S_p^{(l)}-}^{(l)}(\delta_-^{(l)}(S_p^{(l)}))^{1/2}U_-^{(l)}(S_p^{(l)}) + \sigma_{S_p^{(l)}}^{(l)}(\delta_+^{(l)}(S_p^{(l)}))^{1/2}U_+^{(l)}(S_p^{(l)})) \qquad (3.7)$$

where $(\delta_-^{(l)}(S_p^{(l)}), \delta_+^{(l)}(S_p^{(l)}))$ is independent of $U_-^{(l)}(S_p^{(l)})$, $U_+^{(l)}(S_p^{(l)})$. Here, the limit has a mixed normal distribution with random variance

$$\sum_{p:S_p^{(l)}\leq T} (g'(\Delta N^{(l)}(q)_{S_p^{(l)}}))^2 ((\sigma_{S_p^{(l)}-}^{(l)})^2\delta_-^{(l)}(S_p^{(l)}) + (\sigma_{S_p^{(l)}}^{(l)})^2\delta_+^{(l)}(S_p))$$

which depends on $N^{(l)}(q)$ and $\sigma^{(l)}$ as well as on the asymptotics of the observation scheme.

Next, we will motivate the result in the bivariate setting with asynchronous observation times. We again consider a toy example

$$X_t^{toy} = \int_0^t \sigma_s dW_s + \sum_{s\leq t} \Delta N(q)_s, \quad t \geq 0,$$

which is of similar form as the one in the univariate setting. Let $f\colon \mathbb{R}^2 \to \mathbb{R}$ be a twice continuously differentiable function with

$$f(x,y) = \partial_1 f(x,y) = \partial_2 f(x,y) = 0 \text{ for all } (x,y) \in \mathbb{R}^2 \text{ with } xy = 0, \quad (3.8)$$

$$\partial_{kl} f(x,y) = o(|x|^{p_1^{(k,l)}} |y|^{p_2^{(k,l)}}) \text{ as } |xy| \to 0 \quad (3.9)$$

for some $p_1^{(k,l)}, p_2^{(k,l)} \geq 0$ with $p_1^{(k,l)} + p_2^{(k,l)} = 1$ for any $k,l \in \{1,2\}$. Here ∂_l denotes the first order partial derivative with respect to the l-th argument and ∂_{kl} denotes the second order partial derivative where we first differentiate with respect to the l-th argument and then with respect to the k-th argument. By Taylor expansion, compare Theorem 7.2 in [17], we obtain

$$
\begin{aligned}
&\sqrt{n}\big(V(f,\pi_n)_T - B^*(f)_T\big) \\
&= \sqrt{n}\Big(\sum_{i_1,i_2 : t_{i_1,n}^{(1)} \vee t_{i_2,n}^{(2)} \leq T} \big[f(\Delta_{i_1,n}^{(1)} N^{(1)}(q), \Delta_{i_2,n}^{(2)} N^{(2)}(q)) \\
&\quad + \sum_{l=1,2} \partial_l f(\Delta_{i_1,n}^{(1)} N^{(1)}(q), \Delta_{i_2,n}^{(2)} N^{(2)}(q)) \Delta_{i_l,n}^{(l)} C^{toy,(l)} \\
&\quad + \sum_{k,l=1,2} \frac{1}{2} \partial_{kl} f(\Delta_{i_1,n}^{(1)} N^{(1)}(q) + \theta_{i_1,i_2}^n \Delta_{i_1,n}^{(1)} C^{(1)}, \Delta_{i_2,n}^{(2)} N^{(2)}(q) \\
&\qquad\qquad + \theta_{i_1,i_2}^n \Delta_{i_2,n}^{(2)} C^{(2)}) \Delta_{i_k,n}^{(k)} C^{toy,(k)} \Delta_{i_l,n}^{(l)} C^{toy,(l)} \big] \\
&\quad - \sum_{s \leq T} f(\Delta N^{(1)}(q)_s, \Delta N^{(2)}(q)_s) \Big)
\end{aligned}
\quad (3.10)
$$

for some $\theta_{i_1,i_2}^n \in [0,1]$. Hence on the set where any two different jump times of $N(q)$ in $[0,T]$ are further apart than $2|\pi_n|_T$ the expression (3.10) is using (3.8) equal to

$$
\sqrt{n} \sum_{p : S_p \leq T} \sum_{l=1,2} \partial_l f(\Delta_{i_n^{(1)}(S_p),n}^{(1)} N^{(1)}(q), \Delta_{i_n^{(2)}(S_p),n}^{(2)} N^{(2)}(q)) \Delta_{i_n^{(l)}(S_p),n}^{(l)} C^{toy,(l)}
$$

$$(3.11)$$

$$
+ \frac{\sqrt{n}}{2} \sum_{i_1,i_2 : t_{i_1,n}^{(l)} \vee t_{i_2,n}^{(2)} \leq T} \sum_{k,l=1,2} \partial_{kl} f(\Delta_{i_1,n}^{(1)} N^{(1)}(q) + \theta_{i_1,i_2}^n \Delta_{i_1,n}^{(1)} C^{(1)},
$$

$$
\Delta_{i_2,n}^{(2)} N^{(2)}(q) + \theta_{i_1,i_2}^n \Delta_{i_2,n}^{(2)} C^{(2)}) \Delta_{i_k,n}^{(k)} C^{toy,(k)} \Delta_{i_l,n}^{(l)} C^{toy,(l)} \quad (3.12)
$$

where S_p, $p \in \mathbb{N}$, denotes an enumeration of the *common jump times* of $N^{(1)}(q)$ and $N^{(2)}(q)$.

Splitting the sum (3.12) into the summands where

$$(\Delta^{(1)}_{i_1,n} N^{(1)}(q), \Delta^{(2)}_{i_2,n} N^{(2)}(q)) \neq 0$$

and into those summands where increments of $N(q)$ do not contribute similarly as in the univariate setting for (3.4) and (3.5) we obtain that (3.12) is equal to

$$o_{\mathbb{P}}(1) + \sum_{k,l=1}^{2} o_{\mathbb{P}}\big(G^n_{p_1^{(k,l)}+(2-k)+(2-l),\, p_2^{(k,l)}+(k-1)+(l-1)}(T)\big)$$

which vanishes as $n \to \infty$ if we assume that

$$G^n_{p_1^{(k,l)}+(2-k)+(2-l),\, p_2^{(k,l)}+(k-1)+(l-1)}(T), \quad k,l = 1,2,$$

are all $O_{\mathbb{P}}(1)$ as $n \to \infty$. Hence (3.12) is asymptotically negligible.

As in the univariate setting the quantity (3.11) constitutes the terms which are relevant in the limit. It holds

$$\sqrt{n} \sum_{p:S_p \leq T} \sum_{l=1,2} \partial_l f(\Delta^{(1)}_{i_n^{(1)}(S_p),n} N^{(1)}(q), \Delta^{(2)}_{i_n^{(2)}(S_p),n} N^{(2)}(q)) \Delta^{(l)}_{i_n^{(l)}(S_p),n} C^{toy,(l)}$$

$$\overset{\mathcal{L}}{\approx} \sum_{p:S_p \leq T} \Big[\partial_1 f(\Delta N^{(1)}(q)_{S_p}, \Delta N^{(2)}(q)_{S_p}) \Big(\sigma^{(1)}_{S_p-} \sqrt{\mathcal{L}_n(S_p)} U^{(1),-}_{S_p}$$

$$+ \sigma^{(1)}_{S_p} \sqrt{\mathcal{R}_n(S_p)} U^{(1),+}_{S_p}$$

$$+ \sqrt{(\sigma^{(1)}_{S_p-})^2 \mathcal{L}_n^{(1)}(S_p) + (\sigma^{(1)}_{S_p})^2 \mathcal{R}_n^{(1)}(S_p)} U^{(2)}_{S_p} \Big)$$

$$+ \partial_2 f(\Delta N^{(1)}(q)_{S_p}, \Delta N^{(2)}(q)_{S_p}) \Big(\sigma^{(2)}_{S_p-} \rho_{S_p-} \sqrt{\mathcal{L}_n(S_p)} U^{(1),-}_{S_p}$$

$$+ \sigma^{(2)}_{S_p} \rho_{S_p} \sqrt{\mathcal{R}_n(S_p)} U^{(1),+}_{S_p}$$

$$+ \sqrt{(\sigma^{(2)}_{S_p-})^2 (1-(\rho_{S_p-})^2) \mathcal{L}_n(S_p) + (\sigma^{(2)}_{S_p})^2 (1-(\rho_{S_p})^2) \mathcal{R}_n(S_p)} U^{(3)}_{S_p}$$

$$+ \sqrt{(\sigma^{(2)}_{S_p-})^2 \mathcal{L}_n^{(2)}(S_p) + (\sigma^{(2)}_{S_p})^2 \mathcal{R}_n^{(2)}(S_p)} U^{(4)}_{S_p} \Big) \Big] \tag{3.13}$$

where $U^{(1),-}_s, U^{(1),+}_s, U^{(2)}_s, U^{(3)}_s, U^{(4)}_s$, $s \in [0,T]$, are i.i.d. standard normal distributed and we use the random variables

$$\mathcal{L}_n^{(l)}(s) = \max\{t^{(1)}_{i_n^{(1)}(s)-1,n}, t^{(2)}_{i_n^{(2)}(s)-1,n}\} - t^{(l)}_{i_n^{(l)}(s)-1,n}, \quad l=1,2,$$

$$\mathcal{R}_n^{(l)}(s) = t^{(l)}_{i_n^{(l)}(s),n} - \min\{t^{(1)}_{i_n^{(1)}(s),n}, t^{(2)}_{i_n^{(2)}(s),n}\}, \quad l=1,2,$$

$$\mathcal{L}_n(s) = s - \max\{t^{(1)}_{i_n^{(1)}(s)-1,n}, t^{(2)}_{i_n^{(2)}(s)-1,n}\},$$

$$\mathcal{R}_n(s) = \min\{t^{(1)}_{i_n^{(1)}(s),n}, t^{(2)}_{i_n^{(2)}(s),n}\} - s \tag{3.14}$$

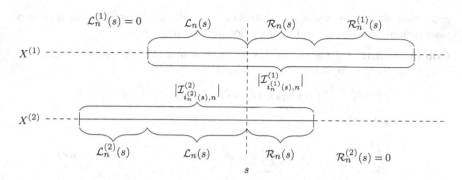

Figure 3.2: Illustration of the terms $\mathcal{L}_n^{(l)}(s), \mathcal{R}_n^{(l)}(s)$, $l = 1, 2$, and $\mathcal{L}_n(s), \mathcal{R}_n(s)$.

to describe lengths of overlapping and non-overlapping parts of the observation intervals containing s; compare Figure 3.2.

As in the univariate situation we then obtain convergence of (3.11) and hence also of $\sqrt{n}(V(f, \pi_n)_T - B^*(f)_T)$ if we assume that the random vectors

$$(n\mathcal{L}_n^{(1)}(s), n\mathcal{R}_n^{(1)}(s), n\mathcal{L}_n^{(2)}(s), n\mathcal{R}_n^{(2)}(s), n\mathcal{L}_n(s), n\mathcal{R}_n(s)), \quad s \in [0, T], \quad (3.15)$$

converge in a suitable sense to random variables

$$(\mathcal{L}^{(1)}(s), \mathcal{R}^{(1)}(s), \mathcal{L}^{(2)}(s), \mathcal{R}^{(2)}(s), \mathcal{L}(s), \mathcal{R}(s)), \quad s \in [0, T]. \quad (3.16)$$

The limit is of the form (3.13) with the only difference that the rescaled interval lengths (3.15) are replaced with their limits (3.16). Again we observe that the limit has a mixed normal distribution with random variance.

If we move from our toy examplse with finite jump activity to general Itô semimartingales X we obtain the limit variables in the central limit theorems for general X by letting $q \to \infty$ in (3.7) and (3.15). To show that the stable convergence is preserved if we take the limit $q \to \infty$ and to formally justify the approximations and estimates sketched above we will need a lot of technical arguments in Section 3.1.2. However, all relevant ideas that lead to the form of the limits in the general central limit theorems can be illustrated using our toy examples.

3.1.1 The Results

In this section we state precise central limit theorems for the non-normalized functionals $V^{(l)}(g, \pi_n)_T$ in the univariate setting and $V(f, \pi_n)_T$ in the bivariate setting based on the previous considerations for our toy examples. As for the motivation we start by presenting the result in the univariate setting. The necessary assumptions are summarized in the following condition.

Condition 3.1. *Let Condition 1.3 be fulfilled and the observation scheme be exogenous. Further it holds* $\sqrt{n}|\pi_n|_T = o_{\mathbb{P}}(1)$ *and*

(i) we have $G_3^{(l),n}(T) = O_{\mathbb{P}}(1)$, $G_4^{(l),n}(T) = O_{\mathbb{P}}(1)$ *as* $n \to \infty$.

(ii) The integral

$$\int_0^T g(x_1, \ldots, x_P)\mathbb{E}\Big[\prod_{p=1}^P h_p(n\delta_{n,-}^{(l)}(x_p), n\delta_{n,+}^{(l)}(x_p))\Big]dx_1 \ldots dx_P \quad (3.17)$$

converges as $n \to \infty$ *to*

$$\int_0^T g(x_1, \ldots, x_P)\prod_{p=1}^P \int_{\mathbb{R}} h_p(y)\Gamma^{univ,(l)}(x_p, dy)dx_1 \ldots dx_P \quad (3.18)$$

for all bounded functions $g\colon [0,T]^P \to \mathbb{R}$, $h_p\colon [0,\infty)^2 \to \mathbb{R}$, $p = 1, \ldots, P$, *and any* $P \in \mathbb{N}$. *Here* $\Gamma^{univ,(l)}(x, dy)$, $x \in [0,T]$, *is a family of probability measures on* $[0,\infty)^2$ *with uniformly bounded first moments.*

Let $Z^{univ,(l)}(s) = (\delta_-^{(l)}(s), \delta_+^{(l)}(s))$, $s \in [0,T]$, be random variables defined on an extended probability space $(\widetilde{\Omega}, \widetilde{\mathcal{F}}, \widetilde{\mathbb{P}})$ whose distribution is given by

$$\widetilde{\mathbb{P}}^{Z^{univ,(l)}(x)}(dy) = \Gamma^{univ,(l)}(x, dy).$$

Further let $(U_-^{(l)}(s), U_+^{(l)}(s))$, $s \in [0,T]$, denote i.i.d. standard normal distributed random variables which are defined on $(\widetilde{\Omega}, \widetilde{\mathcal{F}}, \widetilde{\mathbb{P}})$ as well. The random variables $Z^{univ,(l)}(s)$ are independent of the $(U_-^{(l)}(s), U_+^{(l)}(s))$. All the random variables

$$Z^{univ,(l)}(s), (U_-^{(l)}(s), U_+^{(l)}(s))$$

are independent of each other for different values of $s \in [0,T]$ and independent of the σ-algebra $\mathcal{F}$ which contains the information on the process X and its components.

Theorem 3.2. *Suppose Condition 3.1 is fulfilled. Then we have the* $\mathcal{X}$*-stable convergence*

$$\sqrt{n}(V^{(l)}(g, \pi_n)_T - B^{(l)}(g)_T)$$
$$\xrightarrow{\mathcal{L}-s} \sum_{s \leq T} g'(\Delta X_s^{(l)})(\sigma_{s-}^{(l)}(\delta_-^{(l)}(s))^{1/2}U_-^{(l)}(s) + \sigma_s^{(l)}(\delta_+^{(l)}(s))^{1/2}U_+^{(l)}(s)) \quad (3.19)$$

for all functions $g\colon \mathbb{R} \to \mathbb{R}$ *with* $g(0) = g'(0) = 0$ *and* $g''(x) = o(|x|)$ *as* $x \to 0$.

Here, the notion that a sequence of $\mathcal{F}$-measurable real-valued random variables $(Y_n)_{n \in \mathbb{N}}$ converges $\mathcal{X}$-stably in law to a random variable Y defined on an extended probability space $(\widetilde{\Omega}, \widetilde{\mathcal{F}}, \widetilde{\mathbb{P}})$ means that it holds

$$\mathbb{E}[g(Y_n)Z] \longrightarrow \widetilde{\mathbb{E}}[g(Y)Z]$$

for all bounded functions $g \colon \mathbb{R} \to \mathbb{R}$ and all bounded real-valued $\mathcal{X}$-measurable random variables Z. For more details on stable converge in law consult Appendix B.

Considering the functions $g_p(x) = x^p$ and $\overline{g}_p(x) = |x|^p$ for $p \geq 0$ introduced in Section 2.2 we observe that we get convergence only for $p > 3$. Hence to obtain a central limit theorem we need to impose a stricter condition than was needed to get mere convergence $V^{(l)}(g, \pi_n)_T \xrightarrow{\mathrm{P}} B^{(l)}(g)_T$ where only $p > 2$ was required; compare Corollary 2.2. The assumptions on the function g in the univariate irregular setting are identical to the assumptions which were made in Theorem 5.1.2 of [30] to obtain a central limit theorem in the case of equidistant deterministic observation times. Condition 3.1(ii) is a necessary assumption in the setting of irregular observation schemes because we need that $(\delta_{n,-}^{(l)}(s), \delta_{n,+}^{(l)}(s))$ converges in law as motivated by the toy example. Condition 3.1(i) is an additional structural assumption on the moments of the observation intervals. It is automatically fulfilled in the setting of equidistant observations which is shown in Example 3.3.

In the following examples we discuss Condition 3.1 and Theorem 3.2 for two important observation schemes. The first example shows that our results are a natural generalization of the existing results in the setting of equidistant observation times. In fact we will see that the result from Theorem 3.2 reduces to Theorem 5.1.2 in [30] in this specific setting. In the second example we discuss the genuinely random observation scheme of Poisson sampling and show that our results can be applied in this setting as well.

Example 3.3. *Condition 3.1 is fulfilled in the setting of equidistant observation times* $t_{i,n}^{(l)} = i/n$. *Indeed in that case it holds*

$$G_p^{(l),n}(T) = n^{p/2-1} \sum_{i=1}^{\lfloor nT \rfloor} n^{-p/2} \to T, \quad p \geq 0.$$

Further the outer integral in (3.17) can be interpreted as the expectation with regard to P independent uniformly distributed random variables $S_p^{(l)}$ on $[0, T]$. Then, as the variables $(n\delta_{n,-}^{(l)}(S_p^{(l)}), n\delta_{n,+}^{(l)}(S_p^{(l)}))$ are asymptotically distributed like $(\kappa, 1 - \kappa)$ with $\kappa \sim \mathcal{U}[0,1]$, we obtain the limiting distribution of $(n\delta_{n,-}^{(l)}(s), n\delta_{n,+}^{(l)}(s))$ as $(\delta_-^{(l)}(s), \delta_+^{(l)}(s)) \sim (\kappa^{(l)}(s), 1 - \kappa^{(l)}(s))$ for independent $\kappa^{(l)}(s) \sim \mathcal{U}[0,1]$. Standard *arguments which will be formalized in the proof of Lemma 5.10 further show that* $(n\delta_{n,-}^{(l)}(S_p^{(l)}), n\delta_{n,+}^{(l)}(S_p^{(l)}))$ *and* $(n\delta_{n,-}^{(l)}(S_{p'}^{(l)}), n\delta_{n,+}^{(l)}(S_{p'}^{(l)}))$ *are asymptotically independent for* $p \neq p'$. *Hence Condition 3.1 is fulfilled in the setting of equidistant observation times and we obtain from Theorem 3.2*

$$\sqrt{n}\left(V^{(l)}(g,\pi_n)_T - B^{(l)}(g)_T\right)$$
$$\stackrel{\mathcal{L}-s}{\Longrightarrow} \sum_{s \leq T} g'(\Delta X_s^{(l)})(\sigma_{s-}^{(l)}(\kappa^{(l)}(s))^{1/2}U_-^{(l)}(s) + \sigma_s^{(l)}(1 - \kappa^{(l)}(s))^{1/2}U_+^{(l)}(s)).$$

Here the limit is identical to the limit in Theorem 5.1.2 of [30] which we recover for univariate processes from Theorem 3.2 in the special case of equidistant observations. □

Example 3.4. *Condition 3.1 is also fulfilled in the setting of Poisson sampling where the observation times* $t_{i,n}^{(l)}$ *are recursively defined by* $t_{0,n}^{(l)} = 0$ *and*

$$t_{i,n}^{(l)} = t_{i-1,n}^{(l)} + E_{i,n}^{(l)}/n$$

for i.i.d. $Exp(\lambda_l)$*-distributed random variables* $E_{i,n}^{(l)}$ *for some fixed* $\lambda_l > 0$. *Indeed Condition 3.1(i) holds by Corollary 5.6 and that part (ii) holds follows from Lemma 5.10 with* $f^{(l)}(x,y,z) = (x,y)$.

By the memorylessness of the exponential distribution the variable $n\delta_{n,+}^{(l)}(s)$ *is* $Exp(\lambda_l)$*-distributed and by the discussion following Lemma 5.10 the backward waiting time* $n\delta_{n,-}^{(l)}(s)$ *is asymptotically* $Exp(\lambda_l)$*-distributed as well and independent of* $n\delta_{n,+}^{(l)}(s)$. *Hence we get* $(\delta_-^{(l)}(s), \delta_+^{(l)}(s)) \sim (E_-^{(l)}(s), E_+^{(l)}(s))$ *for i.i.d.* $Exp(\lambda_l)$*-distributed random variables* $E_-^{(l)}(s), E_+^{(l)}(s)$, $s \in [0,T]$, *and (3.19) becomes*

$$\sqrt{n}\left(V^{(l)}(g,\pi_n)_T - B^{(l)}(g)_T\right)$$
$$\stackrel{\mathcal{L}-s}{\Longrightarrow} \sum_{s \leq T} g'(\Delta X_s^{(l)})(\sigma_{s-}^{(l)}(E_-^{(l)}(s))^{1/2}U_-^{(l)}(s) + \sigma_s^{(l)}(E_+^{(l)}(s))^{1/2}U_+^{(l)}(s))$$

in the case of Poisson sampling. □

Next, we present the central limit theorem for $V(f,\pi_n)_T$ in the setting of asynchronous and random observation schemes. The following condition summarizes the necessary assumptions. We use the notation

$$Z_n^{biv}(s) = (n\mathcal{L}_n^{(1)}(s), n\mathcal{R}_n^{(1)}(s), n\mathcal{L}_n^{(2)}(s), n\mathcal{R}_n^{(2)}(s), n\mathcal{L}_n(s), n\mathcal{R}_n(s)). \quad (3.20)$$

Condition 3.5. *Let Condition 1.3 be fulfilled and the observation scheme be exogenous. Additionally, it holds* $\sqrt{n}|\pi_n|_T = o_{\mathbb{P}}(1)$ *and*

(i) it holds $G_{3,0}^n(T) = O_{\mathbb{P}}(1)$, $G_{0,3}^n(T) = O_{\mathbb{P}}(1)$ as $n \to \infty$.

(ii) The integral

$$\int_0^T g(x_1, \dots, x_P) \mathbb{E}\Big[\prod_{p=1}^P h_p(Z_n^{biv}(x_p)) \Big] dx_1 \dots dx_P$$

converges as $n \to \infty$ to

$$\int_0^T g(x_1, \dots, x_P) \prod_{p=1}^P \int_{\mathbb{R}^6} h_p(y) \Gamma^{biv}(x_p, dy) dx_1 \dots dx_P$$

for all bounded functions $g: [0,T]^P \to \mathbb{R}$, $h_p: [0,\infty)^6 \to \mathbb{R}$, $p = 1, \dots, P$, and any $P \in \mathbb{N}$. Here $\Gamma^{biv}(x, dy)$, $x \in [0,T]$, is a family of probability measures on $[0,\infty)^6$ with uniformly bounded first moments.

As for the result in the univariate setting we need an extended probability space $(\widetilde{\Omega}, \widetilde{\mathcal{F}}, \widetilde{\mathbb{P}})$ on which the limit can be defined. On $(\widetilde{\Omega}, \widetilde{\mathcal{F}}, \widetilde{\mathbb{P}})$ we define via

$$Z^{biv}(s) = (\mathcal{L}^{(1)}(s), \mathcal{R}^{(1)}, \mathcal{L}^{(2)}(s), \mathcal{R}^{(2)}, \mathcal{L}(s), \mathcal{R}(s)), \quad s \in [0,T],$$

random variables which are distributed according to $\widetilde{\mathbb{P}}^{Z^{biv}(x)}(dy) = \Gamma^{biv}(x, dy)$. Further let $U_s^{(1),-}, U_s^{(1),+}, U_s^{(2)}, U_s^{(3)}, U_s^{(4)}$ be i.i.d. standard normal distributed random variables defined on $(\widetilde{\Omega}, \widetilde{\mathcal{F}}, \widetilde{\mathbb{P}})$ as well. All these newly introduced random variables should be independent of each other and independent of $\mathcal{F}$. Using these variables we define

$$\Phi_T^{biv}(f) = \sum_{s \le T} \Big[\partial_1 f(\Delta X_s^{(1)}, \Delta X_s^{(2)}) \Big(\sigma_{s-}^{(1)} \sqrt{\mathcal{L}^{(1)}(s)} U_s^{(1),-} + \sigma_s^{(1)} \sqrt{\mathcal{R}(s)} U_s^{(1),+} \Big.$$

$$+ \sqrt{(\sigma_{s-}^{(1)})^2 \mathcal{L}^{(1)}(s) + (\sigma_s^{(1)})^2 \mathcal{R}^{(1)}(s)} U_s^{(2)} \Big)$$

$$+ \partial_2 f(\Delta X_s^{(1)}, \Delta X_s^{(2)}) \Big(\sigma_{s-}^{(2)} \rho_{s-} \sqrt{\mathcal{L}(s)} U_s^{(1),-} + \sigma_s^{(2)} \rho_s \sqrt{\mathcal{R}(s)} U_s^{(1),+}$$

$$+ \sqrt{(\sigma_{s-}^{(2)})^2 (1 - (\rho_{s-})^2) \mathcal{L}(s) + (\sigma_s^{(2)})^2 (1 - (\rho_s)^2) \mathcal{R}(s)} U_s^{(3)}$$

$$+ \sqrt{(\sigma_{s-}^{(2)})^2 \mathcal{L}^{(2)}(s) + (\sigma_s^{(2)})^2 \mathcal{R}^{(2)}(s)} U_s^{(4)} \Big) \Big]$$

Theorem 3.6. *If Condition 3.5 is fulfilled we have the $\mathcal{X}$-stable convergence*

$$\sqrt{n}\big(V(f, \pi_n)_T - B''(f)_T\big) \xrightarrow{\mathcal{L-}\mathbf{s}} \Phi_T^{biv}(f) \tag{3.21}$$

for all functions $f: \mathbb{R}^2 \to \mathbb{R}$ which fulfil (3.8)–(3.9).

In the following we discuss the conditions (3.8)–(3.9) imposed on the function f to obtain a central limit theorem for $V(f, \pi_n)_T$. The assumption $f(x, y) = 0$ whenever $xy = 0$ was already needed in Section 2.1 to obtain convergence of $V(f, \pi_n)_T$ where the limit solely consisted of common jumps. To obtain a central limit theorem we additionally require that the first derivatives fulfil $\partial_1 f(x, y) = \partial_2 f(x, y) = 0$ whenever $xy = 0$. Hence the limit $\Phi_T^{biv}(f)$ depends solely on the common jumps of $X^{(1)}$ and $X^{(2)}$ as well. That the condition on the first derivatives is necessary becomes obvious when looking at (3.10). There in the sum

$$\sqrt{n} \sum_{i_1, i_2 : t_{i_1,n}^{(l)} \vee t_{i_2,n}^{(2)} \leq T} \sum_{l=1,2} \partial_l f(\Delta_{i_1,n}^{(1)} N^{(1)}(q), \Delta_{i_2,n}^{(2)} N^{(2)}(q)) \Delta_{i_l,n}^{(l)} C^{toy,(l)} \quad (3.22)$$

for an idiosyncratic jump of $N^{(1)}(q)$ at time S_p asymptotically the term

$$\partial_l f(\Delta N^{(1)}(q)_{S_p}, 0) \sqrt{n} \Delta_{i_n^{(1)}(S_p),n}^{(l)} C^{toy,(1)}$$

would be included in (3.22) exactly $|\{j : \mathcal{I}_{i_n^{(1)}(S_p),n}^{(l)} \cap \mathcal{I}_{j,n}^{(2)} \neq \emptyset\}|$ times. However, this number in general does not converge. Hence because of the property $\sqrt{n} \Delta_{i_n^{(1)}(S_p),n}^{(l)} C^{toy,(1)} = O_{\mathbb{P}}(1)$ we need $\partial_l f(\Delta N^{(1)}(q)_{S_p}, 0) = 0$ to ensure convergence of (3.22). The condition (3.9) on the second derivatives is e.g. fulfilled for all functions $\overline{f}_{(p_1, p_2)}(x, y) = |x|^{p_1} |y|^{p_2}$ with $p_1 \wedge p_2 > 2$. This requirement corresponds to the condition made on the function f in Theorem 2.3.

As in the univariate setting we discuss Condition 3.5 in the special setting of equidistant synchronous observation times and in the case of Poisson sampling.

Example 3.7. *Condition 3.5 is fulfilled in the setting of equidistant observation times $t_{i,n}^{(l)} = t_{i,n}^{(2)} = i/n$. In fact it holds*

$$G_{p_1,p_2}^n(T) = G_{p_1+p_2}^{(1),n}(T), \quad p_1, p_2 \geq 0,$$
$$\mathcal{L}_n = \delta_{n,-}^{(1)}, \quad \mathcal{R}_n = \delta_{n,+}^{(1)},$$
$$\mathcal{L}_n^{(1)} = \mathcal{L}_n^{(2)} = \mathcal{R}_n^{(1)} = \mathcal{R}_n^{(2)} = 0,$$

due to the synchronicity of the observation times. Hence that Condition 3.5 is fulfilled in the equidistant setting can be shown as in Example 3.3.

Similarly as for (3.19) in Example 3.3 the convergence in (3.21) in this setting is identical to the corresponding statement for bivariate processes in Theorem 5.1.2 of [30]. □

Example 3.8. *Condition 3.5 is also fulfilled in the setting of Poisson sampling. The assumption $\sqrt{n}|\pi_n|_T = o_{\mathbb{P}}(1)$ is fulfilled by (5.2). Further Condition 3.5(i) is fulfilled by Corollary 5.6 and 3.5(ii) follows from Lemma 5.10 where $f^{(3)}$ is chosen to be the projection on the first 3 arguments.*

In the case of Poisson samling the distribution $\Gamma^{biv}(s, dy)$ can be characterized as follows: Let $E_-^{(l)}(s)$, $E_+^{(l)}(s)$, $s \in [0, T]$, be i.i.d. $Exp(\lambda_l)$-distributed random variables for $l = 1, 2$. Then the limit distribution of $nZ_n^{biv}(s)$ is given by

$$Z^{biv}(s) = (\mathcal{L}^{(1)}(s), \mathcal{R}^{(1)}, \mathcal{L}^{(2)}(s), \mathcal{R}^{(2)}, \mathcal{L}(s), \mathcal{R})$$

$$\mathcal{L}(s) = E_-^{(1)}(s) \wedge E_-^{(2)}(s), \quad \mathcal{R}(s) = E_+^{(1)}(s) \wedge E_+^{(2)}(s),$$

$$\mathcal{L}^{(l)}(s) = \left(E_-^{(l)}(s) - E_-^{(1)}(s) \wedge E_-^{(2)}(s)\right)^+,$$

$$\mathcal{R}^{(l)}(s) = \left(E_+^{(l)}(s) - E_+^{(1)}(s) \wedge E_+^{(2)}(s)\right)^+$$

using $\mathcal{L}(s) = \delta_-^{(1)}(s) \wedge \delta_-^{(2)}(s)$, $\mathcal{R}(s) = \delta_-^{(1)}(s) \wedge \delta_-^{(2)}(s)$, the considerations in Example 3.4 and the independence of the two Poisson processes. □

Notes. *The techniques used in this section are based on methods from [8]. In particular Conditions 3.1(ii) and 3.5(ii) are analogous to Assumption 2(ii) and Assumption 4(ii) in [8]. Further, the limiting variables $(\delta_-(s), \delta_+(s))$ and $Z^{biv}(s)$ are also very similar to those in Theorems 2 and 3 of [8]. Our limiting variables are slightly more complicated because we allow for discontinuous volatility processes σ_s.*

3.1.2 The Proofs

Proof of Theorem 3.2. For the sake of clarity we first present the rough structure of the proof while the more technical computations necessary for the individual steps are presented later.

Step 1. First we discretize $\sigma^{(l)}$ and restrict ourselves to the big jumps $\Delta N^{(l)}(q)_s$ of which there are almost surely only finitely many. We define $\sigma^{(l)}(r), C^{(l)}(r)$ for $r \in \mathbb{N}$ by $\sigma^{(l)}(r)_t = \sigma_{(j-1)2^{-r}}^{(l)}$ if $t \in [(j-1)2^{-r}, j2^{-r})$, $C^{(l)}(r)_t = \int_0^t \sigma^{(l)}(r)_s d\overline{W}_s^{(l)}$. Here,

$$\overline{W}_t = (W_t^{(1)}, \rho_t W_t^{(1)} + \sqrt{1 - \rho_t^2} W_t^{(2)})^* \tag{3.23}$$

is defined such that $\overline{W}^{(l)}$ denotes for $l = 1, 2$ the Brownian motion driving the process $X^{(l)}$. Denote by $S_{q,p}^{(l)}$, $p \in \mathbb{N}$, an enumeration of the jump times of $N^{(l)}(q)$. By $\lceil r \rceil - \min\{n \in \mathbb{Z} : n \geq x\}$ we denote for $x \in \mathbb{R}$ the smallest integer which is greater or equal to x. Using the notion of the jump times we modify the discretized processes $\sigma^{(l)}(r), C^{(l)}(r)$ further via

$$\tilde{\sigma}^{(l)}(r,q)_s = \begin{cases} \sigma^{(l)}_{S^{(l)}_{q,p}} & \text{if } s \in [S^{(l)}_{q,p}, \lceil 2^r S^{(l)}_{q,p}\rceil/2^r) \\ \sigma^{(l)}(r)_s & \text{otherwise} \end{cases},$$

$$\widetilde{C}^{(l)}(r,q)_t = \int_0^t \tilde{\sigma}^{(l)}(r,q)_s d\overline{W}^{(l)}_s.$$

Using this notation we then define

$$R^{(l)}(n,q,r) = \sqrt{n} \sum_{i:t_{i,n} \leq T} g'(\Delta^{(l)}_{i,n} N^{(l)}(q)) \Delta^{(l)}_{i,n} \widetilde{C}^{(l)}(r,q).$$

and show

$$\lim_{r\to\infty} \limsup_{q\to\infty} \limsup_{n\to\infty} \mathbb{P}(|\sqrt{n}(V^{(l)}(g,\pi_n)_T - B^{(l)}(g)_T) - R^{(l)}(n,q,r)| > \varepsilon) = 0$$

$$(3.24)$$

for all $\varepsilon > 0$. The proof of (3.24) will require some very technical estimates and will be given after the discussion of the rough structure of the proof.

Step 2. Next we will show the $\mathcal{X}$-stable convergence

$$R^{(l)}(n,q,r) \xrightarrow{\mathcal{L}-s} \Phi_T^{univ,(l)}(g,q,r) := \sum_{S^{(l)}_{q,p}\leq T} g'(\Delta N^{(l)}(q)_{S^{(l)}_{q,p}})$$

$$\times \left(\tilde{\sigma}^{(l)}(r,q)_{S^{(l)}_{q,p}-}(\delta^{(l)}_-(S_{q,p}))^{1/2} U^{(l)}_-(S^{(l)}_{q,p})\right.$$

$$\left. + \tilde{\sigma}^{(l)}(r,q)_{S^{(l)}_{q,p}}(\delta^{(l)}_+(S_{q,p^{(l)}}))^{1/2} U^{(l)}_+(S^{(l)}_{q,p})\right) \quad (3.25)$$

for all $q > 0$ and $r \in \mathbb{N}$. The proof of (3.25) is based on Condition 3.1(ii) and will also be given after the discussion of the rough structure of the proof.

Step 3. Finally we show

$$\lim_{q\to\infty} \limsup_{r\to\infty} \widetilde{\mathbb{P}}(|\Phi_T^{univ,(l)}(g) - \Phi_T^{univ,(l)}(g,q,r)| > \varepsilon) = 0, \quad (3.26)$$

for all $\varepsilon > 0$ where

$$\Phi_T^{univ,(l)}(g) = \sum_{s\leq T} g'(\Delta X^{(l)}_s)(\sigma^{(l)}_{s-}(\delta^{(l)}_-(s))^{1/2} U^{(l)}_-(s) + \sigma^{(l)}_s(\delta^{(l)}_+(s))^{1/2} U^{(l)}_+(s))$$

denotes the limit in (3.19). To show (3.26) note that it holds

$$|\Phi_T^{univ,(l)}(g) - \Phi_T^{univ,(l)}(g,q,r)|$$

$$= \sum_{s \leq T} |g'(\Delta M^{(l)}(q)_s)||\sigma_{s-}^{(l)}(\delta_-^{(l)}(s))^{1/2}U_-^{(l)}(s) + \sigma_s^{(l)}(\delta_+^{(l)}(s))^{1/2}U_+^{(l)}(s)| \quad (3.27)$$

$$+ \sum_{S_{q,p}^{(l)} \leq T} |g'(\Delta N^{(l)}(q)_{S_{q,p}^{(l)}})|(|\sigma_{S_{q,p}^{(l)}-}^{(l)} - \tilde{\sigma}^{(l)}(r,q)_{S_{q,p}^{(l)}-}|(\delta_-^{(l)}(S_{q,p}^{(l)}))^{1/2}|U_-^{(l)}(S_{q,p}^{(l)})|$$

$$+ |\sigma_{S_{q,p}^{(l)}}^{(l)} - \tilde{\sigma}^{(l)}(r,q)_{S_{q,p}^{(l)}}|(\delta_+^{(l)}(S_{q,p}^{(l)}))^{1/2}|U_+^{(l)}(S_{q,p}^{(l)})|). \quad (3.28)$$

Using the boundedness of σ and that the measures $\Gamma^{univ,(l)}(\cdot, dy)$ have uniformly bounded first moments we obtain that the $\mathcal{F}$-conditional expectation of (3.27) is bounded by

$$K \sum_{s \leq T} |g'(\Delta M^{(l)}(q)_s)|. \quad (3.29)$$

Note that $g'(0) = 0$ and $g''(x) = o(|x|)$ imply $g'(x) = o(|x|^2)$ as $x \to 0$. Hence the sum (3.29) is almost surely finite for any q and therefore vanishes as $q \to \infty$. This observation yields using Lemma 2.15 that (3.27) also vanishes in probability as $q \to \infty$. Further (3.28) vanises almost surely as $r \to \infty$ for any $q > 0$ because σ is càdlàg and because of $\tilde{\sigma}^{(l)}(r,q)_{S_{q,p}^{(l)}} = \sigma_{S_{q,p}^{(l)}}^{(l)}$ for any $S_{q,p}^{(l)}$. Hence we have proven (3.26).

Step 4. Combining the results (3.24), (3.25) and (3.26) from steps 1–3 we obtain the claim (3.19) using Lemma B.6. □

Lemma 3.9. *Let Condition 1.3 be satisfied, the processes σ_t, Γ_t be bounded and the observation scheme be exogenous. Then there exists a constant K which is independent of (i,j) such that*

$$\mathbb{E}[(\Delta_{i,n}^{(l)}C)^2(\Delta_{j,n}^{(l')}M(q))^2|\sigma(\mathcal{F}_{t_{i,n}^{(l)} \wedge t_{j,n}^{(l')}}, \mathcal{S})] \leq Keq|\mathcal{I}_{i,n}^{(l)}||\mathcal{I}_{j,n}^{(l')}|, \quad l,l' \in \{1,2\}.$$

$$(3.30)$$

Proof. If $\mathcal{I}_{i,n}^{(l)} \cap \mathcal{I}_{j,n}^{(l')} = \emptyset$ we use iterated expectations and (1.9), (1.10). If the intervals do overlap, we use iterated expectations for the non-overlapping parts to obtain

$$\mathbb{E}[(\Delta_{i,n}^{(l)}C)^2(\Delta_{j,n}^{(l')}M(q))^2|\sigma(\mathcal{F}_{t_{i,n}^{(l)} \wedge t_{j,n}^{(l')}}, \mathcal{S})] \leq Keq(|\mathcal{I}_{i,n}^{(l)}||\mathcal{I}_{j,n}^{(l')}| - |\mathcal{I}_{i,n}^{(l)} \cap \mathcal{I}_{j,n}^{(l')}|^2)$$

$$+ \mathbb{E}[(C_{t_{i,n}^{(l)} \wedge t_{j,n}^{(l')}}^{(l)} - C_{t_{i-1,n}^{(l)} \vee t_{j-1,n}^{(l')}}^{(l)})^2$$

$$\times (M^{(3-l)}(q)_{t_{i,n}^{(l)} \wedge t_{j,n}^{(l')}} - M^{(l')}(q)_{t_{i-1,n}^{(l)} \vee t_{j-1,n}^{(l')}})^2|\sigma(\mathcal{F}_{t_{i,n}^{(l)} \wedge t_{j,n}^{(l')}}, \mathcal{S})].$$

The claim now follows from (8.16) in [32] which is basically (3.30) for $\mathcal{I}_{i,n}^{(l)} = \mathcal{I}_{j,n}^{(l')}$.
$$\square$$

Proof of (3.24). To prove (3.24) we first split
$$|\sqrt{n}(V^{(l)}(g,\pi_n)_T - B^{(l)}(g)_T) - R^{(l)}(n,q,r)|$$
into multiple terms which we will then discuss separately. To this end note that on the set $\Omega^{(l)}(n,q)$ where any two different jumps of $N^{(l)}(q)$ in $[0,T]$ are further apart than $|\pi_n|_T$ it holds

$$|\sqrt{n}(V^{(l)}(g,\pi_n)_T - B^{(l)}(g)_T) - R^{(l)}(n,q,r)|\mathbb{1}_{\Omega^{(l)}(n,q)}$$

$$\leq \left| \sqrt{n} \sum_{S_{q,p}^{(l)} \leq T} \left(g(\Delta_{i_n^{(l)}(S_{q,p}^{(l)}),n}^{(l)} X^{(l)}) - g(\Delta_{i_n^{(l)}(S_{q,p}^{(l)}),n}^{(l)} (X^{(l)} - N^{(l)}(q))) \right. \right.$$

$$\left. \left. - g(\Delta N^{(l)}(q)_{S_{q,p}^{(l)}}) \right) - R^{(l)}(n,q,r) \right| \quad (3.31)$$

$$+ \sqrt{n} \left| \sum_{i:t_{i,n}^{(l)} \leq T} g(\Delta_{i,n}^{(l)}(X^{(l)} - N^{(l)}(q))) - \sum_{s \leq T} g(\Delta M^{(l)}(q)_s) \right|. \quad (3.32)$$

Hence because of $\mathbb{P}(\Omega^{(l)}(n,q)) \to 1$ as $n \to \infty$ for any $q > 0$ it remains to show that (3.31) and (3.32) vanish in the sense of (3.24).

Step 1. We first discuss (3.31). Using a Taylor expansion we obtain

$$g(\Delta_{i_n^{(l)}(S_{q,p}^{(l)}),n}^{(l)} X^{(l)}) = g(\Delta N^{(l)}(q)_{S_{q,p}^{(l)}}) + g'(\Delta N^{(l)}(q)_{S_{q,p}^{(l)}})\Delta_{i_n^{(l)}(S_{q,p}^{(l)}),n}^{(l)}(X^{(l)} - N^{(l)}(q))$$

$$+ \frac{1}{2}g''(\Delta N^{(l)}(q)_{S_{q,p}^{(l)}} + \theta_{q,p}^n \Delta_{i_n^{(l)}(S_{q,p}^{(l)}),n}^{(l)}(X^{(l)} - N^{(l)}(q)))(\Delta_{i_n^{(l)}(S_{q,p}^{(l)}),n}^{(l)}(X^{(l)} - N^{(l)}(q)))^2$$

for some random $\theta_{q,p}^n \in [0,1]$. Hence it remains to show

$$\lim_{r \to \infty} \limsup_{n \to \infty} \mathbb{P}(|Y_1(n,q,r)| + |Y_2(n,q,r)| + |Y_3(n,q,r)| > \varepsilon) = 0 \quad (3.33)$$

for all $\varepsilon > 0$ and any $q > 0$ where

$$Y_1(n,q,r) = \sqrt{n} \sum_{S_{q,p}^{(l)} \leq T} g'(\Delta N^{(l)}(q)_{S_{q,p}^{(l)}})$$

$$\times \Delta_{i_n^{(l)}(S_{q,p}^{(l)}),n}^{(l)}(B^{(l)}(q) + M^{(l)}(q) + C^{(l)} - \widetilde{C}^{(l)}(r,q)),$$

$$Y_2(n,q,r) = \sqrt{n} \sum_{S_{q,p}^{(l)} \leq T} g(\Delta_{i_n^{(l)}(S_{q,p}^{(l)}),n}^{(l)}(X^{(l)} - N^{(l)}(q))),$$

$$Y_3(n,q,r) = \frac{\sqrt{n}}{2} \sum_{S_{q,p}^{(l)} \leq T} g''(\Delta N^{(l)}(q)_{S_{q,p}^{(l)}} + \theta_{q,p}^n \Delta_{i_n^{(l)}(S_{q,p}^{(l)}),n}^{(l)}(X^{(l)} - N^{(l)}(q)))$$

$$\times (\Delta_{i_n^{(l)}(S_{q,p}^{(l)}),n}^{(l)}(X^{(l)} - N^{(l)}(q)))^2.$$

The key ingredient in the proof of (3.33) is to show

$$\lim_{r\to\infty} \limsup_{n\to\infty} \mathbb{P}(\sqrt{n}|\Delta^{(l)}_{i_n^{(l)}(S_{q,p}^{(l)}),n}(B^{(l)}(q) + M^{(l)}(q) + C^{(l)} - \widetilde{C}^{(l)}(r,q))| > \varepsilon) = 0$$

(3.34)

for all $\varepsilon > 0$ and any jump time $S_{q,p}^{(l)}$. (3.34) can be proven similarly as (4.4.23) in [30]. To this end note that a filtration $\mathcal{G}_t(q)$ can be defined such that the $S_{q,p}^{(l)}$ are $\mathcal{G}_0(q)$-measurable and the processes $B^{(l)}(q)$, $M^{(l)}(q)$, $C^{(l)}$, $C^{(l)}(q,r)$ have the same properties as $\mathcal{G}_t(q)$-adapted processes like they have as $\mathcal{F}_t$-adapted processes. Then $S_{q,p}^{(l)}$ is in particular $\mathcal{G}_{t_{i_n^{(l)}(S_{q,p}^{(l)}),n}^{(l)}-1}(q)$-measurable and estimates like in Lemma 1.4 can be used. By inequalities (1.8) and (2.1.34), (2.1.39) from [30] it then can be shown that

$$\lim_{r\to\infty} \limsup_{n\to\infty} \mathbb{E}\big[\sqrt{n}|\Delta^{(l)}_{i_n^{(l)}(S_{q,p}^{(l)}),n}(B^{(l)}(q) + C^{(l)} - \widetilde{C}^{(l)}(r,q))|$$
$$+ \min\{\sqrt{n}|\Delta^{(l)}_{i_n^{(l)}(S_{q,p}^{(l)}),n}M^{(l)}(q))|,c\}\big|\sigma(\mathcal{G}_0(q)\cup \mathcal{S})\big] = 0, \quad c \geq 0,$$

holds. The proof of this convergence makes use of the fact that Condition 3.1(ii) implies the $\mathcal{X}$-stable convergence of $(n\delta_{n,-}^{(l)}(S_{q,p}^{(l)}), n\delta_{n,+}^{(l)}(S_{q,p}^{(l)}))$ which will be shown in the proof of (3.25). For more details consult the arguments leading up to (4.4.23) in [30]. A slightly more formal construction of the filtration $(\mathcal{G}_t(q))_{t\geq 0}$ will be presented in the proof of Corollary 6.4.

Using (3.34) we obtain that $Y_1(n,q,r)$ vanishes as first $n \to \infty$ and then $r \to \infty$ for any $q > 0$ because $N^{(l)}(q)$ almost surely only has finitely many jumps in $[0,T]$.

Further on the set $\Omega^{(l)}(n,m,q,T)$ where $|\Delta^{(l)}_{i_n^{(l)}(S_{q,p}^{(l)}),n}(X^{(l)} - N^{(l)}(q))| < 1/m$ for any $S_{q,p}^{(l)} \leq T$ we obtain

$$|Y_2(n,q,r)|\mathbb{1}_{\Omega^{(l)}(n,m,q,T)} \leq n^{-1} \sum_{S_{q,p}^{(l)}\leq T} K_m|\sqrt{n}\Delta^{(l)}_{i_n^{(l)}(S_{q,p}^{(l)}),n}(X^{(l)} - N^{(l)}(q))|^3$$

because $g(0) = g'(0) = 0$ and $g''(x) = o(|x|)$ imply $g(x) = o(|x|^3)$ as $x \to \infty$. (3.34) then shows that $\sqrt{n}\Delta^{(l)}_{i_n^{(l)}(S_{q,p}^{(l)}),n}(X^{(l)} - N^{(l)}(q))$ and $\sqrt{n}\Delta^{(l)}_{i_n^{(l)}(S_{q,p}^{(l)}),n}C^{(l)}(q,r)$ are asymptotically equivalent and in the proof of (3.25) we will see that the sequence $\sqrt{n}\Delta^{(l)}_{i_n^{(l)}(S_{q,p}^{(l)}),n}C^{(l)}(q,r)$ converges stably in law which yields

$$\sqrt{n}\Delta^{(l)}_{i_n^{(l)}(S_{q,p}^{(l)}),n}C^{(l)}(q,r) = O_{\mathbb{P}}(1),$$

Hence $Y_2(n,q,r)$ vanishes as $n \to \infty$ and then $r \to \infty$ for any $q > 0$ because of $\lim_{n\to\infty} \mathbb{P}(\Omega^{(l)}(n,m,q,T)) = 1$ for any $m > 0$ and $q > 0$.

On the set $\overline{\Omega}^{(l)}(n, M, q, T)$ where

$$|g''(\Delta N^{(l)}(q)_{S_{q,p}^{(l)}} + \theta \Delta_{i_n^{(l)}(S_{q,p}^{(l)}),n}^{(l)}(X^{(l)} - N^{(l)}(q)))| \leq M$$

for any $\theta \in [0,1]$ and all $S_{q,p}^{(l)} \leq T$ we obtain

$$|Y_3(n,q,r)|\mathbb{1}_{\overline{\Omega}^{(l)}(n,m,q,T)} \leq \frac{n^{-1/2}}{2} \sum_{S_{q,p}^{(l)} \leq T} M|\sqrt{n}\Delta_{i_n^{(l)}(S_{q,p}^{(l)}),n}^{(l)}(X^{(l)} - N^{(l)}(q))|^2$$

and hence $Y_3(n, q, r)$ can be discussed similarly as $Y_2(n, q, r)$ because of

$$\lim_{M \to \infty} \limsup_{n \to \infty} \mathbb{P}(\overline{\Omega}(n, M, q, T)) = 1.$$

Combining the above discussions of $Y_i(n, q, r)$, $i = 1, 2, 3$, we have proved (3.33).

Step 2. Next we consider (3.32). Using again a Taylor expansion it remains to prove

$$\lim_{q \to \infty} \limsup_{n \to \infty} \mathbb{P}(|Y_4(n, q)| + |Y_5(n, q)| + |Y_6(n, q)| > \varepsilon) = 0 \qquad (3.35)$$

for any $\varepsilon > 0$ where

$$Y_4(n, q) = \sqrt{n}\Big(\sum_{i:t_{i,n}^{(l)} \leq T} g(\Delta_{i,n}^{(l)} M^{(l)}(q)) - \sum_{s \leq T} g(\Delta M^{(l)}(q)_s)\Big),$$

$$Y_5(n, q) = \sqrt{n} \sum_{i:t_{i,n}^{(l)} \leq T} g'(\Delta_{i,n}^{(l)} M^{(l)}(q))\Delta_{i,n}^{(l)}(B^{(l)}(q) + C^{(l)}),$$

$$Y_6(n, q) = \frac{\sqrt{n}}{2} \sum_{i:t_{i,n}^{(l)} \leq T} g''(\Delta_{i,n}^{(l)} M^{(l)}(q) + \theta_i^n \Delta_{i,n}^{(l)}(B^{(l)}(q) + C^{(l)}))$$

$$\times \left(\Delta_{i,n}^{(l)}(B^{(l)}(q) + C^{(l)})\right)^2$$

for random variables $\theta_i^n \in [0,1]$.

Using Itô's formula, compare Theorem 3.21 in [41], for the processes

$$\left(M^{(l)}(q)_t - M^{(l)}(q)_{t_{i-1,n}^{(l)}}\right)_{t \geq t_{i-1,n}^{(l)}}, \ i, n \in \mathbb{N},$$

we obtain that $Y_4(n,q)$ is equal to

$$\sqrt{n}\Big[\sum_{i:t_{i,n}^{(l)}\le T} \Big(g(0) + \int_{t_{i-1,n}^{(l)}}^{t_{i,n}^{(l)}} g'(M^{(l)}(q)_{s-} - M^{(l)}(q)_{t_{i-1,n}^{(l)}})dM^{(l)}(q)_s$$

$$+ \frac{1}{2}\int_{t_{i-1,n}^{(l)}}^{t_{i,n}^{(l)}} g''(M^{(l)}(q)_{s-} - M^{(l)}(q)_{t_{i-1,n}^{(l)}})d[M^{(l)}(q),M^{(l)}(q)]_s^c$$

$$+ \sum_{s\in(t_{i-1,n}^{(l)},t_{i,n}^{(l)}]}[g(M^{(l)}(q)_s - M^{(l)}(q)_{t_{i-1,n}^{(l)}}) - g(M^{(l)}(q)_{s-} - M^{(l)}(q)_{t_{i-1,n}^{(l)}})$$

$$- g'(M^{(l)}(q)_s - M^{(l)}(q)_{t_{i-1,n}^{(l)}})\Delta M^{(l)}(q)_s]\Big) - \sum_{s\le T} g(\Delta M^{(l)}(q)_s)\Big]$$

$$= \sqrt{n}\sum_{i:t_{i,n}^{(l)}\le T}\int_{t_{i-1,n}^{(l)}}^{t_{i,n}^{(l)}} g'(M^{(l)}(q)_{s-} - M^{(l)}(q)_{t_{i-1,n}^{(l)}})dM^{(l)}(q)_s \qquad (3.36)$$

$$+ \sqrt{n}\sum_{i:t_{i,n}^{(l)}\le T}\sum_{s\in(t_{i-1,n}^{(l)},t_{i,n}^{(l)}]}[g(M^{(l)}(q)_s - M^{(l)}(q)_{t_{i-1,n}^{(l)}})$$

$$- g(M^{(l)}(q)_{s-} - M^{(l)}(q)_{t_{i-1,n}^{(l)}})$$

$$- g(\Delta M^{(l)}(q)_s) - g'(M^{(l)}(q)_s - M^{(l)}(q)_{t_{i-1,n}^{(l)}})\Delta M^{(l)}(q)_s]. \qquad (3.37)$$

$$+ \sqrt{n}\sum_{s\in(t_{i_n^{(l)}(T),n}^{(l)},T]} g(\Delta M^{(l)}(q)_s) \qquad (3.38)$$

where $[Y,Y]_s^c$ denotes the part of the quadratic variation $[Y,Y]_s$ originating from the continous part of a process Y. For $Y = M^{(l)}(q)$ it holds $[M^{(l)}(q),M^{(l)}(q)]_s^c \equiv 0$. Following the ideas used in Step 4) in the proof of Theorem 5.1.2 in [30] we set $T^{(l)}(n,m,i) = \inf\{s > t_{i-1,n}^{(l)} : |M^{(l)}(q)_s - M^{(l)}(q)_{t_{i-1,n}^{(l)}}| > 1/m\}$ and

$$\widehat{\Omega}^{(l)}(n,m,q,T) = \{T^{(l)}(n,m,i) \ge t_{i,n}^{(l)} \text{ for all } i \text{ with } t_{i,n}^{(l)} \le T\}.$$

Then on the set $\widehat{\Omega}^{(l)}(n,m,q,T)$ (3.36) is identical to

$$\sqrt{n}\sum_{i:t_{i,n}^{(l)}\le T}\int_{t_{i-1,n}^{(l)}}^{t_{i,n}^{(l)}\wedge T(n,m,i)} g'(M^{(l)}(q)_{s-} - M^{(l)}(q)_{t_{i-1,n}^{(l)}})dM^{(l)}(q)_s. \qquad (3.39)$$

Using that (3.39) is a sum of martingale differences and the Burkholder-Davis-Gundy inequality (A.1) the $\mathcal{S}$-conditional expectation of the square of (3.36) is, compare (29) in [8], bounded by (by $\delta^{(l)}(s,z)$ we denote the l-th component of $\delta(s,z) \in \mathbb{R}^2$)

$$n \sum_{i:t_{i,n}^{(l)} \leq T} \mathbb{E}\Big[\int_{t_{i-1,n}^{(l)}}^{t_{i,n}^{(l)} \wedge T(n,m,i)} (g'(M^{(l)}(q)_{s-} - M^{(l)}(q)_{t_{i-1,n}^{(l)}}))^2 d[M^{(l)}(q), M^{(l)}(q)]_s \big| \mathcal{S}\Big]$$

$$\leq n \sum_{i:t_{i,n}^{(l)} \leq T} \mathbb{E}\Big[\int_{t_{i-1,n}^{(l)}}^{t_{i,n}^{(l)} \wedge T(n,m,i)} \int_{\mathbb{R}^2} (g'(M^{(l)}(q)_{s-} - M^{(l)}(q)_{t_{i-1,n}^{(l)}}))^2$$

$$\times (\delta^{(l)}(s,z))^2 \mathbb{1}_{\{\gamma(z) \leq 1/q\}} \lambda(dz) ds \big| \mathcal{S}\Big]$$

$$\leq n \sum_{i:t_{i,n}^{(l)} \leq T} Ke_q \mathbb{E}\Big[\int_{t_{i-1,n}^{(l)}}^{t_{i,n}^{(l)} \wedge T(n,m,i)} (g'(M^{(l)}(q)_{s-} - M^{(l)}(q)_{t_{i-1,n}^{(l)}}))^2 ds \big| \mathcal{S}\Big]$$

$$\leq n \sum_{i:t_{i,n}^{(l)} \leq T} Ke_q \mathbb{E}\Big[\int_{t_{i-1,n}^{(l)}}^{t_{i,n}^{(l)} \wedge T(n,m,i)} K_m |M^{(l)}(q)_{s-} - M^{(l)}(q)_{t_{i-1,n}^{(l)}}|^4 ds \big| \mathcal{S}\Big]$$

$$\leq n \sum_{i:t_{i,n}^{(l)} \leq T} Ke_q \int_{t_{i-1,n}^{(l)}}^{t_{i,n}^{(l)}} K_m \mathbb{E}[|M^{(l)}(q)_{s-} - M^{(l)}(q)_{t_{i-1,n}^{(l)}}|^4 | \mathcal{S}] ds$$

$$\leq K_m e_q G_4^{(l),n}(T)$$

where we used $g'(x) = o(|x|^2)$ as $x \to 0$ such that $K_m \to 0$ as $m \to 0$ and inequality (1.10). Hence we have shown that (3.36) vanishes as first $n \to \infty$ and then $q \to \infty$ because of $\lim_{q \to \infty} \limsup_{n \to \infty} \mathbb{P}(\widehat{\Omega}^{(l)}(n,m,q,T)) = 1$ for any $m > 0$.

Further we obtain from (5.1.22) in [30]

$$|g(x+y) - g(x) - g(y) - g'(x)y| \leq K_m |x||y|^2$$

for $x, y \in \mathbb{R}$ with $|x|, |y| \leq 1/m$. Hence on the set $\widehat{\Omega}^{(l)}(n,m,q,T)$ intersected with the set where no jump of $M(q)$ in $[0,T]$ is larger than $1/m$ we obtain that the $\mathcal{S}$-conditional expectation of the absolute value of (3.37) is bounded by

$$\mathbb{E}\Big[\sqrt{n} \sum_{i:t_{i,n}^{(l)} \leq T} \sum_{s \in (t_{i-1,n}^{(l)}, t_{i,n}^{(l)}]} K_m |M^{(l)}(q)_{s-} - M^{(l)}(q)_{t_{i-1,n}^{(l)}}||\Delta M^{(l)}(q)_s|^2 \big| \mathcal{S}\Big]$$

$$= \mathbb{E}\Big[K_m \sqrt{n} \sum_{i:t_{i,n}^{(l)} \leq T} \int_{t_{i-1,n}^{(l)}}^{t_{i,n}^{(l)}} \int_{\mathbb{R}^2} |M^{(l)}(q)_{s-} - M^{(l)}(q)_{t_{i-1,n}^{(l)}}|$$

$$\times (\delta^{(l)}(s,z))^2 \mathbb{1}_{\{\gamma(z) \leq 1/q\}} \mu(dz, ds) \big| \mathcal{S}\Big]$$

$$= \mathbb{E}\Big[K_m \sqrt{n} \sum_{i:t_{i,n}^{(l)} \leq T} \int_{t_{i-1,n}^{(l)}}^{t_{i,n}^{(l)}} \int_{\mathbb{R}^2} |M^{(l)}(q)_{s-} - M^{(l)}(q)_{t_{i-1,n}^{(l)}}|$$

$$\times \, (\delta^{(l)}(s,z))^2 \mathbf{1}_{\{\gamma(z) \leq 1/q\}} \lambda(dz) ds \big| \mathcal{S} \big]$$

$$\leq K_m \sqrt{n} \sum_{i:t_{i,n}^{(l)} \leq T} K e_q \int_{t_{i-1,n}^{(l)}}^{t_{i,n}^{(l)}} \mathbb{E}[|M^{(l)}(q)_{s-} - M^{(l)}(q)_{t_{i-1,n}^{(l)}}| \, \|\mathcal{S}] ds.$$

Here the $\mathcal{S}$-conditional expectation of the expression in the last line is using (1.10) less or equal to $K_m e_q G_3^{(l),n}(T)$. Hence (3.37) also vanishes as first $n \to \infty$ and then $q \to \infty$ by Condition 3.1(i). Finally the $\mathcal{S}$-conditional expectation of the sum (3.38) is using inequality (1.10) bounded by $\sqrt{n} K e_q |\pi_n|_T$ which vanishes by the assumption $\sqrt{n} |\pi_n|_T = o_{\mathbb{P}}(1)$ and we have shown all together that $Y_4(n,q)$ is asymptotically negligible.

Next we discuss $Y_5(n,q)$. On the set $\widehat{\Omega}^{(l)}(n,m,q,T)$ we obtain using the Cauchy-Schwarz inequality, inequalities (1.8)–(1.10) and Lemma 3.9

$$\mathbb{E}[|Y_5(n,q)| \mathbf{1}_{\widehat{\Omega}^{(l)}(n,m,q,T)} | \mathcal{S}]$$

$$\leq \sqrt{n} \sum_{i:t_{i,n}^{(l)} \leq T} K_m \mathbb{E}[(\Delta_{i,n}^{(l)} M^{(l)}(q))^2 |\Delta_{i,n}^{(l)}(B^{(l)}(q) + C^{(l)})| \, | \mathcal{S}]$$

$$\leq K_m \sqrt{n} \sum_{i:t_{i,n}^{(l)} \leq T} \big(\mathbb{E}[(\Delta_{i,n}^{(l)} M^{(l)}(q))^2 | \mathcal{S}]$$

$$\times \, \mathbb{E}[(\Delta_{i,n}^{(l)} M^{(l)}(q))^2 (\Delta_{i,n}^{(l)}(B^{(l)}(q) + C^{(l)}))^2 | \mathcal{S}]\big)^{1/2}$$

$$\leq K_m e_q (K_q (|\pi_n|_T)^{1/2} + 1) G_3^{(l),n}(T)$$

where the expression in the last line vanishes as $n \to \infty$ and then $q \to \infty$ by Condition 3.1(i). Hence $Y_5(n,q)$ is asymptotically negligible by Lemma 2.15 and because of $\lim_{q \to \infty} \limsup_{n \to \infty} \mathbb{P}(\widehat{\Omega}^{(l)}(n,m,q,T)) = 1$ as well.

Finally denote by $\widetilde{\Omega}^{(l)}(n,m,q,T)$ the set where

$$\Delta_{i,n}^{(l)} M^{(l)}(q)_s + \theta \Delta_{i,n}^{(l)}(B^{(l)}(q) + C^{(l)}) < 1/m$$

for all i with $t_{i,n}^{(l)} \leq T$ and all $\theta \in [0,1]$. On this set $|Y_6(n,q)|$ is bounded by

$$\frac{\sqrt{n}}{2} \sum_{i:t_{i,n}^{(l)} \leq T} K_m |\Delta_{i,n}^{(l)} M^{(l)}(q)_s + \theta_i^n \Delta_{i,n}^{(l)}(B^{(l)}(q) + C^{(l)})| (\Delta_{i,n}^{(l)}(B^{(l)}(q) + C^{(l)}))^2$$

$$\leq \frac{\sqrt{n}}{2} \sum_{i:t_{i,n}^{(l)} \leq T} K_m \big(|\Delta_{i,n}^{(l)} M^{(l)}(q)_s| + |\Delta_{i,n}^{(l)}(B^{(l)}(q) + C^{(l)})|\big)$$

$$\times \, \big(\Delta_{i,n}^{(l)}(B^{(l)}(q) + C^{(l)})\big)^2$$

where the $\mathcal{S}$-conditional expectation of the expression in the last line is using the Cauchy-Schwarz inequality and (1.8)–(1.10) bounded by $K_m G_3^{(l),n}(T)$ which vanishes as first $n \to \infty$ and then $m \to \infty$. Hence we have shown (3.35).   □

Proof of (3.25). Step 1. We define the process

$$A^{(l)}(q)_t = \int_0^t \int_{\mathbb{R}^2} \mathbb{1}_{\{\gamma(z) > 1/q\}} \mu(ds, dz), \quad t \geq 0,$$

and denote by $\widetilde{S}_{q,p}^{(l)}$, $p \in \mathbb{N}$, an enumeration of its jump times, compare (4.3.1) in [30]. Further we set

$$Y_n^{(l)}(s) = \big(n\delta_{n,-}^{(l)}(s), n\delta_{n,+}^{(l)}(s), U_{n,-}^{(l)}(s), U_{n,+}^{(l)}(s)\big)^*,$$
$$Y^{(l)}(s) = \big(\delta_-^{(l)}(s), \delta_+^{(l)}(s), U_-^{(l)}(s), U_+^{(l)}(s)\big)^*,$$

with

$$U_{n,-}^{(l)}(s) = (\overline{W}_s^{(l)} - \overline{W}_{t_{i_n^{(l)}(s)-1,n}^{(l)}}^{(l)})/(\delta_{n,-}^{(l)}(s))^{1/2}$$
$$U_{n,+}^{(l)}(s) = (\overline{W}_{t_{i_n^{(l)}(s),n}^{(l)}}^{(l)} - \overline{W}_s^{(l)})/(\delta_{n,+}^{(l)}(s))^{1/2}. \tag{3.40}$$

where $\overline{W}_t$ is defined as in (3.23).

We begin by showing that Condition 3.1(ii) yields the $\mathcal{X}$-stable convergence of all the $Y_n^{(l)}(\widetilde{S}_{q,p}^{(l)})$ to the respective $Y^{(l)}(\widetilde{S}_{q,p}^{(l)})$, i.e. we have to show

$$\mathbb{E}\big[\Lambda f\big((Y_n^{(l)}(S_{q,p}^{(l)}))_{\widetilde{S}_{q,p}^{(l)} \leq T}\big)\big] \to \widetilde{\mathbb{E}}\big[\Lambda f\big((Y^{(l)}(S_{q,p}^{(l)}))_{\widetilde{S}_{q,p}^{(l)} \leq T}\big)\big] \tag{3.41}$$

for all $\mathcal{X}$-measurable bounded random variables Λ and all bounded Lipschitz functions f. We will use the same techniques to prove this convergence as were used in [8] e.g. in the proof of Proposition 3. Similar arguments can also be found in the proof of Lemma 6.2 in [29] and in the proof of Lemma 5.8 in [27].

Denote by $\Omega^{(l)}(q,m,n)$ the subset of Ω on which $|\pi_n|_T < 1/m$ and where two different jumps $\widetilde{S}_{q,p}^{(l)} \leq T$ are further apart than $|\pi_n|_T$. As $\mathbb{P}(\Omega^{(l)}(q,m,n)) \to 1$ for $n \to \infty$ it suffices to prove (3.41) with the indicator $\mathbb{1}_{\Omega^{(l)}(q,m,n)}$ added in both expectations. Further we set

$$B^{(l)}(m) = \bigcup_{\widetilde{S}_{q,p}^{(l)} \leq T} \big(\max\{\widetilde{S}_{q,p}^{(l)} - 1/m, 0\}, \min\{\widetilde{S}_{q,p}^{(l)} + 1/m, T\}\big],$$

$$\overline{W}^{(l)}(m)_t = \int_0^t \mathbb{1}_{B^{(l)}(m)}(s) d\overline{W}^{(l)}(s).$$

Let $\mathcal{G}(m)$ denote the σ-algebra generated by $\overline{W}^{(l)}(m)$ and the jump times $\widetilde{S}_{q,p}^{(l)} \leq T$. By conditioning on $\sigma(\mathcal{G}(m) \cup \mathcal{S})$ we see that for proving (3.41) with the indicator $\mathbb{1}_{\Omega^{(l)}(q,m,n)}$ added in both expectations it is sufficient to consider only $\mathcal{G}(m)$-measurable Λ', as restricted to $\Omega^{(l)}(q,m,n)$ the $Y_n^{(l)}(S_p)$ are $\sigma(\mathcal{G}(m) \cup \mathcal{S})$-measurable. By Lemma 2.1 in [29] we may in particular choose Λ' of the form

$$\Lambda' = \gamma(\overline{W}^{(l)}(m))\kappa\big((S_{q,p}^{(l)})_{\widetilde{S}_{q,p}^{(l)} \leq T}\big)$$

for bounded Lipschitz functions γ and κ.

Because $\overline{W}^{(l)}(m)$ converges to 0 as $m \to \infty$ and because γ, κ, f are bounded we obtain

$$\lim_{m\to\infty} \limsup_{n\to\infty} \Big| \mathbb{E}\big[\gamma(0)\kappa\big((\widetilde{S}_{q,p}^{(l)})_{\widetilde{S}_{q,p}^{(l)} \leq T}\big) f\big((Y_n^{(l)}(\widetilde{S}_{q,p}^{(l)}))_{\widetilde{S}_{q,p}^{(l)} \leq T}\big)\big]$$

$$- \mathbb{E}\big[\gamma(\overline{W}^{(l)}(m))\kappa\big((\widetilde{S}_{q,p}^{(l)})_{\widetilde{S}_{q,p}^{(l)} \leq T}\big) f\big((Y_n^{(l)}(\widetilde{S}_{q,p}^{(l)}))_{\widetilde{S}_{q,p}^{(l)} \leq T}\big)\mathbb{1}_{\Omega^{(l)}(q,m,n)}\big] \Big| = 0$$

and the analogous result for $Y_n^{(l)}(\widetilde{S}_{q,p}^{(l)})$ replaced with $Y^{(l)}(\widetilde{S}_{q,p}^{(l)})$. Hence it remains to prove

$$\mathbb{E}\big[\kappa\big((\widetilde{S}_{q,p}^{(l)})_{\widetilde{S}_{q,p}^{(l)} \leq T}\big) f\big((Y_n^{(l)}(\widetilde{S}_{q,p}^{(l)}))_{\widetilde{S}_{q,p}^{(l)} \leq T}\big)\big]$$

$$\to \widetilde{\mathbb{E}}\big[\kappa\big((\widetilde{S}_{q,p}^{(l)})_{\widetilde{S}_{q,p}^{(l)} \leq T}\big) f\big((Y^{(l)}(\widetilde{S}_{q,p}^{(l)}))_{\widetilde{S}_{q,p}^{(l)} \leq T}\big)\big] \quad (3.42)$$

for all bounded Lipschitz functions κ, f.

Further note that by another density argument, compare again Lemma 2.1 in [30], it suffices to consider functions f of the form

$$f\big((Y_n^{(l)}(\widetilde{S}_{q,p}^{(l)}))_{\widetilde{S}_{q,p}^{(l)} \leq T}\big) = \prod_{p:S_{q,p}^{(l)} \leq T} f_p^{(l)}\big(n\delta_{n,-}^{(l)}(\widetilde{S}_{q,p}^{(l)}), n\delta_{n,+}^{(l)}(\widetilde{S}_{q,p}^{(l)})\big)\tilde{f}_p^{(l)}\big(U_{n,-}^{(l)}(\widetilde{S}_{q,p}^{(l)}), U_{n,+}^{(l)}(\widetilde{S}_{q,p}^{(l)})\big).$$

Then because the $U_{n,-}^{(l)}(\widetilde{S}_{q,p}^{(l)})$, $U_{n,+}^{(l)}(\widetilde{S}_{q,p}^{(l)})$ are i.i.d. $\mathcal{N}(0,1)$-distributed and independent of μ and $(n\delta_{n,-}^{(l)}(\widetilde{S}_{q,p}^{(l)}), n\delta_{n,+}^{(l)}(\widetilde{S}_{q,p}^{(l)}))$, (3.42) becomes

$$\mathbb{E}\big[\kappa\big((\widetilde{S}_{q,p}^{(l)})_{\widetilde{S}_{q,p}^{(l)} \leq T}\big) \prod_{\widetilde{S}_{q,p}^{(l)} \leq T} f_p^{(l)}\big(n\delta_{n,-}^{(l)}(\widetilde{S}_{q,p}^{(l)}), n\delta_{n,+}^{(l)}(\widetilde{S}_{q,p}^{(l)})\big)\big]$$

$$\to \widetilde{\mathbb{E}}\big[\kappa\big((S_{q,p}^{(l)})_{\widetilde{S}_{q,p}^{(l)} \leq T}\big) \prod_{p:\widetilde{S}_{q,p}^{(l)} \leq T} f_p^{(l)}\big(\delta_-^{(l)}(\widetilde{S}_{q,p}^{(l)}), \delta_+^{(l)}(\widetilde{S}_{q,p}^{(l)})\big)\big]. \quad (3.43)$$

However, this is exactly Condition 3.1(ii) as conditional on the event that there are P jumps of $A^{(l)}(q)$ in $[0,T]$ all the $\widetilde{S}_{q,p}^{(l)}$ are independent uniformly distributed on $[0, T]$ by the properties of the Poisson random measure μ, compare Section 1.3.1 [3]. This argument occurs in similar form in [8], compare the paragraph following (42) in that paper. Note that the second expectation in (3.43) can be written in the form (3.18) as the $(\delta_-^{(l)}(s), \delta_+^{(l)}(s))$ are independent of the $\widetilde{S}_{q,p}^{(l)}$ and of each other. Hence we have shown

$$\left((Y_n^{(l)}(\widetilde{S}_{q,p}^{(l)}))_{\widetilde{S}_{q,p}^{(l)} \leq T}\right) \xrightarrow{\mathcal{L}-s} \left((Y^{(l)}(\widetilde{S}_{q,p}^{(l)}))_{\widetilde{S}_{q,p}^{(l)} \leq T}\right). \tag{3.44}$$

Step 2. Denote by $\Omega^{(l)}(q, r, n)$ the subset of Ω where two different jump times $S_{q,p}^{(l)} \neq S_{q,p'}^{(l)}$ of $N^{(l)}(q)$ are further apart than $|\pi_n|_T$ and the jump times $S_{q,p}^{(l)}$ are further away than $|\pi_n|_T$ from the discontinuity points $k/2^r$ of $\tilde{\sigma}(q, r)$. On this set we get

$$R^{(l)}(n, q, r)\mathbb{1}_{\Omega^{(l)}(q,r,n)}$$
$$= \sum_{S_{q,p}^{(l)} \leq T} g'(\Delta N^{(l)}(q)_{S_{q,p}^{(l)}})(\tilde{\sigma}(r,q)_{S_{q,p}^{(l)}-}(n\delta_{n,-}^{(l)}(S_{q,p}^{(l)}))^{1/2}U_{n,-}^{(l)}(S_{q,p}^{(l)})$$
$$+ \tilde{\sigma}(r,q)_{S_{q,p}^{(l)}}(n\delta_{n,+}^{(l)}(S_{q,p}^{(l)}))^{1/2}U_{n,+}^{(l)}(S_{q,p}^{(l)}))\mathbb{1}_{\Omega^{(l)}(q,r,n)}. \tag{3.45}$$

On the other hand from (3.44) and Lemma B.3 we obtain

$$\left(N(q), \tilde{\sigma}(q,r), (\widetilde{S}_{q,p}^{(l)})_{\widetilde{S}_{q,p}^{(l)} \leq T}, (Y_n^{(l)}(\widetilde{S}_{q,p}^{(l)}))_{\widetilde{S}_{q,p}^{(l)} \leq T}\right)$$
$$\xrightarrow{\mathcal{L}-s} \left(N(q), \tilde{\sigma}(q,r), (\widetilde{S}_{q,p}^{(l)})_{\widetilde{S}_{q,p}^{(l)} \leq T}, (Y^{(l)}(\widetilde{S}_{q,p}^{(l)}))_{\widetilde{S}_{q,p}^{(l)} \leq T}\right)$$

which using the continuous mapping theorem for stable convergence in law stated in Lemma B.5 yields

$$\sum_{S_{q,p}^{(l)} \leq T} g'(\Delta N^{(l)}(q)_{S_{q,p}^{(l)}})(\tilde{\sigma}(r,q)_{S_{q,p}^{(l)}-}(n\delta_{n,-}^{(l)}(S_{q,p}^{(l)}))^{1/2}U_{n,-}^{(l)}(S_{q,p}^{(l)})$$
$$+ \tilde{\sigma}(r,q)_{S_{q,p}^{(l)}}(n\delta_{n,+}^{(l)}(S_{q,p}^{(l)}))^{1/2}U_{n,+}^{(l)}(S_{q,p}^{(l)})) \xrightarrow{\mathcal{L}-s} \Phi_T^{univ}(g, q, r). \tag{3.46}$$

Note to this end that the jump times $S_{q,p}^{(l)}$ of $N^{(l)}(q)$ form a subset of the jump times $\widetilde{S}_{q,p}^{(l)}$ of $A^{(l)}(q)$. Finally (3.46) implies (3.25) because of (3.45) and $\mathbb{P}(\Omega(q, r, n)) \to \infty$ as $n \to \infty$ for any $q > 0$ and any $r \in \mathbb{N}$. $\qquad\square$

Proof of Theorem 3.6. The proof of Theorem 3.6 is very similar to the proof of Theorem 3.2. Only in Step 1 the used techniques to derive the necessary estimates are somewhat different due to the more complex asynchronous and bivariate structure.

Step 1. Again we define discretized versions $\tilde{\sigma}^{(l)}(r,q), \widetilde{C}^{(l)}(r,q)$ of σ, C as in Step 1 in the proof of Theorem 3.2. The only difference here is that we use the stopping times $S_{q,p}$, $p \in \mathbb{N}$, defined as the common jump times of $N^{(1)}(q)$ and $N^{(2)}(q)$ in the construction instead of the jump times $S^{(l)}_{q,p}$, $p \in \mathbb{N}$, of $N^{(l)}(q)$. A discretization $\tilde{\rho}(r,q)$ of ρ is obtained analogously.

Using this notation we then define

$$R(n,q,r) = \sqrt{n} \sum_{p:S_{q,p}\leq T} \sum_{l=1,2} \partial_l f(\Delta N^{(1)}_{S_{q,p}}, \Delta N^{(2)}(q)_{S_{q,p}}) \Delta^{(l)}_{i^{(l)}_n(S_{q,p}),n} \widetilde{C}^{(l)}(r,q)$$

and show

$$\lim_{q\to\infty} \limsup_{r\to\infty} \limsup_{n\to\infty} \mathbb{P}(|\sqrt{n}(V(f,\pi_n)_T - B(f)_T) - R(n,q,r)| > \varepsilon) \to 0 \quad (3.47)$$

for all $\varepsilon > 0$. The proof of (3.47) will require some rather technical estimates and is therefore postponed to after the discussion of the general structure of the proof.

Step 2. Next we show the $\mathcal{X}$-stable convergence

$$R(n,q,r) \xrightarrow{\mathcal{L}-s} \Phi^{biv}_T(f,q,r) \quad (3.48)$$

for all $q > 0$ and $r \in \mathbb{N}$ where

$$\Phi^{biv}_T(f,q,r) = \sum_{S_{q,p}\leq T} \Big[\partial_1 f(\Delta N^{(1)}(q)_{S_{q,p}}, \Delta N^{(2)}(q)_{S_{q,p}})$$

$$\times \Big(\tilde{\sigma}^{(1)}(r,q)_{S_{q,p}-}\sqrt{\mathcal{L}(S_{q,p})}U^{(1),-}_{S_{q,p}} + \tilde{\sigma}^{(1)}(r,q)_{S_{q,p}}\sqrt{\mathcal{R}(S_{q,p})}U^{(1),+}_{S_{q,p}}$$

$$+ \sqrt{(\tilde{\sigma}^{(1)}(r,q)_{S_{q,p}-})^2\mathcal{L}^{(1)}(S_{q,p}) + (\tilde{\sigma}^{(1)}(r,q)_{S_{q,p}})^2\mathcal{R}^{(1)}(S_{q,p})}U^{(2)}_{S_{q,p}} \Big)$$

$$+ \partial_2 f(\Delta N^{(1)}(q)_{S_{q,p}}, \Delta N^{(2)}(q)_{S_{q,p}}) \Big(\tilde{\sigma}^{(2)}(r,q)_{S_{q,p}-}\tilde{\rho}(r,q)_{S_{q,p}-}\sqrt{\mathcal{L}(S_{q,p})}U^{(1),-}_{S_{q,p}}$$

$$+ \tilde{\sigma}^{(2)}(r,q)_{S_{q,p}}\tilde{\rho}(r,q)_{S_{q,p}}\sqrt{\mathcal{R}(S_{q,p})}U^{(1),+}_{S_{q,p}}$$

$$+ ((\tilde{\sigma}^{(2)}(r,q)_{S_{q,p}-})^2(1 - (\tilde{\rho}(r,q)_{S_{q,p}-})^2)\mathcal{L}(S_{q,p})$$

$$+ (\tilde{\sigma}^{(2)}(r,q)_{S_{q,p}})^2(1 - (\tilde{\rho}(r,q)_{S_{q,p}})^2)\mathcal{R}(S_{q,p}))^{1/2}U^{(3)}_{S_{q,p}}$$

$$+ \sqrt{\tilde{\sigma}^{(2)}(r,q)_{S_{q,p}-})^2\mathcal{L}^{(2)}(S_{q,p}) + (\tilde{\sigma}^{(2)}(r,q)_{S_{q,p}})^2\mathcal{R}^{(2)}(S_{q,p})}U^{(4)}_{S_{q,p}} \Big) \Big].$$

To this end note that on the set $\Omega(n,q,r)$ where two different jumps of $N(q)$ are further apart than $|\pi_n|_T$ and any jump time of $N(q)$ is further away from $j2^{-r}$ than $k|\pi_n|_T$ for any $j \in \{1,\ldots,\lfloor T2^r \rfloor\}$ it holds

$$R(n,q,r)\mathbb{1}_{\Omega(n,q,r)} = \sqrt{n} \sum_{S_{q,p}\leq T} \Big[\partial_1 f(\Delta N^{(1)}(q)_{S_{q,p}}, \Delta N^{(2)}(q)_{S_{q,p}})$$

$$\times \Big(\tilde{\sigma}^{(1)}(r,q)_{S_{q,p}} \Delta_{\mathcal{L}_n(S_{q,p})} W^{(1)} + \tilde{\sigma}^{(1)}(r,q)_{S_{q,p}} \Delta_{\mathcal{R}_n(S_{q,p})} W^{(1)}$$

$$+ \tilde{\sigma}^{(1)}(r,q)_{S_{q,p}} \Delta_{\mathcal{L}_n^{(1)}(S_{q,p})} W^{(1)} + (\tilde{\sigma}^{(1)}(r,q)_{S_{q,p}} \Delta_{\mathcal{R}_n^{(1)}(S_{q,p})} W^{(1)} \Big)$$

$$+ \partial_2 f(\Delta N^{(1)}(q)_{S_{q,p}}, \Delta N^{(2)}(q)_{S_{q,p}}) \Big(\tilde{\sigma}^{(2)}(r,q)_{S_{q,p}} \tilde{\rho}(r,q)_{S_{q,p}} \Delta_{\mathcal{L}_n(S_{q,p})} W^{(1)}$$

$$+ \tilde{\sigma}^{(2)}(r,q)_{S_{q,p}} \tilde{\rho}(r,q)_{S_{q,p}} \Delta_{\mathcal{R}_n(S_{q,p})} W^{(1)}$$

$$+ \tilde{\sigma}^{(2)}(r,q)_{S_{q,p-}} \sqrt{1 - (\tilde{\rho}(r,q)_{S_{q,p-}})^2} \Delta_{\mathcal{L}_n(S_{q,p})} W^{(2)}$$

$$+ \tilde{\sigma}^{(2)}(r,q)_{S_{q,p}} \sqrt{1 - (\tilde{\rho}(r,q)_{S_{q,p}})^2} \Delta_{\mathcal{R}_n(S_{q,p})} W^{(2)}$$

$$+ \tilde{\sigma}^{(2)}(r,q)_{S_{q,p}} \Delta_{\mathcal{L}_n^{(2)}(S_{q,p})} W^{(2)} + \tilde{\sigma}^{(2)}(r,q)_{S_{q,p}} \Delta_{\mathcal{R}_n^{(2)}(S_{q,p})} W^{(2)} \Big) \Big] \mathbb{1}_{\Omega(n,q,r)}$$

where $\Delta_{\mathcal{L}_n(S_{q,p})} W^{(1)}$ denotes the increment of $W^{(1)}$ over the interval corresponding to $\mathcal{L}_n(S_{q,p})$ and so forth. (3.48) then follows because Condition 3.5(ii) yields i.e. that $\sqrt{n} \Delta_{\mathcal{L}_n(S_{q,p})} W^{(1)}$ converges stably in law to $\mathcal{L}(S_{q,p}) U_{S_{q,p}}^{(1),-}$. The detailed proof of (3.48) is identical to the proof of (3.25).

Step 3. Finally we show

$$\lim_{q\to\infty} \limsup_{r\to\infty} \widetilde{\mathbb{P}}(|\Phi_T^{biv}(f) - \Phi_T^{biv}(f,q,r)| > \varepsilon) = 0, \tag{3.49}$$

for all $\varepsilon > 0$. Here (3.49) can be proven similarly as (3.26).

Step 4. Combining the results (3.47), (3.48) and (3.49) from Steps 1–3 we obtain the claim (3.21) using Lemma B.6.   □

Proof of (3.47). We start as in the proof of (3.24) and observe that on the set $\Omega(n,q)$ where any two different jump times of $N(q)$ in $[0,T]$ are further apart than $2|\pi_n|_T$ it holds

$$|\sqrt{n}(V(f,\pi_n)_T - B(f)_T) - R(n,q,r)| \mathbb{1}_{\Omega(n,q)}$$

$$\leq \Big| \sqrt{n} \sum_{S_{q,p} \leq T} \big(f(\Delta_{i_n^{(1)}(S_{q,p}),n}^{(1)} X^{(1)}, \Delta_{i_n^{(2)}(S_{q,p}),n}^{(2)} X^{(2)})$$

$$- f(\Delta_{i_n^{(1)}(S_{q,p}),n}^{(1)} X^{(1)}, \Delta_{i_n^{(2)}(S_{q,p}),n}^{(2)} (X^{(2)} - N^{(2)}(q)))$$

$$- f(\Delta_{i_n^{(1)}(S_{q,p}),n}^{(1)} (X^{(1)} - N^{(1)}(q)), \Delta_{i_n^{(2)}(S_{q,p}),n}^{(2)} X^{(2)})$$

$$- f(\Delta_{i_n^{(1)}(S_{q,p}),n}^{(1)} (X^{(1)} - N^{(1)}(q)), \Delta_{i_n^{(2)}(S_{q,p}),n}^{(2)} (X^{(2)} - N^{(2)}(q)))$$

$$- f(\Delta N^{(1)}(q)_{S_{q,p}}, \Delta N^{(2)}(q)_{S_{q,p}}) \big) - R(n,q,r) \Big| \tag{3.50}$$

$$+ \sqrt{n} \Big| \sum_{S_{q,p}^{(1)} \leq T} \sum_{j: t_{j,n}^{(2)} \leq T} f(\Delta_{i_n^{(1)}(S_{q,p}^{(1)}),n}^{(1)} X^{(1)}, \Delta_{j,n}^{(2)} (X^{(2)} - N^{(2)}(q))) \mathbb{1}_{i_n^{(1)}(S_{q,p}^{(1)}),j} \Big|$$

$$\tag{3.51}$$

$$+ \sqrt{n} \left| \sum_{S_{q,p}^{(2)} \leq T} \sum_{i:t_{i,n}^{(1)} \leq T} f(\Delta_{i,n}^{(1)}(X^{(1)} - N^{(1)}(q)), \Delta_{i_n^{(2)}(S_{q,p}^{(1)}),n}^{(2)} X^{(2)}) \mathbb{1}_{i,i_n^{(2)}(S_{q,p}^{(1)})} \right|$$

$$\tag{3.52}$$

$$+ \sqrt{n} \left| \sum_{i,j:t_{i,n}^{(1)} \vee t_{j,n}^{(2)} \leq T} f(\Delta_{i,n}^{(1)}(X^{(1)} - N^{(1)}(q)), \Delta_{j,n}^{(2)}(X^{(2)} - N^{(2)}(q))) \mathbb{1}_{i,j} \right.$$

$$\left. - \sum_{s \leq T} f(\Delta M^{(1)}(q)_s, \Delta M^{(2)}(q)_s) \right| \tag{3.53}$$

using the shorthand-notation $\mathbb{1}_{i,j} = \mathbb{1}_{\{\mathcal{I}_{i,n}^{(1)} \cap \mathcal{I}_{j,n}^{(2)} \neq \emptyset\}}$ Hence using $\mathbb{P}(\Omega(n,q)) \to 1$ as $n \to \infty$ for any $q > 0$ it remains to show that (3.50)–(3.53) vanish in the sense of (3.24).

Step 1. Using a Taylor expansion (3.50) becomes equal to

$$\left| \sqrt{n} \sum_{S_{q,p} \leq T} \left(\sum_{l=1,2} \partial_l f(\Delta N^{(1)}(q)_{S_{q,p}}, \Delta N^{(2)}(q)_{S_{q,p}}) \right. \right.$$

$$\times \Delta_{i_n^{(l)}(S_{q,p}),n}^{(l)}(X^{(l)} - N^{(l)}(q) - \tilde{C}^{(l)}(r,q))$$

$$+ \frac{1}{2} \sum_{k,l=1,2} \partial_{kl} f(\Delta N^{(1)}(q)_{S_{q,p}} + \theta_{q,p}^n \Delta_{i_n^{(l)}(S_{q,p}),n}^{(1)}(X^{(l)} - N^{(l)}(q)), \Delta N^{(2)}(q)_{S_{q,p}}$$

$$+ \theta_{q,p}^n \Delta_{i_n^{(2)}(S_{q,p}),n}^{(2)}(X^{(2)} - N^{(2)}(q))) \prod_{\iota=k,l} \Delta_{i_n^{(\iota)}(S_{q,p}),n}^{(\iota)}(X^{(\iota)} - N^{(\iota)}(q))$$

$$- f(\Delta_{i_n^{(1)}(S_{q,p}),n}^{(1)} X^{(1)}, \Delta_{i_n^{(2)}(S_{q,p}),n}^{(2)}(X^{(2)} - N^{(2)}(q)))$$

$$- f(\Delta_{i_n^{(1)}(S_{q,p}),n}^{(1)}(X^{(1)} - N^{(1)}(q)), \Delta_{i_n^{(2)}(S_{q,p}),n}^{(2)} X^{(2)})$$

$$- f(\Delta_{i_n^{(1)}(S_{q,p}),n}^{(1)}(X^{(1)} - N^{(1)}(q)), \Delta_{i_n^{(2)}(S_{q,p}),n}^{(2)}(X^{(2)} - N^{(2)}(q))) \right|. \tag{3.54}$$

The first sum in (3.54) vanishes in the sense of (3.24) because (3.34) also holds under Condition 3.5 (note that $\delta_{n,-}^{(l)} = \mathcal{L}_n(s) + \mathcal{L}_n^{(l)}(s)$, $\delta_{n,+}^{(l)} = \mathcal{R}_n(s) + \mathcal{R}_n^{(l)}(s)$). Further the second sum vanishes similarly as the term $Y_3(n,q,r)$ in the proof of (3.24) also due to (3.34). For the discussion of the last three sums in (3.54) note that (3.8) implies

$$|f(x,y)| = \left| \int_0^x \int_0^{x'} \partial_{11} f(x'',y) dx' dx'' \right| \leq \int_0^x \int_0^{x'} |\partial_{11} f(x'',y)| dx' dx'' \tag{3.55}$$

which then together with $\partial_{11} f(x,y) = o(|x|^{p_1^{(1,1)}} |y|^{p_2^{(1,1)}})$ as $|xy| \to 0$ from (3.9) yields $f(x,y) = o(|x|^{2+p_1^{(1,1)}} |y|^{p_2^{(1,1)}})$ as $|xy| \to 0$. Similarly we obtain

$f(x,y) = o(|x|^{p_1^{(2,2)}} |y|^{2+p_2^{(2,2)}})$ as $|xy| \to 0$. Hence the sums over the expressions in the last three lines of (3.54) vanish also by (3.34), compare the discussion of $Y_2(n,q,r)$ in the proof of (3.24).

Step 2. Denote by $\Omega(N,q,m,n)$ the set where $|\Delta_{i,n}^{(l)} X^{(l)}| \leq N$ and

$$|\Delta_{i,n}^{(l)}(X^{(l)} - N^{(l)}(q))| \leq 1/m$$

for any $t_{i,n}^{(l)} \leq T$, $l = 1,2$. It holds $\mathbb{P}(\Omega(N,q,m,n)) \to 1$ as $N,n \to \infty$ for any $m,q > 0$. On the set $\Omega(N,q,m,n)$ the sums (3.51) and (3.52) are using $f(x,y) = o(|x|^{2+p_1^{(1,1)}} |y|^{p_2^{(1,1)}})$ as $|xy| \to 0$ and $f(x,y) = o(|x|^{p_1^{(2,2)}} |y|^{2+p_2^{(2,2)}})$ as $|xy| \to 0$ bounded by

$$K_{N,m}\sqrt{n} \sum_{l=1,2} \sum_{S_{q,p}^{(l)} \leq T} \sum_{j:t_{j,n}^{(3-l)} \leq T} |\Delta_{i_n^{(1)}(S_{q,p}^{(1)}),n}^{(1)} X^{(1)}|^{p_l^{(3-l,3-l)}}$$

$$\times |\Delta_{j,n}^{(3-l)}(X^{(3-l)} - N^{(3-l)}(q))|^{2+p_{3-l}^{(3-l,3-l)}} \mathbb{1}_{i_n^{(1)}(S_{q,p}^{(1)}),j}.$$

By similar arguments as used in the derivation of (3.34) we can deduce using inequalities 1.8–1.10

$$\mathbb{E}\Big[\sum_{j:t_{j,n}^{(3-l)} \leq T} |\Delta_{j,n}^{(3-l)}(X^{(3-l)} - N^{(3-l)}(q))|^{2+p_{3-l}^{(3-l,3-l)}} \mathbb{1}_{i_n^{(1)}(S_{q,p}^{(1)}),j} \big| \sigma(\mathcal{G}_0(q),\mathcal{S}) \Big]$$

$$\leq K(K_q(|\pi_n|_T)^{1+p_{3-l}^{(3-l,3-l)}} + (|\pi_n|_T)^{p_{3-l}^{(3-l,3-l)}/2} + e_q)3|\pi_n|_T.$$

Hence the sums (3.51) and (3.52) vanish as $n \to \infty$ for any $q > 0$ because $N^{(l)}(q)$, $l = 1,2$, almost surely has only finitely many jumps in $[0,T]$ and because we have $\sqrt{n}|\pi_n|_T = o_\mathbb{P}(1)$ by Condition 3.5(i).

Step 3. Denote by

$$\Delta_{(i,j),n}^{(l,3-l)} M^{(l)}(q) = \big(M^{(l)}(q)_{t_{i,n}^{(l)} \wedge t_{j,n}^{(3-l)}} - M^{(l)}(q)_{t_{i-1,n}^{(l)} \vee t_{j-1,n}^{(3-l)}} \big) \mathbb{1}_{i,j},$$

$$\Delta_{(i,j),n}^{(l\backslash 3-l)} M^{(l)}(q) = \Delta_{i,n}^{(l)} M^{(l)}(q) - \Delta_{(i,j),n}^{(l,3-l)} M^{(l)}(q)$$

the increments of $M^{(l)}(q)$ over the sets $\mathcal{I}_{i,n}^{(l)} \cap \mathcal{I}_{j,n}^{(3-l)}$ respectively $\mathcal{I}_{i,n}^{(l)} \setminus \mathcal{I}_{j,n}^{(3-l)}$. Then using the above notation and a Taylor expansion the sum (3.53) is equal to

$$\sqrt{n}\Big(\sum_{i,j:t_{i,n}^{(1)} \vee t_{j,n}^{(2)} \leq T} f(\Delta_{(i,j),n}^{(1,2)} M^{(1)}(q), \Delta_{(i,j),n}^{(1,2)} M^{(2)}(q))\mathbb{1}_{i,j}$$

$$- \sum_{s \leq T} f(\Delta M^{(1)}(q)_s, \Delta M^{(2)}(q)_s)\Big) \qquad (3.56)$$

$$+ \sqrt{n} \sum_{i,j:t_{i,n}^{(1)} \vee t_{j,n}^{(2)} \leq T} \sum_{l=1,2} \partial_l f(\Delta_{(i,j),n}^{(1,2)} M^{(1)}(q), \Delta_{(i,j),n}^{(1,2)} M^{(2)}(q))$$

$$\times (\Delta_{i,n}^{(l)}(B^{(l)} + C^{(l)}) + \Delta_{(i,j),n}^{(l\backslash 3-l)} M^{(l)}(q)) \mathbb{1}_{i,j} \quad (3.57)$$

$$+ \sqrt{n} \sum_{k,l=1,2} \sum_{i_1,i_2:t_{i_1,n}^{(1)} \vee t_{i_2,n}^{(2)} \leq T} \partial_{kl} f(\Delta_{(i_1,i_2),n}^{(1,2)} M^{(1)}(q)$$

$$+ \theta_{i_1,i_2}^n [\Delta_{i_1,n}^{(1)}(B^{(1)} + C^{(1)}) + \Delta_{(i_1,i_2),n}^{(1\backslash 2)} M^{(1)}(q)],$$

$$\Delta_{(i_1,i_2),n}^{(1,2)} M^{(2)}(q) + \theta_{i_1,i_2}^n [\Delta_{i_2,n}^{(2)}(B^{(2)} + C^{(2)})$$

$$+ \Delta_{(i_2,i_1),n}^{(2\backslash 1)} M^{(2)}(q)]) (\Delta_{i_k,n}^{(k)}(B^{(k)} + C^{(k)}) + \Delta_{(i_k,i_{3-k}),n}^{(k\backslash 3-k)} M^{(k)}(q))$$

$$\times (\Delta_{i_l,n}^{(l)}(B^{(l)} + C^{(l)}) + \Delta_{(i_l,i_{3-l}),n}^{(l\backslash 3-l)} M^{(l)}(q)) \mathbb{1}_{i_1,i_2}. \quad (3.58)$$

Then the $\mathcal{S}$-conditional expectation of the sum corresponding to $l = 1$ in (3.57) is on the set $\Omega(n,m,q)$ where it holds $\Delta_{i,n}^{(l)} M^{(l)}(q) \leq 1/m$, for all $i \in \mathbb{N}$ with $t_{i,n}^{(l)} \leq T$, $l = 1,2$, using iterated expectations, the Cauchy-Schwarz inequality as in the discussion of $Y_5(n,q)$ in the proof of (3.24), Lemma 1.4 and Lemma 3.9 bounded by

$$\sqrt{n} \sum_{i,j:t_{i,n}^{(1)} \vee t_{j,n}^{(2)} \leq T} \mathbb{E}\big[K_m |\Delta_{(i,j),n}^{(1,2)} M^{(1)}(q)|^{1+p_1^{(1,1)}} |\Delta_{(i,j),n}^{(1,2)} M^{(2)}(q)|^{p_2^{(1,1)}} |$$

$$\times (\Delta_{i,n}^{(l)}(B^{(l)} + C^{(l)}) + \Delta_{(i,j),n}^{(l\backslash 3-l)} M^{(l)}(q))| \big| \mathcal{S}\big] \mathbb{1}_{i,j}$$

$$\leq K_m e_q G_{2,1}^n(T).$$

Analogously the $\mathcal{S}$-conditional expectation of the sum in (3.57) corresponding to $l = 2$ can be bounded by $K_m e_q G_{1,2}^n(T)$. Further on the set $\Omega(n,m,q)$ the expression (3.58) is similarly as $Y_6(n,q)$ in the proof of (3.24) bounded by

$$K_m \sum_{k,l=1,2} G_{p_1^{(k,l)}+(2-k)+(2-l),p_2^{(k,l)}+(k-1)+(l-1)}^n(T).$$

Hence (3.57) and (3.58) vanish as $n \to \infty$ for any $q > 0$ because of

$$\lim_{n \to \infty} \mathbb{P}(\Omega(n,m,q)) = 0$$

for any $q, m > 0$, $K_m \to 0$ as $m \to 0$ and because of Condition 3.5(i) and

$$G_{p,3-p}^n(T) \leq G_{3,0}^n(T) + G_{0,3}^n(T), \quad p \in [0,3], \quad (3.59)$$

where the inequality follows from $a^p b^{3-p} \leq a^3 + b^3$ for any $a, b \geq 0$ which can be concluded from Muirhead's inequality; compare Theorem 45 of [19].

Finally (3.56) can be discussed similarly as the term $Y_4(n,q)$ in the proof of (3.24). $\qquad \square$

3.2 Central Limit Theorem for Normalized Functionals

In this section we will discuss how central limit theorems for the normalized functionals $\overline{V}^{(l)}(p, g, \pi_n)_T$ and $\overline{V}(p, f, \pi_n)_T$ introduced in Section 2.2 can be obtained. To illustrate usable methods and arising challenges we are going to discuss the asymptotics of

$$\sqrt{n}\left(\overline{V}(p, f, \pi_n)_T - \int_0^T m_{c_s}(f)dG_p(s)\right) \tag{3.60}$$

under synchronous observations. Hence we will outline how to derive a central limit theorem supporting the law of large numbers from Theorem 2.17. In the end of this section we then briefly discuss why it would be much more difficult to obtain a similar result in the setting of asynchronous observations.

The additional assumptions which are necessary to derive central limit theorems for $\overline{V}^{(l)}(p, g, \pi_n)_T$ and $\overline{V}(p, f, \pi_n)_T$ in the case of synchronous observations are summarized in the following condition. From now on we use the notation from (2.1).

Condition 3.10. *Suppose Condition 1.3 holds, the observation scheme is exogenous and the process X is continuous. Further,*

(i) the volatility process is a continuous 2×2-dimensional semimartingale, i.e.

$$\sigma_t = \sigma_0 + \int_0^t \tilde{b}_s ds + \sum_{l=1,2} \int_0^t \tilde{\sigma}_s^{(l)} dW_s^{(l)} + \int_0^t \tilde{\tau}_s dV_s \tag{3.61}$$

where $\tilde{b}_s, \tilde{\sigma}_s^{(l)}, \tilde{\tau}_s$ are adapted 2×2-dimensional processes with càdlàg paths and V is a 2×2-dimensional standard Brownian motion which is independent of W. The integrals $\int_0^t \tilde{\sigma}_s^{(l)} dW_s^{(l)}$ are to be understood in such a way that each component of $\tilde{\sigma}_s^{(l)}$ is integrated with respect to $W_s^{(l)}$.

(ii) The observation scheme is given by $t_{0,n} = 0$, $n \in \mathbb{N}$, and

$$t_{i,n} = t_{i-1,n} + \frac{E_{i,n}}{n}, \quad i, n \in \mathbb{N},$$

for i.i.d. non-negative random variables $E_{i,n}$, $i, n \in \mathbb{N}$, whose law fulfils $\mathbb{E}[E_{1,1}^k] < \infty$ for all $k \geq 0$.

In the following we will sketch how based on Condition 3.10 a central limit theorem can be obtained for $\overline{V}(p, f, \pi_n)_T$ where the function $f \colon \mathbb{R}^2 \to \mathbb{R}$ is continuously differentiable and positively homogeneous of degree $p \geq 1$ in the setting of synchronous observations. Set $G_p^n(t) = G_p^{(1),n}(t) = G_p^{(2),n}(t)$ in this setting and denote by $G_p(t)$ the (deterministic) limit of $G_p^n(t)$ which always exists under Condition 3.10(ii) by Lemma 5.3. First we decompose (3.60) as follows

$$
\sqrt{n}\Big(\overline{V}(p, f, \pi_n)_T - \int_0^T m_{c_s}(f)dG_p(s)\Big)
$$

$$
= \sqrt{n}n^{p/2-1} \sum_{i:t_{i,n}\leq T} \big(f(\Delta_{i,n}X) - f(\sigma_{t_{i-1,n}}\Delta_{i,n}W)\big) \tag{3.62}
$$

$$
+ \sqrt{n}n^{p/2-1} \sum_{i:t_{i,n}\leq T} \big(f(\sigma_{t_{i-1,n}}\Delta_{i,n}W) - m_{c_{t_{i-1,n}}}(f)|\mathcal{I}_{i,n}|^{p/2}\big) \tag{3.63}
$$

$$
+ \sqrt{n} \sum_{i:t_{i,n}\leq T} m_{c_{t_{i-1,n}}}(f)\big(n^{p/2-1}|\mathcal{I}_{i,n}|^{p/2} - (G_p(t_{i,n}) - G_p(t_{i-1,n}))\big) \tag{3.64}
$$

$$
+ \sqrt{n} \sum_{i:t_{i,n}\leq T} \int_{t_{i-1,n}}^{t_{i,n}} (m_{c_{t_{i-1,n}}}(f) - m_{c_s}(f))dG_p(s) + o_{\mathbb{P}}(1) \tag{3.65}
$$

where $c_s = \sigma_s\sigma_s^*$. Here, (3.62) and (3.65) describe errors which are due to a discretization of σ. Further, the term (3.63) originates from the error in the convergence of the standardized sum $n^{-1}\sum_i n^{p/2}|\mathcal{I}_{i,n}|^{p/2}f(\sigma_{t_{i-1,n}}\Delta_{i,n}W|\mathcal{I}_{i,n}|^{-1/2})$ to corresponding moments of a normal distribution. The term (3.64) represents the error in the convergence of the observation scheme. The $o_{\mathbb{P}}(1)$-term is due to the integral

$$
\sqrt{n} \int_{t_{i_n(T),n}}^T m_{c_s}(f)dG_p(s) = O_{\mathbb{P}}(\sqrt{n}|\mathcal{I}_{i_n(T),n}|) = O_{\mathbb{P}}(n^{-1/2}).
$$

To derive stable convergence of (3.60) we want to use Proposition C.4 or rather a generalized version of it. Hence we have to check (C.1)–(C.5) for

$$
\zeta_i^n = \sqrt{n}\Big(n^{p/2-1}f(\Delta_{i,n}X) - \int_{t_{i-1,n}}^{t_{i,n}} m_{c_s}(f)dG_p(s)\Big).
$$

Instead of checking the conditions (C.1)–(C.5) for ζ_i^n we will check them for the corresponding increments of (3.62)–(3.65) separately.

First we consider (3.63). Setting

$$
\alpha_i^n = n^{(p+1)/2-1}\big(f(\sigma_{t_{i-1,n}}\Delta_{i,n}W) - m_{c_{t_{i-1,n}}}(f)|\mathcal{I}_{i,n}|^{p/2}\big)
$$

we obtain $\mathbb{E}[\alpha_i^n|\mathcal{F}_{t_{i-1,n}},\mathcal{S}]=0$ for all $i,n\in\mathbb{N}$ which yields

$$\sum_{i:t_{i,n}\leq T}\mathbb{E}[\alpha_i^n|\mathcal{F}_{t_{i-1,n}},\mathcal{S}]=0,$$

$$\sum_{i:t_{i,n}\leq T}\mathbb{E}[(\alpha_i^n)^2|\mathcal{F}_{t_{i-1,n}},\mathcal{S}]=n^{p-1}\sum_{i:t_{i,n}\leq T}(m_{c_{t_{i-1,n}}}(f^2)-(m_{c_{t_{i-1,n}}}(f))^2)|\mathcal{I}_{i,n}|^p$$

$$\xrightarrow{\mathbb{P}}\int_0^T(m_{c_s}(f^2)-(m_{c_s}(f))^2)dG_{2p}(s).$$

Denote $m_\Sigma^{(l)}(f,1)=\mathbb{E}[g(\Sigma^{1/2}Z)Z^{(l)}]$ where $Z=(Z^{(1)},\ldots,Z^{(d)})\sim\mathcal{N}(0,I_d)$ for covariance matrizes $\Sigma\in\mathbb{R}^{d\times d}$. Then we derive for $l=1,2$

$$\sum_{i:t_{i,n}\leq T}\mathbb{E}[\alpha_i^n\Delta_{i,n}W^{(l)}|\mathcal{F}_{t_{i-1,n}},\mathcal{S}]$$

$$=n^{(p+1)/2-1}\sum_{i:t_{i,n}\leq T}(m_{c_{t_{i-1,n}}}^{(l)}(f,1)-m_{c_{t_{i-1,n}}}(f)m_{I_1}(\mathrm{id}))|\mathcal{I}_{i,n}|^{(p+1)/2}$$

$$\xrightarrow{\mathbb{P}}\int_0^T m_{c_s}^{(l)}(f,1)dG_{p+1}(s)$$

where we used $m_{I_1}(\mathrm{id})=0$.

Next we look at (3.64). To discuss this term we need Condition 3.10(ii) as here we have to work with the asymptotics of the observation scheme. Note that Condition 3.10(ii) and Lemma 5.3 yield

$$G_k(t)=\frac{\mathbb{E}[E_{1,1}^{k/2}]}{\mathbb{E}[E_{1,1}]}t,\quad\forall t\geq 0,\ \forall k\geq 0.\tag{3.66}$$

Hence we have

$$\sqrt{n}\sum_{i:t_{i,n}\leq T}\mathbb{E}\big[m_{c_{t_{i-1,n}}}(f)\big(n^{p/2-1}|\mathcal{I}_{i,n}|^{p/2}-(G_p(t_{i,n})-G_p(t_{i-1,n}))\big)\big|\mathcal{F}_{t_{i-1,n}}\big]$$

$$=\sqrt{n}\sum_{i:t_{i-1,n}\leq T}m_{c_{t_{i-1,n}}}(f)\mathbb{E}\big[n^{-1}E_{i,n}^{p/2}-\mathbb{E}[E_{1,1}^{p/2}]/\mathbb{E}[E_{1,1}]|\mathcal{I}_{i,n}|]\big]$$

$$=n^{-1/2}\sum_{i:t_{i-1,n}\leq T}m_{c_{t_{i-1,n}}}(f)\mathbb{E}\big[E_{i,n}^{p/2}-\mathbb{E}[E_{i,n}^{p/2}]/\mathbb{E}[E_{i,n}]E_{i,n}\big]$$

$$=0$$

and

$$\sum_{i:t_{i,n}\leq T}\mathbb{E}[(\sqrt{n}m_{c_{t_{i-1,n}}}(f))^2(n^{p/2-1}|\mathcal{I}_{i,n}|^{p/2}-(G_p(t_{i,n})-G_p(t_{i-1,n})))^2|\mathcal{F}_{t_{i-1,n}}]$$

$$=\sum_{i:t_{i,n}\leq T}n(m_{c_{t_{i-1,n}}}(f))^2\mathbb{E}[(n^{p/2-1}|\mathcal{I}_{i,n}|^{p/2}-\mathbb{E}[E_{i,n}^{p/2}]/\mathbb{E}[E_{i,n}]|\mathcal{I}_{i,n}|)^2]$$

$$=\sum_{i:t_{i,n}\leq T}(m_{c_{t_{i-1,n}}}(f))^2\Big[(G_{2p}^n(t_{i,n})-G_{2p}^n(t_{i-1,n}))$$

$$-2\frac{\mathbb{E}[E_{i,n}^{p/2}]}{\mathbb{E}[E_{i,n}]}(G_{p+1}^n(t_{i,n})-G_{p+1}^n(t_{i-1,n}))+\frac{\mathbb{E}[E_{i,n}^{p/2}]^2}{\mathbb{E}[E_{i,n}]^2}(G_4^n(t_{i,n})-G_4^n(t_{i-1,n}))\Big]$$

$$\xrightarrow{\mathbb{P}}\int_0^T(m_{c_s}(f))^2d\Big(G_{2p}-2\frac{G_p}{G_2}G_{p+1}+\frac{(G_p)^2}{(G_2)^2}G_4\Big)(s)$$

where we used (3.66) repeatedly. Here, the fraction $G_p(t)/G_2(t)$ does not depend on t.

Finally we consider the error terms (3.62) and (3.65) which are arising from the discretization of σ. In particular we need to quantify the approximation error due to this discretization and therefore need the structural assumption (3.61) on the form of σ. Further we will use the differentiability of f for the discussion of these terms. Using a Taylor expansion we get

$$\sqrt{n}\sum_{i:t_{i,n}\leq T}\int_{t_{i-1,n}}^{t_{i,n}}(m_{c_{t_{i-1,n}}}(f)-m_{c_s}(f))dG_p(s)$$

$$\approx\sum_{i:t_{i,n}\leq T}\int_{t_{i-1,n}}^{t_{i,n}}(\nabla m_{c_{t_{i-1,n}}}(f))^*\sqrt{n}(\kappa(\sigma_{t_{i-1,n}})-\kappa(\sigma_s))dG_p(s)$$

where

$$\kappa(v)=(v_{1,1},v_{1,2},v_{2,1},v_{2,2})^*\quad\text{for }v=\begin{pmatrix}v_{1,1}&v_{1,2}\\v_{2,1}&v_{2,2}\end{pmatrix}^*\in\mathbb{R}^{2\times2}$$

denotes the operation that tranforms a 2×2-matrix into a 4×1-vector and

$$\nabla m_{x^*x}(f)=\Big(\frac{\partial m_{x^*x}(f)}{\partial x_{1,1}},\frac{\partial m_{x^*x}(f)}{\partial x_{2,1}},\frac{\partial m_{x^*x}(f)}{\partial x_{1,2}},\frac{\partial m_{x^*x}(f)}{\partial x_{2,2}}\Big)$$

denotes the derivative of $m_\Sigma(f)$ with respect to the entries of the root $\Sigma^{1/2}$ of the covariance matrix. Note that f is for $p\geq1$ regular enough, compare Lemma 2.28, such that e.g.

$$\frac{\partial m_{x^*x}(f)}{\partial x_{1,1}}=\frac{\partial}{\partial x_{1,1}}\mathbb{E}[f(xZ)]=\mathbb{E}[\frac{\partial}{\partial x_{1,1}}f(xZ)],\quad Z\sim\mathcal{N}(0,I_2),$$

holds and hence the existence of $\nabla m_{x^*x}(f)$ is assured becaue f is continuously differentiable.

Further, we infer

$$\mathbb{E}[\Delta_{i,n}\sigma|\mathcal{F}_{t_{i-1,n}},\mathcal{S}] = O_{\mathbb{P}}(|\mathcal{I}_{i,n}|), \quad \mathbb{E}[\|\Delta_{i,n}\sigma\|^2|\mathcal{F}_{t_{i-1,n}},\mathcal{S}] = O_{\mathbb{P}}(|\mathcal{I}_{i,n}|)$$

from (3.61), compare Lemma 1.4, which yields

$$\sum_{i:t_{i,n}\leq T} \mathbb{E}[\beta_i^n|\mathcal{F}_{t_{i-1,n}},\mathcal{S}] \xrightarrow{\mathbb{P}} 0, \quad \sum_{i:t_{i,n}\leq T} \mathbb{E}[(\beta_i^n)^2|\mathcal{F}_{t_{i-1,n}},\mathcal{S}] \xrightarrow{\mathbb{P}} 0,$$

where $\beta_i^n = (\nabla m_{c_{t_{i-1,n}}}(f))^* \int_{t_{i-1,n}}^{t_{i,n}} \sqrt{n}(\kappa(\sigma_{t_{i-1,n}}) - \kappa(\sigma_s))dG_p(s)$. Hence (3.65) is asymptotically negligible by Lemma C.3.

Finally we consider (3.62). Note that we can write $X_t = \int_0^t \bar{b}_s ds + C_t$ for some adapted process $\bar{b}_s$ because X is continuous by Condition 3.10. By a Taylor expansion we then obtain

$$\sqrt{n}\left(n^{p/2-1} \sum_{i:t_{i,n}\leq T} f(\Delta_{i,n}X) - n^{p/2-1} \sum_{i:t_{i,n}\leq T} f(\sigma_{t_{i-1,n}}\Delta_{i,n}W)\right)$$

$$\approx n^{(p+1)/2-1} \sum_{i:t_{i,n}\leq T} (\nabla f(\sigma_{t_{i-1,n}}\Delta_{i,n}W))^*$$

$$\times \left(\int_{t_{i-1,n}}^{t_{i,n}} \bar{b}_s ds + \int_{t_{i-1,n}}^{t_{i,n}} \sigma_s dW_s - \sigma_{t_{i-1,n}}\Delta_{i,n}W\right)$$

$$= n^{(p+1)/2-1} \sum_{i:t_{i,n}\leq T} (\nabla f(\sigma_{t_{i-1,n}}\Delta_{i,n}W))^*$$

$$\times \left(\int_{t_{i-1,n}}^{t_{i,n}} \bar{b}_s ds + \int_{t_{i-1,n}}^{t_{i,n}} (\sigma_s - \sigma_{t_{i-1,n}})dW_s\right)$$

$$\approx n^{(p+1)/2-1} \sum_{i:t_{i,n}\leq T} (\nabla f(\sigma_{t_{i-1,n}}\Delta_{i,n}W))^*$$

$$\times \left(\bar{b}_{t_{i-1,n}}|\mathcal{I}_{i,n}| + \int_{t_{i-1,n}}^{t_{i,n}} \left(\sum_{l=1,2} \tilde{\sigma}_{t_{i-1,n}}^{(l)}(W_s^{(l)} - W_{t_{i-1,n}}^{(l)})\right.\right.$$

$$\left.\left. + \tilde{\tau}_{t_{i-1,n}}(V_s - V_{t_{i-1,n}})\right)dW_s\right)$$

where we used (3.61) for the last approximation. Set

$$\gamma_i^n = n^{(p+1)/2-1}(\nabla f(\sigma_{t_{i-1,n}}\Delta_{i,n}W))^*\left(\bar{b}_{t_{i-1,n}}|\mathcal{I}_{i,n}|\right.$$

$$\left. + \int_{t_{i-1,n}}^{t_{i,n}} \left(\sum_{l=1,2} \tilde{\sigma}_{t_{i-1,n}}^{(l)}(W_s^{(l)} - W_{t_{i-1,n}}^{(l)}) + \tilde{\tau}_{t_{i-1,n}}(V_s - V_{t_{i-1,n}})\right)dW_s\right).$$

Using the fact that if f is positively homogeneous of degree $p \geq 1$ and differentiable then ∇f is positively homogeneous of degree $p - 1$, compare Theorem 20.1 in [44], we then derive

$$\sum_{i:t_{i,n} \leq T} \mathbb{E}[\gamma_i^n | \mathcal{F}_{t_{i-1,n}}, \mathcal{S}]$$

$$= n^{(p+1)/2 - 1} \sum_{i:t_{i,n} \leq T} |\mathcal{I}_{i,n}|^{(p+1)/2} \Big(\mathbb{E}[(\nabla f(\sigma_{t_{i-1,n}} \Delta_{i,n} W / |\mathcal{I}_{i,n}|^{1/2}))^* | \mathcal{F}_{t_{i-1,n}}, \mathcal{S}]$$

$$\times \overline{b}_{t_{i-1,n}}$$

$$+ \mathbb{E}[(\nabla f(\sigma_{t_{i-1,n}} \Delta_{i,n} W / |\mathcal{I}_{i,n}|)^{1/2}))^*$$

$$\times \Big(\frac{1}{|\mathcal{I}_{i,n}|} \int_{t_{i-1,n}}^{t_{i,n}} \sum_{l=1,2} \tilde{\sigma}_{t_{i-1,n}}^{(l)} W_s^{(l)} dW_s \Big) | \mathcal{F}_{t_{i-1,n}}, \mathcal{S}] \Big)$$

$$= n^{(p+1)/2 - 1} \sum_{i:t_{i,n} \leq T} |\mathcal{I}_{i,n}|^{(p+1)/2} \Big[(m_{c_{t_{i-1,n}}}(\nabla f))^* \overline{b}_{t_{i-1,n}}$$

$$+ \mathbb{E}[(\nabla f(\sigma_{t_{i-1,n}} \widetilde{V}_1))^* \Big(\frac{1}{|\mathcal{I}_{i,n}|} \sum_{l=1,2} \tilde{\sigma}_{t_{i-1,n}}^{(l)} \int_0^1 \widetilde{V}_s^{(l)} d\widetilde{V}_s \Big) | \mathcal{F}_{t_{i-1,n}}] \Big]$$

$$\overset{\mathbb{P}}{\longrightarrow} \int_0^T (m_{c_s}(\nabla f))^* \overline{b}_s dG_{p+1}(s) + \int_0^T \widetilde{m}_{c_s, \tilde{c}_s^{(1)}, \tilde{c}_s^{(2)}}(\nabla f) dG_{p+1}(s)$$

where $\tilde{c}_s^{(l)} = \tilde{\sigma}_s^{(l)}(\tilde{\sigma}_s^{(l)})^*$ and $\widetilde{V} = (\widetilde{V}^{(1)}, \widetilde{V}^{(2)})^*$ denotes a two-dimensional standard Brownian motion independent of $\mathcal{F}$ and

$$\widetilde{m}_{c_s, \tilde{c}^{(1)}, \tilde{c}^{(2)}}(\nabla f) = \mathbb{E}\Big[(\nabla f(\sigma_s \widetilde{V}_1))^* \Big(\tilde{\sigma}_s^{(1)} \frac{1}{2} \Big(\frac{(\widetilde{V}_1^{(1)})^2 - 1}{\int_0^1 \widetilde{V}_s^{(2)} d\widetilde{V}_s^{(1)}} \Big) + \tilde{\sigma}_s^{(2)} \frac{1}{2} \Big(\frac{\int_0^1 \widetilde{V}_s^{(1)} d\widetilde{V}_s^{(2)}}{(\widetilde{V}_1^{(2)})^2 - 1} \Big) \Big) | \mathcal{F} \Big].$$

Summarizing the investigation of the conditions (C.1)–(C.5) we observe that for (C.1) only (3.62) yields a non-zero contribution. Further, for condition (C.2) we find that the terms (3.63) and (3.64) contribute to the limit while for condition (C.3) only (3.63) yields a non-zero contribution. The term (3.65) does not contribute to the limit at all.

Combining all results from above according to Proposition C.4 we obtain the stable convergence

$$\sqrt{n} \Big(\overline{V}(p, f, \pi_n)_T - \int_0^T m_{c_s}(f) dG_p(s) \Big)$$

$$\overset{\mathcal{L}-s}{\longrightarrow} \int_0^T \big((m_{c_s}(\nabla f))^* \overline{b}_s + \widetilde{m}_{c_s, \tilde{c}_n^{(1)}, \tilde{c}_n^{(2)}}(\nabla f) \big) dG_{p+1}(s)$$

$$+ \sum_{l=1,2} \int_0^T \sqrt{m_{c_s}^{(l)}(f, 1) G'_{p+1}(s)} dW_s^{(l)} + \int_0^T \sqrt{w_s} d\widetilde{W}_s, \quad (3.67)$$

$$w_s = m_{c_s}(f^2)G'_{2p}(s) - \sum_{l=1,2} m_{c_s}^{(l)}(f,1)G'_{p+1}(s)$$

$$+ (m_{c_s}(f))^2 \Big(-\frac{G_p}{G_2}G_{p+1} + \frac{(G_p)^2}{(G_2)^2}G_4 \Big)'(s)$$

where $\widetilde{W}$ denotes a standard Brownian motion which is independent of $\mathcal{F}$. By Lemma 5.3 it is clear that all functions G_p, $p \geq 0$, are differentiable such that the expressions above are well defined. We did not formally check conditions (C.4) and (C.5) but they should follow similarly as in the setting of equidistant deterministic observations which is discussed in [40]. Further it should be checked that no covariations of the terms (3.62)-(3.65) enter the limit.

The structure of the limit (3.67) is apart from the terms originating from the irregular observation scheme similar to Theorem 3.3 in [40]. However, due to the bivariate setting for example the term $\tilde{m}_{c_s,\tilde{c}_s^{(1)},\tilde{c}_s^{(2)}}(\nabla f)$ has a much more complicated structure compared to the corresponding term in the univariate setting. If f is an even function e.g. $f(x_1,x_2) = |x_1|^{p_1}|x_2|^{p_2}$ we have that ∇f is odd and $m_{c_s}^{(l)}(f,1)$, $m_{c_s}(\nabla f)$, $\tilde{m}_{c_s,\tilde{c}_s^{(1)},\tilde{c}_s^{(2)}}(\nabla f)$ all vanish in that case. Hence we obtain

$$\sqrt{n}\Big(\overline{V}(p,f,\pi_n)_T - \int_0^T m_{c_s}(f)dG_p(s)\Big)$$

$$\xrightarrow{\mathcal{L}-s} \int_0^T \sqrt{m_{c_s}(f^2)G'_{2p}(s) + (m_{c_s}(f))^2\big(-G_pG_{p+1}/G_2 + G_p^2 G_4/G_2^2 \big)'(s)}d\widetilde{W}_s$$

for even functions $f\colon \mathbb{R}^2 \to \mathbb{R}$. This corresponds to Theorem 14.3.2 in [30] as there our functions $G_p(t)$ are equal to the integrals

$$\int_0^t m'_{p/2}\theta_s^{(p+1)/2-1}ds$$

using their notation.

In the case of asynchronous observation times we would assume that both the observation times of $X^{(1)}$ and $X^{(2)}$ are of the form as in Condition 3.10(ii). Next, also in the asynchronous setting it would be reasonable to start with a decomposition similar to (3.62)–(3.65). However, it is not straightforward anymore how the volatility process should be discretized. As seen above this discretization error contributes to the limit and hence the choice of the discretization is very important. Additionally, the summands in $\overline{V}(p,f,\pi_n)_T$ in the asynchronous setting do not contain increments over disjoint intervals which makes a direct application of Proposition C.4 infeasible. Further, especially in the situation of Theorem 2.22 where already the limit of $\overline{V}(p,f,\pi_n)_T$ has a structure which is much more complicated than in the synchronous setting it is to be expected that also the asymptotic variances will have a form which is much more involved. However, putting the technical difficulties sketched above aside it is plausible by the exogeneity of the observation scheme and its specific form assumed in Condition 3.10(ii) that a central limit theorem of some kind should indeed hold.

Remark 3.11. *In this remark we discuss the assumptions made in Condition 3.10: The form of the volatility (3.61) was necessary to estimate an error due to the discretization of σ and is also required in the setting of equidistant deterministic observation times; compare (3.18) in [40]. As we here consider a two-dimensional process X we have to consider a matrix-valued process σ. Note that in Section 14.3.1 of [30] a central limit theorem is only derived for one-dimensional processes stating "the result for k ≥ 2 being much more difficult to prove". Maybe we could also allow the volatility process σ to be a discontinuous Itô semimartingale as in Assumption (K) in Section 14.3.1 of [30].*

Condition 3.10(ii) guarantees the convergence of $G_p^n(t)$ for all $p \geq 0$; compare Lemma 5.3. A weaker assumption which would yield a central limit theorem for $G_p^n(t)$, $t \geq 0$, might also be sufficient. □

Notes. *The methods used in this section follow the arguments in Section 3.1 of [40]. Further, we compared our computations to those leading up to Theorem 14.3.2 in [30]. Like this, we made sure that the structure of the limit (3.67) at least coincides in the more simple cases with the results from Theorem 3.3 in [40] and Theorem 14.3.2 in [30].*

Already in the synchronous setting, we observed that the asymptotic variance in (3.67) has a rather complicated form and should therefore be difficult to estimate. To this end note that the Observed Asymptotic Variance *approach discussed in [39] allows to construct asymptotic confidence intervals also without knowing the explicit form of the limit. There, under the assumption that a central limit theorem of some form holds, the asymptotic variance is estimated via a bootstrapping approach directly from the available observations. This has the remarkable advantage that no knowledge of the form of the central limit theorem is required. Hence, if we believe that a central limit theorem also holds in the asynchronous setting corresponding to the situations in Theorems 2.21 or 2.22 we could use their method to estimate the asymptotic variance. However, to formally apply this procedure we have to prove that a central limit theorem of some kind holds which might not be much easier than computing its specific form.*

4 Estimating Asymptotic Laws

In order to make use of the central limit theorems 3.2 and 3.6 the law of the limiting variables

$$\Phi_T^{univ,(l)}(g) = \sum_{s \leq T} g'(\Delta X_s^{(l)})(\sigma_{s-}^{(l)}(\delta_-^{(l)}(s))^{1/2} U_-^{(l)}(s) + \sigma_s^{(l)}(\delta_+^{(l)}(s))^{1/2} U_+^{(l)}(s))$$

and $\Phi_T^{biv}(f)$ or at least a suitable estimate of their law has to be known to the statistician. Denote by $S_p^{(l)}$ the jump times of $X^{(l)}$ and by S_p the common jump times of $X^{(1)}$ and $X^{(2)}$. If we look through the components making up the limits $\Phi_T^{univ,(l)}(g)$ and $\Phi_T^{biv}(f)$, we find that the jump heights of $X^{(l)}$ respectively X, the values and left limits of $\sigma^{(l)}$ respectively $\sigma^{(1)}, \sigma^2)$, ρ at the jump times and the laws $\Gamma^{univ,(l)}(S_p^{(l)}, dy)$, $\Gamma^{biv}(S_p, dy)$ are necessary to describe the asymptotic law but in general unknown to the statistician.

To estimate the jump heights we may use the threshold technique introduced in Section 2.3 to identify „unusually" large increments as jumps. Further we will devote ourselves to the estimation of $\sigma^{(l)}$, $l = 1, 2$, and ρ at specific times in Chapter 6[1]. Hence it remains the task to find suitable estimates for the laws $\Gamma^{univ,(l)}(S_p^{(l)}, dy)$ and $\Gamma^{biv}(S_p, dy)$. First note that without making additional assumptions on the observation scheme it is impossible to complete this task. This is due to the fact that the laws $\Gamma^{univ,(l)}(s, dy)$, $\Gamma^{biv}(s, dy)$ are dependent on the time variable s and the only information on the asymptotic law $\Gamma^{univ,(l)}(s, dy)$, $\Gamma^{biv}(s, dy)$ which we can extract from the observed data are the realizations $(n\delta_{n,-}^{(l)}(s), n\delta_{n,+}^{(l)}(s))$ respectively $Z_n^{biv}(s)$ whose law converges to $\Gamma^{univ,(l)}(s, dy)$, $\Gamma^{biv}(s, dy)$. Here we are given only a single realization according to the approximate asymptotic law no matter how large n is and hence there is no hope to find a consistent estimator for the asymptotic laws.

To circumvent the issue sketched above we impose an asymptotic local homogeneity assumption on the observation scheme π_n. In particular we require that the laws $\Gamma^{univ,(l)}(s', dy)$, $\Gamma^{biv}(s', dy)$ are close to $\Gamma^{univ,(l)}(s, dy)$, $\Gamma^{biv}(s, dy)$ if s' is close to s. In such a situation we can estimate the laws $\Gamma^{univ,(l)}(s, dy)$, $\Gamma^{biv}(s, dy)$ from the realizations $(n\delta_{n,-}^{(l)}(s'), n\delta_{n,+}^{(l)}(s'))$ respectively $Z_n^{biv}(s')$ for multiple s'

---

[1]Although these estimators are already needed in this chapter, I decided to discuss the estimation of $\sigma_s^{(l)}$ and ρ_s in the applications part because these quantities are also of individual interest in certain applications.

© Springer Fachmedien Wiesbaden GmbH, part of Springer Nature 2019
O. Martin, *High-Frequency Statistics with Asynchronous and Irregular Data*,
Mathematische Optimierung und Wirtschaftsmathematik I Mathematical Optimization
and Economathematics, https://doi.org/10.1007/978-3-658-28418-3_4

which are close to s. We are then able to consistently bootstrap the asymptotic laws using this procedure if we pick s' from a shrinking neighbourhood $[s - \varepsilon_n, s + \varepsilon_n]$ of s such that $\varepsilon_n \to 0$ as $n \to \infty$ but also in such a way that the neighbourhood is large enough such that $(n\delta_{n,-}^{(l)}(s'), n\delta_{n,+}^{(l)}(s_n'))$, $(n\delta_{n,-}^{(l)}(s_n''), n\delta_{n,+}^{(l)}(s_n''))$ respectively $Z_n^{biv}(s_n')$, $Z_n^{biv}(s_n'')$ become asymptotically independent for "typical" s_n', s_n'' chosen from $[s - \varepsilon_n, s + \varepsilon_n]$.

As in Chapter 3 we will use the discussion in this chapter concerning the estimation of the laws $\Gamma^{univ,(l)}(S_p^{(l)}, dy)$, $\Gamma^{biv}(S_p, dy)$ to introduce and analyse general methods which will later be used in the applications in Chapters 7–9 as well.

4.1 The Results

To formalize the bootstrap idea sketched above let $(K_n)_{n \in \mathbb{N}}$ and $(M_n)_{n \in \mathbb{N}}$ denote increasing sequences of natural numbers tending to infinity as $n \to \infty$. As in Section 3.1 we first discuss the univariate case because the notation and the results are structurally simpler in that case. We then define

$$\hat{\delta}_{n,m,-}^{(l)}(s) = n\kappa_{n,m}^{(l)}(s) \big| \mathcal{I}_{i_n^{(l)}(s)+V_{n,m}^{(l)}(s),n}^{(l)} \big|,$$
$$\hat{\delta}_{n,m,+}^{(l)}(s) = n(1 - \kappa_{n,m}^{(l)}(s)) \big| \mathcal{I}_{i_n^{(l)}(s)+V_{n,m}^{(l)}(s),n}^{(l)} \big| \tag{4.1}$$

for $m = 1, \ldots, M_n$ where $\kappa_{n,m}^{(l)}(s)$ denotes a $\mathcal{U}[0,1]$-distributed random variable and $V_{n,m}^{(l)}(s)$ is distributed according to

$$\mathbb{P}(V_{n,m}^{(l)}(s) = k | \mathcal{S}) = \big| \mathcal{I}_{i_n^{(l)}(s)+k,n}^{(l)} \big| \Big(\sum_{k'=-K_n}^{K_n} \big| \mathcal{I}_{i_n^{(l)}(s)+k',n}^{(l)} \big| \Big)^{-1}, \tag{4.2}$$

$k \in \{-K_n, \ldots, K_n\}$. See Figure 4.1 for an illustration. The variables $\kappa_{n,m}^{(l)}(s)$ and $V_{n,m}^{(l)}(s)$ are defined on the same extended probability space $(\widetilde{\Omega}, \widetilde{\mathcal{F}}, \widetilde{\mathbb{P}})$ on which $\Phi_T^{univ,(l)}(g)$ is defined. They are apart from the property (4.2) independent of $\mathcal{F}$ and they are conditionally on $\mathcal{F}$ independent of each other for different m and/or different s. Note that by the above construction $(\hat{\delta}_{n,m,-}^{(l)}(s), \hat{\delta}_{n,m,+}^{(l)}(s))$ is equal to

$$\Big(n\delta_{n,-}^{(l)}\big(t_{i_n^{(l)}(s)+V_{n,m}^{(l)}(s)-1,n}^{(l)} + \kappa_{n,m}^{(l)}(s) \big| \mathcal{I}_{i_n^{(l)}(s)+V_{n,m}^{(l)}(s),n}^{(l)} \big| \big),$$
$$n\delta_{n,+}^{(l)}\big(t_{i_n^{(l)}(s)+V_{n,m}^{(l)}(s)-1,n}^{(l)} + \kappa_{n,m}^{(l)}(s) \big| \mathcal{I}_{i_n^{(l)}(s)+V_{n,m}^{(l)}(s),n}^{(l)} \big| \big) \Big)$$

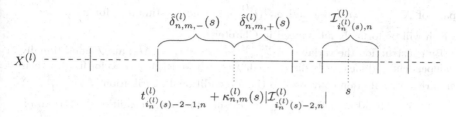

Figure 4.1: Realization of $(\hat{\delta}_{n,m,-}^{(l)}(s), \hat{\delta}_{n,m,+}^{(l)}(s))$ for $V_{n,m}^{(l)}(s) = -2$.

where the argument $t_{i_n^{(l)}(s)+V_{n,m}^{(l)}(s),n}^{(l)} + \kappa_{n,m}^{(l)}(s)\big|\mathcal{I}_{i_n^{(l)}(s)+V_{n,m}^{(l)}(s),n}^{(l)}\big|$ denotes a random variable which is conditionally on $\mathcal{S}$ uniformly distributed on the interval

$$\big[t_{i_n^{(l)}(s)-K_n-1,n}^{(l)}, t_{i_n^{(l)}(s)+K_n,n}^{(l)}\big]. \tag{4.3}$$

Remark 4.1. *The interval (4.3) corresponds to the interval $[s - \varepsilon_n, s + \varepsilon_n]$ used in the introductory sketch and the above construction corresponds to picking s' precisely M_n times uniformly from this interval. To obtain this property we had to weight the probabilities by which $V_{n,m}^{(l)}(s)$ chooses from the $2K_n + 1$ intervals $\mathcal{I}_{i_n^{(l)}(s)+k,n}^{(l)}$, $k \in \{-K_n, \ldots, K_n\}$, surrounding s by the corresponding interval length. This construction relates to Condition 3.1(ii) where the outer integral in (3.17) over the time variables can be interpreted as an expectation with respect to uniform random variables; compare Example 3.3. Also in the limit $\Phi_T^{univ,(l)}(g)$ the variable $(\delta_-^{(l)}(\cdot), \delta_+^{(l)}(\cdot))$ only occurs evaluated at jump times $S_p^{(l)}$ of $N^{(l)}(q)$ and the jump times can under certain conditions be assumed to be uniformly distributed; compare the proof of (3.25).* □

Using the variables defined in (4.1) and i.i.d. standard normal distributed random variables $U_{n,i,m}^{(l),-}, U_{n,i,m}^{(l),+}$ defined on the extended probability space and independent of all previously introduced random variables we define by

$$\widehat{\Phi}_{T,n,m}^{univ,(l)}(g) = \sum_{i:t_{i,n}^{(l)} \leq T} g'(\Delta_{i,n}^{(l)} X^{(l)})\mathbb{1}_{\{|\Delta_{i,n}^{(l)} X^{(l)}| > \beta|\mathcal{I}_{i,n}^{(l)}|^\varpi\}}$$

$$\times \big(\tilde{\sigma}_n^{(l)}(t_{i,n}^{(l)}, -)(\hat{\delta}_{n,m-}^{(l)}(t_{i,n}^{(l)}))^{1/2}U_{n,i,m}^{(l),-} + \tilde{\sigma}_n^{(l)}(t_{i,n}^{(l)}, +)(\hat{\delta}_{n,m,+}^{(l)}(t_{i,n}^{(l)}))^{1/2}U_{n,i,m}^{(l),+}\big),$$

$m = 1, \ldots, M_n$, random variables whose $\mathcal{F}$-conditional distribution approximates the $\mathcal{X}$-conditional distribution of $\Phi_T^{univ,(l)}(g)$. Here, we set $\beta > 0$, $\varpi \in (0, 1/2)$, compare Section 2.3, to separate large increments likely dominated by the jump part of $X^{(l)}$ from small increments likely dominated by the continuous martingale part of $X^{(l)}$. $\tilde{\sigma}_n^{(l)}(t_{i,n}^{(l)}, -)$ and $\tilde{\sigma}_n^{(l)}(t_{i,n}^{(l)}, +)$ denote estimators for $\sigma_{t_{i,n}^{(l)}-}^{(l)}$, $\sigma_{t_{i,n}^{(l)}}^{(l)}$ which will be formally introduced in Chapter 6.

By construction the variables $\widehat{\Phi}_{T,n,m}^{univ,(l)}(g)$, $m = 1, \ldots, M_n$, are $\mathcal{F}$-conditionally independent and identically distributed. Hence if we let $M_n \to \infty$ for fixed n their empirical distribution converges to the $\mathcal{F}$-conditional distribution of $\widehat{\Phi}_{T,n,1}^{univ,(l)}(g)$. On the other hand we defined $\widehat{\Phi}_{T,n,1}^{univ,(l)}(g)$ such that its $\mathcal{F}$-conditional distribution approximates the $\mathcal{X}$-conditional distribution of $\Phi_T^{univ,(l)}(g)$. Note that $\Phi_T^{univ,(l)}(g)$ depends on $\mathcal{F}$ only through $\mathcal{X}$. Hence we can use the empirical distribution on the $\widehat{\Phi}_{T,n,m}^{univ,(l)}(g)$, $m = 1, \ldots, M_n$, to estimate the $\mathcal{X}$-conditional law of the limiting variable $\Phi_T^{univ,(l)}(g)$ which we were interested in from the beginning.

For the construction of asymptotic confidence intervals not the whole distribution of $\Phi_T^{univ,(l)}(g)$ is necessary but only specific quantiles of the distribution. To this end we define

$$\widehat{Q}_{T,n}^{univ,(l)}(g, \alpha) = \widehat{Q}_\alpha(\{\widehat{\Phi}_{T,n,m}^{univ,(l)}(g) | m = 1, \ldots, M_n\})$$

to be the $\lfloor \alpha M_n \rfloor$-largest element of the set $\{\widehat{\Phi}_{T,n,m}^{univ,(l)}(g) | m = 1, \ldots, M_n\}$. Hence by construction $\widehat{Q}_{T,n}^{univ,(l)}(g, \alpha)$ corresponds to the α-quantile of the empirical distribution of the $\widehat{\Phi}_{T,n,m}^{univ,(l)}(g)$, $m = 1, \ldots, M_n$. Then following the above considerations $\widehat{Q}_{T,n}^{univ,(l)}(g, \alpha)$ converges under appropriate conditions to the $\mathcal{X}$-conditional α-quantile $Q_T^{univ,(l)}(g, \alpha)$ of $\Phi_T^{univ,(l)}(g)$ which is defined as the (under certain conditions unique) $\mathcal{X}$-measurable $[-\infty, \infty]$-valued random variable fulfilling

$$\widetilde{\mathbb{P}}(\Phi_T^{univ,(l)}(g) \leq Q_T^{univ,(l)}(g, \alpha) | \mathcal{X})(\omega) = \alpha, \quad \omega \in \Omega. \tag{4.4}$$

In the following condition, we state a precise assumption which yields the local homogeneity that we had to require in the introductory motivation such that our bootstrap procedure is able to consistently estimate the distribution of the limit in Theorem 3.2. Further, some additional structural assumptions have to be made.

Condition 4.2. *Suppose Condition 3.1 is fulfilled, the set $\{s \in [0, T] : g'(\Delta X_s) = 0\}$ is almost surely not empty, the set $\{s \in [0, T] : \sigma_s^{(l)} = 0\}$ is almost surely a Lebesgue null set and $\int_0^T \Gamma^{univ,(l)}(s, \{0\})ds = 0$. Further, let the sequence $(b_n)_{n \in \mathbb{N}}$ used for the estimators $\tilde{\sigma}(\cdot, -)$, $\tilde{\sigma}(\cdot, +)$ fulfil $b_n \to 0$, $|\pi_n|_T / b_n \xrightarrow{\mathbb{P}} 0$ and suppose that $(K_n)_{n \in \mathbb{N}}$ and $(M_n)_{n \in \mathbb{N}}$ are sequences of natural numbers converging to infinity with $K_n / n \to 0$. Additionally:*

(i) It holds

$$\widetilde{\mathbb{P}}\Big(\big|\widetilde{\mathbb{P}}((\hat{\delta}_{n,1,-}^{(l)}(s_p), \hat{\delta}_{n,1,+}^{(l)}(s_p)) \leq x_p, \ p = 1, \dots, P | \mathcal{S})$$

$$- \prod_{p=1}^{P} \widetilde{\mathbb{P}}((\delta_{-}^{(l)}(s_p), \delta_{+}^{(l)}(s_p)) \leq x_p)\big| > \varepsilon\Big) \to 0$$

as $n \to \infty$, for all $\varepsilon > 0$ and any $P \in \mathbb{N}$, $x_p \in \mathbb{R}^2$ and $s_p \in (0, T)$, $p = 1, \dots, P$, with $s_p \neq s_{p'}$ for $p \neq p'$.

(ii) The volatility process $\sigma^{(l)}$ is itself an Itô semimartingale i.e. a process of the form (1.1).

Assuming that the set $\{s \in [0, T] : g'(\Delta X_s) = 0\}$ is almost surely not empty, that the set $\{s \in [0, T] : \sigma_s^{(l)} = 0\}$ is almost surely a Lebesgue null set and that $\int_0^T \Gamma^{univ,(l)}(s, \{0\}) ds = 0$ holds is necessary to guarantee that the limit $\Phi_T^{univ,(l)}(g)$ is almost surely non-singular because on the set where $\Phi_T^{univ,(l)}(g)$ is singular the $\mathcal{X}$-conditional quantile $Q^{univ,(l)}(g, \alpha)$ is not well-defined in the sense of (4.4). Condition 4.2(i) contains the local homogeneity assumption and yields that the $\mathcal{S}$-conditional distribution of $(\hat{\delta}_{n,1,-}^{(l)}(s), \hat{\delta}_{n,1,+}^{(l)}(s))$ converges to $\Gamma^{univ,(l)}(s, dy)$. The sequence $(b_n)_{n \in \mathbb{N}}$ and part (ii) of Condition 4.2 are needed for the consistency of the spot volatility estimators $\tilde{\sigma}^{(l)}(t_{i,n}^{(l)}, -)$, $\tilde{\sigma}^{(l)}(t_{i,n}^{(l)}, +)$.

Under Condition 4.2 we obtain

$$\widehat{Q}_{T,n}^{univ,(l)}(g, \alpha) \overset{\mathbb{P}}{\longrightarrow} Q_T^{univ,(l)}(g, \alpha) \tag{4.5}$$

which is used to obtain the following result.

Theorem 4.3. *Suppose Condition 4.2 is fulfilled and g fulfils the assumptions made in Theorem 3.2. Then we have for $\alpha \in [0, 1]$*

$$\lim_{n \to \infty} \widetilde{\mathbb{P}}\big(\sqrt{n}(V^{(l)}(g, \pi_n)_T - B^{(l)}(g)_T) \leq \widehat{Q}_{T,n}^{univ,(l)}(g, \alpha) | F\big) = \alpha \tag{4.6}$$

for any $F \in \mathcal{X}$ with $\mathbb{P}(F) > 0$.

Here, we are able to consider $\mathcal{X}$-conditional quantiles and to obtain the convergence in (4.6) conditional on sets $F \in \mathcal{X}$ because we are working with the $\mathcal{X}$-stable convergence shown in Theorem 3.2. Using (4.6) we can construct an asymptotic symmetric confidence interval with level α for $B^{(l)}(g)_T$ via

$$[V^{(l)}(g, \pi_n)_T - n^{-1/2}\widehat{Q}_{T,n}^{univ,(l)}(g, (1+\alpha)/2), V^{(l)}(g, \pi_n)_T + n^{-1/2}\widehat{Q}_{T,n}^{univ,(l)}(g, (1+\alpha)/2)]$$

based on the fact that the $\mathcal{F}$-conditional distribution of $\Phi_T^{univ,(l)}(g)$ is symmetrical. Next we consider the bivariate situation where we need to estimate the law of $\Phi_T^{biv}(f)$. Compared to the univariate situation the theory here is a little more complicated because now we have to locate s' in the observation scheme both of $X^{(1)}$ and of $X^{(2)}$ and also because the general structure of the limit in Theorem 3.6 is more complex than in Theorem 3.2. However, the principal ideas are identical and we start again with sequences $(K_n)_{n \in \mathbb{N}}$, $(M_n)_{n \in \mathbb{N}}$ tending to infinity and define for $l = 1, 2$

$$\widehat{\mathcal{L}}_{n,m}^{(l)}(s) = n \Big(\max\{t_{i_n^{(l)}(s)+V_{n,m}^{(l)}(s)-1,n}^{(l)}, t_{i_n^{(3-l)}(s)+V_{n,m}^{(3-l)}(s)-1,n}^{(3-l)}\}$$
$$- t_{i_n^{(3-l)}(s)+V_{n,m}^{(3-l)}(s)-1,n}^{(3-l)} \Big),$$

$$\widehat{\mathcal{R}}_{n,m}^{(l)}(s) = n \Big(t_{i_n^{(3-l)}(s)+V_{n,m}^{(3-l)}(s),n}^{(3-l)}$$
$$- \min\{t_{i_n^{(l)}(s)+V_{n,m}^{(l)}(s),n}^{(l)}, t_{i_n^{(3-l)}(s)+V_{n,m}^{(3-l)}(s),n}^{(3-l)}\} \Big), \quad (4.7)$$

$$\widehat{\mathcal{L}}_{n,m}(s) = n \kappa_{n,m}(s) \big| \mathcal{I}_{i_n^{(1)}(s)+V_{n,m}^{(1)}(s),n}^{(1)} \cap \mathcal{I}_{i_n^{(2)}(s)+V_{n,m}^{(2)}(s),n}^{(2)} \big|,$$

$$\widehat{\mathcal{R}}_{n,m}(s) = n (1 - \kappa_{n,m}(s)) \big| \mathcal{I}_{i_n^{(1)}(s)+V_{n,m}^{(1)}(s),n}^{(1)} \cap \mathcal{I}_{i_n^{(2)}(s)+V_{n,m}^{(2)}(s),n}^{(2)} \big|,$$

$m = 1, \ldots, M_n$, where $\kappa_{n,m}(s)$ denotes a $\mathcal{U}[0,1]$-distributed random variable and $(V_{n,m}^{(1)}(s), V_{n,m}^{(2)}(s))$ is distributed according to

$$\mathbb{P}\big((V_{n,m}^{(1)}(s), V_{n,m}^{(2)}(s)) = (k_1, k_2) \big| \mathcal{S}\big) = \big| \mathcal{I}_{i_n^{(1)}(s)+k_1,n}^{(l)} \cap \mathcal{I}_{i_n^{(2)}(s)+k_2,n}^{(l)} \big|$$

$$\times \Big(\sum_{k_1',k_2'=-K_n}^{K_n} \big| \mathcal{I}_{i_n^{(1)}(s)+k_1',n}^{(l)} \cap \mathcal{I}_{i_n^{(2)}(s)+k_2',n}^{(l)} \big| \Big)^{-1}, \quad k_1, k_2 \in \{-K_n, \ldots, K_n\}.$$

$$(4.8)$$

Again the random variables $\kappa_{n,m}(s)$ and $(V_{n,m}^{(1)}(s), V_{n,m}^{(2)}(s))$ are defined on the extended probability space $(\widetilde{\Omega}, \widetilde{\mathcal{F}}, \widetilde{\mathbb{P}})$ and are apart from the requirement (4.8) independent of $\mathcal{F}$ and of each other for different m and/or different s. See Figure 4.2 for an illustration.

Here, first two intervals $\mathcal{I}_{i_n^{(1)}(s)+k_1,n}^{(l)}$, $\mathcal{I}_{i_n^{(2)}(s)+k_2,n}^{(l)}$ are chosen in the observation scheme of $X^{(1)}$, $X^{(2)}$ with probability proportional to the intersection of these two intervals and then a point in this interval is chosen accordingly to the uniform random variable $\kappa_{n,m}(s)$. Similarly as in the univariate setting stated in Remark 4.1 it holds

$$Z_n^{biv}(U_{n,m}(s)) \overset{\mathcal{L}}{=} (\widehat{\mathcal{L}}_{n,m}^{(1)}(s), \widehat{\mathcal{R}}_{n,m}^{(1)}(s), \widehat{\mathcal{L}}_{n,m}^{(2)}(s), \widehat{\mathcal{R}}_{n,m}^{(2)}(s), \widehat{\mathcal{L}}_{n,m}(s), \widehat{\mathcal{R}}_{n,m}(s))$$
$$=: \widehat{Z}_{n,m}^{biv}(s)$$

for $\mathcal{F}$-conditionally independent random variables $U_{n,m}(s)$, $m = 1, \ldots, M_n$, which are uniformly distributed on the interval

$$\bigcup_{k_1,k_2=-K_n}^{K_n} \left(\mathcal{I}^{(1)}_{i_n^{(1)}(s)+k_1,n} \cap \mathcal{I}^{(2)}_{i_n^{(2)}(s)+k_2,n}\right)$$
$$= \left[t^{(1)}_{i_n^{(1)}(s)-K_n-1,n} \vee t^{(2)}_{i_n^{(2)}(s)-K_n-1,n}, t^{(1)}_{i_n^{(1)}(s)+K_n,n} \wedge t^{(2)}_{i_n^{(2)}(s)+K_n,n}\right].$$

Let $U^{(1),-}_{n,(i,j),m}$, $U^{(1),+}_{n,(i,j),m}$, $U^{(2)}_{n,(i,j),m}$, $U^{(3)}_{n,(i,j),m}$, $U^{(4)}_{n,(i,j),m}$ denote i.i.d. standard normal distributed random variables which are defined on the extended probability space and independent of all previously introduced random variables. Using these random variables and the random variables from (4.7) we proceed similarly as in the univariate setting and define by

$$\widehat{\Phi}^{biv}_{T,n,m}(f) = \sum_{i,j: t^{(1)}_{i,n} \vee t^{(2)}_{j,n} \leq T} \mathbf{1}_{\{|\Delta^{(1)}_{i,n}X^{(1)}| > \beta|\mathcal{I}^{(1)}_{i,n}|^\varpi, |\Delta^{(2)}_{j,n}X^{(2)}| > \beta|\mathcal{I}^{(2)}_{j,n}|^\varpi\}}$$

$$\times \left[\partial_1 f(\Delta^{(1)}_{i,n}X^{(1)}, \Delta^{(2)}_{j,n}X^{(2)})\left(\tilde{\sigma}^{(1)}_n(t^{(1)}_{i,n},-)\sqrt{\widehat{\mathcal{L}}_{n,m}(\tau^n_{i,j})}U^{(1),-}_{n,(i,j),m}\right.\right.$$

$$+ \tilde{\sigma}^{(1)}_n(t^{(1)}_{i,n},+)\sqrt{\widehat{\mathcal{R}}_{n,m}(\tau^n_{i,j})}U^{(1),+}_{n,(i,j),m}$$

$$+ \left.\sqrt{(\tilde{\sigma}^{(1)}_n(t^{(1)}_{i,n},-))^2\widehat{\mathcal{L}}^{(1)}_{n,m}(\tau^n_{i,j}) + (\tilde{\sigma}^{(1)}_n(t^{(1)}_{i,n},+))^2\widehat{\mathcal{R}}^{(1)}_{n,m}(\tau^n_{i,j})}U^{(2)}_{n,(i,j),m}\right)$$

$$+ \partial_2 f(\Delta^{(1)}_{i,n}X^{(1)}, \Delta^{(2)}_{j,n}X^{(2)})\left(\tilde{\sigma}^{(2)}_n(t^{(2)}_{j,n},-)\tilde{\rho}_n(\tau^n_{i,j},-)\sqrt{\widehat{\mathcal{L}}_{n,m}(\tau^n_{i,j})}U^{(1),-}_{n,(i,j),m}\right.$$

$$+ \left.\left.\tilde{\sigma}^{(2)}_n(t^{(2)}_{j,n},+)\tilde{\rho}_n(\tau^n_{i,j},+)\sqrt{\widehat{\mathcal{R}}_{n,m}(\tau^n_{i,j})}U^{(1),+}_{n,(i,j),m}\right.\right.$$

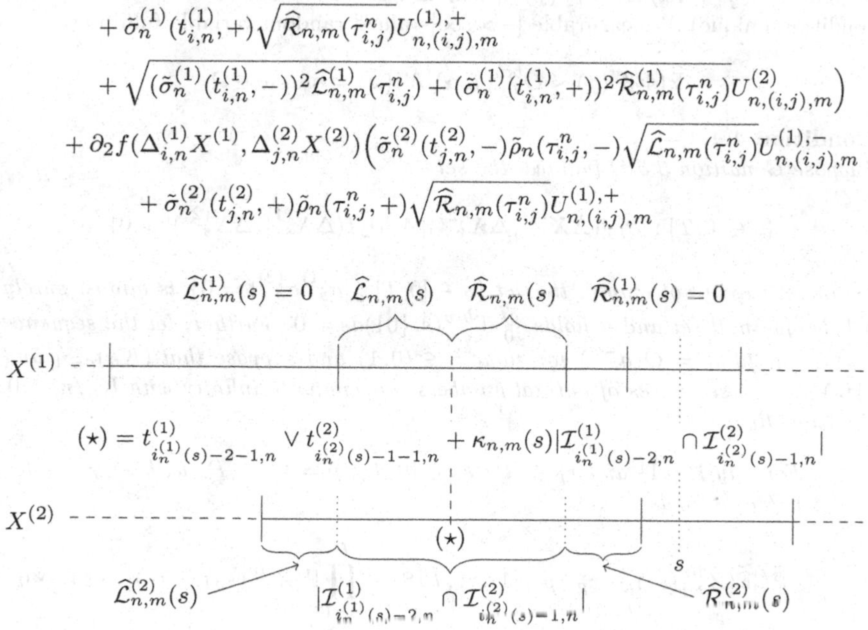

Figure 4.2: Realization of $\widehat{Z}^{biv}_{n,m}(s)$ for $V^{(1)}_{n,m} = -2$, $V^{(2)}_{n,m}(s) = -1$.

$$+ \Big((\tilde{\sigma}_n^{(2)}(t_{j,n}^{(2)}, -))^2 (1 - (\tilde{\rho}_n(\tau_{i,j}^n, -))^2) \widehat{\mathcal{L}}_{n,m}(\tau_{i,j}^n)$$

$$+ (\tilde{\sigma}_n^{(2)}(t_{j,n}^{(2)}, +))^2 (1 - (\tilde{\rho}_n(\tau_{i,j}^n, +))^2) \widehat{\mathcal{R}}_{n,m}(\tau_{i,j}^n) \Big)^{1/2} U_{n,(i,j),m}^{(3)}$$

$$+ \sqrt{(\tilde{\sigma}_n^{(2)}(t_{j,n}^{(2)}, -))^2 \widehat{\mathcal{L}}_{n,m}^{(2)}(\tau_{i,j}^n) + (\tilde{\sigma}_n^{(2)}(t_{j,n}^{(2)}, +))^2 \widehat{\mathcal{R}}_{n,m}^{(2)}(\tau_{i,j}^n)} U_{n,(i,j),m}^{(4)} \Big) \Big],$$

$m = 1, \ldots, M_n$, random variables whose $\mathcal{F}$-conditional distribution approximates the $\mathcal{X}$-conditional distribution of $\Phi_T^{biv}(f)$. Here, we set $\tau_{i,j}^n = t_{i,n}^{(1)} \wedge t_{j,n}^{(2)}$ to shorten notation and $\tilde{\rho}_n(s, -)$, $\tilde{\rho}_n(s, +)$ are consistent estimators for ρ_{s-}, ρ_s which are formally introduced in Chapter 6.

Analogously to the considerations in the univariate situation we define by

$$\widehat{Q}_{T,n}^{biv}(f, \alpha) = \widehat{Q}_\alpha(\{\widehat{\Phi}_{T,n,m}^{biv}(f) | m = 1, \ldots, M_n\})$$

an estimator for the $\mathcal{X}$-conditional α-quantile of $\Phi_T^{biv}(f)$. Again $\widehat{Q}_{T,n}^{biv}(f, \alpha)$ corresponds to the α-quantile of the empirical distribution of the $\widehat{\Phi}_{T,n,m}^{biv}(f)$, $m = 1, \ldots, M_n$.

Then under the upcoming condition $\widehat{Q}_{T,n}^{biv}(f, \alpha)$ converges to the $\mathcal{X}$-conditional α-quantile $Q_T^{biv}(f, \alpha)$ of $\Phi_T^{biv}(f)$ which is defined as the (under the following condition unique) $\mathcal{X}$-measurable $[-\infty, \infty]$-valued random variable fulfilling

$$\widetilde{\mathbb{P}}(\Phi_T^{biv}(f) \le Q_T^{biv}(f, \alpha) | \mathcal{X})(\omega) = \alpha, \quad \omega \in \Omega.$$

Condition 4.4.
Suppose Condition 3.5 is fulfilled, the set

$$\{s \in [0, T] : |\partial_1 f(\Delta X_s^{(1)}, \Delta X_s^{(2)})| + |\partial_2 f(\Delta X_s^{(1)}, \Delta X_s^{(2)})| > 0\}$$

is almost surely not empty, the set $\{s \in [0, T] : |\sigma_s^{(1)} \sigma_s^{(2)}| = 0\}$ is almost surely a Lebesgue null set and it holds $\int_0^T \Gamma^{biv}(s, \{0\}) ds = 0$. Further, let the sequence $(b_n)_{n \in \mathbb{N}}$ fulfil $b_n = O(n^{-\gamma})$ for some $\gamma \in (0, 1)$ and suppose that $(K_n)_{n \in \mathbb{N}}$ and $(M_n)_{n \in \mathbb{N}}$ are sequences of natural numbers converging to infinity with $K_n / n \to 0$. Additionally:

(i) *For any $P \in \mathbb{N}$ and $x_p \in \mathbb{R}^6$, $s_p \in (0, T)$, $p = 1, \ldots, P$, with $s_p \ne s_{p'}$ for $p \ne p'$ it holds*

$$\widetilde{\mathbb{P}}\big(|\widetilde{\mathbb{P}}(\widehat{Z}_{n,1}^{biv}(s_p) \le x_p, \ p = 1, \ldots, P | \mathcal{S}) - \prod_{p=1}^P \widetilde{\mathbb{P}}(Z^{biv}(s_p) \le x_p)| > \varepsilon \big) \to 0$$

as $n \to \infty$, for all $\varepsilon > 0$.

(ii) It holds $(|\pi_n|_T)^{\tilde{\gamma}+\varepsilon} = o_{\mathbb{P}}(n^{-\tilde{\gamma}})$ for any $\tilde{\gamma} > 0$, $\varepsilon > 0$, and we assume $G_{2,0}^n(T) = O_{\mathbb{P}}(1)$, $G_{0,2}^n(T) = O_{\mathbb{P}}(1)$.

(iii) The volatility process σ is itself an $\mathbb{R}^{2\times2}$-valued Itô semimartingale, i.e. a process of similar form as (1.1).

The roles of the specific assumptions in Condition 4.4 are identical to the roles of the corresponding assumptions in Condition 4.2. Part (ii) is an additional structural assumption which is needed in the proof of Theorem 4.5. Under Condition 4.4 we obtain the following result.

Theorem 4.5. Suppose Condition 4.4 is fulfilled and f fulfils the assumptions made in Theorem 3.6 with $p_1^{(1,1)}, p_2^{(2,2)} < 1$. Then we have for $\alpha \in [0,1]$

$$\lim_{n\to\infty} \widetilde{\mathbb{P}}\big(\sqrt{n}(V(f,\pi_n)_T - B^*(f)_T) \le \widehat{Q}_{T,n}^{biv}(f,\alpha)\big|F\big) = \alpha \qquad (4.9)$$

for any $F \in \mathcal{X}$ with $\mathbb{P}(F) > 0$.

Similarly as in the univariate case Theorem 4.5 allows to construct an asymptotic symmetric confidence interval with level α for $B^*(f)_T$ via

$$\big[V(f,\pi_n)_T - n^{-1/2}\widehat{Q}_{T,n}^{biv}(g,(1+\alpha)/2), V(f,\pi_n)_T + n^{-1/2}\widehat{Q}_{T,n}^{biv}(g,(1+\alpha)/2)\big].$$

The assumption $p_1^{(1,1)}, p_2^{(2,2)} < 1$ is not very restrictive. For example in the case $f_{(p,p')}(x,y) = x^p y^{p'}$ we have $\partial_{11} f_{(p,p')}(x,y) = p(p-1)x^{p-2}y^{p'}$, $\partial_{22} f_{(p,p')}(x,y) = p'(p'-1)x^p y^{p'-2}$ and $p_1^{(1,1)}$, $p_2^{(2,2)}$ can be chosen to be zero because we need $p, p' \ge 2$. Hence, usually $p_l^{(l,l)} \le p_{3-l}^{(l,l)}$ which yields $p_l^{(l,l)} < 1$ due to $p_l^{(l,l)} + p_{3-l}^{(l,l)} = 1$. However, this assumption is needed in the proof.

Example 4.6. The assumptions on the observation scheme made in Conditions 4.2 and 4.4 are fulfilled in the setting of equidistant and synchronous observation times $t_{i,n}^{(1)} = t_{i,n}^{(2)} = i/n$. In this specific situation we have seen in Example 3.3 that $(\delta_-^{(l)}(s), \delta_+^{(l)}(s))$ is distributed like $(\kappa^{(l)}(s), 1 - \kappa^{(l)}(s))$ for $\kappa^{(l)}(s) \sim \mathcal{U}[0,1]$. Hence, it holds $\Gamma^{univ,(l)}(s,\{0\}) = 0$ for any $s \in [0,T]$ and because of

$$Z_n^{biv}(s) = (0,0,0,0, n\delta_{n,-}^{(l)}(s), n\delta_{n,+}^{(l)}(s))$$

as derived in Example 3.7 the same holds true for $\Gamma^{biv}(s,\{0\})$.

Further $(\hat{\delta}_{n,1,-}^{(l)}(s_p), \hat{\delta}_{n,1,+}^{(l)}(s_p))$ is distributed like $(\kappa_{n,m}^{(l)}(s_p), 1 - \kappa_{n,m}^{(l)}(s_p))$ for any $s_p \in [0,1]$ because of $|\mathcal{I}_{i,n}^{(l)}| = n^{-1}$ for any $i \in \mathbb{N}$. Hence $(\hat{\delta}_{n,1,-}^{(l)}(s_p), \hat{\delta}_{n,1,+}^{(l)}(s_p))$ has the same distribution as $(\delta_-^{(l)}(s), \delta_+^{(l)}(s))$. Then Condition 4.2(i) is fulfilled *because $(\hat{\delta}_{n,1,-}^{(l)}(s_p), \hat{\delta}_{n,1,+}^{(l)}(s_p))$ and $(\hat{\delta}_{n,1,-}^{(l)}(s_p'), \hat{\delta}_{n,1,+}^{(l)}(s_p'))$ are by construction independent for $s_p \neq s_p'$ as $\kappa_{n,m}^{(l)}(s_p)$ and $\kappa_{n,m}^{(l)}(s_p')$ are independent. These arguments immediately yield that Condition 4.4(i) is fulfilled as well because of $Z_n^{biv}(s) = (0,0,0,0, n\delta_{n,-}^{(l)}(s), n\delta_{n,+}^{(l)}(s))$ as above. Here we observe that in the case of synchronous observations the situation in the bivariate setting is not more complicated than in the univariate setting which intuitively seems reasonable.* □

Example 4.7. *Conditions 4.2 and 4.4 are also fulfilled in the setting of Poisson sampling; compare Definition 5.1. The properties $\Gamma^{univ,(l)}(s,\{0\}) = 0$ and $\Gamma^{biv}(s,\{0\}) = 0$ for any $s \in [0,T]$ follow from the considerations in Remark 5.11 and because $(\hat{\delta}_{n,1,-}^{(l)}(s), \hat{\delta}_{n,1,+}^{(l)}(s))$ and $Z_n^{biv}(s)$ are almost surely different from $0 \in \mathbb{R}^2$ respectively $0 \in \mathbb{R}^4$ for any $n \in \mathbb{N}$.*

That Conditions 4.2(i) and 4.4(i) are fulfilled in the Poisson setting follows from Lemma 5.12. Further part (ii) of Condition 4.4 is fulfilled by (5.2) and Corollary 5.6. □

Remark 4.8. *Instead of choosing realizations $(n\delta_{n,-}^{(l)}(s'), n\delta_{n,+}^{(l)}(s'))$ respectively $Z_n^{biv}(s')$ by picking s' uniformly from the intervals*

$$\left[t_{i_n^{(l)}(s)-K_n-1,n}^{(l)}, t_{i_n^{(l)}(s)+K_n,n}^{(l)} \right],$$

$$\left[t_{i_n^{(1)}(s)-K_n-1,n}^{(1)} \vee t_{i_n^{(2)}(s)-K_n-1,n}^{(2)}, t_{i_n^{(1)}(s)+K_n,n}^{(1)} \wedge t_{i_n^{(2)}(s)+K_n,n}^{(2)} \right]$$

we might also set

$$\left(\hat{\delta}_{n,m,-}^{(l)}(s), \hat{\delta}_{n,m,+}^{(l)}(s)\right) = \left(n\delta_{n,-}^{(l)}(\tilde{\kappa}_{n,m}(s)), n\delta_{n,+}^{(l)}(\tilde{\kappa}_{n,m}(s))\right),$$

$$\widehat{Z}_{n,m}^{biv}(s) = Z_n^{biv}(\tilde{\kappa}_{n,m}(s))$$

for i.i.d. random variables $\tilde{\kappa}_{n,m}(s)$, $m = 1, \ldots, M_n$ which are uniformly distributed on the interval $[s - \varepsilon_n, s + \varepsilon_n]$ for some sequence $(\varepsilon_n)_{n\in\mathbb{N}}$ with $\varepsilon_n \to 0$ and $n\varepsilon_n \to \infty$ as $n \to \infty$, compare the requirement on K_n in Condition 4.2. In this chapter, I chose to discuss the approach where s' is picked from a fixed number of observation intervals because the resulting procedure is easier to implement for practical computations. □

4.2 The Proofs

Conditions 4.2(i) and 4.4(i) yield that the empirical laws on the sets

$$\{(\hat{\delta}_{n,m,-}^{(l)}(s), \hat{\delta}_{n,m,+}^{(l)}(s)) | m = 1, \ldots, M_n\}, \quad \{Z_n^{biv}(s) | m = 1, \ldots, M_n\}$$

converge to the laws of $(\delta_-^{(l)}(s), \delta_+^{(l)}(s))$ and $Z^{biv}(s)$. The following lemma will be the key tool for showing that this convergence already implies the convergence of the empirical laws on $\{\widehat{\Phi}_{T,n,m}^{univ,(l)}(g) | m = 1, \ldots, M_n\}$ and $\{\widehat{\Phi}_{T,n,m}^{biv}(f) | m = 1, \ldots, M_n\}$ to the $\mathcal{X}$-conditional laws of $\Phi_T^{univ,(l)}(g)$ and $\Phi_T^{biv}(f)$. First, we prove a more general statement which then can also be used in Chapters 7–9 for proving similar results in the applications part.

Lemma 4.9. *Suppose we have $A_{n,p} \xrightarrow{\widetilde{\mathbb{P}}} A_p$ for $\mathcal{F}$-measurable random variables $A_{n,p} \in \mathbb{R}^d$, $\mathcal{X}$-measurable random variables $A_p \in \mathbb{R}^d$, and let $S_p \in [0,T]$, $p = 1, \ldots, P$, $p \in \mathbb{N}$, be almost surely distinct $\mathcal{X}$-measurable random variables. For $n \in \mathbb{N}$, $p = 1, \ldots, P$ and $s \in [0,T]$ let $\widehat{Z}_{n,m}^p(s)$, $m = 1, \ldots, M_n$, be $\mathbb{R}^{d'}$-valued random variables which are independent of $\mathcal{X}$, $\mathcal{S}$-conditionally independent and whose $\mathcal{S}$-conditional distributions are identical. Further let $Z^p(s)$, $p = 1, \ldots, P$, be $\mathbb{R}^{d'}$-valued random variables which are independent of $\mathcal{X}$ as well and suppose that for any $x = (x_1, \ldots, x_p) \in \mathbb{R}^{d \times P}$ and any $\varepsilon > 0$ it holds*

$$\widetilde{\mathbb{P}}\Big(\big|\widetilde{\mathbb{P}}(\widehat{Z}_{n,1}^p(s_p) \le x_p, \, p = 1, \ldots, P | \mathcal{S}) - \prod_{p=1}^P \widetilde{\mathbb{P}}(Z^p(s_p) \le x_p)\big| > \varepsilon\Big) \to 0 \quad (4.10)$$

as $n \to \infty$. Then it holds

$$\widetilde{\mathbb{P}}\Big(\Big|\frac{1}{M_n} \sum_{m=1}^{M_n} \mathbb{1}_{\{\varphi((A_{n,p}, \widehat{Z}_{n,m}^p(S_p))_{p=1,\ldots,P}) \le \Upsilon\}} - \widetilde{\mathbb{P}}\big(\varphi((A_p, Z^p(S_p))_{p=1,\ldots,P}) \le \Upsilon | \mathcal{X}\big)\Big| > \varepsilon\Big) \to 0$$

for any $\mathcal{X}$-measurable random variable Υ, any $\varepsilon > 0$ and any continuous function $\varphi : \mathbb{R}^{(d+d') \times P} \to \mathbb{R}$ such that the $\mathcal{X}$-conditional distribution of the random variable $\varphi((A_p, Z^p(S_p))_{p=1,\ldots,P})$ is almost surely continuous.

Proof. First, note that (4.10) implies that

$$\widetilde{\mathbb{P}}\Big(\Big|\frac{1}{M_n} \sum_{m=1}^{M_n} \mathbb{1}_{\{\widehat{Z}_{n,m}^p(s_p) \le x_p, \, p=1,\ldots,P\}} - \widetilde{\mathbb{P}}(Z^p(s_p) \le x_p, \, p = 1, \ldots, P)\Big| > \varepsilon\Big) \quad (4.11)$$

$$\le \widetilde{\mathbb{P}}\Big(\Big|\frac{1}{M_n} \sum_{m=1}^{M_n} \mathbb{1}_{\{\widehat{Z}_{n,m}^p(s_p) \le x_p, \, p=1,\ldots,P\}} - \widetilde{\mathbb{P}}(Z_{n,1}^p(s_p) \le x_p, \, p = 1, \ldots, P | \mathcal{S})\Big| > \frac{\varepsilon}{2}\Big)$$

$$+ \widetilde{\mathbb{P}}\Big(\big|\widetilde{\mathbb{P}}(\widehat{Z}_{n,1}^p(s_p) \le x_p, \, p = 1, \ldots, P | \mathcal{S}) - \widetilde{\mathbb{P}}(Z^p(s_p) \le x_p, \, p = 1, \ldots, P)\big| > \frac{\varepsilon}{2}\Big)$$

converges to zero as $n \to \infty$ for any s_p, x_p, $p = 1, \ldots, P$. In fact, $M_n \to \infty$, the conditional Chebyshev inequality and dominated convergence ensure that the first term vanishes asymptotically because the $(\widehat{Z}_{n,m}^p(s_p))_{p=1,\ldots,P}$ are $\mathcal{S}$-conditionally independent. The second term converges to zero by (4.10).

To shorten notation we set

$$\zeta_n = \frac{1}{M_n} \sum_{m=1}^{M_n} \mathbb{1}_{\{\varphi((A_{n,p}, \widehat{Z}_{n,m}^p(S_p))_{p=1,\dots,P}) \leq \Upsilon\}},$$

$$\zeta = \widetilde{\mathbb{P}}\big(\varphi((A_p, Z^p(S_p))_{p=1,\dots,P}) \leq \Upsilon \,|\, \mathcal{X}\big).$$

The idea for the following steps is to approximate the function φ by piecewise constant functions, use (4.11) to prove the claim for those piecewise constant functions and to show that the convergence is preserved if we take limits. To formalize this approach let $K > 0$, set

$$\square_k(K,r) = \{x \in \mathbb{R}^{d' \times P} \,|\, x_{i,p} \in ((k_{i,p}-1)2^{-r}K, k_{i,p}2^{-r}K], i \leq d', p \leq P\},$$

$k = (k_1, \dots, k_P) \in \mathbb{Z}^{d' \times P}$, $r \in \mathbb{N}$, and define

$\zeta_n(K,r)$

$$= \frac{1}{M_n} \sum_{m=1}^{M_n} \mathbb{1}_{\{\sum_{k \in \{-2^r,\dots,2^r\}^{d' \times P}} \varphi((A_p, k_p 2^{-r}K)_{p \leq P})\mathbb{1}_{\{(\widehat{Z}_{n,m}^p(S_p))_{p \leq P}) \in \square_k(K,r)\}} \leq \Upsilon\}}$$

$$= \sum_{k \in \{-2^r,\dots,2^r\}^{d' \times P}} \mathbb{1}_{\{\varphi((A_p, k_p 2^{-r}K)_{p \leq P}) \leq \Upsilon\}} \frac{1}{M_n} \sum_{m=1}^{M_n} \mathbb{1}_{\{(\widehat{Z}_{n,m}^p(S_p))_{p \leq P}) \in \square_k(K,r)\}},$$

$\zeta(K,r)$

$$= \sum_{k \in \{-2^r,\dots,2^r\}^{d' \times P}} \mathbb{1}_{\{\varphi((A_p, k_p 2^{-r}K)_{p \leq P}) \leq \Upsilon\}} \widetilde{\mathbb{P}}\big((Z^p(S_p))_{p \leq P}) \in \square_k(K,r) \,|\, \mathcal{X}\big),$$

where $\varphi((A_p, k_p 2^{-r}K)_{p \leq P})$ equals $\varphi((A_p, \cdot)_{p=1,\dots,P})$ evaluated at the rightmost vertex of $\square_k(K,r)$.

Using this notation it remains to show

$$\lim_{K \to \infty} \limsup_{r \to \infty} \limsup_{n \to \infty} \widetilde{\mathbb{P}}(|\zeta_n - \zeta_n(K,r)| > \varepsilon) = 0 \quad \forall \varepsilon > 0, \tag{4.12}$$

$$\lim_{n \to \infty} \widetilde{\mathbb{P}}(|\zeta_n(K,r) - \zeta(K,r)| > \varepsilon) = 0 \quad \forall K, \varepsilon > 0 \,\forall r \in \mathbb{N}, \tag{4.13}$$

$$\lim_{K \to \infty} \limsup_{r \to \infty} \widetilde{\mathbb{P}}(|\zeta(K,r) - \zeta| > \varepsilon) = 0 \quad \forall \varepsilon > 0. \tag{4.14}$$

Step 1. We start by showing (4.13). It holds

$$\widetilde{\mathbb{P}}(|\zeta_n(K,r) - \zeta(K,r)| > \varepsilon)$$

$$\leq \sum_{k \in \{-2^r,\dots,2^r\}^{d' \times P}} \mathbb{E}\bigg[\widetilde{\mathbb{P}}\bigg(\bigg|\frac{1}{M_n} \sum_{m=1}^{M_n} \mathbb{1}_{\{(\widehat{Z}_{n,m}^p(S_p))_{p=1,\dots,P}) \in \square_k(K,r)\}}$$

$$- \widetilde{\mathbb{P}}\big((Z^p(S_p))_{p=1,\dots,P}) \in \square_k(K,r) \,|\, \mathcal{X}\big)\bigg| > \varepsilon/(2^{r+1}+1)^{d'P} \,\bigg|\, \mathcal{X}\bigg)\bigg]$$

where each conditional probability vanishes almost surely as $n \to \infty$ by (4.11) because the events

$$\{(\widehat{Z}_{n,m}^p(S_p))_{p=1,\ldots,P}) \in \square_k(K,r)\}, \quad \{(Z^p(S_p))_{p=1,\ldots,P}) \in \square_k(K,r)\}$$

may be written as unions/differences of events of the form

$$\{\widehat{Z}_{n,m}^p(S_p))_{p=1,\ldots,P}) \leq v_{k,i}(K,r)\}, \quad \{Z^p(S_p))_{p=1,\ldots,P}) \leq v_{k,i}(K,r)\}$$

where $v_{k,i}(K,r)$, $i = 1, \ldots, 2^{d'P}$, denote the vertices of the cuboid $\square_k(K,r)$. Note that conditioning on $\mathcal{X}$ here simply has the effect of fixing the S_p. (4.13) then follows by dominated convergence.

Step 2. Next we show (4.12). It holds

$$|\zeta_n - \zeta_n(K,r)|$$

$$\leq \frac{1}{M_n} \sum_{m=1}^{M_n} \mathbb{1}_{\{(\widehat{Z}_{n,m}^p(S_p))_{p=1,\ldots,P} \notin [-K,K]^{d' \times P} \vee (A_p)_{p=1,\ldots,P} \notin [-K,K]^{d \times P}\}}$$

$$+ \sum_{k \in \{-2^r,\ldots,2^r\}^{d' \times P}} \Big| \mathbb{1}_{\{\varphi((A_{n,p}, \widehat{Z}_{n,m}^p(S_p))_{p=1,\ldots,P}) \leq \Upsilon\}}$$

$$- \mathbb{1}_{\{\varphi((A_p, k_p 2^{-r}K)_{p=1,\ldots,P}) \leq \Upsilon\}} \Big|$$

$$\times \frac{1}{M_n} \sum_{m=1}^{M_n} \mathbb{1}_{\{(\widehat{Z}_{n,m}^p(S_p))_{p=1,\ldots,P}) \in \square_k(K,r)\}} \mathbb{1}_{\{(A_p)_{p=1,\ldots,P} \in [-K,K]^{d \times P}\}}. \quad (4.15)$$

The first term in (4.15) becomes arbitrarily small as first $n \to \infty$ and then $K \to \infty$, because for $n \to \infty$ we obtain from (4.11)

$$\frac{1}{M_n} \sum_{m=1}^{M_n} \mathbb{1}_{\{(\widehat{Z}_{n,m}^p(S_p))_{p=1,\ldots,P} \notin [-K,K]^{d' \times P}\}}$$

$$\xrightarrow{\widetilde{\mathbb{P}}} \widetilde{\mathbb{P}}((Z^p(S_p))_{p=1,\ldots,P} \notin [-K,K]^{d' \times P} | \mathcal{X})$$

as in Step 1, where the right-hand side vanishes as $K \to \infty$.

Denote the second term in (4.15) by $\zeta_n'(K,r)$. Then it holds for $\delta > 0$

$$\zeta_n'(K,r) \qquad\qquad\qquad\qquad\qquad\qquad\qquad\qquad\qquad (4.16)$$

$$\leq \mathbb{1}_{\{\|(A_{n,p} - A_p)_{p=1,\ldots,P}\| \geq \delta\}} + \frac{1}{M_n} \sum_{m=1}^{M_n} \mathbb{1}_{\{(\widehat{Z}_{n,m}^p(S_p))_{p=1,\ldots,P} \in [-K,K]^{d' \times P}\}}$$

$$\times \mathbb{1}_{\{|\varphi((A_p, \widehat{Z}_{n,m}^p(S_p))_{p=1,\ldots,P}) - \Upsilon| \leq \rho(K+\delta, \delta, 2^{-r}K)\}} \mathbb{1}_{\{(A_p)_{p=1,\ldots,P} \in [-K,K]^{d \times P}\}}$$

where $\rho(K, a, b)$ is defined as

$$\sup_{(x,y),(x',y') \in [-K,K]^{(d+d') \times P} : \|x-x'\| < a, \|y-y'\|_\infty \leq b} |\varphi((x_p, y_p)_{p=1,\ldots,P}) - \varphi((x_p', y_p')_{p=1,\ldots,P})|.$$

The first summand in (4.16) vanishes as $n \to \infty$ for all $\delta > 0$ since $A_{n,p} \xrightarrow{\text{P}} A_p$. Denoting the second summand in (4.16) by $\zeta_n''(K, r, \delta)$ we further obtain

$$
\begin{aligned}
&\zeta_n''(K, r, \delta) \\
&\leq \frac{1}{M_n} \sum_{m=1}^{M_n} \sum_{k \in \{-2^r, \ldots, 2^r\}^{d' \times P}} \mathbb{1}_{\{\min_{x \in \Box_k(K,r)} |\varphi((A_p, x_p)_{p=1,\ldots,P}) - \Upsilon| \leq \rho(K+\delta, \delta, 2^{-r}K)\}} \\
&\qquad\qquad\qquad\qquad \times \mathbb{1}_{\{(\widehat{Z}_{n,m}^p(S_p))_{p=1,\ldots,P} \in \Box_k(K,r)\}} \mathbb{1}_{\{(A_p)_{p=1,\ldots,P} \in [-K,K]^{d \times P}\}} \\
&\leq \sum_{k \in \{-2^r, \ldots, 2^r\}^{d' \times P}} \Big| \frac{1}{M_n} \sum_{m=1}^{M_n} \mathbb{1}_{\{(\widehat{Z}_{n,m}^p(S_p))_{p=1,\ldots,P} \in \Box_k(K,r)\}} \\
&\qquad\qquad\qquad\qquad\qquad - \widetilde{\mathbb{P}}\big((Z^p(S_p))_{p=1,\ldots,P} \in \Box_k(K,r) \big| \mathcal{X}\big) \Big| \\
&\quad + \sum_{k \in \{-2^r, \ldots, 2^r\}^{d' \times P}} \mathbb{1}_{\{\min_{x \in \Box_k(K,r)} |\varphi((A_p, x_p)_{p=1,\ldots,P}) - \Upsilon| \leq \rho(K+\delta, \delta, 2^{-r}K)\}} \\
&\qquad\qquad \times \widetilde{\mathbb{P}}\big((Z^p(S_p))_{p=1,\ldots,P} \in \Box_k(K,r) \big| \mathcal{X}\big) \mathbb{1}_{\{(A_p)_{p=1,\ldots,P} \in [-K,K]^{d \times P}\}} \quad (4.17)
\end{aligned}
$$

where the first sum vanishes for $n \to \infty$ as shown in Step 1. Denote the second sum in (4.17) by $\zeta_n'''(K, r, \delta)$. Then we finally obtain

$$
\zeta_n'''(K, r, \delta) \leq \widetilde{\mathbb{P}}\big(|\varphi((A_p, Z^p(S_p))_{p=1,\ldots,P}) - \Upsilon| \leq 2\rho(K+\delta, \delta, 2^{-r}K) \big| \mathcal{X}\big)
$$

which converges to zero because $\varphi((A_p, Z^p(S_p))_{p=1,\ldots,P})$ possesses almost surely a continuous $\mathcal{X}$-conditional distribution by assumption and because of

$$
\lim_{\delta \to 0} \limsup_{r \to \infty} \rho(K+\delta, \delta, 2^{-r}K) = 0
$$

for all $K > 0$ as φ is continuous. Hence altogether we have shown

$$
\lim_{K \to \infty} \limsup_{\delta \to 0} \limsup_{r \to \infty} \limsup_{n \to \infty} \widetilde{\mathbb{P}}(|\zeta_n'(K, r)| > \varepsilon)
$$

for all $\varepsilon > 0$ which yields (4.12).

Step 3. It holds

$$
\zeta(K, r) = \widetilde{\mathbb{P}}\Big(\sum_{k \in \{-2^r, \ldots, 2^r\}^{d' \times P}} \varphi((A_p, k_p 2^{-r}K)_{p=1,\ldots,P}) \mathbb{1}_{\{(Z^p(S_p))_{p=1,\ldots,P} \in \Box_k(K,r)\}} \leq \Upsilon \Big| \mathcal{X} \Big).
$$

Hence

$$
\begin{aligned}
|\zeta(K, r) - \zeta| &\leq \widetilde{\mathbb{P}}\big((Z^p(S_p))_{p=1,\ldots,P} \notin [-K,K]^{d' \times P} \big| \mathcal{X}\big) \\
&\quad + \widetilde{\mathbb{P}}\big(|\varphi((A_p, Z^p(S_p))_{p=1,\ldots,P}) - \Upsilon| \leq \tilde{\rho}(K, r, (A_p)_{p=1,\ldots,P}) \big| \mathcal{X}\big) \quad (4.18)
\end{aligned}
$$

where $\tilde{\rho}(K, r, (A_p)_{p=1,\ldots P})$ is defined as

$$
\sup_{y, y' \in [-K,K]^{d' \times P} : \|y - y'\| \leq 2^{-r}K} |\varphi((A_p, y_p)_{p=1,\ldots,P}) - \varphi((A_p, y_p')_{p=1,\ldots,P})|.
$$

The first term on the right-hand side of (4.18) vanishes almost surely as $K \to \infty$. Further it holds

$$\lim_{r \to \infty} \tilde{\rho}(K, r, (A_p)_{p=1,\dots,P}) = 0 \quad \text{almost surely}$$

because $y \mapsto \varphi((A_p, y_p)_{p=1,\dots,P})$ is uniformly continuous on $[-K, K]^{d' \times P}$ for fixed ω. Using this result the second term in (4.18) vanishes almost surely as $r \to \infty$ for any $K > 0$ because the $\mathcal{X}$-conditional distribution of $\varphi((A_p, Z^p(S_p))_{p=1,\dots,P})$ is almost surely continuous by assumption. (4.14) then follows by dominated convergence. $\qquad\square$

The following proposition shows that Condition 4.2 yields that the empirical distribution on $\{\widehat{\Phi}_{T,n,m}^{univ,(l)}(g) | m = 1, \dots, M_n\}$ converges to the $\mathcal{X}$-conditional distribution of $\Phi_T^{univ,(l)}(g)$.

Proposition 4.10. *Suppose that Condition 4.2 is satisfied. Then it holds*

$$\widetilde{\mathbb{P}}\Big(\Big|\frac{1}{M_n} \sum_{m=1}^{M_n} \mathbb{1}_{\{\widehat{\Phi}_{T,n,m}^{univ,(l)}(g) \leq \Upsilon\}} - \mathbb{P}(\Phi_T^{univ,(l)}(g) \leq \Upsilon | \mathcal{X})\Big| > \varepsilon\Big) \to 0 \quad (4.19)$$

for any $\mathcal{X}$-measurable random variable Υ and all $\varepsilon > 0$.

Proof. Step 1. We denote by $S_{q,p}^{(l)}$, $p \in \mathbb{N}$, an enumeration of the jump times of $N^{(l)}(q)$ in $[0, T]$. Here we choose the enumeration such that $(S_{q,p}^{(l)})_{p \in \mathbb{N}}$ is an increasing sequence which is possible because the process $N(q)$ has finite jump activity. For using 4.9 we set

$$A_{n,p} = \Big(g'(\Delta_{i_n^{(l)}(S_{q,p}^{(l)}),n}^{(l)} X^{(l)}) \mathbb{1}_{\{|\Delta_{i_n^{(l)}(S_{q,p}^{(l)}),n}^{(l)} X^{(l)}| > \beta |\mathcal{I}_{i_n^{(l)}(S_{q,p}^{(l)}),n}^{(l)}|^{\varpi}\}}$$

$$, \tilde{\sigma}_n^{(l)}(S_{q,p}^{(l)}, -), \tilde{\sigma}_n^{(l)}(S_{q,p}^{(l)}, +)\Big),$$

$$A_p = \Big(g'(\Delta X_{S_{q,p}^{(l)}}^{(l)}), \sigma_{S_{q,p}^{(l)}-}^{(l)}, \sigma_{S_{q,p}^{(l)}}^{(l)}\Big),$$

$$\widehat{Z}_{n,m}^p(s) = \Big(\hat{\delta}_{n,m,-}^{(l)}(s), \hat{\delta}_{n,m,+}^{(l)}(s), U_{n,i_n^{(l)}(s),m}^{(l),-}, U_{n,i_n^{(l)}(s),m}^{(l),+}\Big),$$

$$Z^p(s) = \Big(\delta_-^{(l)}(s), \delta_+^{(l)}(s), U_-^{(l)}(s), U_+^{(l)}(s)\Big)$$

and define φ via

$$\varphi((A_p, Z^p(S_{q,p}^{(l)}))_{p=1,\dots,P})$$

$$= \sum_{p=1}^{P} g'(\Delta X_{S_{q,p}^{(l)}}^{(l)})\Big(\sigma_{S_{q,p}^{(l)}-}^{(l)} (\delta_-^{(l)}(s))^{1/2} U_-^{(l)}(s) + \sigma_{S_{q,p}^{(l)}}^{(l)} (\delta_+^{(l)}(s))^{1/2} U_+^{(l)}(s)\Big).$$

By Condition 4.2(ii) $A_{n,p} \xrightarrow{\mathbb{P}} A_p$ follows from Corollary 6.4 and (4.10) holds because of Condition 4.2(i) and because $U^{(l),-}_{n,i_n^{(l)}(s),m}$, $U^{(l),+}_{n,i_n^{(l)}(s),m}$ respectively $U^{(l)}_-(s)$, $U^{(l)}_+(s)$ are standard normal distributed and independent of $\hat{\delta}^{(l)}_{n,m,-}(s)$, $\hat{\delta}^{(l)}_{n,m,+}(s)$ respectively $\delta^{(l)}_-(s)$, $\delta^{(l)}_+(s)$. Further the assumptions made in Condition 4.2 guarantee that the $\mathcal{X}$-conditional distribution of $\varphi((A_p, Z^p(S^{(l)}_{q,p}))_{p=1,\dots,P})$ is almost surely continuous for q and P large enough and Lemma 4.9 then proves

$$\widetilde{\mathbb{P}}\left(\left\{\left|\frac{1}{M_n}\sum_{m=1}^{M_n}\mathbb{1}_{\{Y(P,n,m)\leq\Upsilon\}} - \widetilde{\mathbb{P}}\big(Y(P)\leq\Upsilon\big|\mathcal{X}\big)\right| > \varepsilon\right\}\right) \to 0 \qquad (4.20)$$

where

$$
Y(P,n,m) = \sum_{p=1}^{P} g'\big(\Delta^{(l)}_{i_n^{(l)}(S^{(l)}_{q,p}),n}X^{(l)}\big)\mathbb{1}_{\{|\Delta^{(l)}_{i_n^{(l)}(S^{(l)}_{q,p}),n}X^{(l)}|>\beta|\mathcal{I}^{(l)}_{i_n^{(l)}(S^{(l)}_{q,p}),n}|^{\varpi}\}}
$$
$$
\times \mathbb{1}_{\{S^{(l)}_{q,p}\leq T\}}\big(\tilde{\sigma}^{(l)}_n(S^{(l)}_{q,p},-)(\hat{\delta}^{(l)}_{n,m,-}(S^{(l)}_{q,p}))^{1/2}U^{(l),-}_{n,i_n^{(l)}(S^{(l)}_{q,p}),m}
$$
$$
+ \tilde{\sigma}^{(l)}_n(S^{(l)}_{q,p},+)(\hat{\delta}^{(l)}_{n,m,+}(S^{(l)}_{q,p}))^{1/2}U^{(l),+}_{n,i_n^{(l)}(S^{(l)}_{q,p}),m}\big),
$$

$$
Y(P) = \sum_{p=1}^{P} g'\big(\Delta X^{(l)}_{S^{(l)}_{q,p}}\big)\big(\sigma^{(l)}_{S^{(l)}_{q,p}-}(\delta^{(l)}_-(S^{(l)}_{q,p}))^{1/2}U^{(l)}_-(S^{(l)}_{q,p})
$$
$$
+ \sigma^{(l)}_{S^{(l)}_{q,p}}(\delta^{(l)}_+(S^{(l)}_{q,p}))^{1/2}U^{(l)}_+(S^{(l)}_{q,p})\big)\mathbb{1}_{\{S^{(l)}_{q,p}\leq T\}}.
$$

Step 2. Next we prove

$$\lim_{P\to\infty}\limsup_{n\to\infty}\frac{1}{M_n}\sum_{m=1}^{M_n}\widetilde{\mathbb{P}}\big(|Y(P,n,m)-\widehat{\Phi}^{univ,(l)}_{T,n,m}(g)|>\varepsilon\big) = 0 \qquad (4.21)$$

for all $\varepsilon > 0$. Denote by $\Omega^{(l)}(P,r,q,n)$ the set on which there are at most P jumps of $N^{(l)}(q)$ in $[0,T]$, two different jumps of $N^{(l)}(q)$ are further apart than $|\pi_n|_T$ and we have $\Delta^{(l)}_{i,n}(X^{(l)} - N^{(l)}(q)) \leq 1/r$ for all $i \in \mathbb{N}$ with $t^{(l)}_{i,n} \leq T$. Obviously,

$$\lim_{q\to\infty}\limsup_{P\to\infty}\limsup_{n\to\infty}\mathbb{P}(\Omega^{(l)}(P,r,q,n)) = 1$$

for any $r > 0$. On the set $\Omega^{(l)}(P,r,q,n)$ we have

$$\mathbb{E}\big[|Y(P,n,m) - \widehat{\Phi}^{univ,(l)}_{T,n,m}|\,\big|\mathcal{S}\big]\mathbb{1}_{\Omega^{(l)}(P,r,q,n)}$$
$$\leq \sum_{i:t^{(l)}_{i,n}\leq T,\nexists p:S^{(l)}_{q,p}\in\mathcal{I}^{(l)}_{i,n}}\mathbb{E}\big[K_r(\Delta^{(l)}_{i,n}(X^{(l)} - N^{(l)}(q)))^2\mathbb{1}_{\{|\Delta^{(l)}_{i,n}X^{(l)}|>\beta|\mathcal{I}^{(l)}_{i,n}|^{\varpi}\}}}$$

$$\times \big(\tilde{\sigma}_n^{(l)}(t_{i,n}^{(l)},-)(\hat{\delta}_{n,m,-}^{(l)}(t_{i,n}^{(l)}))^{1/2}U_{n,i,m}^{(l),-}$$
$$+ \tilde{\sigma}_n^{(l)}(t_{i,n}^{(l)},+)(\hat{\delta}_{n,m,+}^{(l)}(t_{i,n}^{(l)}))^{1/2}U_{n,i,m}^{(l),+}\big)\big|\mathcal{S}\big]$$
$$\leq \sum_{i:t_{i,n}^{(l)}\leq T} \mathbb{E}\big[K_r(\Delta_{i,n}^{(l)}(X^{(l)}-N^{(l)}(q)))^2\mathbb{1}_{\{|\Delta_{i,n}^{(l)}X^{(l)}|>\beta|\mathcal{I}_{i,n}^{(l)}|^\varpi\}}$$
$$\times \Big(\frac{1}{b_n}\sum_{j\neq i:\mathcal{I}_{j,n}^{(l)}\subset(t_{i,n}^{(l)}-b_n,t_{i,n}^{(l)}+b_n]}(\Delta_{j,n}^{(l)}X^{(l)})^2\Big)^{1/2}$$
$$\times (\hat{\delta}_{n,m,-}^{(l)}(t_{i,n}^{(l)})+\hat{\delta}_{n,m,+}^{(l)}(t_{i,n}^{(l)}))^{1/2}\big|\mathcal{S}\big]. \tag{4.22}$$

Using iterated expectations, Lemma 1.4 and inequality (2.77) we obtain

$$\mathbb{E}\big[(\Delta_{i,n}^{(l)}(X^{(l)}-N^{(l)}(q)))^2\mathbb{1}_{\{|\Delta_{i,n}^{(l)}X^{(l)}|>\beta|\mathcal{I}_{i,n}^{(l)}|^\varpi\}}$$
$$\times \Big(\frac{1}{b_n}\sum_{j\neq i:\mathcal{I}_{j,n}^{(l)}\subset(t_{i,n}^{(l)}-b_n,t_{i,n}^{(l)}+b_n]}(\Delta_{j,n}^{(l)}X^{(l)})^2\Big)^{1/2}\big|\mathcal{S}\big]$$
$$\leq \mathbb{E}\big[(\Delta_{i,n}^{(l)}(X^{(l)}-N^{(l)}(q)))^2\mathbb{1}_{\{|\Delta_{i,n}^{(l)}X^{(l)}|>\beta|\mathcal{I}_{i,n}^{(l)}|^\varpi\}}$$
$$\times \Big(\big(\frac{1}{b_n}\sum_{j\neq i:\mathcal{I}_{j,n}^{(l)}\subset(t_{i,n}^{(l)}-b_n,t_{i,n}^{(l)})}(\Delta_{j,n}^{(l)}X^{(l)})^2\big)^{1/2}$$
$$+ \big(\frac{1}{b_n}\sum_{j\neq i:\mathcal{I}_{j,n}^{(l)}\subset(t_{i,n}^{(l)},t_{i,n}^{(l)}+b_n]}(\Delta_{j,n}^{(l)}X^{(l)})^2\big)^{1/2}\big)\big|\mathcal{S}\big]$$
$$\leq K(K_q|\pi_n|_T + (|\pi_n|_T)^{1/2-\varpi}+e_q)|\mathcal{I}_{i,n}^{(l)}|.$$

Hence as $(\hat{\delta}_{n,m,-}^{(l)}(t_{i,n}^{(l)})+\hat{\delta}_{n,m,+}^{(l)}(t_{i,n}^{(l)}))^{1/2}$ is conditionally on $\mathcal{S}$ independent of all other random variables occuring in the conditional expectation we can bound (4.22) by

$$K(K_q|\pi_n|_T + (|\pi_n|_T)^{1/2-\varpi}+e_q)$$
$$\times \sum_{i:t_{i,n}^{(l)}\leq T}|\mathcal{I}_{i,n}^{(l)}|\mathbb{E}\big[(\hat{\delta}_{n,m,-}^{(l)}(t_{i,n}^{(l)})+\hat{\delta}_{n,m,+}^{(l)}(t_{i,n}^{(l)}))^{1/2}\big|\mathcal{S}\big]$$
$$\leq (K_q(|\pi_n|_T)^{1/2-\varpi}+e_q)\sum_{i:t_{i,n}^{(l)}\leq T}|\mathcal{I}_{i,n}^{(l)}|\mathbb{E}\big[(\hat{\delta}_{n,m,-}^{(l)}(t_{i,n}^{(l)})+\hat{\delta}_{n,m,+}^{(l)}(t_{i,n}^{(l)}))^{1/2}\big|\mathcal{S}\big]$$

Further by resorting the sum we obtain

$$(K_q(|\pi_n|_T)^{1/2-\varpi}+e_q)\sum_{i:t_{i,n}^{(l)}\leq T}|\mathcal{I}_{i,n}^{(l)}|\mathbb{E}\big[(\hat{\delta}_{n,m,-}^{(l)}(t_{i,n}^{(l)})+\hat{\delta}_{n,m,+}^{(l)}(t_{i,n}^{(l)}))^{1/2}\big|\mathcal{S}\big]$$

$$= (K_q(|\pi_n|_T)^{1/2-\varpi} + e_q)\sqrt{n} \sum_{i:t_{i,n}^{(l)}\leq T} |\mathcal{I}_{i,n}^{(l)}| \sum_{k=-K_n}^{K_n} |\mathcal{I}_{i+k,n}^{(l)}|^{3/2}$$

$$\times \left(\sum_{k'=-K_n}^{K_n} |\mathcal{I}_{i+k',n}^{(l)}| \right)^{-1}$$

$$= (K_q(|\pi_n|_T)^{1/2-\varpi} + e_q)\sqrt{n} \sum_{i:t_{i,n}^{(l)}\leq T} |\mathcal{I}_{i,n}^{(l)}|^{3/2} \sum_{k=-K_n}^{K_n} |\mathcal{I}_{i+k,n}^{(l)}|$$

$$\times \left(\sum_{k'=-K_n}^{K_n} |\mathcal{I}_{i+k+k',n}^{(l)}| \right)^{-1}$$

$$\leq (K_q(|\pi_n|_T)^{1/2-\varpi} + e_q)\sqrt{n} \sum_{i:t_{i,n}^{(l)}\leq T} |\mathcal{I}_{i,n}^{(l)}|^{3/2}$$

$$\times \left(\sum_{k=-K_n}^{0} |\mathcal{I}_{i+k,n}^{(l)}|\left(\sum_{k'=-K_n}^{0} |\mathcal{I}_{i+k',n}^{(l)}| \right)^{-1} + \sum_{k=0}^{K_n} |\mathcal{I}_{i+k,n}^{(l)}|\left(\sum_{k'=0}^{K_n} |\mathcal{I}_{i+k',n}^{(l)}| \right)^{-1} \right)$$

$$= 2(K_q(|\pi_n|_T)^{1/2-\varpi} + e_q)G_3^{(l),n}(T). \tag{4.23}$$

Hence (4.22) vanishes as first $n \to \infty$ and then $q \to \infty$ by Condition 3.1(i) and we have proven (4.21).

Step 3. Using dominated convergence, we obtain $Y(P) \xrightarrow{\tilde{\mathbb{P}}} \Phi_T^{univ,(l)}(g)$ as $P \to \infty$. Also, as the $\mathcal{X}$-conditional distribution of $\Phi_T^{univ,(l)}(g)$ is continuous by Condition 4.2, for any choice of $\varepsilon, \eta > 0$ there exists $\delta > 0$ such that

$$\tilde{\mathbb{P}}\big(|\tilde{\mathbb{P}}(\Phi_T^{univ,(l)}(g) \leq \Upsilon|\mathcal{X}) - \tilde{\mathbb{P}}(\Phi_T^{univ,(l)}(g) \pm \delta \leq \Upsilon|\mathcal{X})| > \eta\big) < \varepsilon.$$

Then it is easy to deduce that

$$\tilde{\mathbb{P}}\big(Y(P) \leq \Upsilon|\mathcal{X}\big) \xrightarrow{\tilde{\mathbb{P}}} \tilde{\mathbb{P}}\big(\Phi_T^{univ,(l)}(g) \leq \Upsilon|\mathcal{X}\big) \tag{4.24}$$

holds for $P \to \infty$.

Step 4. For any $\varepsilon > 0$ we have

$$\tilde{\mathbb{E}}\big[|\frac{1}{M_n} \sum_{m=1}^{M_n} \mathbb{1}_{\{Y(P,n,m)\leq \Upsilon\}} - \frac{1}{M_n} \sum_{m=1}^{M_n} \mathbb{1}_{\{\hat{\Phi}_{T,n,m}^{univ,(l)}(g)\leq \Upsilon\}}|\big]$$

$$\leq \tilde{\mathbb{E}}\big[\frac{1}{M_n} \sum_{m=1}^{M_n} \mathbb{1}_{\{|Y(P,n,m)-\hat{\Phi}_{T,n,m}^{univ,(l)}(g)|\geq |Y(P,n,m)-\Upsilon|\}}\big]$$

$$\leq \widetilde{\mathbb{E}}\Big[\frac{1}{M_n}\sum_{m=1}^{M_n}\big(\mathbb{1}_{\{|Y(P,n,m)-\widehat{\Phi}_{T,n,m}^{univ,(l)}(g)|>\varepsilon\}}+\mathbb{1}_{\{|Y(P,n,m)-\Upsilon|\leq\varepsilon\}}\big)\Big].$$

By (4.20) and dominated convergence we obtain

$$\widetilde{\mathbb{E}}\Big[\frac{1}{M_n}\sum_{m=1}^{M_n}\mathbb{1}_{\{|Y(P,n,m)-\Upsilon|\leq\varepsilon\}}\Big]\to\widetilde{\mathbb{P}}(|Y(P)-\Upsilon|\leq\varepsilon), \qquad (4.25)$$

where the right-hand side tends to zero as $\varepsilon\to 0$ using dominated convergence again, because the $\mathcal{X}$-conditional distribution of $Y(P)$ is continuous while Υ is $\mathcal{X}$-measurable. By (4.21) we also have

$$\lim_{P\to\infty}\limsup_{n\to\infty}\widetilde{\mathbb{E}}\Big[\frac{1}{M_n}\sum_{m=1}^{M_n}\mathbb{1}_{\{|Y(P,n,m)-\widehat{\Phi}_{T,n,m}^{univ,(l)}(g)|>\varepsilon\}}\Big]=0 \qquad (4.26)$$

for all $\varepsilon>0$. Thus, using (4.25) and (4.26), we obtain

$$\lim_{P\to\infty}\limsup_{n\to\infty}\widetilde{\mathbb{E}}\Big(\Big|\frac{1}{M_n}\sum_{m=1}^{M_n}\big(\mathbb{1}_{\{Y(P,n,m)\leq\Upsilon\}}-\mathbb{1}_{\{\widehat{\Phi}_{T,n,m}^{univ,(l)}(g)\leq\Upsilon\}}\big)\Big|>\varepsilon\Big)=0$$

$$(4.27)$$

for all $\varepsilon>0$.

Step 5. Finally the claim (4.19) follows from (4.20), (4.24) and (4.27). $\qquad\square$

Proof of Theorem 4.3. Step 1. We first prove (4.5). To this end note that we have for arbitrary $\varepsilon>0$

$$\widetilde{\mathbb{P}}\big(\widehat{Q}_{T,n}^{univ,(l)}(g,\alpha)>Q_T^{univ,(l)}(g,\alpha)+\varepsilon\big)$$

$$=\widetilde{\mathbb{P}}\Big(\frac{1}{M_n}\sum_{m=1}^{M_n}\mathbb{1}_{\{\widehat{\Phi}_{T,n,m}^{univ,(l)}(g)>Q_T^{univ,(l)}(g,\alpha)+\varepsilon\}}>\frac{M_n-(\lfloor\alpha M_n\rfloor-1)}{M_n}\Big)$$

$$\leq\widetilde{\mathbb{P}}\Big(\frac{1}{M_n}\sum_{m=1}^{M_n}\mathbb{1}_{\{\widehat{\Phi}_{T,n,m}^{univ,(l)}(g)>Q_T^{univ,(l)}(g,\alpha)+\varepsilon\}}-\Upsilon(\alpha,\varepsilon)>(1-\alpha)-\Upsilon(\alpha,\varepsilon)\Big)$$

with $\Upsilon(\alpha,\varepsilon)=\widetilde{\mathbb{P}}\big(\Phi_T^{univ,(l)}(g)>Q_T^{univ,(l)}(g,\alpha)+\varepsilon\,|\,\mathcal{X}\big)$. Because the $\mathcal{X}$-conditional distribution of $\Phi_T^{univ,(l)}(g)$ is continuous by Condition 4.2, we have $\Upsilon(\alpha,\varepsilon)<1-\alpha$ almost surely. Then it is easy to deduce

$$\widetilde{\mathbb{P}}\big(\widehat{Q}_{T,n}^{univ,(l)}(g,\alpha)>Q_T^{univ,(l)}(g,\alpha)+\varepsilon\big)\to 0$$

using Proposition 4.10. Similarly we get

$$\widetilde{\mathbb{P}}(\widehat{Q}_{T,n}^{univ,(l)}(g,\alpha) < Q_T^{univ,(l)}(g,\alpha) - \varepsilon) \to 0$$

and combining these two convergences yields (4.5).

Step 2. From Theorem 3.2 and (4.5) we obtain using Proposition B.7(i)

$$(\widehat{Q}_{T,n}^{univ,(l)}(g,\alpha), \sqrt{n}(V^{(l)}(g,\pi_n)_T - B^{(l)}(g)_T)) \overset{\mathcal{L}-s}{\Longrightarrow} (Q_T^{univ,(l)}(g,\alpha), \Phi_T^{univ,(l)}(g)).$$

From this property we obtain using the definition of $\mathcal{X}$-stable convergence, compare (B.4), and (4.4)

$$\begin{aligned}
\widetilde{\mathbb{E}}[\mathbf{1}_F \mathbf{1}_{\{\sqrt{n}(V^{(l)}(g,\pi_n)_T - B^{(l)}(g)_T \leq \widehat{Q}_{T,n}^{univ,(l)}(g,\alpha)\}}] \\
\to \widetilde{\mathbb{E}}[\mathbf{1}_F \mathbf{1}_{\{\Phi_T^{univ,(l)}(g) \leq Q_T^{univ,(l)}(g,\alpha)\}}] \\
= \widetilde{\mathbb{E}}[\mathbf{1}_F \widetilde{\mathbb{P}}(\Phi_T^{univ,(l)}(g) \leq Q_T^{univ,(l)}(g,\alpha)|\mathcal{X})] = \mathbb{P}(F)\alpha
\end{aligned}$$

which is equivalent to (4.6).   □

As for the proof of Theorem 4.3 to prove Theorem 4.5 we will first derive a proposition which will yield that Condition 4.4 implies the convergence of the empirical distribution on the set $\{\widehat{\Phi}_{T,n,m}^{biv}(f)|m = 1,\dots,M_n\}$ to the $\mathcal{X}$-conditional distribution of $\Phi_T^{biv}(f)$.

Proposition 4.11. *Suppose that Condition 4.4 is satisfied. Then it holds*

$$\widetilde{\mathbb{P}}(|\frac{1}{M_n}\sum_{m=1}^{M_n}\mathbf{1}_{\{\widehat{\Phi}_{T,n,m}^{biv}(f)\leq \Upsilon\}} - \widetilde{\mathbb{P}}(\Phi_T^{biv}(f) \leq \Upsilon|\mathcal{X})| > \varepsilon) \to 0 \qquad (4.28)$$

for any $\mathcal{X}$-measurable random variable Υ and all $\varepsilon > 0$.

Proof. Denote by $S_{q,p}$, $p \in \mathbb{N}$, an increasing sequence of stopping times which exhausts the common jump times of $N(q)$ and define

$$\begin{aligned}
Y(P,n,m) = \sum_{p=1}^{P} \mathbf{1}_{\{|\Delta_{i_p,n}^{(1)}X^{(1)}|>\beta|\mathcal{I}_{i_p,n}^{(1)}|^{\varpi}, |\Delta_{j_p,n}^{(2)}X^{(2)}|>\beta|\mathcal{I}_{j_p,n}^{(2)}|^{\varpi}\}}\mathbf{1}_{\{S_{q,p}\leq T\}} \\
\times \Big[\partial_1 f(\Delta_{i_p,n}^{(1)}X^{(1)}, \Delta_{j_p,n}^{(2)}X^{(2)})\Big(\tilde{\sigma}_n^{(1)}(t_{i_p,n}^{(1)},-)\sqrt{\widehat{\mathcal{L}}_{n,m}(\tau_p^n)}U_{n,(i_p,j_p),m}^{(1),-} \\
+ \tilde{\sigma}_n^{(1)}(t_{i_p,n}^{(1)},+)\sqrt{\widehat{\mathcal{R}}_{n,m}(\tau_p^n)}U_{n,(i_p,j_p),m}^{(1),+} \\
+ \sqrt{(\tilde{\sigma}_n^{(1)}(t_{i_p,n}^{(1)},-))^2\widehat{\mathcal{L}}_{n,m}^{(1)}(\tau_p^n) + (\tilde{\sigma}_n^{(1)}(t_{i_p,n}^{(1)},+))^2\widehat{\mathcal{R}}_{n,m}^{(1)}(\tau_p^n)}U_{n,(i_p,j_p),m}^{(2)}\Big)
\end{aligned}$$

$$+ \partial_2 f(\Delta_{i_p,n}^{(1)} X^{(1)}, \Delta_{j_p,n}^{(2)} X^{(2)}) \Big(\tilde{\sigma}_n^{(2)}(t_{j_p,n}^{(2)}, -) \tilde{\rho}_n(\tau_p^n, -) \sqrt{\widehat{\mathcal{L}}_{n,m}(\tau_p^n)} U_{n,(i_p,j_p),m}^{(1),-}$$

$$+ \tilde{\sigma}_n^{(2)}(t_{j_p,n}^{(2)}, +) \tilde{\rho}_n(\tau_p^n, +) \sqrt{\widehat{\mathcal{R}}_{n,m}(\tau_p^n)} U_{n,(i_p,j_p),m}^{(1),+}$$

$$+ \Big((\tilde{\sigma}_n^{(2)}(t_{j_p,n}^{(2)}, -))^2 (1 - (\tilde{\rho}_n(\tau_p^n, -))^2) \widehat{\mathcal{L}}_{n,m}(\tau_p^n)$$

$$+ (\tilde{\sigma}_n^{(2)}(t_{j_p,n}^{(2)}, +))^2 (1 - (\tilde{\rho}_n(\tau_p^n, +))^2) \widehat{\mathcal{R}}_{n,m}(\tau_p^n) \Big)^{1/2} U_{n,(i_p,j_p),m}^{(3)}$$

$$+ \sqrt{(\tilde{\sigma}_n^{(2)}(t_{j_p,n}^{(2)}, -))^2 \widehat{\mathcal{L}}_{n,m}^{(2)}(\tau_p^n) + (\tilde{\sigma}_n^{(2)}(t_{j_p,n}^{(2)}, +))^2 \widehat{\mathcal{R}}_{n,m}^{(2)}(\tau_p^n)} U_{n,(i_p,j_p),m}^{(4)} \Big) \Big]$$

where we set $i_p = i_n^{(1)}(S_{q,p})$, $j_p = i_n^{(2)}(S_{q,p})$ and $\tau_p^n = t_{i_p,n}^{(1)} \wedge t_{j_p,n}^{(2)}$. Further define

$$Y(P) = \sum_{p=1}^{P} \Big[\partial_1 f(\Delta X_{S_{q,p}}^{(1)}, \Delta X_{S_{q,p}}^{(2)}) \Big(\sigma_{S_{q,p}-}^{(1)} \sqrt{\mathcal{L}(S_{q,p})} U_{S_{q,p}}^{(1),-}$$

$$+ \sigma_{S_{q,p}}^{(1)} \sqrt{\mathcal{R}(S_{q,p})} U_{S_{q,p}}^{(1),+}$$

$$+ \sqrt{(\sigma_{S_{q,p}-}^{(1)})^2 \mathcal{L}^{(1)}(S_{q,p}) + (\sigma_{S_{q,p}}^{(1)})^2 \mathcal{R}^{(1)}(S_{q,p})} U_{S_{q,p}}^{(2)} \Big)$$

$$+ \partial_2 f(\Delta X_{S_{q,p}}^{(1)}, \Delta X_{S_{q,p}}^{(2)}) \Big(\sigma_{S_{q,p}-}^{(2)} \rho_{S_{q,p}-} \sqrt{\mathcal{L}(S_{q,p})} U_{S_{q,p}}^{(1),-}$$

$$+ \sigma_{S_{q,p}}^{(2)} \rho_{S_{q,p}} \sqrt{\mathcal{R}(S_{q,p})} U_{S_{q,p}}^{(1),+}$$

$$+ \sqrt{(\sigma_{S_{q,p}-}^{(2)})^2 (1 - (\rho_{S_{q,p}-})^2) \mathcal{L}(S_{q,p}) + (\sigma_{S_{q,p}}^{(2)})^2 (1 - (\rho_{S_{q,p}})^2) \mathcal{R}(S_{q,p})} U_{S_{q,p}}^{(3)}$$

$$+ \sqrt{(\sigma_{S_{q,p}-}^{(2)})^2 \mathcal{L}^{(2)}(S_{q,p}) + (\sigma_{S_{q,p}}^{(2)})^2 \mathcal{R}^{(2)}(S_{q,p})} U_{S_{q,p}}^{(4)} \Big) \Big] \mathbb{1}_{\{S_{q,p} \leq T\}}.$$

Step 1. By choosing $A_{n,p}$, A_p, $\widehat{Z}_{n,m}^p(s)$, $Z^p(s)$ and the function φ appropriately, compare Step 1 in the proof of Proposition 4.10, we obtain the following convergence from Condition 4.4 using Lemma 4.9

$$\lim_{n \to \infty} \widetilde{\mathbb{P}}(\{ \Big| \frac{1}{M_n} \sum_{m=1}^{M_n} \mathbb{1}_{\{Y(P,n,m) \leq \Upsilon\}} - \widetilde{\mathbb{P}}(Y(P) \leq \Upsilon | \mathcal{X}) \Big| > \varepsilon) = 0 \quad (4.29)$$

for any $P \in \mathbb{N}$.

Step 2. Next we prove

$$\lim_{P \to \infty} \limsup_{n \to \infty} \frac{1}{M_n} \sum_{m=1}^{M_n} \widetilde{\mathbb{P}}(\big| Y(P,n,m) - \widehat{\Phi}_{T,n,m}^{biv}(f) \big| > \varepsilon) = 0. \quad (4.30)$$

Denote by $\Omega(P, r, q, n)$ the set on which there are at most P common jumps of $N(q)$ in $[0, T]$, two different jumps of $N(q)$ are further apart than $|\pi_n|_T$ and we have $\Delta_{i,n}^{(l)}(X^{(l)} - N^{(l)}(q)) \leq 1/r$, $l = 1, 2$, for all $i \in \mathbb{N}$ with $t_{i,n}^{(l)} \leq T$. Obviously, $\mathbb{P}(\Omega(P, r, q, n)) \to 1$ for $P, n, q \to \infty$ and any $r > 0$. On the set $\Omega(P, r, q, n)$ we have

$$
\mathbb{E}\big[|Y(P, n, m) - \widehat{\Phi}_{T,n,m}^{biv}(f)|\big|\mathcal{F}\big]\mathbf{1}_{\Omega(P,r,q,n)}
$$

$$
\leq K \sum_{i,j:t_{i,n}^{(1)}\vee t_{j,n}^{(2)}\leq T, \nexists p:S_{q,p}\in\mathcal{I}_{i,n}^{(1)}\cap\mathcal{I}_{j,n}^{(2)}} \mathbf{1}_{\{|\Delta_{i,n}^{(1)}X^{(1)}|>\beta|\mathcal{I}_{i,n}^{(1)}|^{\varpi}, |\Delta_{j,n}^{(2)}X^{(2)}|>\beta|\mathcal{I}_{j,n}^{(2)}|^{\varpi}\}}
$$

$$
\times \Big[|\partial_1 f(\Delta_{i,n}^{(1)}X^{(1)}, \Delta_{j,n}^{(2)}X^{(2)})|\tilde{\sigma}_n^{(1)}(i,n)\sqrt{(\widehat{\mathcal{L}}_{n,m} + \widehat{\mathcal{R}}_{n,m} + \widehat{\mathcal{L}}_{n,m}^{(1)} + \widehat{\mathcal{R}}_{n,m}^{(1)})(\tau_{i,j}^n)}
$$

$$
+ |\partial_2 f(\Delta_{i,n}^{(1)}X^{(1)}, \Delta_{j,n}^{(2)}X^{(2)})|\tilde{\sigma}_n^{(2)}(j,n)\sqrt{(\widehat{\mathcal{L}}_{n,m} + \widehat{\mathcal{R}}_{n,m} + \widehat{\mathcal{L}}_{n,m}^{(2)} + \widehat{\mathcal{R}}_{n,m}^{(2)})(\tau_{i,j}^n)}\Big]
$$

$$
\times \mathbf{1}_{\{\mathcal{I}_{i,n}^{(1)}\cap\mathcal{I}_{j,n}^{(2)}\neq\emptyset\}}\mathbf{1}_{\Omega(P,r,q,n)}
$$

$$
\leq K_r \sum_{i,j:t_{i,n}^{(1)}\vee t_{j,n}^{(2)}\leq T, \nexists p:S_{q,p}\in\mathcal{I}_{i,n}^{(1)}\cap\mathcal{I}_{j,n}^{(2)}} \mathbf{1}_{\{|\Delta_{i,n}^{(1)}X^{(1)}|>\beta|\mathcal{I}_{i,n}^{(1)}|^{\varpi}, |\Delta_{j,n}^{(2)}X^{(2)}|>\beta|\mathcal{I}_{j,n}^{(2)}|^{\varpi}\}}
$$

$$
\times \Big[|\Delta_{i,n}^{(1)}X^{(1)}|^{1+p_1^{(1,1)}}|\Delta_{j,n}^{(2)}X^{(2)}|^{p_2^{(1,1)}}\tilde{\sigma}_n^{(1)}(i,n)
$$

$$
\times \sqrt{(\widehat{\mathcal{L}}_{n,m} + \widehat{\mathcal{R}}_{n,m} + \widehat{\mathcal{L}}_{n,m}^{(1)} + \widehat{\mathcal{R}}_{n,m}^{(1)})(\tau_{i,j}^n)}
$$

$$
+ |\Delta_{i,n}^{(1)}X^{(1)}|^{p_1^{(2,2)}}|\Delta_{j,n}^{(2)}X^{(2)}|^{1+p_2^{(2,2)}}\tilde{\sigma}_n^{(2)}(j,n)
$$

$$
\times \sqrt{(\widehat{\mathcal{L}}_{n,m} + \widehat{\mathcal{R}}_{n,m} + \widehat{\mathcal{L}}_{n,m}^{(2)} + \widehat{\mathcal{R}}_{n,m}^{(2)})(\tau_{i,j}^n)}\Big]
$$

$$
\times \mathbf{1}_{\{\mathcal{I}_{i,n}^{(1)}\cap\mathcal{I}_{j,n}^{(2)}\neq\emptyset\}}\mathbf{1}_{\Omega(P,r,q,n)}
$$

$$
\leq K_r \sum_{i_1,i_2:t_{i_1,n}^{(1)}\vee t_{i_2,n}^{(2)}\leq T} \mathbf{1}_{\{|\Delta_{i_1,n}^{(1)}X^{(1)}|>\beta|\mathcal{I}_{i_1,n}^{(1)}|^{\varpi}, |\Delta_{i_2,n}^{(2)}X^{(2)}|>\beta|\mathcal{I}_{i_2,n}^{(2)}|^{\varpi}\}}
$$

$$
\sum_{l=1,2} \tilde{\sigma}_n^{(l)}(i_l,n)\sqrt{(\widehat{\mathcal{L}}_{n,m} + \widehat{\mathcal{R}}_{n,m} + \widehat{\mathcal{L}}_{n,m}^{(l)} + \widehat{\mathcal{R}}_{n,m}^{(l)})(\tau_{i_1,i_2}^n)}
$$

$$
\times \Big[|\Delta_{i_l,n}^{(l)}(X - N(q))^{(l)}|^{1+p_l^{(l,l)}}|\Delta_{i_{3-l},n}^{(3-l)}X^{(3-l)}|^{p_{3-l}^{(l,l)}}
$$

$$
+ |\Delta_{i_l,n}^{(l)}X^{(l)}|^{1+p_l^{(l,l)}}|\Delta_{i_{3-l},n}^{(3-l)}(X - N(q))^{(3-l)}|^{p_{3-l}^{(l,l)}}\Big]\mathbf{1}_{\{\mathcal{I}_{i_1,n}^{(l)}\cap\mathcal{I}_{i_{3-l},n}^{(3-l)}\neq\emptyset\}} \qquad (4.31)
$$

where we applied the arguments following (3.55) to bound $\partial_l f(\ldots)$, $l = 1, 2$, and used the shorthand notation $\tilde{\sigma}^{(l)}(i,n) = ((\tilde{\sigma}^{(l)}(t_{i,n}^{(l)}, -))^2 + ((\tilde{\sigma}^{(l)}(t_{i,n}^{(l)}, +))^2)^{1/2}$. To distinguish increments over the overlapping interval parts and the increments over the parts which do not overlap we denote by $\Delta_{(j,i),n}^{(3-l,l)}X$ the increment of X over the interval $\mathcal{I}_{i,n}^{(l)} \cap \mathcal{I}_{j,n}^{(3-l)}$ and by $\Delta_{(j,i),n}^{(3-l\setminus l)}X$ the increment of X over $\mathcal{I}_{j,n}^{(3-l)} \setminus \mathcal{I}_{i,n}^{(l)}$ (which might be the sum of the increments over two separate intervals). Then we obtain using Muirhead's inequality as in (2.25), iterated expectations (note that $\tilde{\sigma}^{(l)}(i_l, n)$ does not contain the increment over $\mathcal{I}_{i_l,n}^{(l)}$), Lemma (1.4) and inequality (2.77)

$$\mathbb{E}\big[\tilde{\sigma}_n^{(l)}(i_l, n)|\Delta_{(i_1,i_2),n}^{(1,2)}(X - N(q))^{(l)}|^{1+p_l^{(l,l)}}\frac{\varepsilon}{\varepsilon}|\Delta_{(i_1,i_2),n}^{(1,2)}X^{(3-l)}|^{p_{3-l}^{(l,l)}}$$

$$\times \mathbb{1}_{\{|\Delta_{i_l,n}^{(l)}X^{(l)}|>\beta|\mathcal{I}_{i_l,n}^{(l)}|^{\varpi}\}}\big|\mathcal{S}\big]$$

$$\leq \mathbb{E}\big[\tilde{\sigma}_n^{(l)}(i_l, n)\big(|\Delta_{(i_1,i_2),n}^{(1,2)}(X - N(q))^{(l)}|^2\varepsilon^{-2/(1+p_l^{(l,l)})}\mathbb{1}_{\{|\Delta_{i_l,n}^{(l)}X^{(l)}|>\beta|\mathcal{I}_{i_l,n}^{(l)}|^{\varpi}\}}$$

$$+ |\Delta_{(i_1,i_2),n}^{(1,2)}X^{(3-l)}|^2\varepsilon^{2/p_{3-l}^{(l,l)}}\big)\big|\mathcal{S}\big]$$

$$\leq K\big((K_q|\mathcal{I}_{i_1,n}^{(1)} \cap \mathcal{I}_{i_2,n}^{(2)}| + |\mathcal{I}_{i_l,n}^{(l)}|^{1/2-\varpi} + e_q)\varepsilon^{-2/(1+p_l^{(l,l)})} + K\varepsilon^{2/p_{3-l}^{(l,l)}}\big)$$

$$\times |\mathcal{I}_{i_1,n}^{(1)} \cap \mathcal{I}_{i_2,n}^{(2)}|.$$

Note that $p_{3-l}^{(l,l)} > 0$ because we assumed $p_l^{(l,l)} < 1$ and $p_l^{(l,l)} + p_{3-l}^{(l,l)} = 1$. Analogously the terms with $|\Delta_{i_l,n}^{(l)}X^{(l)}|^{1+p_l^{(l,l)}}|\Delta_{i_3-l,n}^{(3-l)}(X - N(q))^{(3-l)}|^{p_{3-l}^{(l,l)}}$ can be bounded by a similar expression. Hence the $\mathcal{S}$-conditional expectation of the last bound in (4.31) where we replace the increments over $\mathcal{I}_{i_1,n}^{(1)}$, $\mathcal{I}_{i_2,n}^{(2)}$ with increments over $\mathcal{I}_{i_1,n}^{(1)} \cap \mathcal{I}_{i_2,n}^{(2)}$ is less or equal than

$$K_r\theta(n,q,\varepsilon) \sum_{i,j:t_{i,n}^{(1)}\vee t_{j,n}^{(2)}\leq T} |\mathcal{I}_{i,n}^{(1)} \cap \mathcal{I}_{j,n}^{(2)}| \sum_{l=1,2} \mathbb{E}[\sqrt{(\widehat{\mathcal{L}}_{n,m} + \widehat{\mathcal{R}}_{n,m} + \widehat{\mathcal{L}}_{n,m}^{(l)} + \widehat{\mathcal{R}}_{n,m}^{(l)})(\tau_{i,j}^n)}|\mathcal{S}]$$

For some random variable $\theta(n,q,\varepsilon)$ with

$$\lim_{\varepsilon\to 0}\lim\sup_{q\to\infty}\lim\sup_{n\to\infty}\mathbb{P}(|\theta(n,q,\varepsilon)| > \delta) = 0, \quad \forall\delta > 0. \tag{4.32}$$

Further we obtain using a similar index change as in (4.23)

$$\sum_{i,j:t_{i,n}^{(1)}\vee t_{j,n}^{(2)}\leq T} |\mathcal{I}_{i,n}^{(1)} \cap \mathcal{I}_{j,n}^{(2)}| \sum_{l=1,2} \mathbb{E}[\sqrt{(\widehat{\mathcal{L}}_{n,m} + \widehat{\mathcal{R}}_{n,m} + \widehat{\mathcal{L}}_{n,m}^{(l)} + \widehat{\mathcal{R}}_{n,m}^{(l)})(\tau_{i,j}^n)}|\mathcal{S}]$$

$$= \sqrt{n} \sum_{i,j:t_{i,n}^{(1)}\vee t_{j,n}^{(2)}\leq T} |\mathcal{I}_{i,n}^{(1)} \cap \mathcal{I}_{j,n}^{(2)}| \sum_{k_1,k_2=-K_n}^{K_n} \big(|\mathcal{I}_{i+k_1,n}^{(1)}|^{1/2} + |\mathcal{I}_{j+k_2,n}^{(2)}|^{1/2}\big)$$

$$\times |\mathcal{I}_{i+k_1,n}^{(1)} \cap \mathcal{I}_{j+k_2,n}^{(2)}|\Big(\sum_{k_1',k_2'=-K_n}^{K_n} |\mathcal{I}_{i+k_1',n}^{(1)} \cap \mathcal{I}_{j+k_2',n}^{(2)}|\Big)^{-1}$$

$$= O_\mathbb{P}(\sqrt{n}|\pi_n|_T) + \sqrt{n} \sum_{i,j:t_{i,n}^{(1)}\vee t_{j,n}^{(2)}\leq T} (|\mathcal{I}_{i,n}^{(1)}|^{1/2} + |\mathcal{I}_{j,n}^{(2)}|^{1/2})|\mathcal{I}_{i,n}^{(1)} \cap \mathcal{I}_{j,n}^{(2)}|$$

$$\times \sum_{k_1,k_2=-K_n}^{K_n} |\mathcal{I}_{i+k_1,n}^{(1)} \cap \mathcal{I}_{j+k_2,n}^{(2)}| \Big(\sum_{k_1',k_2'=-K_n}^{K_n} |\mathcal{I}_{i+k_1+k_1',n}^{(1)} \cap \mathcal{I}_{j+k_2+k_2',n}^{(2)}| \Big)^{-1}$$

$$\leq O_{\mathbb{P}}(\sqrt{n}|\pi_n|_T) + 4(G_{2,1}^n(T) + G_{1,2}^n(T)) \tag{4.33}$$

where we used

$$\sum_{k_1,k_2=-K_n}^{K_n} |\mathcal{I}_{i+k_1,n}^{(1)} \cap \mathcal{I}_{j+k_2,n}^{(2)}| \Big(\sum_{k_1',k_2'=-K_n}^{K_n} |\mathcal{I}_{i+k_1+k_1',n}^{(1)} \cap \mathcal{I}_{j+k_2+k_2',n}^{(2)}| \Big)^{-1} \leq 4$$

which can be shown similarly as the corresponding estimate in (4.23). By Condition 3.5 it holds $O_{\mathbb{P}}(\sqrt{n}|\pi_n|_T) + 4(G_{2,1}^n(T) + G_{1,2}^n(T)) = O_{\mathbb{P}}(1)$, compare (3.59). Hence the part of the last bound in (4.31) which stems from overlapping interval parts vanishes in the sense of (4.32) by Lemma 2.15.

It remains to discuss the part of the last bound in (4.31) that originates from non-overlapping interval parts. To this end consider the notation

$$Y_{(i,j),n}^{(l)} = \Big(\frac{1}{b_n} \sum_{k \neq i: \mathcal{I}_{k,n}^{(l)} \subset [t_{i,n}^{(l)}-b_n, t_{i,n}^{(l)}+b_n]} (\Delta_{(k,j),n}^{(l,3-l)} X^{(l)})^2 \Big)^{1/2},$$

$$\widetilde{Y}_{(i,j),n}^{(l)} = \Big(\frac{1}{b_n} \sum_{k \neq i: \mathcal{I}_{k,n}^{(l)} \subset [t_{i,n}^{(l)}-b_n, t_{i,n}^{(l)}+b_n]} (\Delta_{(k,j),n}^{(l \backslash 3-l)}, X^{(l)})^2 \Big)^{1/2},$$

which allows to separate increments in $\tilde{\sigma}^{(l)}(i,n)$ over parts of intervals that do overlap with $\mathcal{I}_{j,n}^{(3-l)}$ and parts of increments over intervals that do not overlap. From the Minkowski inequality we then obtain

$$\tilde{\sigma}^{(l)}(i,n) \leq Y_{(i,j),n}^{(l)} + \widetilde{Y}_{(i,j),n}^{(l)}.$$

Let us first consider the last bound in (4.31) with $\tilde{\sigma}^{(l)}(i,n)$ replaced by $Y_{(i,j),n}^{(l)}$. Using iterated expectations, the Cauchy-Schwarz inequality and Lemma 1.4 we obtain

$$\mathbb{E}\big[|\Delta_{i_l,n}^{(l)}(X-N(q))^{(l)}|^{1+p_l^{(l,l)}} |\Delta_{(i_{3-l},i_l),n}^{(3-l\backslash l)} X^{(3-l)}|^{p_{3-l}^{(l,l)}} Y_{(i_l,i_{3-l}),n}^{(l)}$$

$$+ |\Delta_{(i_l,i_{3-l}),n}^{(l\backslash 3-l)}(X-N(q))^{(l)}|^{1+p_l^{(l,l)}} |\Delta_{(i_{3-l},i_l),n}^{(3-l,l)} X^{(3-l)}|^{p_{3-l}^{(l,l)}} Y_{(i_l,i_{3-l}),n}^{(l)} |\mathcal{S}\big]$$

$$\leq K|\mathcal{I}_{i_l,n}^{(l)}|^{(1+p_l^{(l,l)})/2} |\mathcal{I}_{i_{3-l},n}^{(3-l)}|^{p_{3-l}^{(l,l)}/2} |\mathcal{I}_{i_{3-l},n}^{(3-l)}|^{1/2}. \tag{4.34}$$

Then the last bound in (4.31) where we only consider increments over non-overlapping interval parts with $\tilde{\sigma}^{(l)}(i,n)$ replaced by $Y_{(i,j),n}^{(l)}$ is by

$$\sqrt{(\widehat{\mathcal{L}}_{n,m} + \widehat{\mathcal{R}}_{n,m} + \widehat{\mathcal{L}}_{n,m}^{(l)} + \widehat{\mathcal{R}}_{n,m}^{(l)})(\tau_{i_1,i_2}^n)} \leq \sqrt{n}|\pi_n|_T, \tag{4.35}$$

the estimate (4.34) and an anologous inequality for the term where the roles of X and $X - N(q)$ are switched bounded by

$$\frac{\sqrt{n}|\pi_n|_T}{(b_n)^{1/2}} K \sum_{l=1,2} G^n_{1+p_l^{(l,l)},p_{3-l}}(T) \leq \frac{\sqrt{n}|\pi_n|_T}{(b_n)^{1/2}} K\big(G^n_{2,0}(T) + G^n_{0,2}(T)\big), \quad (4.36)$$

compare (3.59). Note that $b_n = O(n^{-\gamma})$, $\gamma \in (0,1)$, yields

$$\frac{\sqrt{n}|\pi_n|_T}{(b_n)^{1/2}} = O(n^{1/2+\gamma/2}|\pi_n|_T).$$

Hence the last bound in (4.36) vanishes as $n \to \infty$ because by Condition 4.4(ii) we have $n^{1/2+\gamma/2}|\pi_n|_T = o_{\mathbb{P}}(1)$ due to $\gamma < 1$ and $G^n_{2,0}(T) = O_{\mathbb{P}}(1)$, $G^n_{0,2}(T) = O_{\mathbb{P}}(1)$.

Finally we consider the last bound in (4.31) where we only consider increments over non-overlapping interval parts and with $\tilde{\sigma}^{(l)}(i,n)$ replaced by $\widetilde{Y}^{(l)}_{(i,j),n}$. The $\mathcal{S}$-conditional expectation of that expression is using the estimate (4.35), iterated expectations and the arguments used in (2.82) bounded by

$$\sqrt{n}|\pi_n|_T K \sum_{i_1,i_2 : t_{i_1,n}^{(1)} \vee t_{i_2,n}^{(2)} \leq T} \sum_{l=1,2} |\mathcal{I}^{(l)}_{i_1,n}|^{\frac{1+p_l^{(l,l)}+(1/2-\varpi)(2-(1+p_l^{(l,l)}))}{2}}$$

$$\times |\mathcal{I}^{(3-l)}_{i_3-l,n}|^{\frac{p_{3-l}^{(l,l)}+(1/2-\varpi)(2-(1+p_l^{(l,l)}))}{2}} \mathbb{1}_{\{\mathcal{I}^{(l)}_{i_1,n} \cap \mathcal{I}^{(3-l)}_{i_3-l,n} \neq \emptyset\}}$$

$$\leq \sqrt{n}(|\pi_n|_T)^{1/2+(1/2-\varpi)((1-p_1^{(1,1)})\wedge(1-p_2^{(2,2)}))/2} \sum_{l=1,2} G^n_{1+p_l^{(l,l)},p_{3-l}}(T)$$

$$\leq \sqrt{n}(|\pi_n|_T)^{1/2+(1/2-\varpi)((1-p_1^{(1,1)})\wedge(1-p_2^{(2,2)}))/2} \big(G^n_{2,0}(T) + G^n_{0,2}(T)\big)$$

where the last bound vanishes as $n \to \infty$ by Condition 4.4(ii) because of $\varpi < 1/2$ and $p_l^{(l,l)} < 1$.

Combining all above arguments we have shown that the $\mathcal{S}$-conditional expectation of the last bound in (4.31) vanishes which yields (4.30) using Lemma 2.15.

Step 3. As in Step 3 in the proof of Proposition 4.10 we deduce

$$\widetilde{\mathbb{P}}\big(Y(P) \leq \Upsilon | \mathcal{X}\big) \xrightarrow{\widetilde{\mathbb{P}}} \widetilde{\mathbb{P}}\big(\Phi_T^{biv}(f) \leq \Upsilon | \mathcal{X}\big) \quad (4.37)$$

for $P \to \infty$.

Step 4. Finally (4.28) follows from (4.29), (4.30) and (4.37); compare Steps 4 and 5 in the proof of Proposition 4.10. $\qquad\square$

Proof of Theorem 4.5. The proof of (4.9) based on Proposition 4.11 is identical to the proof of (4.6) based on Proposition 4.10. $\qquad\square$

5 Observation Schemes

In this chapter, we discuss examples of observation schemes that yield random irregular and asynchronous observations. We will investigate their asymptotics and check whether they fulfil the assumptions made in Chapters 2–4. I chose to collect these results in a separate chapter instead of including them in Chapters 2–4 firstly because it turns out that their proofs are rather technical and require specific arguments unrelated to the arguments in the previous chapters. Further the results in this chapter are of interest also on their own as they provide deeper insights into the nature of specific observation schemes.

The observation schemes we are investigating in this chapter are of the following form: Let $t_{0,n}^{(l)} = 0$ and set

$$t_{i,n}^{(l)} = t_{i-1,n}^{(l)} + r^{(l)}(n)^{-1} E_{i,n}^{(l)}, \ i \geq 1, \tag{5.1}$$

where the $E_{i,n}^{(l)}$, $i, n \in \mathbb{N}$, are i.i.d. random variables with values in $[0, \infty)$ and $r^{(l)} : \mathbb{R} \to [0, \infty)$ are positive functions with $r^{(l)}(n) \to \infty$ as $n \to \infty$ for $l = 1, 2$. Using this construction, the observation times of $X^{(l)}$ stem from a *renewal process* where the waiting times are i.i.d. and decrease with some rate $r^{(l)}(n)$, $l = 1, 2$. Observation schemes as in (5.1) will in the following be called *renewal schemes*. The most prominent example for a member of this class of observation schemes is the case where the $E_{i,n}^{(l)}$ are exponentially distributed and $r^{(l)}(n) = n$. In this setting we have the nice property that the observation scheme is stationary in time. The specific rate $r^{(l)}(n) = n$ is chosen for convenience. As this observation scheme functions as our major example throughout this work we devote it its own name in the following definition. Most of the upcoming results will only be proven for this more specific observation scheme.

Definition 5.1. *We will call the observation scheme* $(\pi_n)_{n \in \mathbb{N}}$ *given by* $t_{0,n}^{(l)} = 0$ *and*

$$t_{i,n}^{(l)} = t_{i-1,n}^{(l)} + \frac{1}{n} E_{i,n}^{(l)}, \ i \geq 1,$$

where the $E_{i,n}^{(l)}$, $i, n \in \mathbb{N}$, *are i.i.d* $Exp(\lambda_l)$-*distributed random variables,* $\lambda_l > 0$, *for* $l = 1, 2$ Poisson sampling. $\qquad \square$

© Springer Fachmedien Wiesbaden GmbH, part of Springer Nature 2019
O. Martin, *High-Frequency Statistics with Asynchronous and Irregular Data*,
Mathematische Optimierung und Wirtschaftsmathematik I Mathematical Optimization
and Economathematics, https://doi.org/10.1007/978-3-658-28418-3_5

This observation scheme has been discussed frequently in the high-frequency statistics literature; see e.g. [23] and [8].

In the Poisson sampling scheme we have the following result for the decay of the moments of $|\pi_n|_T$ as $n \to 0$. This result is taken from Lemma 8 of [23] where a proof can also be found.

Lemma 5.2. *Suppose $(\pi_n)_{n\in\mathbb{N}}$ is the Poisson sampling scheme. Then it holds*

$$\mathbb{E}[(|\pi_n|_T)^q] = o(n^{-\gamma})$$

for any $0 \leq \gamma < q$ and $q \geq 1$.

By Jensen's and Markov's inequalities Lemma 5.2 immediately yields

$$(|\pi_n|_T)^q = o_{\mathbb{P}}(n^{-\gamma}) \tag{5.2}$$

for any $0 \leq \gamma < q$, $q \geq 0$ and in particular it shows $|\pi_n|_T \xrightarrow{\mathbb{P}} 0$ in the case of Poisson sampling.

5.1 The Results

The following lemma is a kind of law of large numbers for moments of the observation interval lengths in the setting where the observation times are generated by renewal processes. The main difference to the classical law of large numbers is that we do not work with a fixed number of random variables at each stage n. But instead, we sum up transformations of random variables as long as their sum does not exceed some threshold t.

Lemma 5.3. *Let the observation scheme be a renewal scheme as in (5.1) and suppose we have $E_{1,1}^{(l)} \in L^2(\Omega)$. Then it holds*

$$r^{(l)}(n)^{-1} \sum_{i:t_{i,n}^{(l)} \leq t} g\big(r^{(l)}(n)|\mathcal{I}_{i,n}^{(l)}|\big) \xrightarrow{\mathbb{P}} \frac{\mathbb{E}[g(E_{1,1}^{(l)})]}{\mathbb{E}[E_{1,1}^{(l)}]} t, \quad t \geq 0,$$

for all functions $g\colon [0,\infty) \to \mathbb{R}$ with $g(E_{1,1}^{(l)}) \in L^1(\Omega)$.

In the specific setting of Poisson sampling, the statement in Lemma 5.3 reads as stated in the following corollary.

Corollary 5.4. *Let $(\pi_n)_{n\in\mathbb{N}}$ be the Poisson sampling scheme introduced in Definition 5.1 and let $E^{(l)}$ be an $Exp(\lambda_l)$-distributed random variable, $\lambda_l > 0$. Then it holds*

$$n^{-1} \sum_{i:t_{i,n}^{(l)} \leq t} g\big(n|\mathcal{I}_{i,n}^{(l)}|\big) \xrightarrow{\mathbb{P}} \mathbb{E}[g(E^{(l)})]\lambda_l t, \quad t \geq 0,$$

for all functions $g\colon [0,\infty) \to \mathbb{R}$ with $g(E^{(l)}) \in L^1(\Omega)$.

In the case of Poisson sampling we also obtain a corresponding result in the bivariate setting where we consider sums which contain observation intervals from both processes $X^{(l)}$, $l = 1, 2$. The following law of large numbers holds for functions $f \colon [0, \infty)^3 \to \mathbb{R}$ evaluated at the lenghts of the rescaled observation intervals

$$\left(n|\mathcal{I}_{i,n}^{(1)}|, n|\mathcal{I}_{j,n}^{(2)}|, n|\mathcal{I}_{i,n}^{(1)} \cap \mathcal{I}_{j,n}^{(2)}|\right)$$

and included in the sum whenever $\mathcal{I}_{i,n}^{(1)} \cap \mathcal{I}_{j,n}^{(2)} \neq \emptyset$. However, unlike in Corollary 5.4, we can in general not find a simple closed form for the limit.

Lemma 5.5. *Let* $(\pi_n)_{n \in \mathbb{N}}$ *be the Poisson sampling scheme introduced in Definition 5.1 and let* $f \colon [0, \infty)^3 \to \mathbb{R}$ *be a function which fulfils*

$$\sum_{i,j : t_{i-1,1}^{(1)} \wedge t_{j-1,1}^{(2)} \in (0,1]} f\left(|\mathcal{I}_{i,1}^{(1)}|, |\mathcal{I}_{j,1}^{(2)}|, |\mathcal{I}_{i,1}^{(1)} \cap \mathcal{I}_{j,1}^{(2)}|\right) \mathbb{1}_{\{\mathcal{I}_{i,1}^{(1)} \cap \mathcal{I}_{j,1}^{(2)} \neq \emptyset\}} \in L^2(\Omega). \tag{5.3}$$

Then the expression

$$\frac{1}{n} \sum_{i,j : t_{i,n}^{(1)} \vee t_{j,n}^{(2)} \leq t} f\left(n|\mathcal{I}_{i,n}^{(1)}|, n|\mathcal{I}_{j,n}^{(2)}|, n|\mathcal{I}_{i,n}^{(1)} \cap \mathcal{I}_{j,n}^{(2)}|\right) \mathbb{1}_{\{\mathcal{I}_{i,n}^{(1)} \cap \mathcal{I}_{j,n}^{(2)} \neq \emptyset\}}$$

converges for $n \to \infty$ *in probability to a deterministic function which is linear in* t.

By applying Lemma 5.5 for certain monomials we obtain the convergence of the functions $G_p^{(l),n}$, G_{p_1,p_2}^n, $H_{k,m,p}^n$ introduced in Section 2.2 in the Poisson setting.

Corollary 5.6. *In the Poisson setting we obtain that the functions* $G_p^{(l),n}(t)$, $G_{p_1,p_2}^n(t)$, $H_{k,m,p}^n(t)$ *defined in (2.39) converge in probability to deterministic linear functions for any* $p, p_1, p_2 \geq 0$ *and* $0 \leq k, m$ *with* $k + m \leq p$.

Example 5.7. *Although in general the limits in Lemma 5.5 and Corollary 5.6 cannot be easily computed, the limit has been computed for certain specific functions. Proposition 1 in [23] e.g. yields*

$$G_{2,2}^n(t) \xrightarrow{\mathbb{P}} \left(\frac{2}{\lambda_1} + \frac{2}{\lambda_2}\right) t,$$

$$H_{0,0,4}^n(t) \xrightarrow{\mathbb{P}} \frac{2}{\lambda_1 + \lambda_2} t.$$

Further, in the case where they cannot be explicitly computed the limits can always naturally be estimated by the observed variables $G_p^{(l),n}(t)$, $G_{p_1,p_2}^n(t)$, $H_{k,m,p}^n(t)$. $\square$

As a further corollary of Lemma 5.5 we obtain the convergence of the functions $G_p^{(l),[k],n}$, $G_{p_1,p_2}^{[k],n}$, $H_{l,m,p}^{[k],n}$ introduced in Section 2.4 in the Poisson setting.

Corollary 5.8. *In the case of Poisson setting we obtain that the functions* $G_p^{(l),[k],n}(t)$, $G_{p_1,p_2}^{[k],n}(t)$, $H_{\iota,m,p}^{[k],n}(t)$ *defined in (2.39) converge in probability to deterministic linear functions for any* $k \in \mathbb{N}$, $p, p_1, p_2 \geq 0$ *and* $\iota, m \geq 0$ *with* $\iota + m \leq p$.

Remark 5.9. *The above results could probably be generalized to hold for more observation schemes than just mere Poisson sampling. One possible extension would be to prove them in the case of inhomogeneous Poisson sampling where the observation times* $t_{i,n}^{(l)}$, $i \in \mathbb{N}$, *follow Poisson point processes on* $[0, \infty)$ *with intensity measure* $n\Lambda^{(l)}$, $l = 1, 2$; *compare Chapter 2 in [45] for a definition of Poisson point processes. If the intensity measure* $\Lambda^{(l)}$ *is sufficiently regular an approximation of* $\Lambda^{(l)}$ *by measures with piece-wise constant Lebesgue-densities might then be sufficient to carry the previously proven results in the situation of Poisson sampling over to the case of inhomogeneous Poisson sampling. To this end note that Poisson sampling, as introduced in Definition 5.1, corresponds to the situation described above with time-constant intensity measures* $n\lambda_l \lambda$, $l = 1, 2$, *where* λ *denotes the Lebesgue measure on* $[0, \infty)$.

Further for a second posssible extension, note that the proof of Lemma (5.5) relies on the stationarity of the Poisson process only through the property that observation times within intervals $[m, m+1]$, $[m', m'+1]$ *are independent and identically distributed. If we instead use intervals of the form* $[m\kappa_n, (m+1)\kappa_n]$ *with* $\kappa_n \to \infty$ *as* $n \to \infty$ *it would be sufficient that the renewal scheme is almost stationary in the long range in the sense that* $(t_{i,1}^{(l)})_{i=1,\ldots,K}$ *and* $(t_{i,1}^{(l)})_{i=M,\ldots,M+K}$ *become independent and that their distributions "converge" in a certain sense as* $M \to \infty$ *for any* $K \in \mathbb{N}$. *Such a property should be fulfilled for "most" renewal schemes due to the i.i.d.-property of the waiting times* $E_{i,n}^{(l)}$. *However proving such a result would require a lot more arguments and is beyond the scope of this work.* $\square$

The above lemmata and corollaries show that the assumptions made in Chapter 2 on the observation scheme are fulfilled in the setting of Poisson sampling. The following result further shows that Conditions 3.1 and 3.5 which were required to obtain central limit theorems in Section 3.1 for the non-normalized functionals from Section 2.1 are fulfilled in the Poisson setting as well. The result is stated in a more general way to also cover the conditions we need to make in Chapters 7-9 to derive central limit theorems for certain selected applications.

Set $\overline{W}_t = (W_t^{(1)}, \rho_t W_t^{(1)} + \sqrt{1 - \rho_t^2} W_t^{(2)})^*$ as in (3.23) such that $\overline{W}^{(l)}$ is the Brownian motion driving the process $X^{(l)}$ for $l = 1, 2$.

Lemma 5.10. *Let* $d \in \mathbb{N}$ *and* $Z_n^{(1)}(s), Z_n^{(2)}(s), Z_n^{(3)}(s)$ *be* $\mathbb{R}^d$-*valued random variables, which can be written as*

$$Z_n^{(l)}(s) = f^{(l)}\big(n(s - t_{i_n^{(l)}(s)-1,n}^{(l)}), n(t_{i_n^{(l)}(s),n}^{(l)} - s), (\sqrt{n}\Delta_{i_n^{(l)}(s)+j,n}\overline{W}^{(l)})_{j \in [k-1]},$$

$$\left(\sqrt{n}\Delta^{(l)}_{i_n^{(l)}(s)+j,k,n}\overline{W}^{(l)}\mathbf{1}_{\{\mathcal{I}^{(3-l)}_{i_n^{(3-l)}(s)+i,k,n}\cap\mathcal{I}^{(l)}_{i_n^{(l)}(s)+j,k,n}\neq\emptyset\}}\right)_{i\in\{0,\dots,k-1\},j\in\mathbb{Z}},\quad l=1,2,$$

$$Z_n^{(3)}(s)=f^{(3)}\Big(\big[n(s-t^{(l)}_{i_n^{(l)}(s)-1,n}),n(t^{(l)}_{i_n^{(l)}(s),n}-s)\big]_{l=1,2},$$

$$n(s-t^{(1)}_{i_n^{(1)}(s)-1,n}\vee t^{(2)}_{i_n^{(2)}(s)-1,n}),n(t^{(1)}_{i_n^{(1)}(s),n}\wedge t^{(2)}_{i_n^{(2)}(s),n}-s),$$

$$\big[(n|\mathcal{I}^{(l)}_{i_n^{(l)}(s)+i,n}|)_{i\in[k]}\big]_{l=1,2},(n|\mathcal{I}^{(1)}_{i_n^{(1)}(s)+i,n}\cap\mathcal{I}^{(2)}_{i_n^{(2)}(s)+j,n}|)_{i,j\in[k]}\Big)$$

with $[k]=\{-k,\dots,k\}$ for Borel-measurable functions $f^{(1)},f^{(2)},f^{(3)}$ and a fixed $k\in\mathbb{N}$.

Then in the situation of Poisson sampling introduced in Definition 5.1 the integral

$$\int_{[0,T]^{P_1+P_2+P_3}}g(x_1,\dots,x_{P_1},x_1',\dots,x_{P_2}',x_1'',\dots,x_{P_3}'')\mathbb{E}\Big[\prod_{p=1}^{P_1}h_p^{(1)}(Z_n^{(1)}(x_p))$$

$$\times\prod_{p=1}^{P_2}h_p^{(2)}(Z_n^{(2)}(x_p'))\prod_{p=1}^{P_3}h_p^{(3)}(Z_n^{(3)}(x_p''))\Big]dx_1\dots dx_{P_1}dx_1'\dots dx_{P_2}'dx_1''\dots dx_{P_3}''$$

$$(5.4)$$

converges for $n\to\infty$ to

$$\int_{[0,T]^{P_1+P_2+P_3}}g(x_1,\dots,x_{P_1},x_1',\dots,x_{P_2}',x_1'',\dots,x_{P_3}'')$$

$$\times\prod_{p=1}^{P_1}\int h_p^{(1)}(y)\Gamma^{(1)}(dy)\prod_{p=1}^{P_2}\int h_p^{(2)}(y)\Gamma^{(2)}(dy)$$

$$\times\prod_{p=1}^{P_3}\int h_p^{(3)}(y)\Gamma^{(3)}(dy)dx_1\dots dx_{P_1}dx_1'\dots dx_{P_2}'dx_1''\dots dx_{P_3}''\qquad(5.5)$$

for all bounded continuous functions $g\colon\mathbb{R}^{P_1+P_2+P_3}\to\mathbb{R}$, $h_p^{(l)}:\mathbb{R}^d\to\mathbb{R}$ and all $P_1,P_2,P_3\in\mathbb{N}$.

Note that the measures $\Gamma^{(1)},\Gamma^{(2)},\Gamma^{(3)}$ implicitly defined in Lemma 5.10 are probability measures on $\mathbb{R}^d$ which do not depend on x_p,x_p',x_p'' like in the general case in Conditions 3.1(ii) and 3.5(ii). This property arises because the Poisson process has stationary increments.

The key idea in the proof of Lemma 5.10 is to observe that due to the stationarity of the increments of the Poisson process and the Brownian motion $\overline{W}$ the law of $Z_n^{(l)}(s)$ depends on s only through the fact that all observation intervals are bounded to the left because all observation times are greater or equal than zero. Further $Z_n^{(l)}(s)$ has the same distribution as $Z_1^{(l)}(ns)$ and the effect on the law of

$Z_1^{(l)}(ns)$ from the property that the observation intervals are bounded to the left becomes asymptotically negligible as $n \to \infty$. Hence $Z_1^{(l)}(ns)$ asymptotically has for $ns \to \infty$ the same distribution as the corresponding random variable which is constructed based on observation times and Brownian motions $\overline{W}^{(l)}$ which are naturally extended on $(-\infty, 0]$. Using this extended observation scheme and extended Brownian motions allows to characterize the limiting distribution.

The following remark shows that in the case of Poisson sampling the limiting laws $\Gamma^{univ}(s, dy)$ and $\Gamma^{biv}(s, dy)$ have no atoms and also that $\Gamma^{univ}(s, dy)$ and $\Gamma^{biv}(s, dy)$ have bounded moments. This property yields that because $\Gamma^{univ}(s, dy)$ and $\Gamma^{biv}(s, dy)$ by Lemma 5.10 do not depend on s the laws described by $\Gamma^{univ}(s, dy)$ and $\Gamma^{biv}(s, dy)$ for $s \in [0, T]$ also have uniformly bounded first moments as required in Conditions 3.1(ii) and 3.5(ii). The computations in the following remark are also valid for the conditions which will be made in Chapters 7–9 where we will also need that the limiting laws do not admit atoms.

Remark 5.11. *In the proof of Lemma 5.10 we obtained*

$$Z_n^{(l)}(s)\mathbb{1}_{\Omega_n^{(l)}(s)} \overset{\mathcal{L}}{=} Z^{(l)}\mathbb{1}_{\widetilde{\Omega}_n^{(l)}(s)}$$

for a random variable $Z^{(l)} \sim \Gamma^{(l)}$ and sequences $(\Omega_n^{(l)}(s))_{n \in \mathbb{N}}$, $(\widetilde{\Omega}_n^{(l)}(s))_{n \in \mathbb{N}}$ with $\Omega_n^{(l)}(s) \uparrow \Omega$, $\widetilde{\Omega}_n^{(l)}(s) \uparrow \widetilde{\Omega}$. Hence if the limit law would admit a part which is singular to the Lebesgue measure, this would also be true for the law of the $Z_n^{(l)}(s)$ for n sufficiently large. In the two choices of the functions $f^{(l)}$ for obtaining Conditions 3.1(ii) and 3.5(ii) it can be shown that $Z_n^{(l)}(s)$ has no atom. Hence in Condition 3.1(ii) and 3.5(ii) we obtain $\Gamma^{univ,(l)}(x, \{0\}) = 0$ and $\Gamma^{biv}(x, \{0\}) = 0$.

Further, by Theorem 6.7 from [9] there exist random variables $\widetilde{Z}^{(l)}$ and $(\widetilde{Z}_n^{(l)})_{n \in \mathbb{N}}$ defined on a probability space $(\Omega', \mathcal{F}', \mathbb{P}')$ with

$$(\widetilde{Z}^{(l)}, (\widetilde{Z}_n^{(l)})_{n \in \mathbb{N}}) \overset{\mathcal{L}}{=} (Z^{(l)}, (Z_n^{(l)})_{n \in \mathbb{N}})$$

and $\widetilde{Z}_n^{(l)} \to \widetilde{Z}$ almost surely. An application of Fatou's lemma then yields

$$\mathbb{E}[\|Z^{(l)}\|^p] = \mathbb{E}'[\|\widetilde{Z}^{(l)}\|^p] \leq \liminf_{n \in \mathbb{N}} \mathbb{E}'[\|\widetilde{Z}_n^{(l)}(s)\|^p] \leq \sup_{n \in \mathbb{N}} \mathbb{E}'[\|\widetilde{Z}_n^{(l)}(s)\|^p] = \sup_{n \in \mathbb{N}} \mathbb{E}[\|Z_n^{(l)}(s)\|^p]$$

for $p \geq 0$. For the cases corresponding to Conditions 3.1(ii) and 3.5(ii) it can be shown that the supremum is finite for all $p \geq 0$ and hence $\Gamma^{univ,(l)}(x, dy)$ and $\Gamma^{biv}(x, dy)$ have finite moments for any $x \in [0, T]$. This yields the requirement of uniformly bounded first and second moments in Condition 3.1(ii) and 3.5(ii). $\square$

Next, we will show that the assumptions made on the observation scheme in Conditions 4.2(i) and 4.4(i) are fulfilled in the case of Poisson sampling. We again derive a more general result which then later can also be used for the applications in Chapters 7–9.

Lemma 5.12. *We consider the situation given in Lemma 5.10. Further define* $\widehat{Z}_{n,1}^{(l)}(s) = Z_n^{(l)}(U_n^{(l)}(s))$ *where we have one of the following three cases*

$$U_n^{(l)}(s) \sim \left[t_{i_n^{(1)}(s)-K_n-1,n}^{(1)}, t_{i_n^{(1)}(s)+K_n,n}^{(1)}\right], \tag{5.6}$$

$$U_n^{(l)}(s) \sim \left[t_{i_n^{(2)}(s)-K_n-1,n}^{(2)}, t_{i_n^{(2)}(s)+K_n,n}^{(2)}\right], \tag{5.7}$$

$$U_n^{(l)}(s) \sim \left[t_{i_n^{(1)}(s)-K_n-1,n}^{(1)} \vee t_{i_n^{(2)}(s)-K_n-1,n}^{(2)}, t_{i_n^{(1)}(s)+K_n,n}^{(1)} \wedge t_{i_n^{(2)}(s)+K_n,n}^{(2)}\right] \tag{5.8}$$

for some sequence $(K_n)_{n\in\mathbb{N}}$ *of natural numbers with* $K_n \to \infty$ *and* $K_n/n \to 0$. *Let the distributions* $\Gamma^{(l)}(dy)$, $l = 1,2,3$, *be continuous and let* $Z^{(l)}(s) \sim \Gamma^{(l)}(s,dy)$ *be random variables independent of* $\mathcal{S}$. *Then it holds*

$$\widetilde{\mathbb{P}}\left(\left|\widetilde{\mathbb{P}}(\widehat{Z}_{n,1}^{(l_p)}(s_p) \leq x_p, \ p=1,\dots,P|\mathcal{S}) - \prod_{p=1}^{P} \widetilde{\mathbb{P}}(Z^{(l)}(s_p) \leq x_p)\right| > \varepsilon\right) \to 0 \tag{5.9}$$

for any $x \in \mathbb{R}^{d\times P}$, $P \in \mathbb{N}$, *and* $s_p \in (0,T)$, $l_p \in \{1,2,3\}$, $p = 1,\dots,P$, *with* $s_p \neq s_{p'}$ *for* $p \neq p'$ *as* $n \to \infty$, *for all* $\varepsilon > 0$.

In the proof of Lemma 5.12 we obtain the following nice property of Poisson processes: Let $(N_t)_{t\geq 0}$ be a Poisson process and let $U_n \sim \mathcal{U}[0,c_n]$, $n \in \mathbb{N}$, be uniformly distributed random variables on $[0,c_n]$ with $c_n \to \infty$ as $n \to \infty$ which are independent of the Poisson process $(N_t)_{t\geq 0}$. Then $(N_{t-U_n} - N_{U_n})_{t\geq U_n}$ has asymptotically the same law as $(N_t)_{t\geq 0}$ and this also holds for the $\sigma(N_t : t \geq 0)$-conditional laws of $(N_{t-U_n} - N_{U_n})_{t\geq U_n}$. Hence taking a fixed realization of a Poisson process and starting it at a uniformly distributed independent random time over an increasing time interval asymptotically yields a Poisson process again. This property is due to the stationarity of the Poisson process and the asymptotic independence of increments of the Poisson process over intervals which are far apart in time.

Remark 5.13. *The statement in Lemma 5.12 can also similarly be proven for* $U_n^{(l)}(s) \sim \mathcal{U}[s - \varepsilon_n, s + \varepsilon_n]$ *where* $(\varepsilon_n)_{n\in\mathbb{N}}$ *denotes some deterministic sequence of positive real numbers with* $\varepsilon_n \to 0$ *and* $n\varepsilon_n \to \infty$; *compare Remark 4.8.* □

5.2 The Proofs

Proof of Lemma 5.3. This proof is based on the proof of Proposition 1 in [23]. Set $m_1^{(l)} = \mathbb{E}[E_{1,1}^{(l)}]$, $m_g^{(l)} = \mathbb{E}[y(E_{1,1}^{(l)})]$ and

$$\lambda^{(l)}(n) = \lceil r^{(l)}(n)T/m_1^{(l)}\rceil.$$

Further define

$$N_T^{(l)}(n) = \sum_{i \in \mathbb{N}} \mathbb{1}_{\{t_{i,n}^{(l)} \le T\}}.$$

Step 1. We prove

$$\mathbb{P}(|N_T^{(l)}(n) - \lambda^{(l)}(n)| > r^{(l)}(n)^{1/2+\varepsilon}) \to 0 \tag{5.10}$$

as $n \to \infty$ for $\varepsilon \in (0, 1/2)$; compare Lemma 9 in [23]. Denote $\nu^{(l)} = Var(E_{1,1}^{(l)})$. The case $\nu^{(l)} = 0$ is simple, so let us assume $\nu^{(l)} > 0$ in the following. It holds

$$N_T^{(l)}(n) - \lambda^{(l)}(n) > r^{(l)}(n)^{1/2+\varepsilon} \Leftrightarrow \sum_{i=1}^{\lambda^{(l)}(n) + r^{(l)}(n)^{1/2+\varepsilon}} |\mathcal{I}_{i,n}^{(l)}| < T$$

$$\Leftrightarrow \frac{\sum_{i=1}^{\lambda^{(l)}(n) + r^{(l)}(n)^{1/2+\varepsilon}} (E_{i,n}^{(l)} - m_1^{(l)})}{\sqrt{(\lambda^{(l)}(n) + r^{(l)}(n)^{1/2+\varepsilon})\nu^{(l)}}} < \frac{r^{(l)}(n)T - (\lambda^{(l)}(n) + r^{(l)}(n)^{1/2+\varepsilon})m_1}{\sqrt{(\lambda^{(l)}(n) + r^{(l)}(n)^{1/2+\varepsilon})\nu^{(l)}}}$$

where the left-hand side is approximately standard normal distributed by the classical central limit theorem while the right-hand side less or equal than

$$\frac{-r^{(l)}(n)^{1/2+\varepsilon} m_1}{\sqrt{(\lambda^{(l)}(n) + r^{(l)}(n)^{1/2+\varepsilon})\nu^{(l)}}} = O\Big(\frac{-r^{(l)}(n)^{1/2+\varepsilon}}{r^{(l)}(n)^{1/2\vee(1/2+\varepsilon)/2}}\Big) \le O(-r^{(l)}(n)^{\varepsilon/2})$$

which tends to $-\infty$ as $n \to \infty$. Hence

$$\mathbb{P}(N_T^{(l)}(n) - \lambda^{(l)}(n) > r^{(l)}(n)^{1/2+\varepsilon}) \to 0. \tag{5.11}$$

Analogously it follows

$$\mathbb{P}(N_T^{(l)}(n) - \lambda^{(l)}(n) < -r^{(l)}(n)^{1/2+\varepsilon}) \to 0$$

which together with (5.11) yields (5.10).

Step 2. By the weak law of large numbers it holds

$$r^{(l)}(n)^{-1} \sum_{i=1}^{\lambda^{(l)}(n)} g\big(r^{(l)}(n)|\mathcal{I}_{i,n}^{(l)}|\big) = \frac{\lambda^{(l)}(n)}{r^{(l)}(n)} \frac{1}{\lambda^{(l)}(n)} \sum_{i=1}^{\lambda^{(l)}(n)} g\big(E_{i,n}^{(l)}\big) \xrightarrow{\mathbb{P}} \frac{T}{m_1^{(l)}} m_g^{(l)}.$$

Hence it remains to prove

$$r^{(l)}(n)^{-1} \sum_{i=\lambda^{(l)}(n) \wedge N_T^{(l)}(n)+1}^{\lambda^{(l)}(n) \vee N_T^{(l)}(n)} g\big(r^{(l)}(n)|\mathcal{I}_{i,n}^{(l)}|\big) \xrightarrow{\mathbb{P}} 0 \tag{5.12}$$

as $n \to \infty$. Because of (5.10) it suffices to prove the convergence (5.12) restricted to the set

$$\Omega^{(l)}(n, \varepsilon) = \{|N_T^{(l)}(n) - \lambda^{(l)}(n)| \leq r^{(l)}(n)^{1/2+\varepsilon}\}.$$

On this set (5.12) is bounded by

$$2r^{(l)}(n)^{(1/2+\varepsilon)-1}\left(\frac{1}{2r^{(l)}(n)^{1/2+\varepsilon}} \sum_{i=\lambda^{(l)}(n)-\lfloor r^{(l)}(n)^{1/2+\varepsilon}\rfloor+1}^{\lambda^{(l)}(n)+\lceil \alpha^{(l)}(n)^{1/2+\varepsilon}\rceil} |g(E_{i,n}^{(l)})|\right)\mathbb{1}_{\Omega^{(l)}(n,\varepsilon)}$$

which converges in probability to zero for $\varepsilon < 1/2$ as $n \to \infty$ because the term in parantheses converges by the law of large numbers in probability to $\mathbb{E}[|g(E_{1,1}^{(l)}|]$. $\square$

Proof of Lemma 5.5. Due to the fact that π_1 and $n\pi_n$ have the same law we obtain

$$\frac{1}{n} \sum_{i,j:t_{i,n}^{(1)}\vee t_{j,n}^{(2)}\leq t} f\left(n|\mathcal{I}_{i,n}^{(1)}|, n|\mathcal{I}_{j,n}^{(2)}|, n|\mathcal{I}_{i,n}^{(1)}\cap \mathcal{I}_{j,n}^{(2)}|\right)\mathbb{1}_{\{\mathcal{I}_{i,n}^{(1)}\cap\mathcal{I}_{j,n}^{(2)}\neq\emptyset\}}$$

$$\stackrel{\mathcal{L}}{=} \frac{1}{n} \sum_{i,j:t_{i,n}^{(1)}\vee t_{j,n}^{(2)}\leq nt} f\left(|\mathcal{I}_{i,1}^{(1)}|, |\mathcal{I}_{j,1}^{(2)}|, |\mathcal{I}_{i,1}^{(1)}\cap \mathcal{I}_{j,1}^{(2)}|\right)\mathbb{1}_{\{\mathcal{I}_{i,1}^{(1)}\cap\mathcal{I}_{j,1}^{(2)}\neq\emptyset\}}$$

$$= \frac{\lfloor nt\rfloor}{n} \frac{1}{\lfloor nt\rfloor} \sum_{m=1}^{\lfloor nt\rfloor} Y_m + O_{\mathbb{P}}(n^{-1})$$

where $\stackrel{\mathcal{L}}{=}$ denotes equality in law and

$$Y_m = \sum_{i,j:t_{i-1,1}^{(1)}\wedge t_{j-1,1}^{(2)}\in(m-1,m]} f\left(|\mathcal{I}_{i,1}^{(1)}|, |\mathcal{I}_{j,1}^{(2)}|, |\mathcal{I}_{i,1}^{(1)}\cap \mathcal{I}_{j,1}^{(2)}|\right)\mathbb{1}_{\{\mathcal{I}_{i,1}^{(1)}\cap\mathcal{I}_{j,1}^{(2)}\neq\emptyset\}}$$

$m \in \mathbb{N}$. Because of the stationarity of the Poisson process, the sequence Y_m, $m \in \mathbb{N}$, is a stationary and by (5.3) square integrable time series. Because further Y_{m_1}, Y_{m_2} become asymptotically independent as $|m_1 - m_2| \to \infty$ it is possible to conclude $(Y_{m_1}, Y_{m_2}) \to 0$ as $|m_1 - m_2| \to \infty$. We then obtain by Theorem 7.1.1 in [10] which is a law of large numbers for stationary processes

$$\frac{1}{n} \sum_{i,j:t_{i,n}^{(1)}\vee t_{j,n}^{(2)}\leq t} f\left(n|\mathcal{I}_{i,n}^{(1)}|, n|\mathcal{I}_{j,n}^{(2)}|, n|\mathcal{I}_{i,n}^{(1)}\cap \mathcal{I}_{j,n}^{(2)}|\right)\mathbb{1}_{\{\mathcal{I}_{i,n}^{(1)}\cap\mathcal{I}_{j,n}^{(2)}\neq\emptyset\}} \stackrel{\mathbb{P}}{\longrightarrow} t\mathbb{E}[Y_1].$$

Hence the limit is linear in t with slope $\mathbb{E}[Y_1]$. $\square$

Proof of Corollary 5.6. The convergence of $G_p^{(l),n}$ follows from Lemma 5.4 because all moments of an exponentially distributed random variable are finite. Lemma 5.5 yields the convergences of G_{p_1,p_2}^n and $H_{k,n,p}^n$ if we can verify (5.3) for the corresponding functions f. Because of

$$0 \leq H_{k,n,p}^n(t) \leq G_{k,p-k}^n(t), \quad t \geq 0,$$

it then suffices to check (5.3) for the function $f_{(p_1,p_2)}(x,y,z) = x^{p_1} y^{p_2}$, $p_1, p_2 \geq 0$. Therefore observe that we obtain using the Cauchy-Schwarz inequality

$$\left(\mathbb{E}\left[\left(\sum_{i,j:t_{i-1,1}^{(1)} \wedge t_{j-1,1}^{(2)} \in (0,1]} |\mathcal{I}_{i,1}^{(1)}|^{p_1} |\mathcal{I}_{j,1}^{(2)}|^{p_2} \mathbb{1}_{\{\mathcal{I}_{i,1}^{(1)} \cap \mathcal{I}_{j,1}^{(2)} \neq \emptyset\}}\right)^2\right]\right)^2$$

$$\leq \left(\mathbb{E}\left[(|\pi_n|_T)^{p_1+p_2}\left(\sum_{i,j:t_{i-1,1}^{(1)} \wedge t_{j-1,1}^{(2)} \leq 1} \mathbb{1}_{\{\mathcal{I}_{i,1}^{(1)} \cap \mathcal{I}_{j,1}^{(2)} \neq \emptyset\}}\right)^2\right]\right)^2$$

$$\leq \mathbb{E}\left[(|\pi_n|_T)^{2(p_1+p_2)}\right]\mathbb{E}\left[\left(\sum_{i,j:t_{i-1,n}^{(1)} \wedge t_{j-1,n}^{(2)} \leq 1} \mathbb{1}_{\{\mathcal{I}_{i,1}^{(1)} \cap \mathcal{I}_{j,1}^{(2)} \neq \emptyset\}}\right)^4\right]. \quad (5.13)$$

Here the first expectation in (5.13) is finite due to Lemma 5.2. Further

$$\sum_{i,j:t_{i-1,1}^{(1)} \wedge t_{j-1,1}^{(2)} \leq 1} \mathbb{1}_{\{\mathcal{I}_{i,1}^{(1)} \cap \mathcal{I}_{j,1}^{(2)} \neq \emptyset\}} = \sum_{l=1,2} \sum_{i=1}^{\infty} \mathbb{1}_{\{t_{i-1,1}^{(l)} \leq 1\}} - 1$$

because for $i \geq 1$ each observation $t_{i,n}^{(l)} \in [0,1]$ except the first one $t_{1,n}^{(1)} \wedge t_{1,n}^{(2)}$ creates a new pair (i,j) with $\mathcal{I}_{i,1}^{(1)} \cap \mathcal{I}_{j,1}^{(2)} \neq \emptyset$. Then the second expectation in (5.13) is finite in the Poisson setting because there the sum $\sum_{i=1}^{\infty} \mathbb{1}_{\{t_{i-1,n}^{(l)} \leq 1\}} - 1$ is Poisson-distributed with parameter λ_l and all moments of the Poisson-distribution are finite. $\square$

Proof of Corollary 5.8. Reproducing the proof of Lemma 5.4 we also obtain the convergence of

$$\frac{1}{n} \sum_{i,j:t_{i,n}^{(1)} \wedge t_{j,n}^{(2)} \leq t} f\left(n|\mathcal{I}_{i,k,n}^{(1)}|, n|\mathcal{I}_{j,k,n}^{(2)}|, n|\mathcal{I}_{i,k,n}^{(1)} \cap \mathcal{I}_{j,k,n}^{(2)}|\right) \mathbb{1}_{\{\mathcal{I}_{i,k,n}^{(1)} \cap \mathcal{I}_{j,k,n}^{(2)} \neq \emptyset\}}$$

whenever the sum is square integrable for $n = 1$. As for Corollary 5.6 it then remains to verify this square integrability condition for appropriate functions f which can be done similarly as in the proof of Corollary 5.6. $\square$

Proof of Lemma 5.10. First we show that any two random variables $Z_n^{(l_1)}(x_1)$ and $Z_n^{(l_2)}(x_2)$ with $l_1, l_2 \in \{1, 2, 3\}$, $x_1, x_2 \in [0, T]$, become asymptotically independent for $x_1 \neq x_2$ which yields that the expectation in (5.4) factorizes in the limit. Let $x_1 < x_2$ and define

$$\Omega_n^-(x_1) = \Big\{ \max_{l=1,2} t_{i_n^{(l)}(\max_{l'=1,2} t_{i_n^{(l')}(x_1)+k}^{(l')})+k,n}^{(l)} \leq \frac{x_1 + x_2}{2} \Big\},$$

$$\Omega_n^+(x_2) = \Big\{ \frac{x_1 + x_2}{2} \leq \min_{l=1,2} t_{i_n^{(l)}(\min_{l'=1,2} t_{i_n^{(l')}(x_2)-k-1}^{(l')})-k-1,n}^{(l)} \Big\},$$

$$\Omega_n(x_1, x_2) = \Omega_n^-(x_1) \cap \Omega_n^+(x_2).$$

Here, $\Omega_n(x_1, x_2)$ describes the subset of Ω on which the set of intervals used for the construction of $Z_n^{(l_1)}(x_1)$ and the set of intervals used for the construction of $Z_n^{(l_2)}(x_2)$ are separated by $(x_1 + x_2)/2$. Then there exist measurable functions g_1, g_2 such that

$$Z_n^{(l_1)}(x_1) \mathbf{1}_{\Omega_n^-(x_1)} = g_1((N_n(t))_{t \in [0,(x_1+x_2)/2]}, (\overline{W}_t^{(l_1)})_{t \in [0,(x_1+x_2)/2]}) \mathbf{1}_{\Omega_n^-(x_1)},$$

$$Z_n^{(l_2)}(x_2) \mathbf{1}_{\Omega_n^+(x_2)} = g_2((N_n(t) - N_n((x_1 + x_2)/2))_{t \in [(x_1+x_2)/2,\infty)},$$

$$(\overline{W}_t^{(l_2)} - \overline{W}_{(x_1+x_2)/2}^{(l_2)})_{t \in [(x_1+x_2)/2,\infty)}) \mathbf{1}_{\Omega_n^+(x_2)}$$

where $N_n(t) = (N_n^{(1)}(t), N_n^{(2)}(t))^*$, $N_n^{(l)}(t) = \sum_{i \in \mathbb{N}} \mathbf{1}_{\{t_{i,n}^{(l)} \leq t\}}$, $l = 1, 2$, denotes the Poisson processes which create the stopping times. These identities yield that the random variables $Z_n^{(l_1)}(x_1) \mathbf{1}_{\Omega_n^-(x_1)}$, $\mathbf{1}_{\Omega_n^-(x_1)}$ are independent from the random variables $Z_n^{(l_2)}(x_2) \mathbf{1}_{\Omega_n^+(x_2)}$, $\mathbf{1}_{\Omega_n^+(,x_2)}$ because the processes $\overline{W}$ and $N_n(t)$, have independent increments. Hence we get

$$\mathbb{E}[h_{p_1}^{(l_1)}(Z_n^{(l_1)}(x_1)) h_{p_2}^{(l_2)}(Z_n^{(l_2)}(x_2)) \mathbf{1}_{\Omega_n^-(x_1)} \mathbf{1}_{\Omega_n^+(x_2)}]$$

$$= \mathbb{E}[h_{p_1}^{(l_1)}(Z_n^{(l_1)}(x_1)) \mathbf{1}_{\Omega_n^-(x_1)}] \mathbb{E}[h_{p_2}^{(l_2)}(Z_n^{(l_2)}(x_2)) \mathbf{1}_{\Omega_n^+(x_2)}]$$

$$= \frac{\mathbb{E}[h_{p_1}^{(l_1)}(Z_n^{(l_1)}(x_1)) \mathbf{1}_{\Omega_n^-(x_1)} \mathbf{1}_{\Omega_n^+(x_2)}]}{\mathbb{E}[\mathbf{1}_{\Omega_n^+(x_2)}]} \frac{\mathbb{E}[h_{p_2}^{(l_2)}(Z_n^{(l_2)}(x_2)) \mathbf{1}_{\Omega_n^+(x_2)} \mathbf{1}_{\Omega_n^-(x_1)}]}{\mathbb{E}[\mathbf{1}_{\Omega_n^-(x_1)}]}$$

$$= \frac{\mathbb{E}[h_{p_1}^{(l_1)}(Z_n^{(l_1)}(x_1)) \mathbf{1}_{\Omega_n(x_1,x_2)}] \mathbb{E}[h_{p_2}^{(l_2)}(Z_n^{(l_2)}(x_2)) \mathbf{1}_{\Omega_n(x_1,x_2)}]}{\mathbb{E}[\mathbf{1}_{\Omega_n(x_1,x_2)}]}$$

which is equivalent to

$$\mathbb{E}[h_{p_1}^{(l_1)}(Z_n^{(l_1)}(x_1)) h_{p_2}^{(l_2)}(Z_n^{(l_2)}(x_2)) | \Omega_n(x_1, x_2)]$$

$$= \mathbb{E}[h_{p_1}^{(l_1)}(Z_n^{(l_1)}(x_1)) | \Omega_n(x_1, x_2)] \mathbb{E}[h_{p_2}^{(l_2)}(Z_n^{(l_2)}(x_2)) | \Omega_n(x_1, x_2)]. \quad (5.14)$$

Further we obtain using Markov's inequality

$$\mathbb{P}(\Omega_n(x_1,x_2)^c) \leq \mathbb{P}(\Omega_n^-(x_1)^c) + \mathbb{P}(\Omega_n^+(x_2)^c)$$

$$= \mathbb{P}\Big(\max_{l=1,2} t^{(l)}_{i_n^{(l)}(\max_{l'=1,2} t^{(l')}_{i_n^{(l')}(x_1)+k})+k,n} > \frac{x_1+x_2}{2}\Big)$$

$$+ \mathbb{P}\Big(\frac{x_1+x_2}{2} > \min_{l=1,2} t^{(l)}_{i_n^{(l)}(\min_{l'=1,2} t^{(l')}_{i_n^{(l')}(x_2)-k-1})-k-1,n}\Big)$$

$$\leq 2K_{\lambda_1,\lambda_2} \frac{1/n}{|x_1-x_2|}$$

for a generic constant K_{λ_1,λ_2} which yields

$$\mathbb{P}(\Omega_n(x_1,x_2)) \geq 1 - \frac{K/n}{|x_1-x_2|}. \tag{5.15}$$

Denote $P = P_1 + P_2 + P_3$, $A_T(\varepsilon) = \{(x_1,\dots,x_P) \in [0,T]^P | \exists i,j : |x_i - x_j| \leq \varepsilon\}$ and $B_T(\sigma) = \{(x_1,\dots x_P) \in [0,T]^P : x_{\sigma(1)} \leq x_{\sigma(2)} \leq \dots \leq x_{\sigma(P)}\}$ where σ denotes a permutation of $\{1,\dots,P\}$. Using (5.14), (5.15), the inequality

$$|\mathbb{E}[X] - \mathbb{E}[X|A]| \leq 2K \frac{1-\mathbb{P}(A)}{\mathbb{P}(A)}$$

which holds for any bounded random variable $X \leq K$ and any set A with $\mathbb{P}(A) > 0$ together with the boundedness of $g, h_p^{(l)}$, $l = 1,2,3$, yields

$$\Big|(5.4) - \int_{[0,T]^{P_1+P_2+P_3}} g(x_1,\dots,x_{P_1},x_1',\dots,x_{P_2}',x_1'',\dots,x_{P_3}'')$$

$$\times \prod_{p=1}^{P_1} \mathbb{E}[h_p^{(1)}(Z_n^{(1)}(x_p))] \prod_{p=1}^{P_2} \mathbb{E}[h_p^{(2)}(Z_n^{(2)}(x_p'))]$$

$$\times \prod_{p=1}^{P_3} \mathbb{E}[h_p^{(3)}(Z_n^{(3)}(x_p''))] dx_1 \dots dx_{P_1} dx_1' \dots dx_{P_2}' dx_1'' \dots dx_{P_3}''\Big|$$

$$\leq K\lambda^{P\otimes}(A_T(\varepsilon))$$

$$+ \sum_{\sigma \in S_P} \Big|(5.4) - \int_{B_T(\sigma)\backslash A_T(\varepsilon)} g(x_1,\dots,x_P)$$

$$\times \mathbb{E}\Big[\prod_{p=1}^{P_1} h_p^{(1)}(Z_n^{(1)}(x_p)) \prod_{p=P_1+1}^{P_1+P_2} h_p^{(2)}(Z_n^{(2)}(x_p))$$

$$\times \prod_{p=P_1+P_2+1}^{P} h_p^{(3)}(Z_n^{(3)}(x_p)) | \bigcap_{i=2}^{P} \Omega_n(x_{\sigma(i-1)},x_{\sigma(i)})) \Big] dx_1 \dots dx_P\Big|$$

$$+ \sum_{\sigma \in S_P} \Bigg| \int_{B_T(\sigma) \setminus A_T(\varepsilon)} g(x_1, \ldots, x_P)$$

$$\times \prod_{l=1}^{3} \prod_{p=\sum_{m<l} P_m+1}^{\sum_{m \leq l} P_m} \mathbb{E}[h_p^{(l)}(Z_n^{(l)}(x_p))| \bigcap_{i=2}^{P} \Omega_n(x_{\sigma(i-1)}, x_{\sigma(i)}))] dx_1 \ldots dx_P$$

$$- \int_{B_T(\sigma) \setminus A_T(\varepsilon)} g(x_1, \ldots, x_P) \prod_{l=1}^{3} \prod_{p=\sum_{m<l} P_m+1}^{\sum_{m \leq l} P_m} \mathbb{E}[h_p^{(l)}(Z_n^{(l)}(x_p))] dx_1 \ldots dx_P \Bigg|$$

$$\leq K \boldsymbol{\lambda}^{P \otimes}(A_T(\varepsilon))$$

$$+ \sum_{\sigma \in S_P} K \int_{B_T(\sigma) \setminus A_T(\varepsilon)} \frac{1 - \mathbb{P}(\bigcap_{i=2}^{P} \Omega_n(x_{\sigma(i-1)}, x_{\sigma(i)}))}{\mathbb{P}(\bigcap_{i=2}^{P} \Omega_n(x_{\sigma(i-1)}, x_{\sigma(i)}))} dx_1 \ldots dx_P$$

$$+ \sum_{\sigma \in S_P} K \int_{B_T(\sigma) \setminus A_T(\varepsilon)} \sum_{j=1}^{P} \frac{1 - \mathbb{P}(\bigcap_{i=2}^{P} \Omega_n(x_{\sigma(i-1)}, x_{\sigma(i)}))}{\mathbb{P}(\bigcap_{i=2}^{P} \Omega_n(x_{\sigma(i-1)}, x_{\sigma(i)}))} dx_1 \ldots dx_P$$

$$\leq K \boldsymbol{\lambda}^{P \otimes}(A_T(\varepsilon))$$

$$+ \sum_{\sigma \in S_P} K \int_{B_T(\sigma) \setminus A_T(\varepsilon)} \frac{(P+1) \sum_{i=2}^{P} \mathbb{P}(\Omega_n(x_{\sigma(i-1)}, x_{\sigma(i)})^c)}{1 - \sum_{i=2}^{P} \mathbb{P}(\Omega_n(x_{\sigma(i-1)}, x_{\sigma(i)})^c)} dx_1 \ldots dx_P$$

$$\leq K \boldsymbol{\lambda}^{P \otimes}(A_T(\varepsilon)) + \binom{P}{2} K(P+1) \frac{PK/n}{\varepsilon} \left(1 - P \frac{K/n}{\varepsilon}\right)^{-1}$$

for all $\varepsilon > 0$ where $\boldsymbol{\lambda}^{P \otimes}$ denotes the Lebesgue measure in $\mathbb{R}^P$. This term vanishes as $n \to \infty$ and then $\varepsilon \to 0$. Next observe that it holds

$$\Bigg| (5.5) - \int_{[0,T]^{P_1+P_2+P_3}} g(x_1, \ldots, x_{P_1}, x_1', \ldots, x_{P_2}', x_1'', \ldots, x_{P_3}'')$$

$$\times \prod_{p=1}^{P_1} \mathbb{E}[h_p^{(1)}(Z_n^{(1)}(x_p))] \prod_{p=1}^{P_2} \mathbb{E}[h_p^{(2)}(Z_n^{(2)}(x_p'))]$$

$$\times \prod_{p=1}^{P_3} \mathbb{E}[h_p^{(3)}(Z_n^{(3)}(x_p''))] dx_1 \ldots dx_{P_1} dx_1' \ldots dx_{P_2}' dx_1'' \ldots dx_{P_3}'' \Bigg|$$

$$\leq K \varepsilon^P + KT^P \sum_{l=1}^{3} \sum_{p=1}^{P_l} \sup_{x \in [\varepsilon, T]} \Bigg| \int h_p^{(l)}(y) \Gamma^{(l)}(dy) - \mathbb{E}[h_p^{(l)}(Z_n^{(l)}(x))] \Bigg|.$$

Hence it remains to show that there exists some measure $\Gamma^{(l)}$ on $\mathbb{R}^d$ with

$$\sup_{x \in [\varepsilon, T]} \Bigg| \int h_p^{(l)}(y) \Gamma^{(l)}(dy) - \mathbb{E}[h_p^{(l)}(Z_n^{(l)}(x))] \Bigg| \to 0 \qquad (5.16)$$

as $n \to \infty$ for all $\varepsilon > 0$.

For proving (5.16) observe that it holds $Z_n^{(l)}(x) \stackrel{\mathcal{L}}{=} Z_1^{(l)}(nx)$. Since the Poisson processes generating the observation times in π_1 and the Brownian motion $\overline{W}^{(l)}$ are independent stationary processes the law of $Z_1^{(l)}(nx)$ depends on n only through the fact that all occurring intervals are bounded to the left by 0. Hence if $\Omega(n, m, x)$ denotes the set on which all intervals needed for the construction of $Z_1^{(l)}(mx)$ are within $[mx - nx, \infty)$, it follows that all members of the sequence $(Z_1^{(l)}(mx) \mathbb{1}_{\Omega(n,m,x)}))_{m \geq n}$ have the same law. This yields because of $\mathbb{P}(\Omega(n, m, x)) = \mathbb{P}(\Omega(n, n, x))$, $m \geq n$, and

$$\lim_{n \to \infty} \mathbb{P}(\Omega(n, n, x)) = 1, \ x > 0,$$

that the sequence $(Z_1^{(l)}(nx))_{n \in \mathbb{N}}$ converges in law for $x > 0$. Hence the sequence $(Z_n^{(l)}(x))_{n \in \mathbb{N}}$ converges in law for $x > 0$ as well. By the stationarity of the processes the law of the limit, which we denote by $\Gamma^{(l)}$, does not depend on x. Finally we obtain the uniform convergence in (5.16) because of $\mathbb{P}(\Omega(n, n, x)) \leq \mathbb{P}(\Omega(n, n, \varepsilon))$ for $x \geq \varepsilon$. $\qquad \square$

Proof of Lemma 5.12. First, note that the variables $\widehat{Z}_{n,m}^{(l_p)}(s_p)$ and $\widehat{Z}_{n,m}^{(l_{p'})}(s_{p'})$ are $\mathcal{S}$-conditionally independent if we are on the set $\Omega(n, s_p, s_{p'})$ on which $\widehat{Z}_{n,m}^{(l_p)}(s_p)$ and $\widehat{Z}_{n,m}^{(l_{p'})}(s_{p'})$ contain no common observation intervals. Without loss of generality let $s_p < s_{p'}$. Using the Markov inequality and the notation (2.28) we get

$$\mathbb{P}(\Omega(n, s_p, s_{p'})^c) \leq \mathbb{P}\Big(\max_{m=1,2} \tau_{n,+}^{(m)} \big(\max_{m'=1,2} t_{i_n^{(m')}(s_p) + K_n, n}^{(m')} \big) \geq s_p + (s_{p'} - s_p)/2 \Big)$$

$$+ \mathbb{P}\Big(\min_{m=1,2} \tau_{n,-}^{(m)} \big(\min_{m'=1,2} t_{i_n^{(m')}(s_{p'}) - K_n - 1, n}^{(m')} \big) \leq s_p + (s_{p'} - s_p)/2 \Big)$$

$$\leq 2 K_{\lambda_1, \lambda_2} \frac{K_n/n}{(s_{p'} - s_p)/2}$$

for a generic constant K_{λ_1, λ_2}. The latter tends to zero as $n \to \infty$ because of $K_n/n \to 0$ by assumption. Hence, we may assume $\widehat{Z}_{n,m}^{(l_p)}(s_p)$ and $\widehat{Z}_{n,m}^{(l_{p'})}(s_{p'})$ to be $\mathcal{S}$-conditionally independent, and it remains to prove (5.9) for $P = 1$. By assumption $Z^{(l)}(s)$ follows a continuous distribution. If we establish weak convergence of the $\mathcal{S}$-conditional distribution of $\widehat{Z}_{n,1}^{(l)}(s)$ to the (unconditional) one of $Z^{(l)}(s)$, then (5.9) follows from the Portmanteau theorem and dominated convergence.

Hence it remains to prove

$$Z_n^{(l)}(U_n^{(l)}(s)) \xrightarrow{\mathcal{L}_\mathcal{S}} Z^{(l)}(s) \qquad (5.17)$$

where $\overset{\mathcal{L}_{\mathcal{S}}}{\longrightarrow}$ denotes convergence of the $\mathcal{S}$-conditional distributions. From here on we will give the proof only in the case (5.7). However the proofs in the cases (5.6) and (5.8) are up to some changes in notation identical. Note that by construction we have

$$Z_n^{(l)}(U_n^{(l)}(s)) = Z_n^{(l)}(U_{n,K_n}^{(l)}(ns)) \overset{\mathcal{L}}{=} Z_1^{(l)}(U_{1,K_n}^{(l)}(ns)),$$

where $U_{n,m}^{(l)}(s) \sim \mathcal{U}[t_{i_n^{(2)}(s)-m-1,n}^{(2)}, t_{i_n^{(2)}(s)+m,n}^{(2)}]$ and hence showing (5.17) is equivalent to showing

$$Z_1^{(l)}(U_{1,K_n}^{(1)}(ns)) \overset{\mathcal{L}_{\mathcal{S}}}{\longrightarrow} Z^{(l)}(s) \tag{5.18}$$

because $Z^{(l)}(s)$ is independent of $\mathcal{S}$.

As shown in the last paragraph of the proof of Lemma 5.10 the laws of $Z_1^{(l)}(ns)$ are equal to the law of $Z^{(l)}(s)$ restricted to sets whose probability tends to 1 as $n \to \infty$. This property yields that as $Z_1^{(l)}(ns)$ is a function of the observation intervals around time ns, and increments of the Brownian motions $\overline{W}^{(l)}$ over these intervals, the distribution of $Z^{(l)}(s)$ can be characterized as follows: Construct two independent Poisson processes on $(-\infty, \infty)$ with intensities λ_l and jump times $\tilde{t}_i^{(l)}$ and two independent Brownian motions $(\widetilde{W}_t^{(l)})_{t \in (-\infty,\infty)}$ with $\widetilde{W}_0^{(l)} = 0$. Here, a Poisson process $(N_t)_{t \in \mathbb{R}}$ on $(-\infty, \infty)$ can be constructed from two independent Poisson processes $(N_t^{(m)})_{t \geq 0}$, $m = 1, 2$, by setting $N_t = N_t^{(1)}$ for $t \geq \tau$ and $N_{\tau-t} = -N_t^{(2)}$ where τ denotes the first jump time of $N^{(1)}$. Then we use the function $f^{(l)}$ from the defintion $Z_1^{(l)}(ns)$ and apply it to the observation intervals based on $\tilde{t}_i^{(l)}$ and increments $\widetilde{W}^{(l)}$ over these intervals around time s. The random variable resulting from this construction has the same law as $Z^{(l)}(s)$. Based on these observations and the fact that the increments of the Brownian motions are merely independent standard normal random variables scaled by the interval length we conclude that for showing (5.18) it is sufficient to show that the $\mathcal{S}$-conditional distribution of the intervals used in the construction of $Z_1^{(l)}(U_{1,K_n}^{(1)}(ns))$ converges to the intervals based on $\tilde{t}_i^{(l)}$ around time s.

To formalize this idea we define for $l = 1, 2$

$$\tilde{t}_{0,n}^{(l)}(s) = U_{1,K_n}^{(l)}(ns),$$

$$\tilde{t}_{k,n}^{(l)}(s) = \inf\{t_{i,1}^{(l)} | t_{i,1}^{(l)} > \tilde{t}_{k-1,n}^{(l)}(s)\}, \quad k \geq 1,$$

$$\tilde{t}_{k,n}^{(l)}(s) = \sup\{t_{i,1}^{(l)} | t_{i,1}^{(l)} < \tilde{t}_{k+1,n}^{(l)}(s)\}, \quad k \leq -1.$$

Since the number of intervals occuring in $Z_1^{(l)}(U_{1,K_n}^{(l)}(ns))$ is bounded in probability and consequently the number of intervals used in the construction of $Z^{(l)}(s)$ sketched above is also bounded in probability for showing (5.18) it suffices to prove

$$((\tilde{t}_{k,n}^{(1)}(s) - \tilde{t}_{k-1,n}^{(1)}(s))_{k=-K+1,\ldots,K}, (\tilde{t}_{k,n}^{(2)}(s) - \tilde{t}_{k-1,n}^{(2)}(s))_{k=-K+1,\ldots,K})$$
$$\xrightarrow{\mathcal{L_S}} ((E_k^{(1)})_{k=-K+1,\ldots,K}, (E_k^{(2)})_{k=-K+1,\ldots,K}) \quad (5.19)$$

for all $K \in \mathbb{N}$ where $E_k^{(1)}, E_k^{(2)}$ are i.i.d. exponentially distributed random variables with parameters λ_1, λ_2, respectively.

We first show

$$(\tilde{t}_{k,n}^{(1)}(s) - \tilde{t}_{k-1,n}^{(1)}(s))_{k=-K+1,\ldots,K} \xrightarrow{\mathcal{L_S}} (E_k^{(1)})_{k=-K+1,\ldots,K}. \quad (5.20)$$

To prove this convergence we consider the $\mathcal{S}$-conditional characteristic function

$$\mathbb{E}\Big[\exp\Big(i\sum_{k=-K+1}^{K} v_k(\tilde{t}_{k,n}^{(1)}(s) - \tilde{t}_{k-1,n}^{(1)}(s))\Big)\Big|\mathcal{S}\Big] \quad (5.21)$$

$$= O_{\mathbb{P}}(K_n^{-1/2}) + \sum_{j=-K_n^*(1)}^{K_n^*(1)} |\mathcal{I}_{i_1^{(1)}(ns)+j,1}^{(1)}| \Big(\sum_{j'=-K_n^*(1)}^{K_n^*(1)} |\mathcal{I}_{i_1^{(1)}(ns)+j',1}^{(1)}|\Big)^{-1}$$

$$\times \exp\Big(i\sum_{k=1}^{K-1} \big(v_{-k}|\mathcal{I}_{i_1^{(1)}(ns)+j-k,1}^{(1)}| + v_{k+1}|\mathcal{I}_{i_1^{(1)}(ns)+j+k,1}^{(1)}|\big)\Big)$$

$$\times \mathbb{E}\Big[\exp\big(iv_0(U_1(ns) - t_{i_1^{(1)}(ns)+j-1,1}^{(1)}(s))$$

$$+ iv_1(t_{i_1^{(1)}(ns)+j,1}^{(1)}(s) - U_1(ns))\big)\Big|\mathcal{S}, U_1(ns) \in \mathcal{I}_{i_1^{(1)}(ns)+j,1}^{(1)}\Big]$$

where we used that the number of observations of $X^{(1)}$ in the interval

$$[t_{i_n^{(2)}(s)-K_n-1,n}^{(2)}, t_{i_n^{(2)}(s)+K_n,n}^{(2)}]$$

equals $(2K_n^*(1) + 1) + O_{\mathbb{P}}(K_n^{1/2})$ with $K_n^*(1) = \lfloor K_n\lambda_1/\lambda_2 \rfloor$. With a random variable $\kappa \sim \mathcal{U}[0,1]$ independent of $\mathcal{S}$ the conditional expectation in the last line equals

$$\mathbb{E}\Big[\exp\big(iv_0\kappa|\mathcal{I}_{i_1^{(1)}(ns)+j,1}^{(1)}| + iv_1(1-\kappa)|\mathcal{I}_{i_1^{(1)}(ns)+j,1}^{(1)}|\big)\Big)\big|\mathcal{S}\Big]$$

$$= \frac{\exp\big(iv_0|\mathcal{I}_{i_1^{(1)}(ns)+j,1}^{(1)}|\big) - \exp\big(iv_1|\mathcal{I}_{i_1^{(1)}(ns)+j,1}^{(1)}|\big)}{i(v_0 - v_1)|\mathcal{I}_{i_1^{(1)}(ns)+j,1}^{(1)}|}.$$

Except for $j = 0$ the length of each observation interval $\mathcal{I}^{(1)}_{i^{(1)}_1(ns)+j,1}$ is exponentially distributed, up to asymptotically negligible boundary effects for $j < 0$, with parameter λ_1. It follows easily that (5.21) has asymptotically the same distribution as

$$\Big(\sum_{j=-K^*_n(1)}^{K^*_n(1)} \widetilde{E}_j \Big)^{-1} \sum_{j=-K^*_n(1)}^{K^*_n(1)} \widetilde{E}_j \exp \Big(i \sum_{k=1}^{K-1} (v_{-k}\widetilde{E}_{j-k} + v_{k+1}\widetilde{E}_{j+k}) \Big)$$
$$\times \frac{\exp(iv_0\widetilde{E}_j) - \exp(iv_1\widetilde{E}_j)}{i(v_0 - v_1)\widetilde{E}_j}$$

for i.i.d. exponentials $\widetilde{E}_j$, $j \in \mathbb{N}$, with parameter λ_1. Expanding by $(2K^*_n + 1)^{-1}$ and using the law of large numbers (note that the summands are independent for $|j - j'| > 2K + 1$), this expression converges almost surely to

$$\mathbb{E}\Big[\lambda_1 \widetilde{E}_0 \exp \Big(i \sum_{k=1}^{K-1} (v_{-k}\widetilde{E}_{j-k} + v_{k+1}\widetilde{E}_{j+k}) \Big)(i(v_0 - v_1)\widetilde{E}_0)^{-1}$$
$$\times \big(\exp(iv_0\widetilde{E}_0) - \exp(iv_1\widetilde{E}_0) \big) \Big]$$
$$= \mathbb{E}\Big[\exp \Big(i \sum_{k=1}^{K-1} (v_{-k}\widetilde{E}_{j-k} + v_{k+1}\widetilde{E}_{j+k}) \Big) \Big]$$
$$\times \int_0^\infty \lambda_1 x \frac{\exp(iv_0 x) - \exp(iv_1 x)}{i(v_0 - v_1)x} \lambda_1 e^{-\lambda_1 x} dx$$
$$= \mathbb{E}\Big[\exp \Big(i \sum_{k=1}^{K-1} (v_{-k}\widetilde{E}_{j-k} + v_{k+1}\widetilde{E}_{j+k}) \Big) \Big] \frac{\lambda_1}{\lambda_1 - iv_0} \frac{\lambda_1}{\lambda_1 - iv_1}$$

which is the characteristic function of a vector of $2K$ independent $\text{Exp}(\lambda_1)$-distributed random variables. This yields (5.20).

Analogously to (5.20) we obtain

$$(\tilde{t}^{(2)}_{k,n}(s) - \tilde{t}^{(2)}_{k-1,n}(s))_{k=-K+1,\dots,K} \xrightarrow{\mathcal{L}s} (E^{(2)}_k)_{k=-K+1,\dots,K}, \qquad (5.22)$$

and finally (5.20) and (5.22) yield (5.19), because the Poisson process has stationary increments and because of the independence of the two processes we have that $\tilde{t}^{(1)}_{k,n}(s) - \tilde{t}^{(1)}_{k-1,n}(s)$ and $\tilde{t}^{(2)}_{k',n}(s) - \tilde{t}^{(2)}_{k'-1,n}(s)$ are asymptotically independent, because dependency only occurs in the $O_{\mathbb{P}}(K_n^{-1/2})$-term of (5.21) which is asymptotically negligible. $\qquad\square$

Part II

Applications

6 Estimating Spot Volatility

Our goal in this chapter is to estimate the *spot volatilities* $\sigma_s^{(1)}$, $\sigma_s^{(2)}$ at some specific time $s \in [0, T]$. In addition to that we would like to estimate the *spot correlation* ρ_s between the two Gaussian processes $C^{(1)}$ and $C^{(2)}$. If we allow σ to be discontinuous we are additionally interested in estimating the left limits $\sigma_{s-}^{(1)}, \sigma_{s-}^{(2)}, \rho_{s-}$.

For finding suitable estimators note that we have

$$\sum_{i:\mathcal{I}_{i,n}^{(l)} \subset (s,s+b]} (\Delta_{i,n}^{(l)} C^{(l)})^2 \xrightarrow{\mathbb{P}} \int_s^{s+b} (\sigma_u^{(l)})^2 du, \quad l = 1, 2,$$

$$\sum_{i,j:\mathcal{I}_{i,n}^{(1)} \cup \mathcal{I}_{j,n}^{(2)} \subset (s,s+b]} \Delta_{i,n}^{(1)} C^{(1)} \Delta_{j,n}^{(2)} C^{(2)} \mathbb{1}_{\{\mathcal{I}_{i,n}^{(1)} \cap \mathcal{I}_{j,n}^{(2)} \neq \emptyset\}} \xrightarrow{\mathbb{P}} \int_s^{s+b} \rho_u \sigma_u^{(1)} \sigma_u^{(2)} du$$

for any $b > 0$ by Corollary 2.19 and Theorem 2.22 (compare also Example 2.23). Using the right-continuity of σ we further get

$$\frac{1}{b} \int_s^{s+b} (\sigma_u^{(l)})^2 du \to (\sigma_s^{(l)})^2, \quad l = 1, 2,$$

$$\frac{1}{b} \int_s^{s+b} \rho_u \sigma_u^{(1)} \sigma_u^{(2)} du \to \rho_s \sigma_s^{(1)} \sigma_s^{(2)}$$

as $b \to 0$. Hence if we choose a deterministic sequence $(b_n)_{n \in \mathbb{N}} \subset (0, \infty)$ such that b_n converges to zero sufficiently slowly we obtain

$$\frac{1}{b_n} \sum_{i:\mathcal{I}_{i,n}^{(l)} \subset (s,s+b_n]} (\Delta_{i,n}^{(l)} C^{(l)})^2 \xrightarrow{\mathbb{P}} (\sigma_s^{(l)})^2, \quad l = 1, 2,$$

$$\frac{1}{b_n} \sum_{i,j:\mathcal{I}_{i,n}^{(1)} \cup \mathcal{I}_{j,n}^{(2)} \subset (s,s+b_n]} \Delta_{i,n}^{(1)} C^{(1)} \Delta_{j,n}^{(2)} C^{(2)} \mathbb{1}_{\{\mathcal{I}_{i,n}^{(1)} \cap \mathcal{I}_{j,n}^{(2)} \neq \emptyset\}} \xrightarrow{\mathbb{P}} \rho_s \sigma_s^{(1)} \sigma_s^{(2)}.$$

In general we do not observe the Gaussian processes $C^{(1)}$, $C^{(2)}$ but only the processes $X^{(1)}$, $X^{(2)}$. However, we will see that asymptotically this distinction here is not important because only jumps in the interval $(s, s + b_n]$ enter the

© Springer Fachmedien Wiesbaden GmbH, part of Springer Nature 2019
O. Martin, *High-Frequency Statistics with Asynchronous and Irregular Data*,
Mathematische Optimierung und Wirtschaftsmathematik I Mathematical Optimization
and Economathematics, https://doi.org/10.1007/978-3-658-28418-3_6

estimation and "large" jumps vanish from the interval $(s, s + b_n]$ as $n \to \infty$ and $b_n \to 0$. Hence we can define consistent estimators via

$$(\tilde{\sigma}_n^{(l)}(s, +))^2 = \frac{1}{b_n} \sum_{i : \mathcal{I}_{i,n}^{(l)} \subset (s, s+b_n]} (\Delta_{i,n}^{(l)} X^{(l)})^2, \quad l = 1, 2,$$

$$\tilde{\rho}_n^*(s, +) = \frac{1}{\tilde{\sigma}_n^{(1)}(s, +)\tilde{\sigma}_n^{(2)}(s, +)} \frac{1}{b_n} \sum_{i,j : \mathcal{I}_{i,n}^{(1)} \cup \mathcal{I}_{j,n}^{(2)} \subset (s, s+b_n]} \Delta_{i,n}^{(1)} X^{(1)} \Delta_{j,n}^{(2)} X^{(2)}$$

$$\times \mathbf{1}_{\{\mathcal{I}_{i,n}^{(1)} \cap \mathcal{I}_{j,n}^{(2)} \neq \emptyset\}}.$$

In the small sample there still might be relatively large jumps of $X^{(l)}$ present in the interval $(s, s + b_n]$. To avoid overestimation of $\sigma_s^{(l)}$ we can use truncated increments to exclude unusually large increments which potentially include jumps from the estimation by using the following estimators

$$(\hat{\sigma}_n^{(l)}(s, +))^2 = \frac{1}{b_n} \sum_{i : \mathcal{I}_{i,n}^{(l)} \subset (s, s+b_n]} (\Delta_{i,n}^{(l)} X^{(l)})^2 \mathbf{1}_{\{|\Delta_{i,n}^{(l)} X^{(l)}| \leq \beta |\mathcal{I}_{i,n}^{(l)}|^{\varpi}\}}, \quad l = 1, 2,$$

$$\hat{\rho}_n^*(s, +) = \frac{1}{\hat{\sigma}_n^{(1)}(s, +)\hat{\sigma}_n^{(2)}(s, +)} \frac{1}{b_n} \sum_{i,j : \mathcal{I}_{i,n}^{(1)} \cup \mathcal{I}_{j,n}^{(2)} \subset (s, s+b_n]} \Delta_{i,n}^{(1)} X^{(1)} \Delta_{j,n}^{(2)} X^{(2)}$$

$$\times \mathbf{1}_{\{\mathcal{I}_{i,n}^{(1)} \cap \mathcal{I}_{j,n}^{(2)} \neq \emptyset\}} \mathbf{1}_{\{|\Delta_{i,n}^{(1)} X^{(1)}| \leq \beta |\mathcal{I}_{i,n}^{(1)}|^{\varpi}\}} \mathbf{1}_{\{|\Delta_{j,n}^{(2)} X^{(2)}| \leq \beta |\mathcal{I}_{j,n}^{(2)}|^{\varpi}\}}$$

for fixed constants $\beta > 0$ and $\varpi \in (0, 1/2)$; compare the discussion at the beginning of Section 2.3.

Analogously we define $\tilde{\sigma}_n^{(l)}(s, -)$, $\tilde{\rho}_n^*(s, -)$ and $\hat{\sigma}_n^{(l)}(s, -)$, $\hat{\rho}_n^*(s, -)$ where we replace the interval $(s, s + b_n]$ in the index of the sums with the interval $(s - b_n, s)$ such that we use increments over intervals to the left of s instead of intervals to the right of s for the estimation.

For practical purposes one might require an estimator for ρ which lies in the interval $[-1, 1]$. Such an estimator can be easily obtained from the estimator introduced above by setting the estimator to 1 if it is larger than 1 and by setting it to -1 if it is less than -1. Therefore we define

$$\tilde{\rho}_n(s, +) = (\tilde{\rho}_n^*(s, +) \wedge 1) \vee (-1), \qquad \tilde{\rho}_n(s, -) = (\tilde{\rho}_n^*(s, -) \wedge 1) \vee (-1),$$

$$\hat{\rho}_n(s, +) = (\hat{\rho}_n^*(s, +) \wedge 1) \vee (-1), \qquad \hat{\rho}_n(s, -) = (\hat{\rho}_n^*(s, -) \wedge 1) \vee (-1),$$

as proposed in Section 3.2 of [22]. Those estimators are also better than the unbounded estimators $\tilde{\rho}_n^*(s, +)$, $\tilde{\rho}_n^*(s, -)$, $\hat{\rho}_n^*(s, +)$, $\hat{\rho}_n^*(s, -)$ in the sense that they are always closer to the true values ρ_s, $\rho_{s-} \in [-1, 1]$ than the unbounded estimators.

6.1 The Results

The following two theorems show that the above constructions indeed yield consistent estimators and clarifies what it means for b_n to converge to zero sufficiently fast. Theorem 6.1 states that if we work with arbitrary observation times that may also be endogenous the estimator $\tilde{\sigma}_n^{(l)}(s,+)$ is consistent for $\sigma_s^{(l)}$, while Theorem 6.2 yields that if we restrict ourselves to exogenous observation times all above estimators are consistent.

Theorem 6.1. *Suppose Condition 1.3 holds and the sequence $(b_n)_{n \in \mathbb{N}}$ fulfils $b_n \to 0$, $\mathbb{E}[(\delta_n^{(l)}(s))^2]/b_n^2 \to 0$ where $\delta_n^{(l)}(s) = \sup\{|\mathcal{I}_{i,n}^{(l)}| : t_{i-1,n}^{(l)} < s + b_n\}$. Then it holds*

$$(\tilde{\sigma}_n^{(l)}(s,+))^2 \xrightarrow{\mathbb{P}} (\sigma_s^{(l)})^2, l = 1, 2. \tag{6.1}$$

The reason why we need a bound on $\delta_n^{(l)}(s)$ here instead of a bound on $|\pi_n|_T$ for some $T > 0$ will become clear in the proof of Theorem (6.1).

Denote by $\Omega(s,-) = \{\sigma_{s-}^{(1)}\sigma_{s-}^{(2)} \neq 0\}$, $\Omega(s,+) = \{\sigma_s^{(1)}\sigma_s^{(2)} \neq 0\}$ the sets on which neither $C^{(1)}$ nor $C^{(2)}$ vanishes to the left respectively to the right of s. Only on these sets it is possible to estimate ρ_s respectively ρ_{s-}.

Theorem 6.2. *Suppose Condition 1.3 holds, the observation scheme is exogenous and we have $b_n \to 0$, $|\pi_n|_T/b_n \xrightarrow{\mathbb{P}} 0$ for some deterministic $T > s$. Then it holds*

$$\begin{aligned}
(\tilde{\sigma}_n^{(l)}(s,+))^2 &\xrightarrow{\mathbb{P}} (\sigma_s^{(l)})^2, & (\tilde{\sigma}_n^{(l)}(s,-))^2 &\xrightarrow{\mathbb{P}} (\sigma_{s-}^{(l)})^2, \\
(\hat{\sigma}_n^{(l)}(s,+))^2 &\xrightarrow{\mathbb{P}} (\sigma_s^{(l)})^2, & (\hat{\sigma}_n^{(l)}(s,-))^2 &\xrightarrow{\mathbb{P}} (\sigma_{s-}^{(l)})^2,
\end{aligned} \tag{6.2}$$

for $l = 1, 2$,

$$\tilde{\rho}_n(s,+)\mathbb{1}_{\Omega(s,+)} \xrightarrow{\mathbb{P}} \rho_s\mathbb{1}_{\Omega(s,+)}, \quad \tilde{\rho}_n(s,-)\mathbb{1}_{\Omega(s,-)} \xrightarrow{\mathbb{P}} \rho_{s-}\mathbb{1}_{\Omega(s,-)}, \tag{6.3}$$

and if we further assume $(b_n)^{-1}(G_{1,1}^n(s+b_n) - G_{1,1}^n(s-b_n)) = O_{\mathbb{P}}(1)$, see (2.39) for the definition of $G_{1,1}^n$, we also have

$$\hat{\rho}_n(s,+)\mathbb{1}_{\Omega(s,+)} \xrightarrow{\mathbb{P}} \rho_s\mathbb{1}_{\Omega(s,+)}, \quad \hat{\rho}_n(s,-)\mathbb{1}_{\Omega(s,-)} \xrightarrow{\mathbb{P}} \rho_{s-}\mathbb{1}_{\Omega(s,-)}. \tag{6.4}$$

Although the convergence $(\tilde{\sigma}_n^{(l)}(s,+))^2 \xrightarrow{\mathbb{P}} (\sigma_s^{(l)})^2$ has already been proven in Theorem 6.1 for the more general setting of possibly endogenous observation times we include it in Theorem 6.2 as the condition $|\pi_n|_T/b_n \xrightarrow{\mathbb{P}} 0$ is slightly weaker than $\mathbb{E}[\delta^{(l)}(s)]/b_n \to 0$, $l = 1, 2$.

It is not obvious whether the condition

$$(b_n)^{-1}(G_{1,1}^n(s+b_n) - G_{1,1}^n(s-b_n)) = O_{\mathbb{P}}(1)$$

is really necessary to obtain the convergences of $\hat{\rho}_n(s,+)$ and $\hat{\rho}_n(s,-)$. However, by the method of proof used here, I didn't manage to derive the needed estimates without it.

Example 6.3. *In the setting of Poisson sampling introduced in Definition 5.1 the assumptions made for Theorems 6.1 and 6.2 are fulfilled. We have $b_n \to 0$, $\mathbb{E}[\delta_n^{(l)}(s)]/b_n \to 0$ and $|\pi_n|_T/b_n \xrightarrow{\mathbb{P}} 0$ by Lemma 5.2 for $b_n = n^{-\alpha}$ with $\alpha \in (0,1)$. Further $(b_n)^{-1}(G_{1,1}^n(s+b_n) - G_{1,1}^n(s-b_n)) = O_{\mathbb{P}}(1)$ holds because*

$$\mathbb{E}[(b_n)^{-1}(G_{1,1}^n(s+b_n) - G_{1,1}^n(s-b_n))$$
$$= \mathbb{E}[(nb_n)^{-1}(G_{1,1}^1(n(s+b_n)) - G_{1,1}^1(n(s-b_n)))]$$

is bounded by some constant; compare the proof of Corollary 5.6. □

In some of the upcoming applications when testing for jumps we are not only interested in estimating the spot volatility and spot correlation at fixed deterministic times, but also at random times e.g. we would like to estimate $\sigma_{S_p}^{(1)}$, $\sigma_{S_p}^{(2)}$, ρ_{S_p} where $S_p = \inf\{s \geq 0 : \|\Delta X_s\| > 1\}$ denotes the first time where X has a jump whose norm is larger than 1. This is also possible and the results are given in the following corollary.

Corollary 6.4. *Let τ be an $(\mathcal{F}_t)_{t \geq 0}$-stopping time and suppose that we have $\mathbb{E}[(\delta_n^{(l)}(T))^2]/b_n^2 \to 0$, then it holds*

$$(\bar{\sigma}_n^{(l)}(\tau,+))^2 \mathbb{1}_{\{\tau < T\}} \xrightarrow{\mathbb{P}} (\sigma_\tau^{(l)})^2 \mathbb{1}_{\{\tau < T\}}, \quad l = 1, 2, \tag{6.5}$$

and if the observation scheme is exogenous and $|\pi_n|_T/b_n \xrightarrow{\mathbb{P}} 0$ is fulfilled, then it further holds

$$(\tilde{\sigma}_n^{(l)}(\tau,+))^2 \mathbb{1}_{\{\tau < T\}} \xrightarrow{\mathbb{P}} (\sigma_\tau^{(l)})^2 \mathbb{1}_{\{\tau < T\}}, \quad l = 1, 2,$$
$$(\hat{\sigma}_n^{(l)}(\tau,+))^2 \mathbb{1}_{\{\tau < T\}} \xrightarrow{\mathbb{P}} (\sigma_\tau^{(l)})^2 \mathbb{1}_{\{\tau < T\}}, \quad l = 1, 2,$$
$$\tilde{\rho}_n^{(l)}(\tau,+) \mathbb{1}_{\Omega(s,+)} \mathbb{1}_{\{\tau < T\}} \xrightarrow{\mathbb{P}} \rho_\tau \mathbb{1}_{\Omega(\tau,+)} \mathbb{1}_{\{\tau < T\}},$$

and if we further have $(b_n)^{-1}(G_{1,1}^n(s+b_n) - G_{1,1}^n(s-b_n)) = O_{\mathbb{P}}(1)$ for any $s < T$, then it also holds

$$\hat{\rho}_n^{(l)}(\tau,+) \mathbb{1}_{\Omega(s,+)} \mathbb{1}_{\{\tau < T\}} \xrightarrow{\mathbb{P}} \rho_\tau \mathbb{1}_{\Omega(\tau,+)} \mathbb{1}_{\{\tau < T\}}. \tag{6.6}$$

Suppose additionally that $(\sigma_t)_{t\geq 0}$ is an Itô semimartingale and the stopping time τ fulfils $\|\Delta X_\tau\| + \|\Delta \sigma_\tau\| > 0$ almost surely. Then it holds

$$(\tilde{\sigma}_n^{(l)}(\tau, -))^2 \mathbb{1}_{\{\tau < T\}} \xrightarrow{\mathbb{P}} (\sigma_{\tau-}^{(l)})^2 \mathbb{1}_{\{\tau < T\}}, \quad l = 1, 2,$$

$$(\hat{\sigma}_n^{(l)}(\tau, -))^2 \mathbb{1}_{\{\tau < T\}} \xrightarrow{\mathbb{P}} (\sigma_{\tau-}^{(l)})^2 \mathbb{1}_{\{\tau < T\}}, \quad l = 1, 2, \tag{6.7}$$

$$\tilde{\rho}_n^{(l)}(\tau, -) \mathbb{1}_{\Omega(s, -)} \mathbb{1}_{\{\tau < T\}} \xrightarrow{\mathbb{P}} \rho_{\tau-} \mathbb{1}_{\Omega(\tau-)} \mathbb{1}_{\{\tau < T\}},$$

and if we assume again that $(b_n)^{-1}(G_{1,1}^n(s + b_n) - G_{1,1}^n(s - b_n)) = O_{\mathbb{P}}(1)$ for any $s < T$ it also holds

$$\hat{\rho}_n^{(l)}(\tau, -) \mathbb{1}_{\Omega(s, -)} \mathbb{1}_{\{\tau < T\}} \xrightarrow{\mathbb{P}} \rho_{\tau-} \mathbb{1}_{\Omega(\tau, -)} \mathbb{1}_{\{\tau < T\}}. \tag{6.8}$$

The assumption $(b_n)^{-1}(G_{1,1}^n(s + b_n) - G_{1,1}^n(s - b_n)) = O_{\mathbb{P}}(1)$ is again only needed for the convergences of $\hat{\rho}_n^{(l)}(\tau, +)$ and $\hat{\rho}_n^{(l)}(\tau, -)$. Further it is not clear whether the additional assumption for the convergence for the left limits is necessary, but otherwise a different technique of proof would be needed. One might additionally argue that the condition $\|\Delta X_\tau\| + \|\Delta \sigma_\tau\| > 0$ is rather mild as on the set $\Delta \sigma_\tau = 0$ the estimators $\tilde{\sigma}_n^{(l)}(\tau, +)$, $\hat{\sigma}_n^{(l)}(\tau, +)$ are consistent for $\sigma_\tau^{(l)} = \sigma_{\tau-}^{(l)}$; compare the paragraph following Theorem 9.3.2 in [30].

Notes. *The results in this chapter are inspired by Theorem 9.3.2 in [30]. There similar results are proved in the setting of equidistant and synchronous observation times. In Section 13.3 of [30], a central limit theorem for the spot volatility estimators is derived in the same setting of equidistant and synchronous observation times. Probably up to additional terms due to the irregularity of the stopping times a similar central limit theorem could also be derived in our setting. However, as we are not directly interested in estimating spot volatility but will use the results in this chapter only to estimate variances of limiting variables in central limit theorems for other statistics we do not need a central limit theorem for the spot volatility estimators.*

The estimation of spot volatilities, spot covariances and spot correlation based on high-frequency data has been vastly discussed in the literature. However, I could not find an approach which works based on asynchronous observations for models with discontinuous X and σ. Selected literature apart from the textbook [30] which relates to our approach is briefly summarized in the following.

In [15] a kernel based estimator for the spot volatility σ_s in a model with continuous X and σ is derived based on equidistant and synchronous observations. They derive a central limit theorem for their estimator $\hat{\Gamma}_s$ as well as a central limit theorem for the uniform error bound $\sup_{s \in [0, T]} \|\hat{\Gamma}_s - \sigma_s\|$.

[1] discuss an estimator for spot covariations $\kappa_s = \rho_s \sigma_s^{(1)} \sigma_s^{(2)}$ based on asynchronous observations. However they include a synchronization step and work with the synchronized data instead of the asynchronous data itself.

Corollary 3.2 in [22] states how to estimate ρ_t under the assumption that $\rho_t \equiv \rho$ is constant based on asynchronous and exogenous observations using our approach presented in this chapter for fixed b_n.

[7] discuss estimators for spot volatilities, spot covariations and spot correlations based on asynchronously observed high-frequency data under the presence of microstructure noise using an approach different from ours based on spectral statistics. In their model, they require X to be continuous (which might be relaxed in their opinion) and σ to be Hölder-continuous.

6.2 The Proofs

Proof of Theorem 6.1. Denote by $\Omega(n, q)$ the set such that $N(q)$ realizes no jump in $(s, s+b_n]$. Using (2.49) for $f(x) = x^2$ and (1.13)–(1.16) we obtain

$$\mathbb{E}\big[|(\tilde{\sigma}_n^{(l)}(s,+))^2 - \frac{1}{b_n} \sum_{i:t_{i-1,n}^{(l)} \in (s,s+b_n)} (\Delta_{i,n}^{(l)} C^{(l)})^2|\mathbb{1}_{\Omega(n,q)}\big]$$

$$\leq \mathbb{E}\big[\frac{1}{b_n} \sum_{i:t_{i-1,n}^{(l)} \in (s,s+b_n)} |(\Delta_{i,n}^{(l)}(X^{(l)} - N^{(l)}(q)))^2 - (\Delta_{i,n}^{(l)} C^{(l)})^2|\mathbb{1}_{\Omega(n,q)}\big]$$

$$+ \frac{1}{b_n}\mathbb{E}[(\Delta_{i_n^{(l)}(s+b_n),n}^{(l)} X^{(l)})^2|]$$

$$\leq \mathbb{E}\big[\frac{1}{b_n} \sum_{i:t_{i-1,n}^{(l)} \in (s,s+b_n)} (K_\varepsilon(\Delta_{i,n}^{(l)}(B^{(l)} + M^{(l)}(q)))^2 + \theta(\varepsilon)(\Delta_{i,n}^{(l)} C^{(l)})^2)\big]$$

$$+ K\mathbb{E}\big[\frac{\delta_n^{(l)}(s)}{b_n}\big]$$

$$\leq \mathbb{E}\big[\frac{1}{b_n} \sum_{i\in\mathbb{N}} \mathbb{E}[K_\varepsilon(\Delta_{i,n}^{(l)} B^{(l)})^2 + K_\varepsilon(\Delta_{i,n}^{(l)} M^{(l)}(q))^2 + \theta(\varepsilon)(\Delta_{i,n}^{(l)} C^{(l)})^2|\mathcal{F}_{t_{i-1,n}^{(l)}}]$$

$$\times \mathbb{1}_{\{t_{i-1,n}^{(l)} \in (s,s+b_n)\}}\big] + K\mathbb{E}[\delta_n^{(l)}(s)/b_n]$$

$$\leq \mathbb{E}\big[\frac{1}{b_n} \sum_{i:t_{i-1,n}^{(l)} \in (s,s+b_n)} \mathbb{E}[K_\varepsilon K_q|\mathcal{I}_{i,n}^{(l)}|^2 + K_\varepsilon e_q|\mathcal{I}_{i,n}^{(l)}| + \theta(\varepsilon)K|\mathcal{I}_{i,n}^{(l)}||\mathcal{F}_{t_{i-1,n}^{(l)}}]\big]$$

$$+ K\mathbb{E}\big[\frac{\delta_n^{(l)}(s)}{b_n}\big]$$

$$= \mathbb{E}\big[\frac{1}{b_n} \sum_{i:t_{i-1,n}^{(l)} \in (s,s+b_n)} (K_\varepsilon K_q|\mathcal{I}_{i,n}^{(l)}|^2 + K_\varepsilon e_q|\mathcal{I}_{i,n}^{(l)}| + \theta(\varepsilon)K|\mathcal{I}_{i,n}^{(l)}|)\big] + K\mathbb{E}\big[\frac{\delta_n^{(l)}(s)}{b_n}\big]$$

$$\leq \mathbb{E}[(K_\varepsilon K_q \delta_n^{(l)}(s) + K_\varepsilon e_q + K\theta(\varepsilon))(b_n + \delta_n^{(l)}(s))/b_n] + K\mathbb{E}[\delta_n^{(l)}(s)/b_n] \quad (6.9)$$

which vanishes as $n, q \to \infty$ and $\varepsilon \to 0$ since it holds

$$\mathbb{E}[\delta_n^{(l)}(s)/b_n] \le (\mathbb{E}[(\delta_n^{(l)}(s))^2]/b_n^2)^{1/2} \to 0 \tag{6.10}$$

by assumption. Hence because of $\mathbb{P}(\Omega(n, q)) \to 0$ as $n \to \infty$ for any $q > 0$ the convergence (6.1) is equivalent to

$$\frac{1}{b_n} \sum_{i: t_{i-1,n}^{(l)} \in (s, s+b_n)} (\Delta_{i,n}^{(l)} C^{(l)})^2 \xrightarrow{\mathbb{P}} (\sigma_s^{(l)})^2, \quad l = 1, 2. \tag{6.11}$$

For proving (6.11) we want to apply Lemma C.2 with

$$\zeta_i^n = \frac{1}{b_n} (\Delta_{i_n^{(l)}(s)+i,n}^{(l)} C^{(l)})^2 - \frac{1}{b_n} (\sigma_s^{(l)})^2 |\mathcal{I}_{i_n^{(l)}(s)+i,n}^{(l)}|, \quad \mathcal{G}_i^n = \mathcal{F}_{t_{i_n^{(l)}(s)+i,n}^{(l)}} \tag{6.12}$$

We obtain using the conditional Itô isometry, compare Theorem 3.20 in [12], and the fact that σ is bounded

$$\mathbb{E}\Big[\sum_{i: t_{i_n^{(l)}(s)+i-1,n}^{(l)} \in (s,s+b_n)} |\mathbb{E}[\zeta_i^n | \mathcal{G}_i^n]|\Big]$$

$$= \frac{1}{b_n} \mathbb{E}\Big[\sum_{i: t_{i-1,n}^{(l)} \in (s,s+b_n)} \mathbb{E}\Big[\int_{t_{i-1,n}^{(l)}}^{t_{i,n}^{(l)}} |(\sigma_u^{(l)})^2 - (\sigma_s^{(l)})^2 | du | \mathcal{F}_{t_{i-1,n}^{(l)}} \Big]\Big]$$

$$\le \frac{1}{b_n} \mathbb{E}\Big[\sum_{i: t_{i-1,n}^{(l)} \in (s,s+b_n)} \mathbb{E}\Big[\varepsilon |\mathcal{I}_{i,n}^{(l)}| + K |\mathcal{I}_{i,n}^{(l)}| \mathbb{1}_{\{\sup_{u \in \mathcal{I}_{i,n}^{(l)}} |(\sigma_u^{(l)})^2 - (\sigma_s^{(l)})^2| > \varepsilon\}} |\mathcal{F}_{t_{i-1,n}^{(l)}} \Big]\Big]$$

$$\le \frac{1}{b_n} \mathbb{E}\Big[(b_n + \delta_n^{(l)}(s))(\varepsilon + K \mathbb{1}_{\{\sup_{u \in (s,s+b_n+\delta_n^{(l)}(s))} |(\sigma_u^{(l)})^2 - (\sigma_s^{(l)})^2| > \varepsilon\}})\Big]$$

$$\le \varepsilon + K \mathbb{P}\Big(\sup_{u \in (s,s+b_n+\delta_n^{(l)}(s))} |(\sigma_u^{(l)})^2 - (\sigma_s^{(l)})^2| > \varepsilon\Big) + (K + \varepsilon) \mathbb{E}[\delta_n^{(l)}(s)/b_n] \tag{6.13}$$

where the first term vanishes as $\varepsilon \to 0$, the second term vanishes as $n \to \infty$ for all $\varepsilon > 0$ because $\sigma^{(l)}$ is right-continuous and the third term vanishes as $n \to \infty$ by (6.10). Further we get

$$\mathbb{E}\Big[\sum_{i: t_{i_n^{(l)}(s)+i-1,n}^{(l)} \in (s,s+b_n)} \mathbb{E}[(\zeta_i^n)^2 | \mathcal{G}_i^n]\Big]$$

$$< \frac{2}{b_n^2} K \mathbb{E}[\delta_n^{(l)}(s)(b_n + \delta_n^{(l)}(s))] + \frac{2}{b_n^2} \mathbb{E}[(\sigma_s^{(l)})^4 \delta_n^{(l)}(s)(b_n + \delta_n^{(l)}(s))] \to 0 \tag{6.14}$$

where we used $(a-b)^2 \leq 2a^2 + 2b^2$, inequality (1.14), the boundedness of $\sigma^{(l)}$ and the assumption $\mathbb{E}[(\delta_n^{(l)}(s))^2]/b_n^2 \to 0$. Hence we obtain $\sum_i \mathbb{E}[\zeta_i^n|\mathcal{G}_i^n] \xrightarrow{\mathbb{P}} 0$ and $\sum_i \mathbb{E}[(\zeta_i^n)^2|\mathcal{G}_i^n] \xrightarrow{\mathbb{P}} 0$ from (6.13) and (6.14) which yields

$$\sum_{\substack{i:t_{i_n^{(l)}(s)+i-1,n}^{(l)} \in (s,s+b_n)}} \zeta_i^n \xrightarrow{\mathbb{P}} 0$$

by Lemma C.2. The convergence (6.1) then follows because of

$$\sum_{\substack{i:t_{i_n^{(l)}(s)+i-1,n}^{(l)} \in (s,s+b_n)}} \zeta_i^n = (\tilde{\sigma}_n^{(l)}(s,+))^2 - (\sigma_s^{(l)})^2 + O_\mathbb{P}(\delta_n^{(l)}(s)/b_n).$$

and because $\mathbb{E}[(\delta_n^{(l)}(s))^2]/b_n^2 \to 0$ implies $\delta_n^{(l)}(s)/b_n \xrightarrow{\mathbb{P}} 0$. $\square$

The following Lemma is needed in the proof of (6.3) and (6.4). The techniques used in the proof are taken from the proof of Theorem 3.1 in [22].

Lemma 6.5. *Let $Y^{(1)}, Y^{(2)}$ be martingales which are adapted to the filtration $(\mathcal{F}_t)_{t\geq 0}$. Further let $(s,t]$ be a deterministic interval and the observation scheme be exogenous. If the inequalities*

$$\mathbb{E}[(Y_{t'}^{(1)} - Y_{s'}^{(1)})^2|\mathcal{F}_{s'}] \leq K_{Y^{(1)}}(t'-s'), \quad \mathbb{E}[(Y_{t'}^{(2)} - Y_{s'}^{(2)})^2|\mathcal{F}_{s'}] \leq K_{Y^{(2)}}(t'-s'),$$
$$(6.15)$$

$$\mathbb{E}\big[(Y_{t'}^{(1)} - Y_{s'}^{(1)})^2(Y_{t'}^{(2)} - Y_{s'}^{(2)})^2|\mathcal{F}_{s'}\big] \leq K_{Y^{(1)},Y^{(2)}}(t'-s')^2, \qquad (6.16)$$

hold for all $s \leq s' \leq t' \leq t$ then we have

$$\mathbb{E}[(\sum_{i,j:\mathcal{I}_{i,n}^{(1)} \cup \mathcal{I}_{j,n}^{(2)} \subset (s,t]} \Delta_{i,n}^{(1)}Y^{(1)}\Delta_{j,n}^{(2)}Y^{(2)} \mathbb{1}_{\{\mathcal{I}_{i,n}^{(1)} \cap \mathcal{I}_{j,n}^{(2)} \neq \emptyset\}})^2|\sigma(\mathcal{F}_s, \mathcal{S})] \leq \widetilde{K}_{Y^{(1)},Y^{(2)}}(t-s)^2$$

where $\widetilde{K}_{Y^{(1)},Y^{(2)}} = 9K_{Y^{(1)}}K_{Y^{(2)}} + 3K_{Y^{(1)},Y^{(2)}}$.

Proof. To shorten notation set $\mathbb{1}_{i,j} := \mathbb{1}_{\{\mathcal{I}_{i,n}^{(1)} \cap \mathcal{I}_{j,n}^{(2)} \neq \emptyset\}}$. Note that it holds

$$\Big(\sum_{i,j:\mathcal{I}_{i,n}^{(1)} \cup \mathcal{I}_{j,n}^{(2)} \subset (s,t]} \Delta_{i,n}^{(1)}Y^{(1)}\Delta_{j,n}^{(2)}Y^{(2)} \mathbb{1}_{i,j}\Big)^2$$

$$= \sum_{i,j,i',j':\mathcal{I}_{i,n}^{(1)} \cup \mathcal{I}_{j,n}^{(2)} \cup \mathcal{I}_{i',n}^{(1)} \cup \mathcal{I}_{j',n}^{(2)} \subset (s,t]} \Delta_{i,n}^{(1)}Y^{(1)}\Delta_{j,n}^{(2)}Y^{(2)}\Delta_{i',n}^{(1)}Y^{(1)}\Delta_{j',n}^{(2)}Y^{(2)} \mathbb{1}_{i,j}\mathbb{1}_{i',j'}$$

$$= \sum_{i=i',j=j'} (\Delta_{i,n}^{(1)}Y^{(1)})^2(\Delta_{j,n}^{(2)}Y^{(2)})^2 \mathbb{1}_{i,j} \qquad (6.17)$$

$$+ \sum_{i=i',j\neq j'} (\Delta_{i,n}^{(1)}Y^{(1)})^2 \Delta_{j,n}^{(2)}Y^{(2)} \Delta_{j',n}^{(2)}Y^{(2)} \mathbb{1}_{i,j}\mathbb{1}_{i,j'} \tag{6.18}$$

$$+ \sum_{i\neq i',j=j'} \Delta_{i,n}^{(1)}Y^{(1)} \Delta_{i',n}^{(1)}Y^{(1)} (\Delta_{j,n}^{(2)}Y^{(2)})^2 \mathbb{1}_{i,j}\mathbb{1}_{i',j} \tag{6.19}$$

$$+ \sum_{i\neq i',j\neq j'} \Delta_{i,n}^{(1)}Y^{(1)} \Delta_{i',n}^{(1)}Y^{(1)} \Delta_{j,n}^{(2)}Y^{(2)} \Delta_{j',n}^{(2)}Y^{(2)} \mathbb{1}_{i,j}\mathbb{1}_{i',j'}. \tag{6.20}$$

In the following we discuss the terms (6.17)–(6.20) separately.

For discussing (6.17) we denote by $\Delta_{(i,j),n}^{(1,2)}Y^{(l)}$ the increment of $Y^{(l)}$, $l = 1, 2$, over the interval $\mathcal{I}_{i,n}^{(1)} \cap \mathcal{I}_{j,n}^{(2)}$ and by $\Delta_{(i,\{j\}),n}^{(l\backslash 3-l)}Y^{(l)}$ the increment of $Y^{(l)}$ over the interval $\mathcal{I}_{i,n}^{(l)} \setminus \mathcal{I}_{j,n}^{(3-l)}$ (which might also be the sum of increments over two distinct intervals). Using this notation, iterated expectations and the inequalities (6.15), (6.16) we obtain that the $\sigma(\mathcal{F}_s, \mathcal{S})$-conditional expectation of (6.17) is equal to

$$\sum_{i,j} \mathbb{E}[(\Delta_{(i,j),n}^{(1,2)}Y^{(1)} + \Delta_{(i,\{j\}),n}^{(1\backslash 2)}Y^{(1)})^2 (\Delta_{(i,j),n}^{(1,2)}Y^{(2)} + \Delta_{(j,\{i\}),n}^{(2\backslash 1)}Y^{(2)})^2 |\sigma(\mathcal{F}_s, \mathcal{S})]\mathbb{1}_{i,j}$$

$$= \sum_{i,j} \mathbb{E}[((\Delta_{(i,j),n}^{(1,2)}Y^{(1)})^2 + (\Delta_{(i,\{j\}),n}^{(1\backslash 2)}Y^{(1)})^2)(\Delta_{(j,\{i\}),n}^{(2\backslash 1)}Y^{(2)})^2$$

$$+ (\Delta_{(i,\{j\}),n}^{(1\backslash 2)}Y^{(1)})^2(\Delta_{(i,j),n}^{(1,2)}Y^{(2)})^2 |\sigma(\mathcal{F}_s, \mathcal{S})]\mathbb{1}_{i,j}$$

$$+ \sum_{i,j} \mathbb{E}[(\Delta_{(i,j),n}^{(1,2)}Y^{(1)})^2(\Delta_{(i,j),n}^{(1,2)}Y^{(2)})^2 |\sigma(\mathcal{F}_s, \mathcal{S})]$$

$$\leq \sum_{i,j} K_{Y^{(1)}} K_{Y^{(2)}} (|\mathcal{I}_{i,n}^{(1)}||\mathcal{I}_{j,n}^{(2)} \setminus \mathcal{I}_{i,n}^{(1)}| + |\mathcal{I}_{i,n}^{(1)} \setminus \mathcal{I}_{j,n}^{(2)}||\mathcal{I}_{i,n}^{(1)} \cap \mathcal{I}_{j,n}^{(2)}|)\mathbb{1}_{i,j}$$

$$+ \sum_{i,j} K_{Y^{(1)},Y^{(2)}} |\mathcal{I}_{i,n}^{(1)} \cap \mathcal{I}_{j,n}^{(2)}|^2$$

$$= (K_{Y^{(1)}} K_{Y^{(2)}} + K_{Y^{(1)},Y^{(2)}}) \sum_{i,j:\mathcal{I}_{i,n}^{(1)}\cup\mathcal{I}_{j,n}^{(2)}\subset(s,t]} |\mathcal{I}_{i,n}^{(1)}||\mathcal{I}_{j,n}^{(2)}|\mathbb{1}_{i,j}$$

$$\leq 3(K_{Y^{(1)}} K_{Y^{(2)}} + K_{Y^{(1)},Y^{(2)}})(|\pi_n|_t \wedge (t-s))(t-s).$$

For treating (6.18) we additionally denote by $\Delta_{(i,\{j,j'\}),n}^{(1\backslash 2)}Y^{(1)}$ the increment of $Y^{(1)}$ over the interval $\mathcal{I}_{i,n}^{(1)}\setminus(\mathcal{I}_{j,n}^{(2)}\cup\mathcal{I}_{j',n}^{(2)})$ (which might also be the sum of increments over up to three distinct intervals). We then obtain by using iterated expectations, the fact that $Y^{(1)}, Y^{(2)}$ are martingales, the Cauchy-Schwarz inequality and (6.15) that the $\sigma(\mathcal{F}_s, \mathcal{S})$-conditional expectation of (6.18) is bounded by

$$\sum_{i,j\neq j'} |\mathbb{E}[(\Delta_{(i,j),n}^{(1,2)}Y^{(1)} + \Delta_{(i,j'),n}^{(1,2)}Y^{(1)} + \Delta_{(i,\{j,j'\}),n}^{(1\backslash 2)}Y^{(1)})^2$$

$$\times \Delta_{(i,j),n}^{(1,2)}Y^{(2)} \Delta_{(i,j'),n}^{(1,2)}Y^{(2)} |\sigma(\mathcal{F}_s, \mathcal{S})]|\mathbb{1}_{i,j}\mathbb{1}_{i,j'}$$

$$= 2 \sum_{i,j \neq j'} |\mathbb{E}[\Delta^{(1,2)}_{(i,j),n} Y^{(1)} \Delta^{(1,2)}_{(i,j),n} Y^{(2)} \Delta^{(1,2)}_{(i,j'),n} Y^{(1)} \Delta^{(1,2)}_{(i,j'),n} Y^{(2)} | \sigma(\mathcal{F}_s, \mathcal{S})]|$$

$$\leq 2 \sum_{i,j \neq j'} K_{Y^{(1)}} K_{Y^{(2)}} |\mathcal{I}^{(1)}_{i,n} \cap \mathcal{I}^{(2)}_{j,n}| |\mathcal{I}^{(1)}_{i,n} \cap \mathcal{I}^{(2)}_{j',n}|$$

$$\leq 2 K_{Y^{(1)}} K_{Y^{(2)}} \sum_{i: \mathcal{I}^{(1)}_{i,n} \subset (s,t]} |\mathcal{I}^{(1)}_{i,n}|^2 \leq 2 K_{Y^{(1)}} K_{Y^{(2)}} (|\pi_n|_t \wedge (t-s))(t-s).$$

By symmetry (6.19) can be treated similarly as (6.18). Hence it remains to discuss (6.20). Using the notation introduced above we obtain that the $\sigma(\mathcal{F}_s, \mathcal{S})$-conditional expectation of (6.19) is equal to

$$\sum_{i \neq i', j \neq j'} \mathbb{E}[(\Delta^{(1,2)}_{(i,j),n} Y^{(1)} + \Delta^{(1,2)}_{(i,j'),n} Y^{(1)} + \Delta^{(1 \setminus 2)}_{(i,\{j,j'\}),n} Y^{(1)})$$

$$\times (\Delta^{(1,2)}_{(i',j),n} Y^{(1)} + \Delta^{(1,2)}_{(i',j'),n} Y^{(1)} + \Delta^{(1 \setminus 2)}_{(i',\{j,j'\}),n} Y^{(1)})$$

$$\times (\Delta^{(1,2)}_{(i,j),n} Y^{(2)} + \Delta^{(1,2)}_{(i',j),n} Y^{(2)} + \Delta^{(2 \setminus 1)}_{(j,\{i,i'\}),n} Y^{(2)})$$

$$\times (\Delta^{(1,2)}_{(i,j'),n} Y^{(2)} + \Delta^{(1,2)}_{(i',j'),n} Y^{(2)} + \Delta^{(2 \setminus 1)}_{(j',\{i,i'\}),n} Y^{(2)}) | \sigma(\mathcal{F}_s, \mathcal{S})] \mathbb{1}_{i,j} \mathbb{1}_{i',j'}$$

$$= \sum_{i \neq i', j \neq j'} \mathbb{E}[(\Delta^{(1,2)}_{(i,j),n} Y^{(1)} + \Delta^{(1,2)}_{(i,j'),n} Y^{(1)})(\Delta^{(1,2)}_{(i',j),n} Y^{(1)} + \Delta^{(1,2)}_{(i',j'),n} Y^{(1)})$$

$$\times (\Delta^{(1,2)}_{(i,j),n} Y^{(2)} + \Delta^{(1,2)}_{(i',j),n} Y^{(2)})(\Delta^{(1,2)}_{(i,j'),n} Y^{(2)} + \Delta^{(1,2)}_{(i',j'),n} Y^{(2)}) | \sigma(\mathcal{F}_s, \mathcal{S})]$$

$$\times \mathbb{1}_{i,j} \mathbb{1}_{i',j'}.$$

Here terms including factors like $\Delta^{(1 \setminus 2)}_{(i,\{j,j'\}),n} Y^{(1)}$ vanish using iterated expectations because $Y^{(1)}$ is a martingale and $\mathcal{I}^{(1)}_{i,n} \setminus (\mathcal{I}^{(2)}_{j,n} \cup \mathcal{I}^{(2)}_{j',n})$ overlaps with none of the other intervals for $i \neq i'$. Expanding the product above and observing that the expectation vanishes for all products where one of i, i', j, j' occurs exactly once yields that the $\sigma(\mathcal{F}_s, \mathcal{S})$-conditional expectation of (6.20) is equal to

$$\sum_{i \neq i', j \neq j'} \mathbb{E}[\Delta^{(1,2)}_{(i,j),n} Y^{(1)} \Delta^{(1,2)}_{(i',j'),n} Y^{(1)} \Delta^{(1,2)}_{(i,j),n} Y^{(2)} \Delta^{(1,2)}_{(i',j'),n} Y^{(2)}$$

$$+ \Delta^{(1,2)}_{(i,j),n} Y^{(1)} \Delta^{(1,2)}_{(i',j'),n} Y^{(1)} \Delta^{(1,2)}_{(i',j),n} Y^{(2)} \Delta^{(1,2)}_{(i,j'),n} Y^{(2)}$$

$$+ \Delta^{(1,2)}_{(i,j'),n} Y^{(1)} \Delta^{(1,2)}_{(i',j),n} Y^{(1)} \Delta^{(1,2)}_{(i,j),n} Y^{(2)} \Delta^{(1,2)}_{(i',j'),n} Y^{(2)}$$

$$+ \Delta^{(1,2)}_{(i,j'),n} Y^{(1)} \Delta^{(1,2)}_{(i',j),n} Y^{(1)} \Delta^{(1,2)}_{(i',j),n} Y^{(2)} \Delta^{(1,2)}_{(i,j'),n} Y^{(2)} | \sigma(\mathcal{F}_s, \mathcal{S})] \mathbb{1}_{i,j} \mathbb{1}_{i',j'}$$

$$= \sum_{i \neq i', j \neq j'} \mathbb{E}[\Delta^{(1,2)}_{(i,j),n} Y^{(1)} \Delta^{(1,2)}_{(i',j'),n} Y^{(1)} \Delta^{(1,2)}_{(i,j),n} Y^{(2)} \Delta^{(1,2)}_{(i',j'),n} Y^{(2)}$$

$$+ \Delta^{(1,2)}_{(i,j'),n} Y^{(1)} \Delta^{(1,2)}_{(i',j),n} Y^{(1)} \Delta^{(1,2)}_{(i',j),n} Y^{(2)} \Delta^{(1,2)}_{(i,j'),n} Y^{(2)} | \sigma(\mathcal{F}_s, \mathcal{S})] \mathbb{1}_{i,j} \mathbb{1}_{i',j'}$$

where the second identity holds as one of the intervals $\mathcal{I}_{i,n}^{(1)} \cap \mathcal{I}_{j,n}^{(2)}$, $\mathcal{I}_{i,n}^{(1)} \cap \mathcal{I}_{j',n}^{(2)}$, $\mathcal{I}_{i',n}^{(1)} \cap \mathcal{I}_{j,n}^{(2)}$, $\mathcal{I}_{i',n}^{(1)} \cap \mathcal{I}_{j',n}^{(2)}$ always has to be empty for $i \neq i'$, $j \neq j'$. We then further obtain using iterated expectations, the conditional Cauchy-Schwarz inequality and (6.16) that the $\sigma(\mathcal{F}_s, \mathcal{S})$-conditional expectation of (6.20) is bounded by

$$K_{Y^{(1)}} K_{Y^{(2)}} \sum_{i \neq i', j \neq j'} (|\mathcal{I}_{i,n}^{(1)} \cap \mathcal{I}_{j,n}^{(2)}||\mathcal{I}_{i',n}^{(1)} \cap \mathcal{I}_{j',n}^{(2)}| + |\mathcal{I}_{i,n}^{(1)} \cap \mathcal{I}_{j',n}^{(2)}||\mathcal{I}_{i',n}^{(1)} \cap \mathcal{I}_{j,n}^{(2)}|) \mathbb{1}_{i,j} \mathbb{1}_{i',j'}$$

$$\leq 2 K_{Y^{(1)}} K_{Y^{(2)}} \sum_{i : \mathcal{I}_{i,n}^{(1)} \subset (s,t]} |\mathcal{I}_{i,n}^{(1)}| \sum_{i' : \mathcal{I}_{i',n}^{(1)} \subset (s,t]} |\mathcal{I}_{i',n}^{(1)}| \leq 2 K_{Y^{(1)}} K_{Y^{(2)}} (t-s)^2.$$

We conclude the proof by combining the bounds for (6.17)–(6.20) and using Lemma 2.15. $\qquad\square$

Proof of Theorem 6.2. Without loss of generalization we may assume $s + b_n \leq T$ as $(b_n)_{n \in \mathbb{N}}$ is a deterministic sequence tending to zero and $s < T$.

We first prove $(\tilde{\sigma}_n^{(l)}(s, +))^2 \overset{\mathbb{P}}{\longrightarrow} (\sigma_s^{(l)})^2$. The proof is similar to the proof of Theorem 6.1 where the main difference is that we consider $\mathcal{S}$-conditional expectations. Indeed we obtain analogously to (6.9) using the inequalities (2.49), (1.8)–(1.10) and the fact that $|\pi_n|_T$ is $\mathcal{S}$-measurable

$$\mathbb{E}\big[|(\tilde{\sigma}_n^{(l)}(s, +))^2 - \frac{1}{b_n} \sum_{i : \mathcal{I}_{i,n}^{(l)} \subset (s, s+b_n]} (\Delta_{i,n}^{(l)} C^{(l)})^2 \mathbb{1}_{\Omega(n,q)} |\mathcal{S}\big]$$

$$\leq \mathbb{E}\big[\frac{1}{b_n} \sum_{i : \mathcal{I}_{i,n}^{(l)} \subset (s, s+b_n]} \big(K_\varepsilon (\Delta_{i,n}^{(l)} (B^{(l)} + M^{(l)}(q))^2 + \theta(\varepsilon)(\Delta_{i,n}^{(l)} C^{(l)})^2\big) |\mathcal{S}\big]$$

$$\leq K_\varepsilon K_q |\pi_n|_T + K_\varepsilon e_q + K\theta(\varepsilon)$$

which vanishes as $n, q \to \infty$ and $\varepsilon \to 0$. $\Omega(n,q)$ is defined as in the proof of Theorem 6.1 to be the set on which $N(q)$ realizes no jump within the interval $(s, s+b_n]$. Hence because of $\mathbb{P}(\Omega(n,q)) \to 1$ as $n \to \infty$ it suffices to prove

$$\frac{1}{b_n} \sum_{i : \mathcal{I}_{i,n}^{(l)} \subset (s, s+b_n]} (\Delta_{i,n}^{(l)} C^{(l)})^2 \overset{\mathbb{P}}{\longrightarrow} (\sigma_s^{(l)})^2. \tag{6.21}$$

Therefore we define ζ_i^n as in (6.12) and set $\mathcal{G}_i^n = \sigma(\mathcal{F}_{t_{i_n^{(l)}(s)+i,n}^{(l)}}, \mathcal{S})$. We then obtain similarly as in the proof of Theorem 6.1

$$\sum_{i : \mathcal{I}_{i_n^{(l)}(s)+i,n}^{(l)} \subset (s, s+b_n]} |\mathbb{E}[\zeta_i^n | \mathcal{G}_i^n]| \leq \varepsilon + K\mathbb{1}_{\{\sup_{u \in (s, s+b_n]} |(\sigma_u^{(l)})^2 - (\sigma_s^{(l)})^2| > \varepsilon\}} \overset{\mathbb{P}}{\longrightarrow} 0,$$

$$\tag{6.22}$$

$$\sum_{\substack{i:\mathcal{I}^{(l)}_{i^{(l)}_n(s)+i,n}\subset(s,s+b_n]}} \mathbb{E}[(\zeta^n_i)^2|\mathcal{G}^n_i] \le K\frac{(|\pi_n|_T)^2}{(b_n)^2} \xrightarrow{\mathbb{P}} 0.$$

Hence using Lemma C.2 we get $\sum_i \zeta^n_i \xrightarrow{\mathbb{P}} 0$ which yields (6.21) because of

$$\sum_{\substack{i:\mathcal{I}^{(l)}_{i^{(l)}_n(s)+i,n}\subset(s,s+b_n]}} \zeta^n_i = \frac{1}{b_n}\sum_{\substack{i:\mathcal{I}^{(l)}_{i,n}\subset(s,s+b_n]}} (\Delta^{(l)}_{i,n}C^{(l)})^2 - (\sigma^{(l)}_s)^2 + O_{\mathbb{P}}(|\pi_n|_T/b_n).$$

For proving $(\hat{\sigma}^{(l)}_n(s,+))^2 \xrightarrow{\mathbb{P}} (\sigma^{(l)}_s)^2$ we derive

$$\mathbb{E}\big[|(\tilde{\sigma}^{(l)}_n(s,+))^2 - (\hat{\sigma}^{(l)}_n(s,+))^2|\mathbf{1}_{\Omega(n,q)}|\mathcal{S}\big]$$

$$\le \frac{1}{b_n}\sum_{\substack{i:\mathcal{I}^{(l)}_{i,n}\subset(s,s+b_n]}} \mathbb{E}[(\Delta^{(l)}_{i,n}(X^{(l)} - N^{(l)}(q)))^2\mathbf{1}_{\{|\Delta^{(l)}_{i,n}X^{(l)}|>\beta|\mathcal{I}^{(l)}_{i,n}|^\varpi\}}|\mathcal{S}]$$

$$\le \frac{3}{b_n}\sum_{\substack{i:\mathcal{I}^{(l)}_{i,n}\subset(s,s+b_n]}} \big(\mathbb{E}[(\Delta^{(l)}_{i,n}B^{(l)}(q))^2|\mathcal{S}]$$

$$+ \mathbb{E}[(\Delta^{(l)}_{i,n}C^{(l)})^2\mathbf{1}_{\{|\Delta^{(l)}_{i,n}X^{(l)}|>\beta|\mathcal{I}^{(l)}_{i,n}|^\varpi\}}|\mathcal{S}] + \mathbb{E}[(\Delta^{(l)}_{i,n}M^{(l)}(q))^2|\mathcal{S}]\big)$$

$$\le K_q|\pi_n|_T + K(|\pi_n|_T)^{1/2-\varpi} + Ke_q \qquad (6.23)$$

where we used Lemma 1.4 and (2.77). Because (6.23) vanishes as first $n \to \infty$ and then $q \to \infty$ and $\mathbb{P}(\Omega(n,q)) \to 1$ as $n \to \infty$ for any $q > 0$ we obtain that the convergences in (6.1) for $\tilde{\sigma}^{(l)}_n(s,+)$ and $\hat{\sigma}^{(l)}_n(s,+)$ are equivalent. Hence we get $(\hat{\sigma}^{(l)}_n(s,+))^2 \xrightarrow{\mathbb{P}} (\sigma^{(l)}_s)^2$ because we have already shown that the convergence $(\tilde{\sigma}^{(l)}_n(s,+))^2 \xrightarrow{\mathbb{P}} (\sigma^{(l)}_s)^2$ holds.

The proofs for the convergences of $\tilde{\sigma}^{(l)}_n(s,-)$ and $\hat{\sigma}^{(l)}_n(s,-)$ are identical to the proofs for $\tilde{\sigma}^{(l)}_n(s,+)$ and $\hat{\sigma}^{(l)}_n(s,+)$. We only have to change the indices of the sums and as we condition on $\mathcal{S}$ all arguments remain valid. Hence alltogether we have shown (6.2).

For proving (6.3) and (6.4) first denote

$$\tilde{\kappa}_n(s,+) = \frac{1}{b_n}\sum_{\substack{i,j:\mathcal{I}^{(1)}_{i,n}\cup\mathcal{I}^{(2)}_{j,n}\in(s,s+b_n]}} \Delta^{(1)}_{i,n}X^{(1)}\Delta^{(2)}_{j,n}X^{(2)}\mathbf{1}_{\{\mathcal{I}^{(1)}_{i,n}\cap\mathcal{I}^{(2)}_{j,n}\ne\emptyset\}}$$

$$\hat{\kappa}_n(s,+) = \frac{1}{b_n}\sum_{\substack{i,j:\mathcal{I}^{(1)}_{i,n}\cup\mathcal{I}^{(2)}_{j,n}\in(s,s+b_n]}} \Delta^{(1)}_{i,n}X^{(1)}\Delta^{(2)}_{j,n}X^{(2)}\mathbf{1}_{\{\mathcal{I}^{(1)}_{i,n}\cap\mathcal{I}^{(2)}_{j,n}\ne\emptyset\}}$$

$$\times \mathbf{1}_{\{|\Delta^{(1)}_{i,n}X^{(1)}|\le\beta|\mathcal{I}^{(1)}_{i,n}|^\varpi\}}\mathbf{1}_{\{|\Delta^{(2)}_{j,n}X^{(2)}|\le\beta|\mathcal{I}^{(2)}_{j,n}|^\varpi\}}$$

and we define $\tilde{\kappa}_n(s,-)$, $\hat{\kappa}_n(s,-)$ analogously. Then (6.3) and (6.4) follow from (6.2),

$$
\begin{aligned}
\tilde{\kappa}_n(s,+) &\xrightarrow{\mathbb{P}} \rho_s \sigma_s^{(1)} \sigma_s^{(2)}, & \tilde{\kappa}_n(s,-) &\xrightarrow{\mathbb{P}} \rho_{s-} \sigma_{s-}^{(1)} \sigma_{s-}^{(2)}, \\
\hat{\kappa}_n(s,+) &\xrightarrow{\mathbb{P}} \rho_s \sigma_s^{(1)} \sigma_s^{(2)}, & \hat{\kappa}_n(s,-) &\xrightarrow{\mathbb{P}} \rho_{s-} \sigma_{s-}^{(1)} \sigma_{s-}^{(2)}
\end{aligned}
\tag{6.24}
$$

and the continuous mapping theorem for convergence in probability. Hence it remains to prove (6.24).

Like in the proof of (6.2) we will first consider $\tilde{\kappa}_n(s,+) \xrightarrow{\mathbb{P}} \rho_s \sigma_s^{(1)} \sigma_s^{(2)}$ and to this end we will prove

$$
\tilde{\kappa}^{(l)}(s,+) - \frac{1}{b_n} \sum_{i,j:\mathcal{I}_{i,n}^{(1)} \cup \mathcal{I}_{j,n}^{(2)} \subset (s,s+b_n]} \Delta_{i,n}^{(1)} C^{(1)} \Delta_{j,n}^{(2)} C^{(2)} \mathbb{1}_{\{\mathcal{I}_{i,n}^{(1)} \cap \mathcal{I}_{j,n}^{(2)} \neq \emptyset\}} \xrightarrow{\mathbb{P}} 0.
\tag{6.25}
$$

In fact it holds

$$
\tilde{\kappa}^{(l)}(s,+) - \frac{1}{b_n} \sum_{i,j:\mathcal{I}_{i,n}^{(1)} \cup \mathcal{I}_{j,n}^{(2)} \subset (s,s+b_n]} \Delta_{i,n}^{(l)} C^{(1)} \Delta_{j,n}^{(l)} C^{(2)} \mathbb{1}_{\{\mathcal{I}_{i,n}^{(1)} \cap \mathcal{I}_{j,n}^{(2)} \neq \emptyset\}}
$$

$$
= \sum_{l=1,2} \frac{1}{b_n} \sum_{i,j:\mathcal{I}_{i,n}^{(l)} \cup \mathcal{I}_{j,n}^{(3-l)} \subset (s,s+b_n]} \Delta_{i,n}^{(l)} B^{(l)}(q) \Delta_{j,n}^{(3-l)} X^{(3-l)} \mathbb{1}_{\{\mathcal{I}_{i,n}^{(1)} \cap \mathcal{I}_{j,n}^{(2)} \neq \emptyset\}} \tag{6.26}
$$

$$
- \frac{1}{b_n} \sum_{i,j:\mathcal{I}_{i,n}^{(1)} \cup \mathcal{I}_{j,n}^{(2)} \subset (s,s+b_n]} \Delta_{i,n}^{(1)} B^{(1)}(q) \Delta_{j,n}^{(2)} B^{(2)}(q) \mathbb{1}_{\{\mathcal{I}_{i,n}^{(1)} \cap \mathcal{I}_{j,n}^{(2)} \neq \emptyset\}} \tag{6.27}
$$

$$
+ \frac{1}{b_n} \sum_{i,j:\mathcal{I}_{i,n}^{(1)} \cup \mathcal{I}_{j,n}^{(2)} \subset (s,s+b_n]} \left(\Delta_{i,n}^{(1)} N^{(1)}(q) \Delta_{j,n}^{(2)} (X^{(2)} - B^{(2)}(q)) \right.
$$

$$
\left. + \Delta_{i,n}^{(1)} (C^{(1)} + M^{(1)}(q)) \Delta_{j,n}^{(2)} N^{(2)}(q) \right) \mathbb{1}_{\{\mathcal{I}_{i,n}^{(1)} \cap \mathcal{I}_{j,n}^{(2)} \neq \emptyset\}} \tag{6.28}
$$

$$
+ \sum_{l=1,2} \frac{1}{b_n} \sum_{i,j:\mathcal{I}_{i,n}^{(l)} \cup \mathcal{I}_{j,n}^{(3-l)} \subset (s,s+b_n]} \Delta_{i,n}^{(l)} M^{(l)}(q) \Delta_{j,n}^{(3-l)} C^{(3-l)} \mathbb{1}_{\{\mathcal{I}_{i,n}^{(l)} \cap \mathcal{I}_{j,n}^{(3-l)} \neq \emptyset\}}
$$

$$
\tag{6.29}
$$

$$
+ \frac{1}{b_n} \sum_{i,j:\mathcal{I}_{i,n}^{(1)} \cup \mathcal{I}_{j,n}^{(2)} \subset (s,s+b_n]} \Delta_{i,n}^{(1)} M^{(1)}(q) \Delta_{j,n}^{(2)} M^{(2)}(q) \mathbb{1}_{\{\mathcal{I}_{i,n}^{(1)} \cap \mathcal{I}_{j,n}^{(2)} \neq \emptyset\}}. \tag{6.30}
$$

where all terms (6.26)–(6.30) vanish as first $n \to \infty$ and then $q \to \infty$ which we will show in the following.

The $\mathcal{S}$-conditional expectation of the absolute value of (6.26) plus (6.27) is using (1.8), (1.12) and Jensen's inequality bounded by

$$\sum_{l=1,2} \frac{1}{b_n} \sum_{i:t_{i,n}^{(l)} \in (s,s+b_n]} K_q |\mathcal{I}_{i,n}^{(l)}|$$

$$\times \mathbb{E}\Big[\big(X^{(3-l)}_{t_{i_n^{(3-l)}(t_{i,n}^{(l)})\wedge i_n^{(3-l)}(s+b_n),n}} - X^{(3-l)}_{t_{i_n^{(3-l)}(t_{i-1,n}^{(l)})-1)\vee(i_n^{(3-l)}(s)-1),n}}\big)^2\Big|\mathcal{S}\Big]^{1/2}$$

$$+ \frac{1}{b_n} \sum_{i,j:\mathcal{I}_{i,n}^{(1)}\cup\mathcal{I}_{j,n}^{(2)}\subset(s,s+b_n]} K_q |\mathcal{I}_{i,n}^{(1)}||\mathcal{I}_{j,n}^{(2)}| \mathbb{1}_{\{\mathcal{I}_{i,n}^{(1)}\cap\mathcal{I}_{j,n}^{(2)}\neq\emptyset\}}$$

$$\leq (3|\pi_n|_T)^{1/2} \sum_{l=1,2} \frac{1}{b_n} \sum_{i:\mathcal{I}_{i,n}^{(l)}\in(s,s+b_n]} K_q|\mathcal{I}_{i,n}^{(l)}| + K_q|\pi_n|_T \frac{b_n}{b_n}$$

$$\leq K_q(|\pi_n|_T)^{1/2}$$

which vanishes as $n \to \infty$. Further, (6.28) vanishes as $n \to \infty$ because of

$$\lim_{n\to\infty} \mathbb{P}(\exists t \in (s,s+b_n] : \Delta N(q)_t \neq 0) = 0.$$

Using (1.9), (1.10) and (3.30) we may apply Lemma 6.5 for (6.29) which yields

$$\mathbb{E}[(6.29)^2|\mathcal{S}] \leq K e_q.$$

This vanishes as $q \to \infty$. Next consider the following decomposition of (6.30) using the notation from the proof of Lemma 6.5

$$(6.30) = \frac{1}{b_n} \sum_{i,j:\mathcal{I}_{i,n}^{(1)}\cup\mathcal{I}_{j,n}^{(2)}\subset(s,s+b_n]} \Delta^{(1,2)}_{(i,j),n}M^{(1)}(q)\Delta^{(1,2)}_{(i,j),n}M^{(2)}(q)$$

$$+ \sum_{l=1,2} \frac{1}{b_n} \sum_{i,j:\mathcal{I}_{i,n}^{(l)}\cup\mathcal{I}_{j,n}^{(3-l)}\subset(s,s+b_n]} \Delta^{(l\backslash3-l)}_{(i,\{j\}),n}M^{(l)}(q)\Delta^{(l,3-l)}_{(i,j),n}M^{(3-l)}(q)$$

$$\times \mathbb{1}_{\{\mathcal{I}_{i,n}^{(l)}\cap\mathcal{I}_{j,n}^{(3-l)}\neq\emptyset\}}$$

$$+ \frac{1}{b_n} \sum_{i,j:\mathcal{I}_{i,n}^{(1)}\cup\mathcal{I}_{j,n}^{(2)}\subset(s,s+b_n]} \Delta^{(1\backslash2)}_{(i,j),n}M^{(1)}(q)\Delta^{(2\backslash1)}_{(j,i),n}M^{(2)}(q)\mathbb{1}_{\{\mathcal{I}_{i,n}^{(1)}\cap\mathcal{I}_{j,n}^{(2)}\neq\emptyset\}}$$

where the $\mathcal{S}$-conditional expectation of the first sum can be bounded by e_q using (1.10) because of $|\Delta^{(1,2)}_{(i,j),n}M^{(1)}(q)\Delta^{(1,2)}_{(i,j),n}M^{(2)}(q)| \leq \|\Delta^{(1,2)}_{(i,j),n}M(q)\|^2$. Further the expectation of the square of the second and third sum can be bounded as in the proof of Lemma 6.5 by $K(e_q)^2$ because the condition (6.16) is only needed for the treatment of the sum involving terms of the form $(\Delta^{(1,2)}_{(i,j),n}M^{(1)}(q))^2(\Delta^{(1,2)}_{(i,j),n}M^{(2)}(q))^2$ which we here treat separately.

Hence we have proven (6.25) which yields that for proving $\tilde{\kappa}_n(s,+) \xrightarrow{\mathbb{P}} \rho_s \sigma_s^{(1)} \sigma_s^{(2)}$ it suffices to show

$$\frac{1}{b_n} \sum_{i,j:\mathcal{I}_{i,n}^{(1)} \cup \mathcal{I}_{j,n}^{(2)} \subset (s,s+b_n]} \Delta_{i,n}^{(1)} C^{(1)} \Delta_{j,n}^{(2)} C^{(2)} \mathbb{1}_{\{\mathcal{I}_{i,n}^{(1)} \cap \mathcal{I}_{j,n}^{(2)} \neq \emptyset\}} \xrightarrow{\mathbb{P}} \rho_s \sigma_s^{(1)} \sigma_s^{(2)}. \quad (6.31)$$

For proving (6.31) define

$$\zeta_i^n = \frac{1}{b_n} \sum_{j,k:\mathcal{I}_{j,n}^{(1)} \cup \mathcal{I}_{k,n}^{(2)} \subset (s+(i-1)b_n/r_n, s+ib_n/r_n]} \Delta_{j,n}^{(1)} C^{(1)} \Delta_{k,n}^{(2)} C^{(2)} \mathbb{1}_{\{\mathcal{I}_{j,n}^{(1)} \cap \mathcal{I}_{k,n}^{(2)} \neq \emptyset\}}$$
$$- \frac{1}{r_n} \rho_s \sigma_s^{(1)} \sigma_s^{(2)}$$

and $\mathcal{G}_i^n = \sigma(\mathcal{F}_{s+ib_n/r_n}, \mathcal{S})$ for some deterministic sequence $(r_n)_{n\in\mathbb{N}} \subset \mathbb{N}$ with $r_n \to \infty$ and $r_n |\pi_n|_T / b_n \xrightarrow{\mathbb{P}} 0$. Such a sequence $(r_n)_{n\in\mathbb{N}}$ always exists because of $|\pi_n|_T / b_n \xrightarrow{\mathbb{P}} 0$. Denote

$$L(n,i) = \{(j,k) : \mathcal{I}_{j,n}^{(1)} \cup \mathcal{I}_{k,n}^{(2)} \subset (s+(i-1)b_n/r_n, s+ib_n/r_n]\}.$$

We then obtain using iterated expectations, the fact that $C^{(1)}, C^{(2)}$ are martingales and a form of the conditional Itô isometry, compare Theorem 3.20 in [12], for two different integrals

$$\sum_{i=1}^{r_n} \mathbb{E}[\zeta_i^n | \mathcal{G}_{i-1}^n]$$
$$= \frac{1}{b_n} \sum_{i=1}^{r_n} \sum_{(j,k)\in L(n,i)} \mathbb{E}[\Delta_{(j,k),n}^{(1,2)} C^{(1)} \Delta_{(j,k),n}^{(1,2)} C^{(2)} | \mathcal{G}_{i-1}^n] - \rho_s \sigma_s^{(1)} \sigma_s^{(2)}$$
$$= \frac{1}{b_n} \sum_{i=1}^{r_n} \mathbb{E}[\int_{s+(i-1)b_n/r_n)}^{s+ib_n/r_n} (\rho_u \sigma_u^{(1)} \sigma_u^{(2)} - \rho_s \sigma_s^{(1)} \sigma_s^{(2)}) du \,|| \mathcal{G}_{i-1}^n] + O_\mathbb{P}(\frac{r_n |\pi_n|_T}{b_n}).$$

As argued for (6.13) and (6.22) this vanishes in probability as $n \to \infty$ because $\rho, \sigma^{(1)}, \sigma^{(2)}$ are right-continuous and bounded and because of the condition on r_n. Note that we obtain from the Cauchy-Schwarz inequality and (1.9)

$$\mathbb{E}[(C_t^{(1)} - C_s^{(1)})^2 (C_t^{(2)} - C_s^{(2)})^2 | \mathcal{F}, \mathcal{S}]$$
$$\leq (\mathbb{E}[(C_t^{(1)} - C_s^{(1)})^4 | \mathcal{F}, \mathcal{S}] \mathbb{E}[(C_t^{(2)} - C_s^{(2)})^4 | \mathcal{F}, \mathcal{S}])^{1/2} \leq K(t-s)^2$$

for any $t \geq s \geq 0$. Hence an application of Lemma 6.5 yields

$$\sum_{i=1}^{r_n} \mathbb{E}[(\zeta_i^n)^2 | \mathcal{G}_{i-1}^n]$$

$$\leq \frac{2}{b_n^2} \sum_{i=1}^{r_n} \mathbb{E}\Big[\Big(\sum_{(j,k)\in L(n,i)} \Delta_{j,n}^{(l)} C^{(1)} \Delta_{k,n}^{(l)} C^{(2)} \mathbb{1}_{\{\mathcal{I}_{j,n}^{(1)} \cap \mathcal{I}_{k,n}^{(2)} \neq \emptyset\}} \Big)^2 \Big| \mathcal{G}_{i-1}^n\Big] + r_n \frac{K}{r_n^2}$$

$$\leq \frac{2}{b_n^2} \sum_{i=1}^{r_n} K \frac{b_n^2}{r_n^2} + \frac{K}{r_n} = \frac{K}{r_n}$$

which vanishes as $r_n \to 0$ for $n \to \infty$. Lemma C.2 then yields

$$\sum_{i=1}^{r_n} \zeta_i^n = \frac{1}{b_n} \sum_{i,j:\mathcal{I}_{i,n}^{(1)} \cup \mathcal{I}_{j,n}^{(2)} \subset (s,s+b_n]} \Delta_{i,n}^{(1)} C^{(1)} \Delta_{j,n}^{(2)} C^{(2)} \mathbb{1}_{\{\mathcal{I}_{i,n}^{(1)} \cap \mathcal{I}_{j,n}^{(2)} \neq \emptyset\}}$$

$$- \rho_s \sigma_s^{(1)} \sigma_s^{(2)} + O_{\mathbb{P}}(r_n |\pi_n|_T / b_n) \overset{\mathbb{P}}{\longrightarrow} 0$$

which is equivalent to (6.25) and hence implies $\tilde{\kappa}_n(s,+) \overset{\mathbb{P}}{\longrightarrow} \rho_s \sigma_s^{(1)} \sigma_s^{(2)}$.

For proving the convergence of $\hat{\kappa}_n(s,+)$ we will proceed as in (6.23) and show that $\tilde{\kappa}_n(s,+) - \hat{\kappa}_n(s,+)$ vanishes asymptotically. Because the terms containing increments of $N^{(l)}(q)$ vanish as in (6.28) we have

$$|\tilde{\kappa}_n(s,+) - \hat{\kappa}_n(s,+)|$$

$$\leq \frac{1}{b_n} \sum_{i,j:\mathcal{I}_{i,n}^{(1)} \cup \mathcal{I}_{j,n}^{(2)} \subset (s,s+b_n]} |\Delta_{i,n}^{(1)}(X^{(1)} - N^{(1)}(q)| |\Delta_{j,n}^{(2)}(X^{(2)} - N^{(2)}(q))|$$

$$\times \mathbb{1}_{\{\mathcal{I}_{i,n}^{(1)} \cap \mathcal{I}_{j,n}^{(2)} \neq \emptyset\}} \mathbb{1}_{\{|\Delta_{i,n}^{(1)} X^{(1)}| > \beta |\mathcal{I}_{i,n}^{(1)}|^{\varpi} \vee |\Delta_{j,n}^{(2)} X^{(2)}| > \beta |\mathcal{I}_{j,n}^{(2)}|^{\varpi}\}} + o_{\mathbb{P}}(1)$$

$$\leq \frac{1}{b_n} \sum_{i,j:\mathcal{I}_{i,n}^{(1)} \cup \mathcal{I}_{j,n}^{(2)} \subset (s,s+b_n]} |\Delta_{i,n}^{(1)}(X^{(1)} - N^{(1)}(q)| |\Delta_{j,n}^{(2)}(X^{(2)} - N^{(2)}(q))|$$

$$\times \mathbb{1}_{\{\mathcal{I}_{i,n}^{(1)} \cap \mathcal{I}_{j,n}^{(2)} \neq \emptyset\}} \big(\mathbb{1}_{\{|\Delta_{i,n}^{(1)} X^{(1)}| > \beta |\mathcal{I}_{i,n}^{(1)}|^{\varpi}\}} + \mathbb{1}_{\{|\Delta_{j,n}^{(2)} X^{(2)}| > \beta |\mathcal{I}_{j,n}^{(2)}|^{\varpi}\}} \big) + o_{\mathbb{P}}(1).$$
$$(6.32)$$

Then (6.32) is using the Cauchy-Schwarz inequality, inequalities (1.8)-(1.10) and (2.77) bounded by

$$K(K_q(|\pi_n|_T)^{1/2} + (|\pi_n|_T)^{(1/2-\varpi)/2} + (e_q)^{1/2})$$

$$\times \frac{1}{b_n} \sum_{i,j:\mathcal{I}_{i,n}^{(1)} \cup \mathcal{I}_{j,n}^{(2)} \subset (s,s+b_n]} |\mathcal{I}_{i,n}^{(1)}|^{1/2} |\mathcal{I}_{j,n}^{(2)}|^{1/2} \mathbb{1}_{\{\mathcal{I}_{i,n}^{(1)} \cap \mathcal{I}_{j,n}^{(2)} \neq \emptyset\}}$$

$$\leq K(K_q(|\pi_n|_T)^{1/2} + (|\pi_n|_T)^{(1/2-\varpi)/2} + (e_q)^{1/2}) \frac{1}{b_n} \big(G_{1,1}^n(s+b_n) - G_{1,1}^n(s) \big)$$

which vanishes using the assumption $(b_n)^{-1}(G_{1,1}^n(s+b_n) - G_{1,1}^n(s)) = O_{\mathbb{P}}(1)$. Hence we have shown

$$\tilde{\kappa}_n(s,+) - \hat{\kappa}_n(s,+) \xrightarrow{\ \mathbb{P}\ } 0.$$

The convergence of $\hat{\kappa}_n(s,+)$ in (6.24) then follows from the convergence of $\tilde{\kappa}_n(s,+)$ which we have already proven. The proof for the convergences of $\tilde{\kappa}_n(s,-)$ and $\hat{\kappa}_n(s,-)$ is similar. $\qquad\square$

Proof of Corollary 6.4. The proofs of (6.5)–(6.6) are identical to the proofs in Theorem 6.1 and 6.2. To see this observe that if we replace s with τ in the proof of the convergences for $\tilde{\sigma}_n^{(l)}(s,+)$, $\tilde{\kappa}_n(s,+)$ the occurring random times $t_{i_n^{(l)}(\tau)+i,n}^{(l)}$ respectively $\tau + ib_n/r_n$ remain stopping times and hence all arguments remain valid. Also we may always work with $\mathcal{F}_\tau$-conditional expectations. However if we proceed similarly with the proofs for $\tilde{\sigma}_n^{(l)}(s,-)$, $\tilde{\kappa}_n(s,-)$ we obtain the random times $t_{i_n^{(l)}(\tau)-i,n}^{(l)}$ respectively $\tau - ib_n/r_n$ which are not necessarily stopping times with respect to the initial filtration anymore.

For still being able to use the arguments from the proof of Theorem 6.1 we will look at a different filtration. The following arguments are taken from Step 6 in the proof of Theorem 9.3.2 from [30]. First note that if σ is itself an Itô semimartingale with respect to the filtration $(\mathcal{F}_t)_{t\geq 0}$ i.e. σ is of the form

$$\sigma_t = \sigma_0 + \int_0^t b_s^\sigma \, ds + \int_0^t \sigma_s^\sigma \, dW_s^\sigma + \int_0^t \int_{\mathbb{R}^2} \delta^\sigma(s,z) \mathbb{1}_{\{\|\delta^\sigma(s,z)\|\leq 1\}}(\mu^\sigma - \nu^\sigma)(ds,dz)$$

$$+ \int_0^t \int_{\mathbb{R}^2} \delta^\sigma(s,z) \mathbb{1}_{\{\|\delta^\sigma(s,z)\|>1\}}\mu^\sigma(ds,dz),$$

for appropriate random variables $b^\sigma, \sigma^\sigma, W^\sigma, \delta^\sigma, \mu^\sigma, \nu^\sigma$, then $Y = (X, \varepsilon)$ is an Itô semimartingale as well. We denote the jump measure of this 6-dimensional Itô semimartingale by μ^Y. Further denote by $\mu^Y\big|_{A_\varepsilon}$ the restriction of μ^Y to $[0,\infty) \times A_\varepsilon$, $A_\varepsilon = [-\varepsilon, \varepsilon]^6$, and denote by $Y^\varepsilon = (X^\varepsilon, \sigma^\varepsilon)$ the process which is obtained from Y by replacing μ^Y and ν^Y by $\mu^Y\big|_{A_\varepsilon}$ and $\nu^Y\big|_{A_\varepsilon}$ (the drift b^Y also hast to be modified). Finally define a new filtration via

$$\mathcal{F}_t^\varepsilon = \sigma(\mathcal{F}_t, \mu^Y\big|_{A_\varepsilon^c})$$

i.e. $(\mathcal{F}_t^\varepsilon)_{t\geq 0}$ is the smallest filtration which dominates $(\mathcal{F}_t)_{t\geq 0}$ such that $\mu^Y\big|_{A_\varepsilon^c}$ is known at time 0. By the property of Poisson random measures then $\mu^Y\big|_{A_\varepsilon}$,

$\mu^Y\big|_{A_\varepsilon^c}$ are independent and Y^ε is an Itô semimartingale also with respect to the filtration $(\mathcal{F}_t^\varepsilon)_{t\geq 0}$. Here the Grigelionis representations of Y and Y^ε only differ in the fact that μ^Y and ν^Y are replaced by $\mu^Y\big|_{A_\varepsilon}$, $\mu^Y\big|_{A_\varepsilon^c}$ and b^Y is modified appropriately. In particular the continuous martingale part of X^ε is equal to $\int_0^t \sigma_s dW_s$ for all $\varepsilon > 0$. Denote by S_p^ε, $p \in \mathbb{N}$, an enumeration of the times with $\mu^Y\big|_{A_\varepsilon^c}(\{S_p^\varepsilon\} \times \mathbb{R}) = 1$. Then the stopping times S_p^ε are $\mathcal{F}_0^\varepsilon$-measurable and hence practically constant with respect to the filtration $(\mathcal{F}_t^\varepsilon)_{t\geq 0}$. Hence we can use the arguments from the proof of Theorem 6.2 to prove

$$\tilde{\sigma}_n^{(l)}(X^\varepsilon, S_p^\varepsilon, -) \xrightarrow{\mathbb{P}} \sigma_{S_p^\varepsilon-}^{(l)}, \quad \tilde{\kappa}_n^{(l)}(X^\varepsilon, S_p^\varepsilon, -) \xrightarrow{\mathbb{P}} \rho_{S_p^\varepsilon-}\sigma_{S_p^\varepsilon-}^{(1)}\sigma_{S_p^\varepsilon-}^{(2)}, \quad (6.33)$$

where the extra argument for the estimators indicates that they are based on increments of X^ε instead of X. Finally increments of X and X^ε only differ if one of the remaining jump times $S_{p'}^\varepsilon$ lies within the interval $(S_p^\varepsilon - b_n, S_p^\varepsilon)$ over which the estimator is computed. Hence as there are only finitely many stopping times S_p^ε within a bounded interval we have $\tilde{\sigma}_n^{(l)}(X, S_p^\varepsilon, -) = \tilde{\sigma}_n^{(l)}(X^\varepsilon, S_p^\varepsilon, -)$ and $\tilde{\kappa}_n^{(l)}(X, S_p^\varepsilon, -) = \tilde{\kappa}_n^{(l)}(X^\varepsilon, S_p^\varepsilon, -)$ with probability approaching 1 as $n \to \infty$ and (6.33) yields

$$\tilde{\sigma}_n^{(l)}(X, S_p^\varepsilon, -) \xrightarrow{\mathbb{P}} \sigma_{S_p^\varepsilon-}^{(l)}, \quad \tilde{\kappa}_n^{(l)}(X, S_p^\varepsilon, -) \xrightarrow{\mathbb{P}} \rho_{S_p^\varepsilon-}\sigma_{S_p^\varepsilon-}^{(1)}\sigma_{S_p^\varepsilon-}^{(2)} \quad (6.34)$$

for any jump time S_p^ε. By assumption we have $\|\Delta Y_\tau\| > 0$ almost surely which yields

$$\mathbb{P}(\tau \in \{S_p^\varepsilon | \varepsilon \in \mathbb{Q} \cap (0, \infty), p \in \mathbb{N}\}) = 1.$$

Hence we also get

$$\tilde{\sigma}_n^{(l)}(X, \tau, -) \xrightarrow{\mathbb{P}} \sigma_{\tau-}^{(l)}, \quad \tilde{\kappa}_n^{(l)}(X, \tau, -) \xrightarrow{\mathbb{P}} \rho_{\tau-}\sigma_{\tau-}^{(1)}\sigma_{\tau-}^{(2)}$$

from (6.34). Using these results we obtain (6.7) and (6.8) because

$$\tilde{\sigma}_n^{(l)}(X^\varepsilon, S_p^\varepsilon, -) - \hat{\sigma}_n^{(l)}(X^\varepsilon, S_p^\varepsilon, -) \xrightarrow{\mathbb{P}} 0,$$
$$\tilde{\kappa}_n^{(l)}(X^\varepsilon, S_p^\varepsilon, -) - \hat{\kappa}_n^{(l)}(X^\varepsilon, S_p^\varepsilon, -) \xrightarrow{\mathbb{P}} 0$$

follow as in the proof of Theorem 6.2. $\square$

7 Estimating Quadratic Covariation

Historically the *integrated volatility* or *realized volatility* of the process $X^{(l)}$

$$\int_0^t (\sigma_s^{(l)})^2 ds$$

and the *integrated co-volatility* of the processes $X^{(1)}$ and $X^{(2)}$

$$\int_0^t \rho_s \sigma_s^{(1)} \sigma_s^{(2)} ds$$

were among the first quantities that have been investigated in high-frequency statistics. They function as measures for how much a continuous process fluctuates or how much two continuous processes fluctuate "together" and are therefore used to measure e.g. financial risk in terms of how volatile price processes are. The estimation of realized volatility dates back to [4] and [6]. While the spot volatilities and spot correlation investigated in Chapter 6 describe the size and dependence of „instantaneous" fluctuations, the integrated volatility and co-volatility are measures for the cummulative or average amount of fluctuations over a certain time interval.

When working with stochastic processes that allow for jumps the *quadratic variation process* of the semimartingale $X^{(l)}$ which is defined by

$$[X^{(l)}, X^{(l)}]_t = \int_0^t (\sigma_s^{(l)})^2 ds + \sum_{s \leq t} (\Delta X_s^{(l)})^2$$

and the *covariation process* of the processes $X^{(1)}$ and $X^{(2)}$ defined by

$$[X^{(1)}, X^{(2)}]_t = \int_0^t \rho_s \sigma_s^{(1)} \sigma_s^{(2)} ds + \sum_{s \leq t} \Delta X_s^{(1)} \Delta X_s^{(2)}$$

replace integrated volatility and co-volatility as measures for how much processes fluctuate (together). Hence it is of interest to find methods for estimating these quantities as well. The quadratic variation and covariation processes also play an important role in stochastic analysis.

© Springer Fachmedien Wiesbaden GmbH, part of Springer Nature 2019
O. Martin, *High-Frequency Statistics with Asynchronous and Irregular Data*,
Mathematische Optimierung und Wirtschaftsmathematik | Mathematical Optimization
and Economathematics, https://doi.org/10.1007/978-3-658-28418-3_7

Further in the setting of processes with jumps it is also of interest to quantify how large the contributions are that stem from the continuous parts respectively from the jump parts of $X^{(1)}$ and $X^{(2)}$. Hence we are looking for estimators for the integrated (co-)volatility also if X is discontinuous and further we would like to find estimators for the *jump parts*

$$\sum_{s \leq t} (\Delta X_s^{(l)})^2 \quad \text{and} \quad \sum_{s \leq t} \Delta X_s^{(1)} \Delta X_s^{(2)}$$

in the quadratic variation respectively the covariation process.

7.1 Consistency Results

In the univariate and also in the bivariate but synchronous setting we have the following classical result; compare e.g. Theorem 23 in [41].

Theorem 7.1. *Suppose Condition 1.3 is fulfilled. Then it holds*

$$V^{(l)}(g_2, \pi_n)_T = \sum_{i: t_{i,n}^{(l)} \leq T} (\Delta_{i,n}^{(l)} X^{(l)})^2 \xrightarrow{\mathbb{P}} [X^{(l)}, X^{(l)}]_T, \quad l = 1, 2,$$

where $g_2(x) = x^2$, $x \in \mathbb{R}$. If the observation scheme is further synchronous it also holds

$$V(f_{(1,1)}, \pi_n)_T = \sum_{i,j: t_{i,n}^{(1)} \vee t_{j,n}^{(2)} \leq T} \Delta_{i,n}^{(1)} X^{(1)} \Delta_{j,n}^{(2)} X^{(2)} \mathbb{1}_{\{\mathcal{I}_{i,n}^{(1)} \cap \mathcal{I}_{j,n}^{(2)} \neq \emptyset\}}$$

$$= \sum_{i: t_{i,n} \leq T} \Delta_{i,n} X^{(1)} \Delta_{i,n} X^{(2)} \xrightarrow{\mathbb{P}} [X^{(1)}, X^{(2)}]_T$$

with $f_{(1,1)}(x, y) = xy$, $(x, y) \in \mathbb{R}^2$, where we used the notation from (2.1).

When looking at the proofs of the convergences in Theorem 7.1 we observe that the continuous parts i.e. realized volatility and co-volatility of $[X^{(l)}, X^{(l)}]_T$ and $[X^{(1)}, X^{(2)}]_T$ originate from increments of the continuous part of X and the jump parts originate from increments of the jump part of X. The information on the continuous part is mostly captured in small increments of X while information on the jump part is contained in the comparably large increments. In Section 2.3 we introduced appropriate thresholds to separate comparably small respectively large increments and thereby separate the contributions of the continuous and jump parts in certain functionals. Using results from Section 2.3 we are able to obtain consistent estimators for the continuous and jump parts in the quadratic variation process. In the univariate setting Corollary 2.34 and part a) of Theorem 2.38 yield the following result.

Theorem 7.2. *Suppose Condition 1.3 is fulfilled and the observation scheme is exogenous. Then it holds*

$$V_-^{(l)}(g_2, \pi_n, (\beta, \varpi))_T \xrightarrow{\mathbb{P}} \int_0^T (\sigma_s^{(l)})^2 ds$$

$$V_+^{(l)}(g_2, \pi_n, (\beta, \varpi))_T \xrightarrow{\mathbb{P}} \sum_{s \leq T} (\Delta X_s^{(l)})^2$$

for $\beta > 0$ and $\varpi \in (0, 1/2)$.

In the setting of asynchronous observations, a consistent estimator for the quadratic covariation process can be found from Theorem 3 in [8]. Further, we obtain similarly as in the univariate setting also in the bivariate setting consistent estimators for the continuous and jump parts in the quadratic covariation from Theorem 2.35 and part b) of Theorem 2.38.

Theorem 7.3. *Suppose Condition 1.3 is fulfilled and the observation scheme is exogenous. Then we have*

$$V(f_{(1,1)}, \pi_n)_T \xrightarrow{\mathbb{P}} [X^{(1)}, X^{(2)}]_T. \tag{7.1}$$

If additionally it holds $G_{1,1}^n(T) = O_{\mathbb{P}}(1)$, we also have

$$V_-(f_{(1,1)}, \pi_n, (\beta, \varpi))_T \xrightarrow{\mathbb{P}} \int_0^T \rho_s \sigma_s^{(1)} \sigma_s^{(2)} ds$$

$$V_+(f_{(1,1)}, \pi_n, (\beta, \varpi))_T \xrightarrow{\mathbb{P}} \sum_{o \leq T} \Delta X_s^{(1)} \Delta X_s^{(2)} \tag{7.2}$$

for $\beta > 0$ and $\varpi \in (0, 1/2)$.

A result similar to (7.2) is stated in Corollary 3 of [34]. In that paper also a central limit theorem for an estimator for the integrated co-volatility is given under the additional assumption that $\sum_{s \leq T} \|\Delta X_s\|^{\beta'} < \infty$ holds almost surely for some $\beta' \in [0, 1)$ i.e. that the generalized Blumenthal-Getoor index is less than 1.

Further there exists the following remarkable result which even holds for endogenous observation times which is due to [35] and based on results from [21].

Theorem 7.4. *Suppose Condition 1.3 is fulfilled and the processes $X^{(1)}$ and $X^{(2)}$ are of finite jump activity on $[0, T]$ i.e. they almost surely have only finitely many jumps in the compact time interval $[0, T]$. Then it holds*

$$V(f_{(1,1)}, \pi_n)_T \xrightarrow{\mathbb{P}} [X^{(1)}, X^{(2)}]_T.$$

In the setting of finite jump activity processes further a consistent estimator for the realized
the estimator for the covariation and for the co-volatility also lead to an estimator for

$$\sum_{s \leq T} \Delta X_s^{(1)} \Delta X_s^{(2)}.$$

Contrary to our approach they use truncated increments via a global threshold and not a local threshold depending on the length of the observation interval as in Section 2.3; compare the Notes at the end of Section 2.3.1.

7.2 Central Limit Theorems

In this section, we construct central limit theorems accompanying the convergences in the previous section. We are going to restrict ourselves to finding central limit theorems in the setting of exogenous observation times. Central limit theorems have also been found for specific endogenous observation schemes, compare e.g. [18] and [47], but they are notoriously difficult to derive as in the case of endogenous observations the asymptotic variances also tend to depend on the covariance structure between the process and its components and the observation times. Further we are going to derive central limit theorems only for the estimators leading to the quadratic variation $[X^{(l)}, X^{(l)}]_T$ respectively the covariation $[X^{(1)}, X^{(2)}]_T$ at time $T > 0$. To be able to obtain feasibility of the central limit theorems in Section 7.3 we will develop stable central limit theorems; compare Chapter 3 and Appendix B.

We will start by developing a central limit theorem for the estimator $V(f_{(1,1)}, \pi_n)_T$ in the univariate setting. To motivate the structure of the asymptotic variance we first consider the following simple toy example

$$X_t^{toy,(l)} = \sigma^{(l)} W_t^{(l)} + \sum_{s \leq t} \Delta N^{(l)}(q)_s, \quad t \geq 0,$$

for some $q > 0$. Here, the volatility $\sigma^{(l)}$ is assumed to be constant and it is assumed that $X^{toy,(l)}$ almost surely has only finitely many jumps in any compact time interval. Then on the set where any two jumps of $X^{toy,(l)}$ are further apart than $|\pi_n|_T$ it holds

$$V^{(l)}(g_2, \pi_n)_T - [X^{toy,(l)}, X^{toy,(l)}]_T$$
$$= \sum_{i:t_{i,n}^{(l)} \leq T} (\sigma^{(l)} \Delta_{i,n}^{(l)} W^{(l)} + \Delta_{i,n}^{(l)} N^{(l)}(q))^2 - (\sigma^{(l)})^2 T - \sum_{s \leq t} (\Delta N^{(l)}(q)_s)^2$$

$$= (\sigma^{(l)})^2 \sum_{i:t_{i,n}^{(l)} \leq T} ((\Delta_{i,n}^{(l)} W^{(l)})^2 - |\mathcal{I}_{i,n}|)$$

$$+ 2\sigma^{(l)} \sum_{i:t_{i,n}^{(l)} \leq T} \Delta_{i,n}^{(l)} W^{(l)} \Delta_{i,n}^{(l)} N^{(l)}(q) + O_{\mathbb{P}}(|\pi_n|_T). \qquad (7.3)$$

The first term in this decomposition has mean zero and it holds

$$\mathbb{E}\left[\left(\sqrt{n}(\sigma^{(l)})^2 \sum_{i:t_{i,n}^{(l)} \leq T} ((\Delta_{i,n}^{(l)} W^{(l)})^2 - |\mathcal{I}_{i,n}|)\right)^2 \bigg| \mathcal{S}\right]$$

$$= (\sigma^{(l)})^4 n \sum_{i:t_{i,n}^{(l)} \leq T} |\mathcal{I}_{i,n}|^2 \mathbb{E}[(U^2 - 1)^2|\mathcal{S}] = 2(\sigma^{(l)})^4 G_4^{(l),n}(T)$$

where U denotes a standard normal distributed random variable independent of $\mathcal{S}$. Further if we denote by $i_n^{(l)}(s)$ the index of the interval characterized by $s \in \mathcal{I}_{i_n^{(l)}(s),n}^{(l)}$, then for the second term in the decomposition (7.3) it holds

$$\sqrt{n}2\sigma^{(l)} \sum_{i:t_{i,n}^{(l)} \leq T} \Delta_{i,n}^{(l)} W^{(l)} \Delta_{i,n}^{(l)} N^{(l)}(q) \stackrel{\mathcal{L}}{\approx} 2\sigma^{(l)} \sum_{s \leq T} \sqrt{n|\mathcal{I}_{i,n}^{(l)}(s)|} U_s \Delta N^{(l)}(q)_s$$

where U_s are i.i.d. standard normal random variables independent of $\mathcal{F}$. Here we have asymptotic independence of the increments of the Brownian motion and the jumps which is intuitively due to the fact that increments over very small intervals asymptotically cannot have a huge influence on the global jump behaviour of the process. If we assume that $G_4^{(l),n}(T)$ converges in probability to some $G_4^{(l)}(T)$ and the $(n|\mathcal{I}_{i,n}^{(l)}(s)|)^{1/2}$ converge suitably in law to random variables $\eta(s)$ it is possible to conclude

$$\sqrt{n}(V^{(l)}(g_2, \pi_n)_T - [X^{toy,(l)}, X^{toy,(l)}]_T)$$

$$\stackrel{\mathcal{L}-s}{\longrightarrow} \sqrt{2(\sigma^{(l)})^4 G_4^{(l)}(T)} U + 2\sigma^{(l)} \sum_{s \leq T} \sqrt{\eta(s)} U_s \Delta N^{(l)}(q)_s$$

for i.i.d. standard normal random variables U, U_s independent of $\mathcal{F}$.

The structure of the limit for general Itô semimartingales is similar to the limit in the toy example. In particular the contribution of the processes $B(q)$ and $M(q)$ to the limit vanishes as $q \to \infty$. However, the structure of the limit becomes more complicated if we allow for time-varying volatility processes $\sigma_s^{(l)}$. We first recall the following notation

$$(\delta_{n,-}^{(l)}(s), \delta_{n,+}^{(l)}(s)) = \left(s - t_{i_n^{(l)}(s)-1,n}^{(l)}, t_{i_n^{(l)}(s),n}^{(l)} - s\right)$$

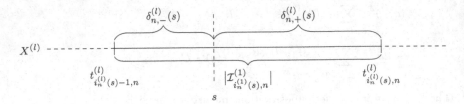

Figure 7.1: Illustration of the terms $\delta_{n,-}^{(l)}(s)$ and $\delta_{n,+}^{(l)}(s)$.

introduced in Section 3.1, compare (3.6), to describe the distances of s to previous and upcoming observation times. For an illustration see Figure 7.1. The necessary assumptions required for obtaining a central limit theorem are collected in the following condition; compare Assumption 2 in [8].

Condition 7.5. *Suppose Condition 1.3 is fulfilled, the observation scheme is exogenous and it further holds:*

(i) *We have* $\mathbb{E}[(|\pi_n|_T)^q] = o(n^{-\alpha})$ *for any* $q \geq \alpha$ *and* $0 < \alpha < q$.

(ii) *It holds* $G_4^{(l),n}(t) \xrightarrow{\mathbb{P}} G_4^{(l)}(t)$, $t \in [0,T]$, *for some increasing (possibly random) function* $G_4^{(l)} : [0,T] \to [0,\infty)$.

(iii) *The integral*

$$\int_0^T g(x_1,\ldots,x_P)\mathbb{E}\Big[\prod_{p=1}^{P} h_p(n\delta_{n,-}^{(l)}(s), n\delta_{n,+}^{(l)}(s))\Big]dx_1\ldots dx_P$$

converges as $n \to \infty$ *to*

$$\int_0^T g(x_1,\ldots,x_P)\prod_{p=1}^{P} h_p(y)\Gamma^{VAR,(l)}(x_p,dy)dx_1\ldots dx_P$$

for all bounded functions $g : [0,T]^P \to \mathbb{R}$, $h_p : [0,\infty)^2 \to \mathbb{R}$, $p = 1,\ldots,P$, *and any* $P \in \mathbb{N}$. *Here* $\Gamma^{VAR,(l)}(x,dy)$, $x \in [0,T]$, *is a family of probability measures on* $[0,\infty)^2$ *with uniformly bounded first moments.*

Part (iii) of Condition 7.5 is the specific assumption that is needed such that the two components of $n|\mathcal{I}_{i_n^{(l)}(s),n}^{(l)}|$ which correspond to the interval parts before and after s converge in the required sense. It is identical to part (ii) of Condition 3.1 and in particular it holds $\Gamma^{VAR,(l)} = \Gamma^{univ,(l)}$. Next we consider an extension of the probability space $(\Omega, \mathcal{F}, \mathbb{P})$ to a probability space $(\widetilde{\Omega}, \widetilde{\mathcal{F}}, \widetilde{\mathbb{P}})$ on which we can define i.i.d. standard normal distributed random variables U, U_s, $s \in [0, T]$, and random variables $(\delta_-^{(l)}(s), \delta_+^{(l)}(s))$, $s \in [0, T]$, which are distributed according to $\Gamma^{VAR,(l)}(s, dy)$. Further, the random variables U, U_s and $(\delta_-^{(l)}(s), \delta_+^{(l)}(s))$, $s \in [0, T]$, are supposed to be independent of each other and of $\mathcal{F}$. Using these newly introduced random variables we can state the central limit theorem for the estimator of the quadratic variation; compare Theorem 2 in [8].

Theorem 7.6. *Assume that Condition 7.5 is fulfilled. Then we have the following $\mathcal{X}$-stable convergence*

$$\sqrt{n}\big(V^{(l)}(g_2, \pi_n)_T - [X^{(l)}, X^{(l)}]_T\big) \xrightarrow{\mathcal{L}-s} \sqrt{2}\Big(\int_0^T (\sigma_s^{(l)})^4 dG_4^{(l)}(s)\Big)^{1/2} U$$

$$+ 2 \sum_{p:S_p^{(l)} \leq T} \Delta X_{S_p^{(l)}} \sqrt{(\sigma_{S_p^{(l)}-}^{(l)})^2 \delta_-^{(l)}(S_p^{(l)}) + (\sigma_{S_p}^{(l)})^2 \delta_+^{(l)}(S_p^{(l)})} U_{S_p^{(l)}} \quad (7.4)$$

where $(S_p^{(l)})_{p\in\mathbb{N}}$ denotes an enumeration of the jump times of $X^{(l)}$.

To motivate the structure of the central limit theorem for the estimator of the covariation we again consider a toy example

$$X_t^{toy} = \sigma W_t + \sum_{s \leq T} \Delta N(q)_s$$

where the volatility matrix

$$\sigma = \begin{pmatrix} \sigma^{(1)} & 0 \\ \rho\sigma^{(2)} & \sqrt{1-\rho^2}\sigma^{(2)} \end{pmatrix}$$

is constant and X has finite jump activity. Then on the set where any two jumps of X are further apart than $2|\pi_n|_T$ it holds

$$V(f_{(1,1)}, \pi_n)_T - [X^{toy,(1)}, X^{toy,(2)}]_T$$

$$= \sum_{i,j:t_{i,n}^{(1)} \vee t_{j,n}^{(2)} \leq T} (\sigma^{(1)}\Delta_{i,n}^{(1)}W^{(1)} + \Delta_{i,n}^{(1)}N^{(1)}(q))$$

$$\times (\rho\sigma^{(2)}\Delta_{j,n}^{(2)}W^{(1)} + \sqrt{1-\rho^2}\sigma^{(2)}\Delta_{j,n}^{(2)}W^{(2)} + \Delta_{j,n}^{(2)}N^{(2)}(q))\mathbb{1}_{\{\mathcal{I}_{i,n}^{(1)} \cap \mathcal{I}_{j,n}^{(2)} \neq \emptyset\}}$$

$$- \rho\sigma^{(1)}\sigma^{(2)}T - \sum_{s \leq T} \Delta N^{(1)}(q)_s \Delta N^{(2)}(q)_s$$

$$= \sigma^{(1)}\sigma^{(2)} \sum_{i,j:t_{i,n}^{(1)} \vee t_{j,n}^{(2)} \leq T} [\Delta_{i,n}^{(1)}W^{(1)}(\rho\Delta_{j,n}^{(2)}W^{(1)} + \sqrt{1-\rho^2}\Delta_{j,n}^{(2)}W^{(2)})$$

$$- \rho |\mathcal{I}_{i,n}^{(1)} \cap \mathcal{I}_{j,n}^{(2)}|] \mathbb{1}_{\{\mathcal{I}_{i,n}^{(1)} \cap \mathcal{I}_{j,n}^{(2)} \neq \emptyset\}}$$

$$+ \sigma^{(2)} \sum_{i,j:t_{i,n}^{(1)} \vee t_{j,n}^{(2)} \leq T} \Delta_{i,n}^{(1)} N^{(1)}(q)(\rho \Delta_{j,n}^{(2)} W^{(1)} + \sqrt{1-\rho^2} \Delta_{j,n}^{(2)} W^{(2)})$$

$$\times \mathbb{1}_{\{\mathcal{I}_{i,n}^{(1)} \cap \mathcal{I}_{j,n}^{(2)} \neq \emptyset\}}$$

$$+ \sigma^{(1)} \sum_{i,j:t_{i,n}^{(1)} \vee t_{j,n}^{(2)} \leq T} \Delta_{i,n}^{(1)} W^{(1)} \Delta_{j,n}^{(2)} N^{(2)}(q) \mathbb{1}_{\{\mathcal{I}_{i,n}^{(1)} \cap \mathcal{I}_{j,n}^{(2)} \neq \emptyset\}} + O_{\mathbb{P}}(|\pi_n|_T).$$

$$(7.5)$$

The first term in the above decomposition has mean zero and the $\mathcal{S}$-conditional expectation of its square equals

$$\mathbb{E}\Big[\Big(\sqrt{n} \sum_{i,j:t_{i,n}^{(1)} \vee t_{j,n}^{(2)} \leq T} [\Delta_{i,n}^{(1)} W^{(1)}(\rho \Delta_{j,n}^{(2)} W^{(1)} + \sqrt{1-\rho^2} \Delta_{j,n}^{(2)} W^{(2)})$$

$$- \rho |\mathcal{I}_{i,n}^{(1)} \cap \mathcal{I}_{j,n}^{(2)}|] \mathbb{1}_{\{\mathcal{I}_{i,n}^{(1)} \cap \mathcal{I}_{j,n}^{(2)} \neq \emptyset\}}\Big)^2 \Big| \mathcal{S}\Big]$$

$$= G_{2,2}^n(T) + \rho^2 H_{0,0,4}^n(T) + \rho^2 \widetilde{H}^n(T), \quad (7.6)$$

compare (2.39) for the definition of the functions $G_{2,2}^n(t)$ and $H_{0,0,4}^n(t)$, where

$$\widetilde{H}^n(t) = n \sum_{l=1,2} \sum_{i,j,j':t_{i,n}^{(l)} \vee t_{j,n}^{(3-l)} \vee t_{j',n}^{(3-l)} \leq t} |\mathcal{I}_{i,n}^{(l)} \cap \mathcal{I}_{j,n}^{(3-l)}||\mathcal{I}_{i,n}^{(l)} \cap \mathcal{I}_{j',n}^{(3-l)}|, \quad t \in [0,T].$$

The calculations leading to (7.6) are presented in Section 7.4. Regarding the last two sums in (7.5) it holds similarly as in the univariate situation

$$\sqrt{n}\sigma^{(2)} \sum_{i:t_{i,n}^{(l)} \leq T} \Delta_{i,n}^{(1)} N^{(1)} \sum_{j:t_{j,n}^{(2)} \leq T} (\rho \Delta_{j,n}^{(2)} W^{(1)} + \sqrt{1-\rho^2} \Delta_{j,n}^{(2)} W^{(2)}) \mathbb{1}_{\{\mathcal{I}_{i,n}^{(1)} \cap \mathcal{I}_{j,n}^{(2)} \neq \emptyset\}}$$

$$+ \sqrt{n}\sigma^{(1)} \sum_{j:t_{j,n}^{(2)} \leq T} \Delta_{j,n}^{(2)} N^{(2)}(q) \sum_{i:t_{i,n}^{(l)} \leq T} \Delta_{i,n}^{(1)} W^{(1)} \mathbb{1}_{\{\mathcal{I}_{i,n}^{(1)} \cap \mathcal{I}_{j,n}^{(2)} \neq \emptyset\}}$$

$$\stackrel{\mathcal{L}}{\approx} \sum_{s \leq T} \Big[\Delta N^{(1)}(q)_s \sigma^{(2)} \big(\rho(\eta_n^{(1,2)}(s))^{1/2} U_s^{(1)} + \rho(\eta_n^{(2\backslash 1)}(s))^{1/2} U_s^{(2)}$$

$$+ \sqrt{1-\rho^2}(\eta_n^{(1,2)}(s) + \eta_n^{(2\backslash 1)}(s))^{1/2} U_s^{(3)}\big)$$

$$+ \Delta N^{(2)}(q)_s \sigma^{(1)} \big((\eta_n^{(1,2)}(s))^{1/2} U_s^{(1)} + (\eta_n^{(1\backslash 2)}(s))^{1/2} U_s^{(4)}\big)\Big]$$

for i.i.d. standard normal random variables $U_s^{(1)}, U_s^{(2)}, U_s^{(3)}, U_s^{(4)}$, $s \in [0,T]$ and random variables

$$\eta_n^{(1,2)}(s) = n|\mathcal{I}_{i_n^{(1)}(s),n}^{(1)} \cup \mathcal{I}_{i_n^{(2)}(s),n}^{(2)}|,$$

$$\eta_n^{(3-l\backslash l)}(s) = n \sum_{\substack{j \neq i_n^{(3-l)}(s):\mathcal{I}_{j,n}^{(3-l)} \cup \mathcal{I}_{i_n^{(l)}(s),n}^{(l)} \neq \emptyset}} |\mathcal{I}_{j,n}^{(3-l)} \setminus \mathcal{I}_{i_n^{(l)}(s),n}^{(l)}|, \quad l = 1,2.$$

If we assume that $G_{2,2}^n(T)$, $H_{0,0,4}^n(T)$, $\widetilde{H}^n(T)$ converge in probability to some $G_{2,2}(T)$, $H_{0,0,4}(T)$, $\widetilde{H}(T)$ and that the random variables

$$(\eta_n^{(1,2)}(s), \eta_n^{(2\backslash 1)}(s), \eta_n^{(1\backslash 2)}(s))$$

converges suitably in law to random variables $(\eta^{(1,2)}(s), \eta^{(2\backslash 1)}(s), \eta^{(1\backslash 2)}(s))$ it is possible to conclude

$$\sqrt{n}\big(V(f_{(1,1)}, \pi_n)_T - [X^{toy,(1)}, X^{toy,(2)}]_T\big)$$
$$\xrightarrow{\mathcal{L}\text{-}s} \big((\sigma^{(1)}\sigma^{(2)})^2[G_{2,2}(T) + \rho^2 H_{0,0,4}(T) + \rho^2 \widetilde{H}(T)]\big)^{1/2} U$$
$$+ \sum_{s \leq T} [\Delta N^{(1)}(q)_s \sigma^{(2)}\big(\rho(\eta_n^{(1,2)}(s))^{1/2} U_s^{(1)} + \rho(\eta^{(2\backslash 1)}(s))^{1/2} U_s^{(2)}$$
$$+ \sqrt{1-\rho^2}(\eta^{(1,2)}(s) + \eta^{(2\backslash 1)}(s))^{1/2} U_s^{(3)}\big)$$
$$+ \Delta N^{(2)}(q)_s \sigma^{(1)}\big((\eta_n^{(1,2)}(s))^{1/2} U_s^{(1)} + (\eta^{(1\backslash 2)}(s))^{1/2} U_s^{(4)}\big)] \qquad (7.7)$$

where $U, U_s^{(1)}, U_s^{(2)}, U_s^{(3)}, U_s^{(4)}$, $s \in [0,T]$, denote i.i.d. standard normal random variables independent of $\mathcal{J}$.

To describe the limit for general Itô semimartingales with potentially discontinuous volatility process σ_s we need to introduce some additional notation. Using the notation from (2.28) we define the random variables

$$\mathcal{L}_n^{COV,(l)}(s) = n\Big(\min\{\tau_{n,-}^{(1)}(s), \tau_{n,-}^{(2)}(s)\} - \tau_{n,-}^{(l)}(\min\{\tau_{n,-}^{(1)}(s), \tau_{n,-}^{(2)}(s)\})\Big), \quad l = 1,2,$$

$$\mathcal{R}_n^{COV,(l)}(s) = n\Big(\tau_{n,+}^{(l)}(\max\{\tau_{n,+}^{(1)}(s), \tau_{n,+}^{(2)}(s)\}) - \max\{\tau_{n,+}^{(1)}(s), \tau_{n,+}^{(2)}(s)\}\Big), \quad l = 1,2,$$

$$\mathcal{L}_n^{COV}(s) = n\Big(s - \min\{\tau_{n,-}^{(1)}(s), \tau_{n,-}^{(2)}(s)\}\Big),$$

$$\mathcal{R}_n^{COV}(s) = n\Big(\max\{\tau_{n,+}^{(1)}(s), \tau_{n,+}^{(2)}(s)\} - s\Big),$$

$$\mathcal{Z}_n^{COV}(s)$$
$$= (\mathcal{L}_n^{COV,(1)}(s), \mathcal{R}_n^{COV,(1)}(s), \mathcal{L}_n^{COV,(2)}(s), \mathcal{R}_n^{COV,(2)}(s), \mathcal{L}_n^{COV}(s), \mathcal{R}_n^{COV}(s)).$$
$$\qquad (7.8)$$

For an illustration see Figure 7.2. Note that these random variables differ from those introduced in (3.14); compare also Figure 3.2. This is necessary because for the function $f_{(1,1)}$ idiosyncratic jumps contribute in the central limit theorem in (7.7) while in Section 3.1 we only looked at functions f where $\partial_1(x,y) = \partial_2(x,y) = 0$ for any $(x,y) \in \mathbb{R}^2$ with $|xy| = 0$ and hence only common jumps contributed in the central limit theorem 3.6.

The necessary assumptions required to obtain a central limit theorem for the estimator of the quadratic covariation are collected in the following condition; compare Assumption 4 in [8].

Condition 7.7. *Suppose Condition 1.3 and 7.5(i) are fulfilled, the observation scheme is exogenous and it further holds:*

(i) *The functions $G_{2,2}^n(t)$, $H_{0,0,4}^n(t)$, $\widetilde{H}^n(t)$ converge pointwise in $t \in [0,T]$ in probability to increasing functions $G_{2,2}(t)$, $H_{0,0,4}(t)$, $\widetilde{H}(t)$.*

(ii) *The integral*

$$\int_0^T g(x_1, \ldots, x_P) \mathbb{E}\Big[\prod_{p=1}^P h_p(Z_n^{COV}(x_p)) \Big] dx_1 \ldots dx_P$$

converges as $n \to \infty$ to

$$\int_0^T g(x_1, \ldots, x_P) \prod_{p=1}^P h_p(y) \Gamma^{COV}(x_p, dy) dx_1 \ldots dx_P$$

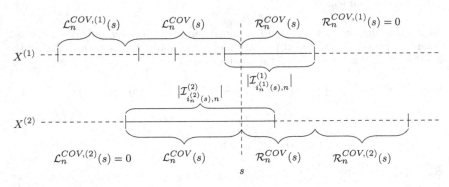

Figure 7.2: Illustration of the terms $\mathcal{L}_n^{COV,(l)}(s)$, $\mathcal{R}_n^{COV,(l)}(s)$, $l = 1, 2$, and $\mathcal{L}_n^{COV}(s)$, $\mathcal{R}_n^{COV}(s)$.

for all bounded functions $g \colon [0,T]^P \to \mathbb{R}$, $h_p \colon [0,\infty)^6 \to \mathbb{R}$, $p = 1,\ldots,P$, and any $P \in \mathbb{N}$. Here $\Gamma^{COV}(x,dy)$, $x \in [0,T]$, is a family of probability measures on $[0,\infty)^6$ with uniformly bounded first moments.

Unlike in the univariate setting here part (ii) of Condition 7.7 is slightly different from the corresponding condition in Condition 3.5 because the variables Z_n^{COV} and Z_n^{biv} differ.

To describe the limit in the upcoming central limit theorem we denote

$$\mathcal{V}_t^{COV} = \int_0^t (\sigma_s^{(1)}\sigma_s^{(2)})^2 dG_{2,2}(s) + \int_0^t (\rho_s\sigma_s^{(1)}\sigma_s^{(2)})^2 dH_{0,0,4}(s) + \int_0^t (\rho_s\sigma_s^{(1)}\sigma_s^{(2)})^2 d\widetilde{H}(s),$$

$t \in [0,T]$. Further let

$$Z^{COV}(s) = (\mathcal{L}^{COV,(1)}(s), \mathcal{R}^{COV,(1)}(s), \mathcal{L}^{COV,(2)}(s), \mathcal{R}^{COV,(2)}(s), \mathcal{L}^{COV}(s), \mathcal{R}^{COV}(s))$$

be distributed according to $\Gamma^{COV}(s,dy)$ and defined on an extended probability space $(\widetilde{\Omega}, \widetilde{\mathcal{F}}, \widetilde{\mathbb{P}})$. $U, U_s^{(1),-}, U_s^{(1),+}, U_s^{(2)}, U_s^{(3)}, U_s^{(4)}$, $s \in [0,T]$, are i.i.d. standard normal distributed random variables defined on the extended probability space as well. The random variables $Z^{COV}(s)$ and $U, U_s^{(1),-}, U_s^{(1),+}, U_s^{(2)}, U_s^{(3)}, U_s^{(4)}$ are all independent of each other and independent of $\mathcal{F}$. Using these newly introduced random variables we define

$$
\begin{aligned}
\mathcal{Z}_t^{COV} = \sum_{p:S_p \leq t} \Big[&\Delta X_{S_p}^{(1)} \Big(\sigma_{S_p-}^{(2)} \rho_{S_p-} \sqrt{\mathcal{L}^{COV}(S_p)} U_{S_p}^{(1),-} \\
&+ \sigma_{S_p}^{(2)} \rho_{S_p} \sqrt{\mathcal{R}^{COV}(S_p)} U_{S_p}^{(1),+} \\
&+ \sqrt{(\sigma_{S_p-}^{(2)})^2 (1 - (\rho_{S_p-})^2) \mathcal{L}^{COV}(S_p) + (\sigma_{S_p}^{(2)})^2 (1 - (\rho_{S_p})^2) \mathcal{R}^{COV}(S_p)} U_{S_p}^{(3)} \\
&+ \sqrt{(\sigma_{S_p-}^{(2)})^2 \mathcal{L}^{COV,(2)}(S_p) + (\sigma_{S_p}^{(2)})^2 \mathcal{R}^{(2)}(S_p)} U_{S_p}^{(4)} \Big) \\
&+ \Delta X_{S_p}^{(2)} \Big(\sigma_{S_p-}^{(1)} \sqrt{\mathcal{L}^{COV}(S_p)} U_{S_p}^{(1),-} + \sigma_{S_p}^{(1)} \sqrt{\mathcal{R}^{COV}(S_p)} U_{S_p}^{(1),+} \\
&+ \sqrt{(\sigma_{S_p-}^{(1)})^2 \mathcal{L}^{COV,(1)}(S_p) + (\sigma_{S_p}^{(1)})^2 \mathcal{R}^{COV,(1)}(S_p)} U_{S_p}^{(2)} \Big) \Big]
\end{aligned}
$$

where $(S_p)_{p \in \mathbb{N}}$ denotes an enumeration of the jump times of X. Using the processes $\mathcal{V}_t^{COV}$ and $\mathcal{Z}_t^{COV}$ we are able to state the following central limit theorem; compare Theorem 3 in [8].

Theorem 7.8. *If Condition 7.7 is fulfilled we have the following $\mathcal{X}$-stable convergence*

$$\sqrt{n}\big(V(f_{(1,1)}, \pi_n)_T - [X^{(1)}, X^{(2)}]_T \big) \xrightarrow{\mathcal{L}-s} \Phi_T^{COV} := (\mathcal{V}_T^{COV})^{1/2} U + \mathcal{Z}_T^{COV}. \tag{7.9}$$

The notation of the continuous term in the limit of Theorem 7.8 differs slightly from the notation of the corresponding term Theorem 3 of [8]. I chose a slightly different notation here to be closer to the notation used in Section 2.2.

Example 7.9. *Conditions 7.5 and 7.7 are fulfilled in the case of Poisson sampling introduced in Definition 5.1. Condition 7.5 holds by Lemma 5.2. Further Condition 7.5(ii) and 7.7(i) follow from Corollary 5.6. Note that the convergence of $\widetilde{H}^n(t)$ does not directly follow from Corollary 5.6, but can be proven similarly. That Conditions 7.5(iii) and 7.7(ii) are fulfilled in the Poisson setting follows from Lemma 5.10; compare Examples 3.4 and 3.8.* $\square$

Notes. *In this section I mainly presented results from [8]. However, I stated slightly modified versions of their theorems which are also valid in the setting of discontinuous volatility processes σ and thereby extend the applicability of their central limit theorems.*

7.3 Constructing Asymptotic Confidence Intervals

To turn the central limit theorems (7.4) and (7.9) into feasible central limit theorems which can be used in statistical applications we have to estimate the asymptotic variances.

First of all we unfortunately find that the results from Section 2.2 and Section 2.3 do not yield consistent estimators for the terms

$$\mathcal{V}_T^{VAR,(l)} = 2 \int_0^T (\sigma_s^{(l)})^4 dG_4^{(l)}(s), \quad l = 1, 2,$$

and $\mathcal{V}_T^{COV}$ although the limits in the results there resemble these terms. Indeed Corollary 2.19 and Theorem 2.38 yield that the quantities $\overline{V}^{(l)}(4, g_4, \pi_n)_T$ and $\overline{V}_-^{(l)}(4, g_4, \pi_n, (\beta, \varpi))_T$ converge to $3\mathcal{V}_T^{VAR,(l)}/2$, but only for continuous processes $X^{(l)}$. Further the limits of the two functionals $\overline{V}(4, f_{(2,2)}, \pi_n)_T$ and $\overline{V}_-(4, f_{(2,2)}, \pi_n, (\beta, \varpi))_T$ have some resemblance to $\mathcal{V}_T^{COV}$, but they do not include the integral with respect to $\widetilde{H}$ and these functionals converge only for continuous X as well.

However, we are able to estimate the functions $G_4^{(l)}(t)$, $G_{2,2}(t)$, $H_{0,0,4}(t)$ and $\widetilde{H}(t)$ pointwise in $t \in [0, T]$ by the sums $G_4^{(l),n}(t)$, $G_{2,2}^n(t)$, $H_{0,0,4}^n(t)$ and $\widetilde{H}^n(t)$. This fact is trivial from the definition of these functions. Further we know from

Chapter 6 consistent estimators for the spot volatilities $\sigma_t^{(l)}$ and spot correlations ρ_t, $t \in [0, T]$. Next we can approximate $\mathcal{V}_T^{VAR,(l)}$ by Riemann-sums of the form

$$\mathcal{V}_T^{VAR,(l)}(r) = \sum_{k=1}^{\lfloor T/r \rfloor} (\sigma_{(k-1)/r}^{(l)})^4 \big(G_4^{(l)}(k/r) - G_4^{(l)}((k-1)/r) \big)$$

and it holds $\mathcal{V}_T^{VAR,(l)}(r) \to \mathcal{V}_T^{VAR,(l)}$ pointwise in ω as $r \to \infty$. Further

$$\mathcal{V}_T^{VAR,(l)}(r,n) = \sum_{k=1}^{\lfloor T/r \rfloor} (\tilde{\sigma}_n^{(l)}((k-1)/r,+))^4 \big(G_4^{(l),n}(k/r) - G_4^{(l),n}((k-1)/r) \big)$$

yields a consistent estimator for $\mathcal{V}_T^{VAR,(l)}(r)$. Note that instead of the estimators $\tilde{\sigma}_n^{(l)}(s,+)$ for the spot volatility we could also use the estimators $\hat{\sigma}_n^{(l)}(s,+)$. Based on the above considerations there has to exist some sequence of positive real numbers $(r_n)_{n \in \mathbb{N}}$ with $r_n \to \infty$ as $n \to \infty$ and

$$\mathcal{V}_T^{VAR,(l)}(r_n,n) \xrightarrow{\mathbb{P}} \mathcal{V}_T^{VAR,(l)}.$$

Similarly we obtain a sequence of random variables $\mathcal{V}_T^{COV}(r_n,n)$ which consistently estimates $\mathcal{V}_T^{COV}$. To further investigate these estimators we have to specify the choice of r_n. However, this choice depends on the convergence rates of $\tilde{\sigma}_n^{(l)}(s,-)$, $\tilde{\sigma}_n^{(l)}(s,+)$, $\tilde{\rho}_n(s,-)$, $\tilde{\rho}_n(s,+)$ as well as on the convergence rates of the functions $G_4^{(l),n}(t), G_{2,2}^n(t), H_{0,0,4}^n(t), \widetilde{H}^n(t)$ which are even themselves difficult to determine. For this reason we will not further investigate the estimators $\mathcal{V}_T^{VAR,(l)}(r_n,n)$ and $\mathcal{V}_T^{COV}(r_n,n)$.

To estimate the asymptotic distribution of

$$\mathcal{Z}_T^{VAR,(l)} = 2 \sum_{p:S_p^{(l)} \leq T} \sqrt{(\sigma_{S_p^{(l)}-}^{(l)})^2 \delta_-^{(l)}(S_p^{(l)}) + (\sigma_{S_p^{(l)}}^{(l)})^2 \delta_+^{(l)}(S_p^{(l)})} U_{S_p^{(l)}} \Delta X_{S_p^{(l)}}$$

and of $\mathcal{Z}_T^{COV}$ we can use the bootstrap method introduced in Chapter 4. First we discuss the estimation of the law of $\mathcal{Z}_T^{VAR,(l)}$ for which in a first step we have to estimate the law of $(\delta_-^{(l)}(s), \delta_+^{(l)}(s))$. To this end let $(K_n)_{n \in \mathbb{N}}$ and $(M_n)_{n \in \mathbb{N}}$ denote increasing sequences of natural numbers and define

$$(\hat{\delta}_{n,m,-}^{(l)}(s), \hat{\delta}_{n,m,+}^{(l)}(s)), \quad m = 1, \dots, M_n,$$

as in (4.1). Using these estimators for realizations of $(\delta_-^{(l)}(s), \delta_+^{(l)}(s))$ we define for $\beta > 0$ and $\varpi \in (0, 1/2)$

$$\widehat{\Phi}_{T,n,m}^{VAR,(l)} = (\widehat{\mathcal{V}}_{T,n}^{VAR,(l)})^{1/2} U_{n,m} + 2 \sum_{i:t_{i,n}^{(l)} \leq T} \Delta_{i,n}^{(l)} X^{(l)} \mathbb{1}_{\{|\Delta_{i,n}^{(l)} X^{(l)}| > \beta |\mathcal{I}_{i,n}^{(l)}|^{\varpi}\}}$$

$$\times \sqrt{(\tilde{\sigma}_n^{(l)}(t_{i,n}^{(l)}, -))^2 \hat{\delta}_{n,m-}^{(l)}(t_{i,n}^{(l)}) + (\tilde{\sigma}_n^{(l)}(t_{i,n}^{(l)}, +))^2 \hat{\delta}_{n,m,+}^{(l)}(t_{i,n}^{(l)})} U_{n,i,m}$$

where $U_{n,m}, U_{n,i,m}$ are i.i.d. standard normal distributed random variables independent of all previously introduced random variables and $\widehat{\mathcal{V}}_{T,n}^{VAR,(l)}$ is a suitable estimator for $\mathcal{V}_T^{VAR,(l)}$. Further we denote by

$$\widehat{Q}_{T,n}^{VAR,(l)}(\alpha) = \widehat{Q}_\alpha(\{\widehat{\Phi}_{T,n,m}^{VAR,(l)} | m = 1, \ldots, M_n\})$$

the $\lfloor \alpha M_n \rfloor$-largest element of the set $\{\widehat{\Phi}_{T,n,m}^{VAR} | m = 1, \ldots, M_n\}$. Then $\widehat{Q}_{T,n}^{VAR,(l)}(\alpha)$ converges under appropriate conditions to the $\mathcal{X}$-conditional α-quantile of

$$(\mathcal{V}_T^{VAR,(l)})^{1/2} U + \mathcal{Z}_T^{VAR,(l)}$$

which is the limit in Theorem 7.6. The $\mathcal{X}$-conditional α-quantile $Q^{VAR,(l)}(\alpha)$ is defined as the (under certain conditions unique) $\mathcal{X}$-measurable random variable fulfilling

$$\widetilde{\mathbb{P}}((\mathcal{V}_T^{VAR,(l)})^{1/2} U + \mathcal{Z}_T^{VAR,(l)} \leq Q^{VAR,(l)}(\alpha) | \mathcal{X})(\omega) = \alpha, \quad \omega \in \Omega.$$

To be able to consistently estimate an asymptotic confidence interval we need the following structural assumptions.

Condition 7.10. *Suppose Condition 7.5 is fulfilled, it holds $\int_0^T |\sigma_s^{(l)}| ds > 0$ almost surely and $t \mapsto G_4^{(l)}(t)$ is strictly increasing on $[0, T]$. Further, let the sequence $(b_n)_{n \in \mathbb{N}} \subset (0, \infty)$ fulfil $b_n \to 0$, $|\pi_n|_T / b_n \xrightarrow{\mathbb{P}} 0$ and suppose that $(K_n)_{n \in \mathbb{N}}$ and $(M_n)_{n \in \mathbb{N}}$ are sequences of natural numbers converging to infinity and $K_n/n \to 0$. Additionally:*

(i) It holds

$$\widetilde{\mathbb{P}}\Big(\Big| \widetilde{\mathbb{P}}((\hat{\delta}_{n,1,-}^{(l)}(s_p), \hat{\delta}_{n,1,+}^{(l)}(s_p)) \leq x_p, \; p = 1, \ldots, P | \mathcal{S})$$

$$- \prod_{p=1}^{P} \widetilde{\mathbb{P}}((\delta_-^{(l)}(s_p), \delta_+^{(l)}(s_p)) \leq x_p) \Big| > \varepsilon \Big) \to 0$$

as $n \to \infty$, for all $\varepsilon > 0$ and any $x = (x_1, \ldots, x_P) \in \mathbb{R}^{2 \times P}$, $P \in \mathbb{N}$, and $s_p \in (0, T)$, $p = 1, \ldots, P$.

(ii) The volatility process $\sigma^{(l)}$ is itself an Itô semimartingale i.e. a process of the form (1.1).

Part (i) of Condition 7.10 yields that the empirical distribution on the random variables

$$(\hat{\delta}^{(l)}_{n,m,-}(s_p), \hat{\delta}^{(l)}_{n,m,+}(s_p)), \quad m = 1, \ldots, M_n,$$

converges to the distribution of $(\delta^{(l)}_-(s_p), \delta^{(l)}_+(s_p))$, while part (ii) is needed for the convergence of the estimators $\bar{\sigma}^{(l)}_n(S^{(l)}_p, -)$. Under Condition 7.10 we obtain the following result.

Theorem 7.11. *Suppose Condition 7.10 is fulfilled and it holds $\widehat{\mathcal{V}}^{VAR,(l)}_{T,n} \overset{\mathbb{P}}{\longrightarrow} \mathcal{V}^{VAR,(l)}_T$ for some sequence of $\mathcal{F}$-measurable random variables $\widehat{\mathcal{V}}^{COV}_{T,n}$, $n \in \mathbb{N}$. Then we have for $\alpha \in [0,1]$*

$$\lim_{n \to \infty} \widetilde{\mathbb{P}}([X^{(l)}, X^{(l)}]_T \in C^{VAR,(l)}_{T,n}(\alpha) | F) = \alpha \qquad (7.10)$$

for any $F \in \mathcal{X}$ with $\mathbb{P}(F) > 0$ where $C^{VAR,(l)}_{T,n}(\alpha)$ is defined as

$$[V^{(l)}(g_2, \pi_n)_T - n^{-1/2}\widehat{Q}^{VAR,(l)}_{T,n}((1+\alpha)/2), V^{(l)}(g_2, \pi_n)_T + n^{-1/2}\widehat{Q}^{VAR,(l)}_{T,n}((1+\alpha)/2)].$$

We proceed similarly to estimate an asymptotic confidence interval for the covariation $[X^{(1)}, X^{(2)}]_T$. Let $(M_n)_{n \in \mathbb{N}}$ be an increasing sequence of natural numbers and define

$$Z^{COV}_{n,m}(s) = (\widehat{\mathcal{L}}^{COV,(l)}_{n,m}(s), \widehat{\mathcal{R}}^{COV,(l)}_{n,m}(s), \widehat{\mathcal{L}}^{COV,(2)}_{u,m}(s), \widehat{\mathcal{R}}^{COV,(2)}_{n,m}(s), \widehat{\mathcal{L}}^{COV}_{n,m}(s), \widehat{\mathcal{R}}^{COV}_{n,m}(s)),$$

$m = 1, \ldots, M_n$, via

$$Z^{COV}_{n,m}(s) = Z^{COV}_n(\kappa_{n,m}(s)), \quad \kappa_{n,m}(s) \sim \mathcal{U}[s - b_n, s + b_n], \qquad (7.11)$$

where $\kappa_{n,m}(s)$ is apart from the above property independent of $\mathcal{F}$ and $(b_n)_{n \in \mathbb{N}}$ is a sequence of non-negative real numbers with $b_n \to 0$, $nb_n \to \infty$; compare Remark 4.8. Further we define by

$$\bar{\sigma}^{(l)}(s, -) = \Big(\frac{1}{b_n} \sum_{i:\mathcal{I}^{(l)}_{i,n} \subset (s - b_n, s - |\pi_n|_T)} (\Delta^{(l)}_{i,n} X^{(l)})^2 \Big)^{\frac{1}{2}},$$

$$\bar{\sigma}^{(l)}(s, +) = \boldsymbol{\delta}^{(l)}(s, +) = \Big(\frac{1}{b_n} \sum_{i:\mathcal{I}^{(l)}_{i,n} \subset (s, s + b_n)} (\Delta^{(l)}_{i,n} X^{(l)})^2 \Big)^{\frac{1}{2}},$$

estimators for $\sigma_{s-}^{(l)}$, $\sigma_s^{(l)}$ where $\overline{\sigma}^{(l)}(s,-)$ is a slightly modified version of the estimator $\tilde{\sigma}^{(l)}(s,-)$. The estimators $\tilde{\sigma}^{(l)}(s,-)$, $\tilde{\sigma}^{(l)}(s,+)$ have been introduced and discussed in Chapter 6. Using these random variables we define

$$
\widehat{\Phi}_{T,n,m}^{COV} = (\widehat{\mathcal{V}}_{T,n}^{COV})^{1/2} U_{n,m}
$$

$$
+ \sum_{i,j:t_{i,n}^{(1)} \vee t_{j,n}^{(2)} \leq T} \mathbb{1}_{\{\mathcal{I}_{i,n}^{(1)} \cap \mathcal{I}_{j,n}^{(2)} \neq \emptyset\}} \left[\Delta_{i,n}^{(1)} X^{(1)} \mathbb{1}_{\{|\Delta_{i,n}^{(1)} X^{(1)}| > \beta |\mathcal{I}_{i,n}^{(1)}|^{\varpi}\}} \right.
$$

$$
\times \left(\overline{\sigma}_n^{(2)}(t_{i,n}^{(1)},-) \tilde{\rho}_n(\tau_{i,j}^n,-) \sqrt{\widehat{\mathcal{L}}_{n,m}^{COV}(\tau_{i,j}^n)} U_{n,(i,j),m}^{(1),-} \right.
$$

$$
+ \overline{\sigma}_n^{(2)}(t_{i,n}^{(1)},+) \tilde{\rho}_n(\tau_{i,j}^n,+) \sqrt{\widehat{\mathcal{R}}_{n,m}^{COV}(\tau_{i,j}^n)} U_{n,(i,j),m}^{(1),+}
$$

$$
+ \left((\overline{\sigma}_n^{(2)}(t_{i,n}^{(1)},-))^2 (1 - (\tilde{\rho}_n(\tau_{i,j}^n,-))^2) \widehat{\mathcal{L}}_{n,m}^{COV}(\tau_{i,j}^n) \right.
$$

$$
\left. + (\overline{\sigma}_n^{(2)}(t_{i,n}^{(1)},+))^2 (1 - (\tilde{\rho}_n(\tau_{i,j}^n,+))^2) \widehat{\mathcal{R}}_{n,m}^{COV}(\tau_{i,j}^n) \right)^{1/2} U_{n,(i,j),m}^{(3)}
$$

$$
\left. + \sqrt{(\overline{\sigma}_n^{(2)}(t_{i,n}^{(1)},-))^2 \widehat{\mathcal{L}}_{n,m}^{COV,(2)}(\tau_{i,j}^n) + (\overline{\sigma}_n^{(2)}(t_{i,n}^{(1)},+))^2 \widehat{\mathcal{R}}_{n,m}^{COV,(2)}(\tau_{i,j}^n)} U_{n,(i,j),m}^{(4)} \right)
$$

$$
+ \Delta_{j,n}^{(2)} X^{(2)} \mathbb{1}_{\{|\Delta_{j,n}^{(2)} X^{(2)}| > \beta |\mathcal{I}_{j,n}^{(2)}|^{\varpi}\}} \left(\overline{\sigma}_n^{(1)}(t_{j,n}^{(2)},-) \sqrt{\widehat{\mathcal{L}}_{n,m}^{COV}(\tau_{i,j}^n)} U_{n,(i,j),m}^{(1),-} \right.
$$

$$
+ \overline{\sigma}_n^{(1)}(t_{j,n}^{(2)},+) \sqrt{\widehat{\mathcal{R}}_{n,m}^{COV}(\tau_{i,j}^n)} U_{n,(i,j),m}^{(1),+}
$$

$$
\left. \left. + \sqrt{(\overline{\sigma}_n^{(1)}(t_{j,n}^{(2)},-))^2 \widehat{\mathcal{L}}_{n,m}^{COV,(1)}(\tau_{i,j}^n) + (\overline{\sigma}_n^{(1)}(t_{j,n}^{(2)},+))^2 \widehat{\mathcal{R}}_{n,m}^{COV,(1)}(\tau_{i,j}^n)} U_{n,(i,j),m}^{(2)} \right) \right]
$$

$$
\times \mathbb{1}_{\{|\Delta_{i,n}^{(1)} X^{(1)}| = \max_{k:\mathcal{I}_{k,n}^{(1)} \cap \mathcal{I}_{j,n}^{(2)} \neq \emptyset} |\Delta_{k,n}^{(1)} X^{(1)}|\}}
$$

$$
\times \mathbb{1}_{\{|\Delta_{j,n}^{(2)} X^{(2)}| = \max_{k:\mathcal{I}_{k,n}^{(2)} \cap \mathcal{I}_{i,n}^{(1)} \neq \emptyset} |\Delta_{k,n}^{(2)} X^{(2)}|\}}
$$

where $\tau_{i,j}^n = t_{i,n}^{(1)} \wedge t_{j,n}^{(2)}$ and $U_{n,m}$, $U_{n,(i,j),m}^{(1),-}$, $U_{n,(i,j),m}^{(1),+}$, $U_{n,(i,j),m}^{(2)}$, $U_{n,(i,j),m}^{(3)}$, $U_{n,(i,j),m}^{(4)}$ are i.i.d. standard normal distributed random variables independent of all previously introduced random variables and $\widehat{\mathcal{V}}_{T,n}^{COV}$ is a suitable estimator for $\mathcal{V}_T^{COV}$. Here, the structure of the second part in $\widehat{\Phi}_{T,n,m}^{COV}$ is fundamentally different from the structure of the estimator $\widehat{\Phi}_{T,n,m}^{biv}(f)$ used in Chapter 4 because for $\widehat{\Phi}_{T,n,m}^{COV}$ idiosyncratic jumps also have to be estimated as they are contained in the variable $\mathcal{Z}_T^{COV}$ which is part of the limit Φ_T^{COV}. Therefore summands are included whenever $|\Delta_{i,n}^{(1)} X^{(1)}| > \beta |\mathcal{I}_{i,n}^{(1)}|^{\varpi}$ or $|\Delta_{j,n}^{(2)} X^{(2)}| > \beta |\mathcal{I}_{j,n}^{(2)}|^{\varpi}$ is fulfilled. The indicator over the set $|\Delta_{i,n}^{(1)} X^{(1)}| = \max_{k:\mathcal{I}_{k,n}^{(1)} \cap \mathcal{I}_{j,n}^{(2)} \neq \emptyset} |\Delta_{k,n}^{(1)} X^{(1)}|$ ensures that no increment $\Delta_{j,n}^{(2)} X^{(2)}$ is included more than once in the sum and the indicator over the set $|\Delta_{j,n}^{(2)} X^{(2)}| = \max_{k:\mathcal{I}_{k,n}^{(2)} \cap \mathcal{I}_{i,n}^{(1)} \neq \emptyset} |\Delta_{k,n}^{(2)} X^{(2)}|$ ensures that no increment

$\Delta_{i,n}^{(1)} X^{(1)}$ is included more than once. Note that the maxima are almost surely unique if we assume that the volatility processes $\sigma^{(1)}$, $\sigma^{(2)}$ do not vanish. Thereby we make sure that no jump (which is estimated by an increment exceeding the threshold) enters the sum multiple times. Further we use the estimator $\overline{\sigma}^{(l)}(s, -)$ instead of $\tilde{\sigma}^{(l)}(s, -)$ such that in the construction of $\overline{\sigma}^{(l)}(t_{i,n}^{(3-l)}, -)$ no increments over intervals $\mathcal{I}_{j,n}^{(l)}$ are used which overlap with $\mathcal{I}_{i,n}^{(3-l)}$. Otherwise, as we evaluate $\overline{\sigma}^{(l)}(t_{i,n}^{(3-l)}, -)$ whenever $X^{(l)}$ jumps in $\mathcal{I}_{i,n}^{(l)}$, for co-jumps of $X^{(l)}$ and $X^{(3-l)}$ the jump might also enter the estimation of $\sigma^{(l)}$ and hence $\tilde{\sigma}^{(l)}(t_{i,n}^{(3-l)}, -)$ might diverge. The estimator $\tilde{\sigma}^{(l)}(t_{i,n}^{(3-l)}, +)$ already has this desired property.

Based on the variables $\widehat{\Phi}_{T,n,m}^{COV}$, $m = 1, \ldots, M_n$, we define via

$$\widehat{Q}_{T,n}^{COV}(\alpha) = \widehat{Q}_\alpha(\{\widehat{\Phi}_{T,n,m}^{COV} | m = 1, \ldots, M_n\})$$

an estimator for the $\mathcal{X}$-conditional α-quantile $Q^{COV}(\alpha)$ of $(\mathcal{V}_T^{COV})^{1/2} U + \mathcal{Z}_T^{COV}$. To be able to consistently estimate an asymptotic confidence interval we need the following structural assumptions.

Condition 7.12. *Suppose Condition 7.5 is fulfilled, it holds $\int_0^T |\sigma_s^{(1)} \sigma_s^{(2)}| ds > 0$ almost surely and the function $t \mapsto G_{2,2}(t)$ is strictly increasing. Further, let the sequence $(b_n)_{n \in \mathbb{N}} \subset (0, \infty)$ fulfil $b_n \to 0$, $|\pi_n|_T / b_n \xrightarrow{\mathbb{P}} 0$, $n(|\pi_n|_T)^2 / b_n \xrightarrow{\mathbb{P}} 0$ and suppose that $(M_n)_{n \in \mathbb{N}}$ is a sequence of natural numbers converging to infinity. Additionally:*

(i) For any $x \in \mathbb{R}^{6 \times P}$, $P \in \mathbb{N}$, and $s_p \in (0, T)$, $p = 1, \ldots, P$, it holds

$$\widetilde{\mathbb{P}}\big(|\widetilde{\mathbb{P}}(\widetilde{Z}_{n,1}^{COV}(s_p) \le x_p, \ p = 1, \ldots, P | \mathcal{S}) - \prod_{p=1}^{P} \widetilde{\mathbb{P}}(Z^{COV}(s_p) \le x_p)| > \varepsilon\big) \to 0$$

as $n \to \infty$, for all $\varepsilon > 0$.

(ii) The volatility process σ is itself an Itô semimartingale, i.e. a process of the form (1.1).

Under Condition 7.12 we are able to state the following result which yields an asymptotically valid confidence interval for $[X^{(1)}, X^{(2)}]_T$.

Theorem 7.13. *Suppose Condition 7.12 is fulfilled and it holds $\widehat{\mathcal{V}}_{T,n}^{COV} \xrightarrow{\mathbb{P}} \mathcal{V}_T^{COV}$ for some sequence of $\mathcal{F}$-measurable random variables $\widehat{\mathcal{V}}_{T,n}^{COV}$, $n \in \mathbb{N}$. Then we have for $\alpha \in [0, 1]$*

$$\lim_{n \to \infty} \widetilde{\mathbb{P}}\big([X^{(1)}, X^{(2)}]_T \in C_{T,n}^{COV}(\alpha) | F\big) = \alpha \tag{7.12}$$

for any $F \in \mathcal{X}$ with $\mathbb{P}(F) > 0$ where $C_{T,n}^{COV}(\alpha)$ is defined as

$$[V(f_{(1,1)}, \pi_n)_T - n^{-1/2}\widehat{Q}_{T,n}^{COV}((1+\alpha)/2), V(f_{(1,1)}, \pi_n)_T + n^{-1/2}\widehat{Q}_{T,n}^{COV}((1+\alpha)/2)].$$

Example 7.14. *In the Poisson setting we have the convergences $|\pi_n|_T/b_n \to 0$ and $n(|\pi_n|_T)^2/b_n \to 0$ in probability whenever $b_n = O(n^{-\alpha})$ for some $\alpha \in (0,1)$ by (5.2). Note also that $|\pi_n|_T/b_n \xrightarrow{\mathbb{P}} 0$ implies $nb_n \to \infty$ because of $|\mathcal{I}_{i,n}^{(l)}| = O_{\mathbb{P}}(n^{-1})$ as $n \to \infty$ and any $i \in \mathbb{N}$, $l = 1, 2$. Further Conditions 7.10(i) and 7.12(i) are fulfilled by Lemma 5.12; compare also Remark 5.13 regarding the proof of Condition 7.12(i).* $\qquad\square$

7.4 The Proofs

Proof of (7.6). We compute

$$\mathbb{E}\Big[\Big(\sqrt{n}\sum_{i,j:t_{i,n}^{(1)}\vee t_{j,n}^{(2)}\leq T}[\Delta_{i,n}^{(1)}W^{(1)}(\rho\Delta_{j,n}^{(2)}W^{(1)} + \sqrt{1-\rho^2}\Delta_{j,n}^{(2)}W^{(2)})$$

$$- \rho|\mathcal{I}_{i,n}^{(1)}\cap\mathcal{I}_{j,n}^{(2)}|]\mathbb{1}_{\{\mathcal{I}_{i,n}^{(1)}\cap\mathcal{I}_{j,n}^{(2)}\neq\emptyset\}}\Big)^2\Big|\mathcal{S}\Big]$$

$$= n\mathbb{E}\Big[\sum_{i,j:t_{i,n}^{(1)}\vee t_{j,n}^{(2)}\leq T}[\Delta_{i,n}^{(1)}W^{(1)}(\rho\Delta_{j,n}^{(2)}W^{(1)} + \sqrt{1-\rho^2}\Delta_{j,n}^{(2)}W^{(2)}) - \rho|\mathcal{I}_{i,n}^{(1)}\cap\mathcal{I}_{j,n}^{(2)}|]$$

$$\times \sum_{i',j':t_{i',n}^{(1)}\vee t_{j',n}^{(2)}\leq T}[\Delta_{i',n}^{(1)}W^{(1)}(\rho\Delta_{j',n}^{(2)}W^{(1)} + \sqrt{1-\rho^2}\Delta_{j',n}^{(2)}W^{(2)})$$

$$- \rho|\mathcal{I}_{i',n}^{(1)}\cap\mathcal{I}_{j',n}^{(2)}|]$$

$$\times \mathbb{1}_{\{\mathcal{I}_{i,n}^{(1)}\cap\mathcal{I}_{j,n}^{(2)}\neq\emptyset\}}\mathbb{1}_{\{\mathcal{I}_{i',n}^{(1)}\cap\mathcal{I}_{j',n}^{(2)}\neq\emptyset\}}\Big|\mathcal{S}\Big]$$

$$= n\mathbb{E}\Big[\sum_{i,j:t_{i,n}^{(1)}\vee t_{j,n}^{(2)}\leq T}\sum_{i',j':t_{i',n}^{(1)}\vee t_{j',n}^{(2)}\leq T}[\Delta_{i,n}^{(1)}W^{(1)}(\rho\Delta_{j,n}^{(2)}W^{(1)} + \sqrt{1-\rho^2}\Delta_{j,n}^{(2)}W^{(2)})$$

$$\times \Delta_{i',n}^{(1)}W^{(1)}(\rho\Delta_{j',n}^{(2)}W^{(1)} + \sqrt{1-\rho^2}\Delta_{j',n}^{(2)}W^{(2)}) - \rho^2|\mathcal{I}_{i,n}^{(1)}\cap\mathcal{I}_{j,n}^{(2)}||\mathcal{I}_{i',n}^{(1)}\cap\mathcal{I}_{j',n}^{(2)}|]$$

$$\times \mathbb{1}_{\{\mathcal{I}_{i,n}^{(1)}\cap\mathcal{I}_{j,n}^{(2)}\neq\emptyset\}}\mathbb{1}_{\{\mathcal{I}_{i',n}^{(1)}\cap\mathcal{I}_{j',n}^{(2)}\neq\emptyset\}}\Big|\mathcal{S}\Big]$$

$$= (1-\rho^2)G_{2,2}^n(T)$$

$$+ n\rho^2\mathbb{E}\Big[\sum_{i,j:t_{i,n}^{(1)}\vee t_{j,n}^{(2)}\leq T}\sum_{i',j':t_{i',n}^{(1)}\vee t_{j',n}^{(2)}\leq T}[\Delta_{i,n}^{(1)}W^{(1)}\Delta_{j,n}^{(2)}W^{(1)}\Delta_{i',n}^{(1)}W^{(1)}\Delta_{j',n}^{(2)}W^{(1)}$$

$$- |\mathcal{I}_{i,n}^{(1)}\cap\mathcal{I}_{j,n}^{(2)}||\mathcal{I}_{i',n}^{(1)}\cap\mathcal{I}_{j',n}^{(2)}|]\mathbb{1}_{\{\mathcal{I}_{i,n}^{(1)}\cap\mathcal{I}_{j,n}^{(2)}\neq\emptyset\}}\mathbb{1}_{\{\mathcal{I}_{i',n}^{(1)}\cap\mathcal{I}_{j',n}^{(2)}\neq\emptyset\}}\Big|\mathcal{S}\Big]. \qquad (7.13)$$

The last equality holds because for terms including increments of W^2 only expectations of those terms including $\Delta_{j,n}^{(2)} W^{(2)} \Delta_{j',n}^{(2)} W^{(2)}$ with $j = j'$ and consequently also containing $\Delta_{i,n}^{(1)} W^{(1)} \Delta_{i',n}^{(1)} W^{(1)}$ for $i = i'$ do not vanish. Next, we discuss the second sum in (7.13). For the terms with $i = i'$ and $j = j'$ it holds

$$\mathbb{E}\big[\Delta_{i,n}^{(1)} W^{(1)} \Delta_{j,n}^{(2)} W^{(1)} \Delta_{i',n}^{(1)} W^{(1)} \Delta_{j',n}^{(2)} W^{(1)} - |\mathcal{I}_{i,n}^{(1)} \cap \mathcal{I}_{j,n}^{(2)}||\mathcal{I}_{i',n}^{(1)} \cap \mathcal{I}_{j',n}^{(2)}|\,\big|\mathcal{S}\big]$$

$$= \mathbb{E}\big[(\Delta_{i,n}^{(1)} W^{(1)})^2 (\Delta_{j,n}^{(2)} W^{(1)})^2 - |\mathcal{I}_{i,n}^{(1)} \cap \mathcal{I}_{j,n}^{(2)}|^2\,\big|\mathcal{S}\big]$$

$$= 3|\mathcal{I}_{i,n}^{(1)} \cap \mathcal{I}_{j,n}^{(2)}|^2 + |\mathcal{I}_{i,n}^{(1)} \setminus \mathcal{I}_{j,n}^{(2)}||\mathcal{I}_{i,n}^{(1)} \cap \mathcal{I}_{j,n}^{(2)}|$$

$$\quad + |\mathcal{I}_{i,n}^{(1)} \cap \mathcal{I}_{j,n}^{(2)}||\mathcal{I}_{j,n}^{(2)} \setminus \mathcal{I}_{i,n}^{(1)}| + |\mathcal{I}_{i,n}^{(1)} \setminus \mathcal{I}_{j,n}^{(2)}||\mathcal{I}_{j,n}^{(2)} \setminus \mathcal{I}_{i,n}^{(1)}| - |\mathcal{I}_{i,n}^{(1)} \cap \mathcal{I}_{j,n}^{(2)}|^2$$

$$= |\mathcal{I}_{i,n}^{(1)} \cap \mathcal{I}_{j,n}^{(2)}|^2 + |\mathcal{I}_{i,n}^{(1)}||\mathcal{I}_{j,n}^{(2)}|.$$

To derive these identities we separated the increments of $W^{(1)}$ into increments over overlapping and non-overlapping intervals. In the case $i = i'$, $j \neq j'$ we obtain

$$\mathbb{E}\big[\Delta_{i,n}^{(1)} W^{(1)} \Delta_{j,n}^{(2)} W^{(1)} \Delta_{i',n}^{(1)} W^{(1)} \Delta_{j',n}^{(2)} W^{(1)} - |\mathcal{I}_{i,n}^{(1)} \cap \mathcal{I}_{j,n}^{(2)}||\mathcal{I}_{i',n}^{(1)} \cap \mathcal{I}_{j',n}^{(2)}|\,\big|\mathcal{S}\big]$$

$$= \mathbb{E}\big[(\Delta_{i,n}^{(1)} W^{(1)})^2 \Delta_{j,n}^{(2)} W^{(1)} \Delta_{j',n}^{(2)} W^{(1)} - |\mathcal{I}_{i,n}^{(1)} \cap \mathcal{I}_{j,n}^{(2)}||\mathcal{I}_{i,n}^{(1)} \cap \mathcal{I}_{j',n}^{(2)}|\,\big|\mathcal{S}\big]$$

$$= \mathbb{E}\big[(\Delta_{i \cap j} W^{(1)} + \Delta_{i \cap j'} W^{(1)} + \Delta_{i \setminus \{j,j'\}} W^{(1)})^2 \Delta_{j,n}^{(2)} W^{(1)} \Delta_{j',n}^{(2)} W^{(1)}$$

$$\quad - |\mathcal{I}_{i,n}^{(1)} \cap \mathcal{I}_{j,n}^{(2)}||\mathcal{I}_{i,n}^{(1)} \cap \mathcal{I}_{j',n}^{(2)}|\,\big|\mathcal{S}\big]$$

$$= 2|\mathcal{I}_{i,n}^{(1)} \cap \mathcal{I}_{j,n}^{(2)}||\mathcal{I}_{i,n}^{(1)} \cap \mathcal{I}_{j',n}^{(2)}| - |\mathcal{I}_{i,n}^{(1)} \cap \mathcal{I}_{j,n}^{(2)}||\mathcal{I}_{i,n}^{(1)} \cap \mathcal{I}_{j',n}^{(2)}|$$

$$= |\mathcal{I}_{i,n}^{(1)} \cap \mathcal{I}_{j,n}^{(2)}||\mathcal{I}_{i,n}^{(1)} \cap \mathcal{I}_{j',n}^{(2)}|$$

where $\Delta_{i \cap j} W^{(1)}, \Delta_{i \cap j'} W^{(1)}, \Delta_{i \setminus \{j,j'\}} W^{(1)}$ denote the increments of $W^{(1)}$ over the intervals $\mathcal{I}_{i,n}^{(1)} \cap \mathcal{I}_{j,n}^{(2)}$, $\mathcal{I}_{i,n}^{(1)} \cap \mathcal{I}_{j',n}^{(2)}$, $\mathcal{I}_{i,n}^{(1)} \setminus (\mathcal{I}_{j,n}^{(2)} \cup \mathcal{I}_{j',n}^{(2)})$. By symmetry we obtain the same result in the case $i \neq i'$, $j = j'$. In the case $i \neq i'$, $j \neq j'$ we obtain using similar decompositions

$$\mathbb{E}\big[\Delta_{i,n}^{(1)} W^{(1)} \Delta_{j,n}^{(2)} W^{(1)} \Delta_{i',n}^{(1)} W^{(1)} \Delta_{j',n}^{(2)} W^{(1)} - |\mathcal{I}_{i,n}^{(1)} \cap \mathcal{I}_{j,n}^{(2)}||\mathcal{I}_{i',n}^{(1)} \cap \mathcal{I}_{j',n}^{(2)}|\,\big|\mathcal{S}\big]$$

$$\quad \times \mathbb{1}_{\{\mathcal{I}_{i,n}^{(1)} \cap \mathcal{I}_{j,n}^{(2)} \neq \emptyset\}} \mathbb{1}_{\{\mathcal{I}_{i',n}^{(1)} \cap \mathcal{I}_{j',n}^{(2)} \neq \emptyset\}}$$

$$= \big(|\mathcal{I}_{i,n}^{(1)} \cap \mathcal{I}_{j,n}^{(2)}||\mathcal{I}_{i',n}^{(1)} \cap \mathcal{I}_{j',n}^{(2)}| + |\mathcal{I}_{i,n}^{(1)} \cap \mathcal{I}_{j',n}^{(2)}||\mathcal{I}_{i',n}^{(1)} \cap \mathcal{I}_{j,n}^{(2)}|$$

$$\quad - |\mathcal{I}_{i,n}^{(1)} \cap \mathcal{I}_{j,n}^{(2)}||\mathcal{I}_{i',n}^{(1)} \cap \mathcal{I}_{j',n}^{(2)}|\big)$$

$$\quad \times \mathbb{1}_{\{\mathcal{I}_{i,n}^{(1)} \cap \mathcal{I}_{j,n}^{(2)} \neq \emptyset\}} \mathbb{1}_{\{\mathcal{I}_{i',n}^{(1)} \cap \mathcal{I}_{j',n}^{(2)} \neq \emptyset\}}$$

$$= |\mathcal{I}_{i,n}^{(1)} \cap \mathcal{I}_{j',n}^{(2)}||\mathcal{I}_{i',n}^{(1)} \cap \mathcal{I}_{j,n}^{(2)}| \mathbb{1}_{\{\mathcal{I}_{i,n}^{(1)} \cap \mathcal{I}_{j,n}^{(2)} \neq \emptyset\}} \mathbb{1}_{\{\mathcal{I}_{i',n}^{(1)} \cap \mathcal{I}_{j',n}^{(2)} \neq \emptyset\}}$$

which is always equal to zero because the intervals $\mathcal{I}_{i,n}^{(1)}$, $\mathcal{I}_{i',n}^{(1)}$ cannot at the same time both overlap with both other intervals $\mathcal{I}_{j,n}^{(2)}$, $\mathcal{I}_{j',n}^{(2)}$.

Putting the results from those four cases together we obtain

$$n\rho^2 \mathbb{E}\Big[\sum_{i,j:t_{i,n}^{(1)}\vee t_{j,n}^{(2)}\leq T}\ \sum_{i',j':t_{i',n}^{(1)}\vee t_{j',n}^{(2)}\leq T} [\Delta_{i,n}^{(1)}W^{(1)}\Delta_{j,n}^{(2)}W^{(1)}\Delta_{i',n}^{(1)}W^{(1)}\Delta_{j',n}^{(2)}W^{(1)}$$

$$- |\mathcal{I}_{i,n}^{(1)}\cap\mathcal{I}_{j,n}^{(2)}||\mathcal{I}_{i',n}^{(1)}\cap\mathcal{I}_{j',n}^{(2)}|]\mathbb{1}_{\{\mathcal{I}_{i,n}^{(1)}\cap\mathcal{I}_{j,n}^{(2)}\neq\emptyset\}}\mathbb{1}_{\{\mathcal{I}_{i',n}^{(1)}\cap\mathcal{I}_{j',n}^{(2)}\neq\emptyset\}}\Big|\mathcal{S}\Big]$$

$$= G_{2,2}^n(T) + \rho^2(H_{0,0,4}^n(T) + \tilde{H}^n(T))$$

which yields the claim.   □

Proof of Theorem 7.3. The proof of (7.1) is very similar to the proof of (6.25) and of (6.31). It holds

$$V(f_{(1,1)},\pi_n)_T = \sum_{l=1,2}\ \sum_{i,j:t_{i,n}^{(l)}\vee t_{j,n}^{(3-l)}\leq T} \Delta_{i,n}^{(l)}B^{(l)}(q)\Delta_{j,n}^{(3-l)}X^{(3-l)}\mathbb{1}_{\{\mathcal{I}_{i,n}^{(l)}\cap\mathcal{I}_{j,n}^{(3-l)}\neq\emptyset\}}$$

$$\tag{7.14}$$

$$- \sum_{i,j:t_{i,n}^{(1)}\vee t_{j,n}^{(2)}\leq T} \Delta_{i,n}^{(1)}B^{(1)}(q)\Delta_{j,n}^{(2)}B^{(2)}(q)\mathbb{1}_{\{\mathcal{I}_{i,n}^{(1)}\cap\mathcal{I}_{j,n}^{(2)}\neq\emptyset\}} \tag{7.15}$$

$$+ \sum_{i,j:t_{i,n}^{(1)}\vee t_{j,n}^{(2)}\leq T} \Delta_{i,n}^{(1)}M^{(1)}(q)\Delta_{j,n}^{(2)}M^{(2)}(q)\mathbb{1}_{\{\mathcal{I}_{i,n}^{(1)}\cap\mathcal{I}_{j,n}^{(2)}\neq\emptyset\}} \tag{7.16}$$

$$+ \sum_{l=1,2}\ \sum_{i,j:t_{i,n}^{(l)}\vee t_{j,n}^{(3-l)}\leq T} \Delta_{i,n}^{(l)}C^{(l)}\Delta_{j,n}^{(3-l)}M^{(3-l)}(q)\mathbb{1}_{\{\mathcal{I}_{i,n}^{(l)}\cap\mathcal{I}_{j,n}^{(3-l)}\neq\emptyset\}} \tag{7.17}$$

$$+ \sum_{l=1,2}\ \sum_{i,j:t_{i,n}^{(l)}\vee t_{j,n}^{(3-l)}\leq T} \Delta_{i,n}^{(l)}(M^{(l)}(q)+C^{(l)})\Delta_{j,n}^{(3-l)}N^{(3-l)}(q)\mathbb{1}_{\{\mathcal{I}_{i,n}^{(l)}\cap\mathcal{I}_{j,n}^{(3-l)}\neq\emptyset\}}$$

$$\tag{7.18}$$

$$+ \sum_{i,j:t_{i,n}^{(1)}\vee t_{j,n}^{(2)}\leq T} \Delta_{i,n}^{(1)}C^{(1)}\Delta_{j,n}^{(2)}C^{(2)}\mathbb{1}_{\{\mathcal{I}_{i,n}^{(1)}\cap\mathcal{I}_{j,n}^{(2)}\neq\emptyset\}} \tag{7.19}$$

$$+ \sum_{i,j:t_{i,n}^{(1)}\vee t_{j,n}^{(2)}\leq T} \Delta_{i,n}^{(1)}N^{(1)}(q)\Delta_{j,n}^{(2)}N^{(2)}(q)\mathbb{1}_{\{\mathcal{I}_{i,n}^{(1)}\cap\mathcal{I}_{j,n}^{(2)}\neq\emptyset\}}. \tag{7.20}$$

The $\mathcal{S}$-conditional expectation of the absolute value of the term (7.14) is using similar arguments as for (6.26) bounded by $K_q(|\pi_n|_T)^{1/2}T$ which vanishes as

$n \to \infty$ for any $q > 0$. Further, the $\mathcal{S}$-conditional expectation of the absolute value of the term (7.15) is using (1.8) bounded by

$$\sum_{i,j:t_{i,n}^{(1)} \vee t_{j,n}^{(2)} \leq T} K_q^2 |\mathcal{I}_{i,n}^{(1)}| |\mathcal{I}_{j,n}^{(2)}| \mathbb{1}_{\{\mathcal{I}_{i,n}^{(l)} \cap \mathcal{I}_{j,n}^{(3-l)} \neq \emptyset\}} \leq 3K_q^2 |\pi_n|_T T$$

which vanishes as well for $n \to \infty$ and any $q > 0$. The term (7.16) can be treated similarly as (6.29) to find an upper bound for the $\mathcal{S}$-conditional expectation of the square of (7.16) of the form $K(e_q)^2 T^2$ which vanishes as $q \to \infty$. Further the expectation of the square of (7.17) is using (1.9), (1.10), (3.30) and Lemma 6.5 bounded uniformly in n by $Ke_q T^2$ which vanishes as $q \to \infty$. The sum (7.18) vanishes as $n \to \infty$ for any $q > 0$ because $N(q)$ almost surely has only finitely many jumps and because of

$$\sum_{i:t_{i,n}^{(l)} \leq T} \Delta_{i,n}^{(l)} (M^{(l)}(q) + C^{(l)}) \mathbb{1}_{\{S_p \in \mathcal{I}_{i,n}^{(l)}\}} \to 0$$

almost surely as $n \to \infty$ for any jump time S_p of $N(q)$.

Finally we consider the terms (7.19) and (7.20) which make up the limit. First recall that

$$\sum_{i,j:t_{i,n}^{(1)} \vee t_{j,n}^{(2)} \leq T} \Delta_{i,n}^{(1)} C^{(1)} \Delta_{j,n}^{(2)} C^{(2)} \mathbb{1}_{\{\mathcal{I}_{i,n}^{(1)} \cap \mathcal{I}_{j,n}^{(2)} \neq \emptyset\}} \xrightarrow{\mathbb{P}} \int_0^T \rho_s \sigma_s^{(1)} \sigma_s^{(2)} ds.$$

follows from Theorem 2.22 as bas been shown in Example 2.23. Further regarding (7.20) we observe

$$\sum_{i,j:t_{i,n}^{(1)} \vee t_{j,n}^{(2)} \leq T} \Delta_{i,n}^{(1)} N^{(1)}(q) \Delta_{j,n}^{(2)} N^{(2)}(q) \mathbb{1}_{\{\mathcal{I}_{i,n}^{(1)} \cap \mathcal{I}_{j,n}^{(2)} \neq \emptyset\}}$$

$$\xrightarrow{\mathbb{P}} \sum_{s \leq T} \Delta N^{(1)}(q)_s \Delta N^{(2)}(q)_s$$

as $N(q)$ almost surely only has finitely many jumps in $[0, T]$ which are separated by the observation scheme for $|\pi_n|_T$ small enough.

Combining the discussion of (7.14)–(7.20) above we obtain

$$\lim_{q \to \infty} \limsup_{n \to \infty} \mathbb{P}(|V(f_{(1,1)}, \pi_n)_T - [X^{(1)}, X^{(2)}]_T| > \delta) = 0$$

for any $\delta > 0$ which yields (7.1).

The convergences in (7.2) follow from part b) of Theorem 2.38 respectively from part b) of Theorem 2.35. $\square$

Theorems 7.6 and 7.8 can be proven very similarly to Theorems 2 and 3 in [8]. The only difference is that we need a slightly more involved discretization of σ due to possible co-jumps of X and σ. This slightly more involved discretization has already been used in the proofs in Section 3.1. To illustrate the differences we lay out the rough structure of the proof of Theorem 7.6 but refer to [8] for the arguments which are similar.

Proof of Theorem 7.6. Denote

$$
\Phi_T^{VAR,(l)} = \sqrt{2}\left(\int_0^T (\sigma_s^{(l)})^4 dG_4^{(l)}(s) \right)^{1/2} U
$$
$$
+ 2 \sum_{p: S_p^{(l)} \leq T} \sqrt{(\sigma_{S_p^{(l)}-}^{(l)})^2 \delta_-^{(l)}(S_p^{(l)}) + (\sigma_{S_p^{(l)}}^{(l)})^2 \delta_+^{(l)}(S_p^{(l)})} \, U_{S_p^{(l)}} \Delta X_{S_p^{(l)}} \quad (7.21)
$$

and using the notation from the proof of Theorem 3.2 we define

$$
R^{(l)}(n,q,r) = \sqrt{n}\left(\sum_{i:t_{i,n} \leq T} \Delta_{i,n}^{(l)} C^{(l)}(r) - \int_0^T (\sigma^{(l)}(r)_s)^2 ds \right)
$$
$$
+ \sqrt{n} \sum_{i:t_{i,n} \leq T} \Delta_{i,n}^{(l)} N^{(l)}(q) \Delta_{i,n}^{(l)} \widetilde{C}^{(l)}(r,q).
$$

$$
\Phi_T^{VAR,(l)}(q,r) = \sqrt{2}\left(\int_0^T (\sigma^{(l)}(r)_s^{(l)})^4 dG_4^{(l)}(s) \right)^{1/2} U
$$
$$
+ 2 \sum_{p: S_{q,p}^{(l)} \leq T} \sqrt{(\tilde{\sigma}^{(l)}(r,q)_{S_p^{(l)}-})^2 \delta_-^{(l)}(S_{q,p}^{(l)}) + (\tilde{\sigma}^{(l)}(r,q)_{S_{q,p}^{(l)}})^2 \delta_+^{(l)}(S_{q,p}^{(l)})}
$$
$$
\times U_{S_p^{(l)}} \Delta N(q)_{S_{q,p}^{(l)}}
$$

Following the structure of the proof of Theorem 3.2 we then show

$$
\lim_{q \to \infty} \limsup_{r \to \infty} \limsup_{n \to \infty} \mathbb{P}(|\sqrt{n}(V^{(l)}(g_2,\pi_n)_T - [X^{(l)},X^{(l)}]_T) - R^{(l)}(n,q,r)| > \varepsilon) = 0,
$$
$$
(7.22)
$$

$$
\lim_{q \to \infty} \limsup_{r \to \infty} \widetilde{\mathbb{P}}(|\Phi_T^{VAR,(l)} - \Phi_T^{VAR,(l)}(q,r)| > \varepsilon) = 0, \quad \forall \varepsilon > 0, \quad (7.23)
$$

$$
R^{(l)}(n,q,r) \xrightarrow{\mathcal{L}_s} \Phi_T^{VAR,(l)}(q,r), \quad \forall q,r > 0. \quad (7.24)
$$

Here, (7.22) can be proven similarly as the corresponding statement in the proof of Theorem 2 in [8]. The statement (7.23) follows from

$$
\widetilde{\mathbb{E}}[|\Phi_T^{VAR,(l)} - \Phi_T^{VAR,(l)}(q,r)|^2 |\mathcal{X}] \xrightarrow{\mathbb{P}} 0 \quad (7.25)
$$

and Lemma 2.15. (7.25) can be shown as in Step 3 of the proof of Theorem 3.2 using the boundedness of σ and that the measures $\Gamma^{VAR,(l)}(\cdot, dy)$ have uniformly bounded first moments. Further for proving (7.24), observe that the stable convergence of the second term in $R^{(l)}(n, q, r)$ to the second term in $\Phi_T^{VAR,(l)}(q, r)$ follows as in the proof of (3.25) while the convergence of the first term in $R^{(l)}(n, q, r)$ to the first term in $\Phi_T^{VAR,(l)}(q, r)$ and the asymptotic independence of the two terms in $R^{(l)}(n, q, r)$ is shown in the proof of Proposition 3 in [8].

Finally (7.4) follows from (7.22)–(7.24) and Lemma B.6. $\qquad\square$

Proof of Theorem 7.8. Theorem 7.8 can be proven in the same way as Theorem 7.6 above where we use arguments from the proof of Theorem 3 in [8] and the proofs in Section 3.1 in the right places. $\qquad\square$

As in Chapter 4 we first consider two propositions for the proofs of Theorem 7.11 and 7.13.

Proposition 7.15. *Suppose that Condition 7.10 is satisfied and let $\Phi_T^{VAR,(l)}$ be defined as in (7.21). Then it holds*

$$\widetilde{\mathbb{P}}\big(\big|\frac{1}{M_n}\sum_{m=1}^{M_n}\mathbb{1}_{\{\widehat{\Phi}_{T,n,m}^{VAR,(l)}\leq\Upsilon\}} - \widetilde{\mathbb{P}}\big(\Phi_T^{VAR,(l)}\leq\Upsilon|\mathcal{X}\big)\big| > \varepsilon\big) \to 0 \qquad (7.26)$$

for any $\mathcal{X}$-measurable random variable Υ and all $\varepsilon > 0$.

Proof. Denote by $S_{q,p}^{(l)}$, $p \in \mathbb{N}$, an increasing sequence of stopping times which exhausts the jump times of $N^{(l)}(q)$ and define

$$Y(P, n, m) = (\widehat{\mathcal{V}}_{T,n}^{VAR,(l)})^{1/2}U_{n,m}$$

$$+ 2\sum_{p=1}^{P}\Delta_{i_n^{(l)}(S_{q,p}^{(l)}),n}^{(l)}X^{(l)}\mathbb{1}_{\{|\Delta_{i_n^{(l)}(S_{q,p}^{(l)}),n}^{(l)}X^{(l)}|>\beta|\mathcal{I}_{i_n^{(l)}(S_{q,p}^{(l)}),n}^{(l)}|^{\varpi}\}}$$

$$\times\big(\widetilde{\sigma}_n^{(l)}(t_{i_n^{(l)}(S_{q,p}^{(l)}),n}^{(l)},-)\big)^2\widehat{\delta}_{n,m-}^{(l)}(t_{i_n^{(l)}(S_{q,p}^{(l)}),n}^{(l)})$$

$$+ \big((\widetilde{\sigma}_n^{(l)}(t_{i_n^{(l)}(S_{q,p}^{(l)}),n}^{(l)},+)\big)^2\widehat{\delta}_{n,m,+}^{(l)}(t_{i_n^{(l)}(S_{q,p}^{(l)}),n}^{(l)})\big)^{1/2}$$

$$\times U_{n,i_n^{(l)}(S_{q,p}^{(l)}),m}\mathbb{1}_{\{S_{q,p}^{(l)}\leq T\}},$$

$$Y(P) = (\mathcal{V}_T^{VAR,(l)})^{1/2}U$$

$$+ 2\sum_{p=1}^{P}\Delta X_{S_{q,p}^{(l)}}\sqrt{(\sigma_{S_{q,p}^{(l)}-}^{(l)})^2\delta_-^{(l)}(S_{q,p}^{(l)}) + (\sigma_{S_p^{(l)}}^{(l)})^2\delta_+^{(l)}(S_{q,p}^{(l)})}U_{S_{q,p}^{(l)}}\mathbb{1}_{\{S_{q,p}^{(l)}\leq T\}}.$$

Step 1. Then Lemma 4.9 with $A_{n,1} = \widehat{\mathcal{V}}^{VAR,(l)}_{T,n}$, $A_1 = \mathcal{V}^{VAR,(l)}_T$, $\widehat{Z}^1_{n,m}(s) = U_{n,m}$, $Z^1(s) = U$ and

$$A_{n,p+1} = \big(\Delta^{(l)}_{i^{(l)}_n(S^{(l)}_{q,p}),n} X^{(l)} \mathbf{1}_{\{|\Delta^{(l)}_{i^{(l)}_n(S^{(l)}_{q,p}),n} X^{(l)}|>\beta|\mathcal{I}^{(l)}_{i^{(l)}_n(S^{(l)}_{q,p}),n}|^{\varpi}\}},$$
$$\tilde{\sigma}^{(l)}_n(S^{(l)}_{q,p},-), \tilde{\sigma}^{(l)}_n(S^{(l)}_{q,p},+)\big),$$
$$A_{p+1} = \big(\Delta X^{(l)}_{S^{(l)}_{q,p}}, \sigma^{(l)}_{S^{(l)}_{q,p}-}, \sigma^{(l)}_{S^{(l)}_{q,p}}\big),$$
$$\widehat{Z}^{p+1}_{n,m}(s) = \big(\hat{\delta}^{(l)}_{n,m,-}(s), \hat{\delta}^{(l)}_{n,m,+}(s), U_{n,i^{(l)}_n(s),m}\big),$$
$$Z^{p+1}(s) = \big(\delta^{(l)}_-(s), \delta^{(l)}_+(s), U_s\big),$$

$p = 1, \dots P$, and the function φ defined such that

$$\varphi((A_p, Z^p(S^{(l)}_{q,p-1}))_{p=1,\dots,P+1}) = Y(P)$$
$$\varphi((A_{n,p}, Z^p_{n,m}(S^{(l)}_{q,p-1}))_{p=1,\dots,P+1}) = Y(P,n,m)$$

(set $S^{(l)}_{q,0} = 0$) yields

$$\widetilde{\mathbb{P}}(\{|\frac{1}{M_n}\sum_{m=1}^{M_n}\mathbf{1}_{\{Y(P,n,m)\leq\Upsilon\}} - \widetilde{\mathbb{P}}(Y(P)\leq\Upsilon|\mathcal{X})|>\varepsilon)\to 0 \qquad (7.27)$$

as $n\to\infty$ for any $P\in\mathbb{N}$.

Step 2. Next we prove

$$\lim_{P\to\infty}\limsup_{n\to\infty}\frac{1}{M_n}\sum_{m=1}^{M_n}\widetilde{\mathbb{P}}(|Y(P,n,m)-\widehat{\Phi}^{VAR,(l)}_{T,n,m}|>\varepsilon)=0 \qquad (7.28)$$

for all $\varepsilon > 0$. Denote by $\Omega(P,q,n)$ the set on which there are at most P jumps of $N^{(l)}(q)$ in $[0,T]$ and two different jumps of $N^{(l)}(q)$ are further apart than $|\pi_n|T$. Obviously, $\mathbb{P}(\Omega(P,q,n))\to 1$ for $P,n\to\infty$ and any $q>0$. On the set $\Omega(P,q,n)$ we have

$$\mathbb{E}\big[|Y(P,n,m)-\widehat{\Phi}^{VAR,(l)}_{T,n,m}|^2\mathbf{1}_{\Omega(P,q,n)}|\mathcal{S}\big]$$
$$\leq \sum_{t^{(l)}_{i,n}\leq T, \nexists p:S^{(l)}_{q,p}\in\mathcal{I}^{(l)}_{i,n}} \mathbb{E}\big[(\Delta^{(l)}_{i,n}(X^{(l)}-N^{(l)}(q)))^2\mathbf{1}_{\{|\Delta^{(l)}_{i,n}X^{(l)}|>\beta|\mathcal{I}^{(l)}_{i,n}|^{\varpi}\}}$$
$$\times ((\tilde{\sigma}^{(l)}_n(t^{(l)}_{i,n},-))^2\hat{\delta}^{(l)}_{n,m,-}(t^{(l)}_{i,n}) + (\tilde{\sigma}^{(l)}_n(t^{(l)}_{i,n},+))^2\hat{\delta}^{(l)}_{n,m,+}(t^{(l)}_{i,n}))(U_{n,i,m})^2|\mathcal{S}\big]$$
$$\leq \sum_{t^{(l)}_{i,n}\leq T} \mathbb{E}\big[(\Delta^{(l)}_{i,n}(X^{(l)}-N^{(l)}(q)))^2\mathbf{1}_{\{|\Delta^{(l)}_{i,n}X^{(l)}|>\beta|\mathcal{I}^{(l)}_{i,n}|^{\varpi}\}}$$

$$\times \Big(\frac{1}{b_n} \sum_{j\neq i:\mathcal{I}_{j,n}^{(l)}\subset(t_{i,n}^{(l)}-b_n,t_{i,n}^{(l)}+b_n]} (\Delta_{j,n}^{(l)}X^{(l)})^2\Big)(\hat{\delta}_{n,m,-}^{(l)}(t_{i,n}^{(l)}) + \hat{\delta}_{n,m,+}^{(l)}(t_{i,n}^{(l)}))|\mathcal{S}]$$

$$(7.29)$$

where the expectation of terms with $\Delta_{i,n}^{(l)}(X^{(l)} - N^{(l)}(q))\Delta_{j,n}^{(l)}(X^{(l)} - N^{(l)}(q))$, $i \neq j$ vanishes because of

$$\mathbb{E}[U_{n,i,m}U_{n,j,m}|\sigma(\mathcal{F}, \{(\hat{\delta}_{n,m,-}^{(l)}(s), \hat{\delta}_{n,m,+}^{(l)}(s))|s \in [0,T]\})] = \mathbb{E}[U_{n,i,m}U_{n,j,m}] = 0$$

for $i \neq j$. The expression in the last line of (7.29) is using iterated expectations, Lemma 1.4 and inequality (2.77), compare Step 1 in the proof of Proposition 4.10 for details, bounded by

$$(K_q|\pi_n|_T + (|\pi_n|_T)^{1/2-\varpi} + e_q) \sum_{t_{i,n}^{(l)}\leq T} |\mathcal{I}_{i,n}^{(l)}| \sum_{k=-K_n}^{K_n} n|\mathcal{I}_{i+k,n}^{(l)}|^2 \Big(\sum_{k'=-K_n}^{K_n} |\mathcal{I}_{i+k',n}^{(l)}|\Big)^{-1}$$

$$\leq 2(K_q|\pi_n|_T + (|\pi_n|_T)^{1/2-\varpi} + e_q)G_4^{(l),n}(T).$$

Hence (7.28) follows from Condition 7.5(ii) and Lemma 2.15.

Step 3. As in Step 3 in the proof of Proposition 4.10 we deduce

$$\widetilde{\mathbb{P}}(Y(P) \leq \Upsilon|\mathcal{X}) \xrightarrow{\widetilde{\mathbb{P}}} \widetilde{\mathbb{P}}(\Phi_T^{VAR,(l)} \leq \Upsilon|\mathcal{X}) \qquad (7.30)$$

for $P \to \infty$ and as in that proof we obtain (7.26) from (7.27), (7.28) and (7.30). $\square$

Proposition 7.16. *Suppose that Condition 7.12 is satisfied and set* $\Phi_T^{COV} = (\mathcal{V}_T^{COV})^{1/2}U + \mathcal{Z}_T^{COV}$. *Then it holds*

$$\widetilde{\mathbb{P}}\Big(\Big|\frac{1}{M_n}\sum_{m=1}^{M_n} \mathbb{1}_{\{\hat{\Phi}_{T,n,m}^{COV}\leq\Upsilon\}} - \widetilde{\mathbb{P}}(\Phi_T^{COV}\leq\Upsilon|\mathcal{X})\Big| > \varepsilon\Big) \to 0 \qquad (7.31)$$

for any $\mathcal{X}$-*measurable random variable* Υ *and all* $\varepsilon > 0$.

Proof. The structure of this proof is identical to the structure of the proof of Proposition 7.15. We denote by $S_{q,p}$, $p \in \mathbb{N}$, an increasing sequence of stopping times which exhausts the jump times of $N(q)$ and define

$$Y(P,n,m) = (\widehat{\mathcal{V}}_{T,n}^{COV})^{1/2}U_{n,m}$$

$$+ \sum_{p=1}^{P}\mathbb{1}_{\{S_{q,p}\leq T\}} \sum_{i,j:t_{i,n}^{(1)}\vee t_{j,n}^{(2)}\leq T} \mathbb{1}_{\{i=i_*^{(1)}(S_{q,p}) \text{ or } j=i_*^{(2)}(S_{q,p})\}}$$

$$\times \mathbb{1}_{\{\mathcal{I}_{i,n}^{(1)}\cap\mathcal{I}_{j,n}^{(2)}\neq\emptyset\}}\Big[\Delta_{i,n}^{(1)}X^{(1)}\mathbb{1}_{\{|\Delta_{i,n}^{(1)}X^{(1)}|>\beta|\mathcal{I}_{i,n}^{(1)}|^{\varpi}\}}$$

$$
\times \Big(\overline{\sigma}_n^{(2)}(t_{i,n}^{(1)},-) \tilde{\rho}_n(\tau_{i,j}^n,-) \sqrt{\widehat{\mathcal{L}}_{n,m}(\tau_{i,j}^n)} U_{n,(i,j),m}^{(1),-}
$$

$$
+ \overline{\sigma}_n^{(2)}(t_{i,n}^{(1)},+) \tilde{\rho}_n(\tau_{i,j}^n,+) \sqrt{\widehat{\mathcal{R}}_{n,m}(\tau_{i,j}^n)} U_{n,(i,j),m}^{(1),+}
$$

$$
+ \Big((\overline{\sigma}_n^{(2)}(t_{i,n}^{(1)},-))^2 (1 - (\tilde{\rho}_n(\tau_{i,j}^n,-))^2) \widehat{\mathcal{L}}_{n,m}(\tau_{i,j}^n)
$$

$$
+ (\overline{\sigma}_n^{(2)}(t_{i,n}^{(1)},+))^2 (1 - (\tilde{\rho}_n(\tau_{i,j}^n,+))^2) \widehat{\mathcal{R}}_{n,m}(\tau_{i,j}^n) \Big)^{1/2} U_{n,(i,j),m}^{(3)}
$$

$$
+ \sqrt{(\overline{\sigma}_n^{(2)}(t_{i,n}^{(1)},-))^2 \widehat{\mathcal{L}}_{n,m}^{(2)}(\tau_{i,j}^n) + (\overline{\sigma}_n^{(2)}(t_{i,n}^{(1)},+))^2 \widehat{\mathcal{R}}_{n,m}^{(2)}(\tau_{i,j}^n)} U_{n,(i,j),m}^{(4)} \Big)
$$

$$
+ \Delta_{j,n}^{(2)} X^{(2)} \mathbb{1}_{\{|\Delta_{j,n}^{(2)} X^{(2)}| > \beta |\mathcal{I}_{j,n}^{(2)}|^\varpi\}} \Big(\overline{\sigma}_n^{(1)}(t_{j,n}^{(2)},-) \sqrt{\widehat{\mathcal{L}}_{n,m}(\tau_{i,j}^n)} U_{n,(i,j),m}^{(1),-}
$$

$$
+ \overline{\sigma}_n^{(1)}(t_{j,n}^{(2)},+) \sqrt{\widehat{\mathcal{R}}_{n,m}(\tau_{i,j}^n)} U_{n,(i,j),m}^{(1),+}
$$

$$
+ \sqrt{(\overline{\sigma}_n^{(1)}(t_{j,n}^{(2)},-))^2 \widehat{\mathcal{L}}_{n,m}^{(1)}(\tau_{i,j}^n) + (\overline{\sigma}_n^{(1)}(t_{j,n}^{(2)},+))^2 \widehat{\mathcal{R}}_{n,m}^{(1)}(\tau_{i,j}^n)} U_{n,(i,j),m}^{(2)} \Big) \Big]
$$

$$
\times \mathbb{1}_{\{|\Delta_{i,n}^{(1)} X^{(1)}| = \max_{k:\mathcal{I}_{k,n}^{(1)} \cap \mathcal{I}_{j,n}^{(2)} \neq \emptyset} |\Delta_{k,n}^{(1)} X^{(1)}|\}}
$$

$$
\times \mathbb{1}_{\{|\Delta_{j,n}^{(2)} X^{(2)}| = \max_{k:\mathcal{I}_{k,n}^{(2)} \cap \mathcal{I}_{i,n}^{(1)} \neq \emptyset} |\Delta_{k,n}^{(2)} X^{(2)}|\}}.
$$

Further we define $Y(P)$ similarly as Φ_T^{COV} with the only difference that in the term $\mathcal{Z}_T^{COV}$ we do not sum over all jump times $S_p \leq T$, but only over the jump times $S_{q,p} \leq T$. Suppose that at $S_{q,p}$ there is a common jump of $X^{(1)}$ and $X^{(2)}$. Then for $|\pi_n|_T$ small enough we have that $\Delta_{i_n^{(1)}(S_{q,p}),n}^{(1)} X^{(1)}$ is the largest increment among all increments over intervals $\mathcal{I}_{k,n}^{(1)}$ which overlap with $\mathcal{I}_{i_n^{(2)}(S_{q,p}),n}^{(2)}$ and $\Delta_{i_n^{(2)}(S_{q,p}),n}^{(2)} X^{(2)}$ is the largest increment among all increments over intervals $\mathcal{I}_{k,n}^{(2)}$ which overlap with $\mathcal{I}_{i_n^{(1)}(S_{q,p}),n}^{(1)}$. Hence exactly for $i_p^n = i_n^{(1)}(S_{q,p})$ and $j_p^n = i_n^{(2)}(S_{q,p})$ we obtain a summand which is asymptotically not equal to zero. Next suppose that only $X^{(1)}$ jumps at $S_{q,p}$. As argued above only summands with $i = i_n^{(1)}(S_{q,p})$ will not vanish in the limit. Further for exactly one j with $\mathcal{I}_{j,n} \cap \mathcal{I}_{i_n^{(1)}(S_{q,p}),n}^{(1)} \neq \emptyset$ we have $|\Delta_{j,n}^{(2)} X^{(2)}| = \max_{k:\mathcal{I}_{k,n}^{(2)} \cap \mathcal{I}_{i,n}^{(1)} \neq \emptyset} |\Delta_{k,n}^{(2)} X^{(2)}|$ because $\sigma^{(l)}$ does not vanish by Condition 7.12 and the size of this $|\Delta_{j,n}^{(2)} X^{(2)}|$ vanishes as $n \to \infty$ because $X^{(2)}$ does not jump at $S_{q,p}$. Furthermore by construction the $\mathcal{F}$-conditional distribution of

$$
\Delta_{i_p^n,n}^{(1)} X^{(1)} \mathbb{1}_{\{|\Delta_{i_p^n,n}^{(1)} X^{(1)}| > \beta |\mathcal{I}_{i_p^n,n}^{(1)}|^\varpi\}}
$$

$$
\times \Big(\overline{\sigma}_n^{(2)}(t_{i_p^n,n}^{(1)},-) \tilde{\rho}_n(\tau_{i_p^n,j}^n,-) \sqrt{\widehat{\mathcal{L}}_{n,m}(\tau_{i_p^n,j}^n)} U_{n,(i_p^n,j),m}^{(1),-}
$$

$$+ \overline{\sigma}_n^{(2)}(t_{i_p^n,n}^{(1)}, +)\tilde{\rho}_n(\tau_{i_p^n,j}^n, +)\sqrt{\widehat{\mathcal{R}}_{n,m}(\tau_{i_p^n,j}^n)}U_{n,(i_p^n,j),m}^{(1),+}$$

$$+ \left((\overline{\sigma}_n^{(2)}(t_{i_p^n,n}^{(1)}, -))^2(1 - (\tilde{\rho}_n(\tau_{i_p^n,j}^n, -))^2)\widehat{\mathcal{L}}_{n,m}(\tau_{i_p^n,j}^n)\right.$$

$$+ (\overline{\sigma}_n^{(2)}(t_{i_p^n,n}^{(1)}, +))^2(1 - (\tilde{\rho}_n(\tau_{i_p^n,j}^n, +))^2)\widehat{\mathcal{R}}_{n,m}(\tau_{i_p^n,j}^n)\right)^{1/2}U_{n,(i_p^n,j),m}^{(3)}$$

$$+ \sqrt{(\overline{\sigma}_n^{(2)}(t_{i_p^n,n}^{(1)}, -))^2\widehat{\mathcal{L}}_{n,m}^{(2)}(\tau_{i_p^n,j}^n) + (\overline{\sigma}_n^{(2)}(t_{i_p^n,n}^{(1)}, +))^2\widehat{\mathcal{R}}_{n,m}^{(2)}(\tau_{i_p^n,j}^n)}U_{n,(i_p^n,j),m}^{(4)}\right)$$

$$+ \Delta_{j,n}^{(2)}X^{(2)}\mathbb{1}_{\{|\Delta_{j,n}^{(2)}X^{(2)}|>\beta|\mathcal{I}_{j,n}^{(2)}|^{\varpi}\}}\left(\overline{\sigma}_n^{(1)}(t_{j,n}^{(2)}, -)\sqrt{\widehat{\mathcal{L}}_{n,m}(\tau_{i_p^n,j}^n)}U_{n,(i_p^n,j),m}^{(1),-}\right.$$

$$+ \overline{\sigma}_n^{(1)}(t_{j,n}^{(2)}, +)\sqrt{\widehat{\mathcal{R}}_{n,m}(\tau_{i_p^n,j}^n)}U_{n,(i_p^n,j),m}^{(1),+}$$

$$+ \sqrt{(\overline{\sigma}_n^{(1)}(t_{j,n}^{(2)}, -))^2\widehat{\mathcal{L}}_{n,m}^{(1)}(\tau_{i_p^n,j}^n) + (\overline{\sigma}_n^{(1)}(t_{j,n}^{(2)}, +))^2\widehat{\mathcal{R}}_{n,m}^{(1)}(\tau_{i_p^n,j}^n)}U_{n,(i_p^n,j),m}^{(2)}\right)$$

asymptotically does not depend on j for all $j \in \mathbb{N}$ with $\mathcal{I}_{i_p^n,n}^{(1)} \cap \mathcal{I}_{j,n}^{(2)} \neq \emptyset$ because j only has an influence on the correlation between the term involving $\Delta_{i_p^n,n}^{(1)}X^{(1)}$ and the term involving $\Delta_{j,n}^{(2)}X^{(2)}$ which becomes irrelevant as the second term vanishes anyways. Hence based on these considerations and similar arguments for idiosyncratic jumps in $X^{(2)}$ we obtain that $Y(P,n,m)$ is asymptotically equivalent to

$$Y'(P,n,m) = (\widehat{\mathcal{V}}_{T,n}^{COV})^{1/2}U_{n,m}$$

$$+ \sum_{p=1}^{P}\mathbb{1}_{\{S_{q,p}\leq T\}}\sum_{i,j:t_{i,n}^{(1)}\vee t_{j,n}^{(2)}\leq T}\mathbb{1}_{\{i=i_n^{(1)}(S_{q,p})\text{ and }j=i_n^{(2)}(S_{q,p})\}}$$

$$\times \left[\Delta_{i,n}^{(1)}X^{(1)}\mathbb{1}_{\{|\Delta_{i,n}^{(1)}X^{(1)}|>\beta|\mathcal{I}_{i,n}^{(1)}|^{\varpi}\}}\right.$$

$$\times \left(\overline{\sigma}_n^{(2)}(t_{i,n}^{(1)}, -)\tilde{\rho}_n(\tau_{i,j}^n, -)\sqrt{\widehat{\mathcal{L}}_{n,m}(\tau_{i,j}^n)}U_{n,(i,j),m}^{(1),-}\right.$$

$$+ \overline{\sigma}_n^{(2)}(t_{i,n}^{(1)}, +)\tilde{\rho}_n(\tau_{i,j}^n, +)\sqrt{\widehat{\mathcal{R}}_{n,m}(\tau_{i,j}^n)}U_{n,(i,j),m}^{(1),+}$$

$$+ \left((\overline{\sigma}_n^{(2)}(t_{i,n}^{(1)}, -))^2(1 - (\tilde{\rho}_n(\tau_{i,j}^n, -))^2)\widehat{\mathcal{L}}_{n,m}(\tau_{i,j}^n)\right.$$

$$+ (\overline{\sigma}_n^{(2)}(t_{i,n}^{(1)}, +))^2(1 - (\tilde{\rho}_n(\tau_{i,j}^n, +))^2)\widehat{\mathcal{R}}_{n,m}(\tau_{i,j}^n)\right)^{1/2}U_{n,(i,j),m}^{(3)}$$

$$+ \sqrt{(\overline{\sigma}_n^{(2)}(t_{i,n}^{(1)}, -))^2\widehat{\mathcal{L}}_{n,m}^{(2)}(\tau_{i,j}^n) + (\overline{\sigma}_n^{(2)}(t_{i,n}^{(1)}, +))^2\widehat{\mathcal{R}}_{n,m}^{(2)}(\tau_{i,j}^n)}U_{n,(i,j),m}^{(4)}\right)$$

$$+ \Delta_{j,n}^{(2)}X^{(2)}\mathbb{1}_{\{|\Delta_{j,n}^{(2)}X^{(2)}|>\beta|\mathcal{I}_{j,n}^{(2)}|^{\varpi}\}}\left(\overline{\sigma}_n^{(1)}(t_{j,n}^{(2)}, -)\sqrt{\widehat{\mathcal{L}}_{n,m}(\tau_{i,j}^n)}U_{n,(i,j),m}^{(1),-}\right.$$

$$+ \overline{\sigma}_n^{(1)}(t_{j,n}^{(2)}, +)\sqrt{\widehat{\mathcal{R}}_{n,m}(\tau_{i,j}^n)}U_{n,(i,j),m}^{(1),+}$$

$$+ \sqrt{(\overline{\sigma}_n^{(1)}(t_{j,n}^{(2)}, -))^2 \widehat{\mathcal{L}}_{n,m}^{(1)}(\tau_{i,j}^n) + (\overline{\sigma}_n^{(1)}(t_{j,n}^{(2)}, +))^2 \widehat{\mathcal{R}}_{n,m}^{(1)}(\tau_{i,j}^n)} U_{n,(i,j),m}^{(2)}\Big)\Big]$$

in the sense that they have for $m = 1, \ldots, M_n$ asymptotically identical common $\mathcal{F}$-conditional distributions.

Step 1. Note that it holds $\overline{\sigma}^{(3-l)}(t_{i_n^{(l)}(S_{q,p})}^{(l)}, -) = \tilde{\sigma}^{(3-l)}(S_{q,p}, -) + o_{\mathbb{P}}(1)$. Hence, using Lemma 4.9, Corollary 6.4 and Condition 7.12 we obtain similarly as in Step 2 of the proof of Proposition

$$\widetilde{\mathbb{P}}(\{|\frac{1}{M_n} \sum_{m=1}^{M_n} \mathbb{1}_{\{Y'(P,n,m) \leq \Upsilon\}} - \widetilde{\mathbb{P}}(Y(P) \leq \Upsilon | \mathcal{X})| > \varepsilon) \to 0$$

as $n \to \infty$ for any $P \in \mathbb{N}$ which then yields

$$\widetilde{\mathbb{P}}(\{|\frac{1}{M_n} \sum_{m=1}^{M_n} \mathbb{1}_{\{Y(P,n,m) \leq \Upsilon\}} - \widetilde{\mathbb{P}}(Y(P) \leq \Upsilon | \mathcal{X})| > \varepsilon) \to 0 \qquad (7.32)$$

based on the above considerations.

Step 2. Next we prove

$$\lim_{P \to \infty} \limsup_{n \to \infty} \frac{1}{M_n} \sum_{m=1}^{M_n} \widetilde{\mathbb{P}}(|Y(P,n,m) - \widehat{\Phi}_{T,n,m}^{COV}| > \varepsilon) = 0 \qquad (7.33)$$

for all $\varepsilon > 0$. Denote by $\Omega(P,q,n)$ the set on which there are at most P jumps of $N(q)$ in $[0,T]$ and two different jumps of $N(q)$ are further apart than $|\pi_n|_T$. Obviously, $\mathbb{P}(\Omega(P,q,n)) \to 1$ for $P, n \to \infty$ and any $q > 0$. On the set $\Omega(P,q,n)$ we obtain similarly as in Step 2 of the proof of Proposition 7.16 the following estimate

$$\mathbb{E}\big[|Y(P,n,m) - \widehat{\Phi}_{T,n,m}^{COV}|^2 \mathbb{1}_{\Omega(P,q,n)} | \mathcal{F}\big]$$

$$\leq K \sum_{l=1,2} \sum_{i_l, i_{3-l}: t_{i_l,n}^{(l)} \vee t_{i_{3-l},n}^{(3-l)} \leq T} (\Delta_{i_l,n}^{(l)}(X^{(l)} - N^{(l)}(q)))^2 \mathbb{1}_{\{|\Delta_{i_l,n}^{(l)} X^{(l)}| > \beta | \mathcal{I}_{i_l,n}^{(l)}|^{\varpi}\}}$$

$$\times \Big(\frac{1}{b_n} \sum_{k: \mathcal{I}_{k,n}^{(3-l)} \subset (t_{i_l,n}^{(l)} - b_n, t_{i_l,n}^{(l)} - |\pi_n|_T] \cup (t_{i_l,n}^{(l)}, t_{i_l,n}^{(l)} + b_n]} (\Delta_{k,n}^{(3-l)} X^{(3-l)})^2\Big)$$

$$\times \mathbb{1}_{\{\mathcal{I}_{i_l,n}^{(l)} \cap \mathcal{I}_{i_{3-l},n}^{(3-l)} \neq \emptyset\}} \prod_{l=1,2} \mathbb{1}_{\{|\Delta_{i_l,n}^{(l)} X^{(l)}| = \max_{k: \mathcal{I}_{k,n}^{(l)} \cap \mathcal{I}_{i_{3-l},n}^{(3-l)} \neq \emptyset} |\Delta_{k,n}^{(l)} X^{(l)}|\}}$$

$$\times \mathbb{E}[\widehat{\mathcal{L}}_{n,m}^{COV}(\tau_{i_l,i_{3-l}}^n) + \widehat{\mathcal{R}}_{n,m}^{COV}(\tau_{i_l,i_{3-l}}^n)$$

$$+ \widehat{\mathcal{L}}_{n,m}^{COV,(3-l)}(\tau_{i_l,i_{3-l}}^n) + \widehat{\mathcal{R}}_{n,m}^{COV,(3-l)}(\tau_{i_l,i_{3-l}}^n)|\mathcal{F}]. \qquad (7.34)$$

From the definition of $\widehat{Z}_{n,m}^{COV}(s)$ in (7.11) we obtain

$$
\mathbb{E}[\widehat{\mathcal{L}}_{n,m}^{COV}(\tau_{i_l,i_{3-l}}^n) + \widehat{\mathcal{R}}_{n,m}^{COV}(\tau_{i_l,i_{3-l}}^n)
$$
$$
+ \widehat{\mathcal{L}}_{n,m}^{COV,(3-l)}(\tau_{i_l,i_{3-l}}^n) + \widehat{\mathcal{R}}_{n,m}^{COV,(3-l)}(\tau_{i_l,i_{3-l}}^n)|\mathcal{F}]
$$
$$
\leq \frac{1}{2b_n} \sum_{i_l',i_{3-l}':|\tau_{i',i_{3-l}'}^n - \tau_{i,i_{3-l}}^n|\leq b_n} \mathcal{M}_n^{(3-l)}(\tau_{i_l',i_{3-l}'}^n)|\mathcal{I}_{i',n}^{(1)} \cap \mathcal{I}_{i_{3-l}',n}^{(2)}| + n|\pi_n|_T \frac{|\pi_n|_T}{2b_n}
$$

$$(7.35)$$

where

$$
\mathcal{M}_n^{(l)}(s) = n \sum_{i:t_{i,n}^{(l)}\leq T} |\mathcal{I}_{i,n}^{(l)}|\mathbb{1}_{\{\mathcal{I}_{i,n}^{(l)}\cap\mathcal{I}_{i_n^{(3-l)}(s),n}^{(3-l)} \neq \emptyset\}}, \quad l = 1,2.
$$

Using the estimate (7.35) we find that we can split the discussion of (7.34) into two parts. First we discuss (7.34) with the $\mathcal{F}$-conditional expectation replaced by $|\pi_n|_T/(2b_n)$. This expression is bounded by

$$
\frac{Kn(|\pi_n|_T)^2}{2b_n} \sum_{l=1,2} \sum_{i:t_{i,n}^{(l)}\leq T} (\Delta_{i,n}^{(l)}(X^{(l)} - N^{(l)}(q)))^2
$$
$$
\times \Big(\frac{1}{b_n} \sum_{k:\mathcal{I}_{k,n}^{(3-l)}\subset(t_{i,n}^{(l)}-b_n,t_{i,n}^{(l)}-|\pi_n|_T)\cup(t_{i,n}^{(l)},t_{i,n}^{(l)}+b_n]} (\Delta_{k,n}^{(3-l)}X^{(3-l)})^2 \Big) \quad (7.36)
$$

because each increment $(\Delta_{i,n}^{(l)}(X^{(l)} - N^{(l)}(q)))^2$ occurs in the sum at most once. The expression (7.36) vanishes as $n \to \infty$ due to $|\pi_n|_T/b_n \xrightarrow{\mathbb{P}} 0$ by Condition 7.12 because the $\mathcal{S}$-conditional expectation of the sum in (7.36) is bounded in probability using iterated expectations and Lemma 1.4. Note to this end that by construction no interval $\mathcal{I}_{k,n}^{(3-l)} \subset (t_{i,n}^{(l)} - b_n, t_{i,n}^{(l)} - |\pi_n|_T) \cup (t_{i,n}^{(l)}, t_{i,n}^{(l)} + b_n]$ overlaps with $\mathcal{I}_{i,n}^{(l)}$.

Next, we discuss (7.34) where we replace the expectation by the sum over the $\mathcal{M}_n^{(3-l)}(\tau_{i_l',i_{3-l}'}^n)$ from (7.35). By an index change as in (4.33) this expression is equal to

$$
K \sum_{l=1,2} \sum_{i_l,i_{3-l}:t_{i_l,n}^{(l)} \vee t_{i_{3-l},n}^{(3-l)}\leq T} \mathcal{M}_n^{(3-l)}(\tau_{i_l,i_{3-l}}^n)|\mathcal{I}_{i_l,n}^{(1)} \cap \mathcal{I}_{i_{3-l},n}^{(2)}|
$$
$$
\times \frac{1}{2b_n} \sum_{i_l',i_{3-l}':|\tau_{i',i_{3-l}'}^n - \tau_{i,i_{3-l}}^n|\leq b_n} (\Delta_{i_l',n}^{(l)}(X^{(l)} - N^{(l)}(q)))^2 \mathbb{1}_{\{|\Delta_{i_l',n}^{(l)}X^{(l)}|>\beta|\mathcal{I}_{i_l',n}^{(l)}|^\varpi\}}
$$

$$\times \Big(\frac{1}{b_n} \sum_{k: \mathcal{I}_{k,n}^{(3-l)} \subset (t_{i_l,n}^{(l)} - b_n, t_{i_l,n}^{(l)} - |\pi_n|_T] \cup (t_{i_l,n}^{(l)}, t_{i_l,n}^{(l)} + b_n]} (\Delta_{k,n}^{(3-l)} X^{(3-l)})^2 \Big)$$

$$\times \mathbb{1}_{\{\mathcal{I}_{i_l',n}^{(l)} \cap \mathcal{I}_{i_{3-l},n}^{(3-l)} \neq \emptyset\}}$$

$$\times \prod_{l=1,2} \mathbb{1}_{\{|\Delta_{i_l',n}^{(l)} X^{(l)}| = \max_{k: \mathcal{I}_{k,n}^{(l)} \cap \mathcal{I}_{i_{3-l}',n}^{(3-l)} \neq \emptyset} |\Delta_{k,n}^{(l)} X^{(l)}|\}}$$

$$\leq K \sum_{l=1,2} \sum_{i_l, i_{3-l}: t_{i_l,n}^{(l)} \vee t_{i_{3-l},n}^{(3-l)} \leq T} \mathcal{M}_n^{(3-l)} (\tau_{i_l, i_{3-l}}^n) |\mathcal{I}_{i_l,n}^{(1)} \cap \mathcal{I}_{i_{3-l},n}^{(2)}|$$

$$\times \frac{1}{2b_n} \sum_{i_l': |t_{i_l',n}^{(l)} - t_{i_l,n}^{(l)}| \leq b_n + 2|\pi_n|_T} (\Delta_{i_l',n}^{(l)} (X^{(l)} - N^{(l)}(q)))^2 \mathbb{1}_{\{|\Delta_{i_l',n}^{(l)} X^{(l)}| > \beta |\mathcal{I}_{i_l',n}^{(l)}|^{\varpi}\}}$$

$$\times \Big(\frac{1}{b_n} \sum_{k: \mathcal{I}_{k,n}^{(3-l)} \subset (t_{i_l',n}^{(l)} - b_n, t_{i_l',n}^{(l)} - |\pi_n|_T] \cup (t_{i_l',n}^{(l)}, t_{i_l',n}^{(l)} + b_n]} (\Delta_{k,n}^{(3-l)} X^{(3-l)})^2 \Big)$$

$$\times \mathbb{1}_{\{\mathcal{I}_{i_l',n}^{(l)} \cap \mathcal{I}_{i_{3-l},n}^{(3-l)} \neq \emptyset\}} \tag{7.37}$$

where we again used that each increment $(\Delta_{i_l',n}^{(l)} (X^{(l)} - N^{(l)}(q)))^2$ occurs in the sum at most once. Using iterated expectations, Lemma 1.4 and inequality (2.77) similarly as for (7.36) we obtain that the $\mathcal{S}$-conditional expectation of (7.37) is bounded by

$$K \sum_{l=1,2} \sum_{i_l, i_{3-l}: t_{i_l,n}^{(l)} \vee t_{i_{3-l},n}^{(3-l)} \leq T} \mathcal{M}_n^{(3-l)} (\tau_{i_l, i_{3-l}}^n) |\mathcal{I}_{i_l,n}^{(l)} \cap \mathcal{I}_{i_{3-l},n}^{(3-l)}|$$

$$\times \frac{1}{2b_n} \sum_{i_l': |t_{i_l',n}^{(l)} - t_{i_l,n}^{(l)}| \leq b_n + |\pi_n|_T} (K_q |\pi_n|_T + (|\pi_n|_T)^{1/2 - \varpi} + e_q) |\mathcal{I}_{i_l',n}^{(l)}| \frac{2b_n}{b_n}$$

$$\leq K \frac{b_n + |\pi_n|_T}{b_n} (K_q |\pi_n|_T + (|\pi_n|_T)^{1/2 - \varpi} + e_q)$$

$$\times \sum_{l=1,2} n \sum_{i_l, i_{3-l}: t_{i_l,n}^{(l)} \vee t_{i_{3-l},n}^{(3-l)} \leq T} |\mathcal{I}_{i_l,n}^{(l)} \cap \mathcal{I}_{i_{3-l},n}^{(3-l)}|$$

$$\times \sum_{j: t_{j,n}^{(3-l)} \leq T} |\mathcal{I}_{j,n}|^{(3-l)} \mathbb{1}_{\{\mathcal{I}_{j,n}^{(3-l)} \cap \mathcal{I}_{i_l,n}^{(l)} \neq \emptyset\}}$$

$$\leq K \Big(\frac{b_n + |\pi_n|_T}{b_n} \Big)^2 (K_q |\pi_n|_T + (|\pi_n|_T)^{1/2 - \varpi} + e_q) G_{2,2}^n(T)$$

which vanishes as first $n \to \infty$ and then $q \to \infty$. Hence, alltogether we have shown that the $\mathcal{S}$-conditional expectation of (7.34) vanishes as $n \to \infty$ and then $q \to \infty$ which yields (7.33) using Lemma 2.15.

Step 3. As in Step 3 in the proof of Proposition 4.10 we deduce

$$\widetilde{\mathbb{P}}(Y(P) \leq \Upsilon | \mathcal{X}) \xrightarrow{\widetilde{\mathbb{P}}} \widetilde{\mathbb{P}}(\Phi_T^{COV} \leq \Upsilon | \mathcal{X}) \tag{7.38}$$

for $P \to \infty$ and as in that proof we obtain (7.31) from (7.32), (7.33) and (7.38). $\quad\square$

Proof of Theorems 7.11 and 7.13. Analogously as in the proof of Theorem 4.3, Propositions 7.15 and 7.16 yield that under Conditions 7.10 respectively 7.12 it holds

$$\lim_{n \to \infty} \widetilde{\mathbb{P}}(\sqrt{n}([X^{(l)}, X^{(l)}]_T - V^{(l)}(g_2, \pi_n)_T) \leq \widehat{Q}_{T,n}^{VAR,(l)}(\alpha) | \mathcal{F}) = \alpha,$$

$$\lim_{n \to \infty} \widetilde{\mathbb{P}}(\sqrt{n}([X^{(1)}, X^{(2)}]_T - V(f_{(1,1)}, \pi_n)_T) \leq \widehat{Q}_{T,n}^{COV}(\alpha) | \mathcal{F}) = \alpha$$

for any $\alpha \in [0, 1]$. These convergences together with the fact that the $\mathcal{F}$-conditional distributions of $\Phi_T^{VAR,(l)}$ and Φ_T^{COV} are symmetrical yield (7.10) and (7.12). $\quad\square$

8 Testing for the Presence of Jumps

When choosing a suitable continuous time process e.g. to model an economic or financial time series one of the first steps is to decide whether a stochastic model is sufficient that produces continuous paths or whether jumps have to be incorporated. To this end, one is faced with the problem to infer from observed data (which is usually only available at discrete time points) whether the underlying model is continuous or allows for jumps. Sometimes this problem is relatively simple to solve e.g. in the situation when very large jumps are easy to identify in a visualization of the time series data. However, when only small but very frequent jumps are present the data might look very similar to observations of continuous time processes. Further, the method of visual inspection becomes infeasible when working with very large amounts of data. Hence automated methods and ideally mathematically precise tests are needed to tackle this problem. The decision whether to use models with or without jumps is not only of theoretical interest as to which model best fits reality but is also of great practical relevance „since models with and without jumps do have quite different mathematical properties and financial consequences (for option hedging, portfolio optimization, etc.)"([2], page 184, second paragraph).

Next, we formalize the problem discussed above under the assumption that the data is generated by discrete time observations of an Itô semimartingale $X^{(l)}$ as in (1.1) at random and irregular times $t_{i,n}^{(l)}$, $i \in \mathbb{N}_0$. Here, we are working within a high-frequency setting, where we observe one realized path $X_t^{(l)}(\omega)$, $t \in [0,T]$, up to a fixed time horizon $T > 0$ and consider asymptotics as the mesh of the observation times tends to zero. Hence as we are only considering one specific realization it is impossible to decide whether the underlying model in principle allows for jumps or not. This is due to the fact that the model might allow for jumps but no jumps have been realized for the ω which we observe. However, we are able to decide asymptotically whether jumps in the realized path $X_t^{(l)}(\omega)$, $t \in [0,T]$, are present or not we are observing this path at finer and finer grids. Hence in this chapter, we will *not construct statistical tests that allow to decide whether the underlying model allows for jumps* but we will construct a statistical *test which allows to decide whether jumps are present in a realized path* or not. So mathematically we are looking for a test based on the observations $X_{t_{i,n}^{(l)}}^{(l)}(\omega)$,

© Springer Fachmedien Wiesbaden GmbH, part of Springer Nature 2019
O. Martin, *High-Frequency Statistics with Asynchronous and Irregular Data*,
Mathematische Optimierung und Wirtschaftsmathematik | Mathematical Optimization and Economathematics, https://doi.org/10.1007/978-3-658-28418-3_8

$i \in \mathbb{N}_0$, which allows to decide to which of the following two subsets of Ω

$$\Omega_T^{J,(l)} = \{\exists t \in [0,T] : \Delta X_t^{(l)} \neq 0\},$$
$$\Omega_T^{C,(l)} = (\Omega_T^{J,(l)})^c = \{\Delta X_t^{(l)} = 0 \ \forall t \in [0,T]\}$$

ω belongs. Here, $\Omega_T^{J,(l)}$ is the set of all ω for which the path of X up to T has at least one jump and $\Omega_T^{C,(l)}$ is the set of all ω for which the path of X is continuous on $[0,T]$.

The test constructed in the following section can be understood as a generalization of tests described in [2], in Section 11.4 of [30] or in Chapter 10 of [3] which are based on equidistant and deterministic observation times to the setting of irregular and asynchronous observation times.

8.1 Theoretical Results

To decide whether $\omega \in \Omega_T^{J,(l)}$ or $\omega \in \Omega_T^{C,(l)}$ is true we need to find statistics that behave differently on $\Omega_T^{J,(l)}$ and $\Omega_T^{C,(l)}$. Recall that by Corollary 2.2 it holds

$$V^{(l)}(\bar{g}_p, \pi_n)_T \xrightarrow{\mathbb{P}} B^{(l)}(\bar{g}_p)_T$$

for $\bar{g}_p(x) = |x|^p$, $p > 2$, where $(B^{(l)}(\bar{g}_p)_T)(\omega) = \sum_{s \leq T} |\Delta X_s^{(l)}(\omega)|^p$ is equal to zero if and only if $\omega \in \Omega_T^{C,(l)}$. Hence $V^{(l)}(\bar{g}_p, \pi_n)_T$ converges to a strictly positive value on $\Omega_T^{J,(l)}$ and to zero on $\Omega_T^{C,(l)}$. Then based on the asymptotics of $V^{(l)}(\bar{g}_p, \pi_n)_T$ it is in principle possible to decide whether $\omega \in \Omega_T^{J,(l)}$ or $\omega \in \Omega_T^{C,(l)}$. In Chapter 3 we found that a central limit theorem for $V^{(l)}(\bar{g}_p, \pi_n)_T$ can only be derived for $p > 3$ and for mathematical reasons it turns out that it is especially convenient to work with $p = 4$ in which case we have $\bar{g}_4(x) = g_4(x) = x^4$.

Further, as we are going to construct a statistical test under the null hypothesis that jumps do exist, i.e. $\omega \in \Omega_T^{J,(l)}$, it is advantegeous to work with a statistic that converges to a fixed value on $\Omega_T^{J,(l)}$ and not to an arbitrary limit greater than zero. To this end we define the statistic

$$\Psi_{k,T,n}^{J,(l)} = \frac{V^{(l)}(g_4, [k], \pi_n)_T}{k V^{(l)}(g_4, \pi_n)_T},$$

$k \geq 2$; compare (2.88) for the definition of $V^{(l)}(g_4, [k], \pi_n)_T$. Here, the functional $V^{(l)}(g_4, [k], \pi_n)_T$ relies on increments of $X^{(l)}$ over the intervals $(t_{i-k,n}^{(l)}, t_{i,n}^{(l)}]$ while $V^{(l)}(g_4, \pi_n)_T$ relies on increments over the original observation intervals $(t_{i-1,n}^{(l)}, t_{i,n}^{(l)}]$. Hence by evaluating the ratio of the two we are comparing the expression $V^{(l)}(g_4, \pi_n)_T$ based on data sampled at different frequencies, once at the original frequency and once at $1/k$ times the original frequency. As the increments of continuous Itô semimartingales and Itô semimartingales with jumps scale differently with the length of the observation interval, compare the discussion at the beginning of Section 2.2, we will see that the limit of $\Phi_{k,T,n}^{J,(l)}$ is different on $\Omega_T^{J,(l)}$ and on $\Omega_T^{C,(l)}$.

Remark 8.1. *In the setting of equidistant observation times* $t_{i,n}^{(l)} = i/n$ *our statistic becomes*

$$\Psi_{k,T,n}^{J,(l)} = \frac{\sum_{i=k}^{\lfloor nT \rfloor} g(\Delta_{i,k,n}^{(l)} X^{(l)})}{k \sum_{i=1}^{\lfloor T/n \rfloor} g(\Delta_{i,n}^{(l)} X^{(l)})}. \tag{8.1}$$

On the contrary in [2] a test is constructed based on the statistic

$$\widetilde{\Psi}_{k,T,n}^{J,(l)} = \frac{\sum_{i=1}^{\lfloor nT/k \rfloor} g(\Delta_{ik,k,n}^{(l)} X^{(l)})}{\sum_{i=1}^{\lfloor T/n \rfloor} g(\Delta_{i,n}^{(l)} X^{(l)})} \tag{8.2}$$

where at the lower observation frequency n/k *only increments over certain observation intervals* $\mathcal{I}_{ik,k,n}$ *enter the estimation. Intuitively it seems that using the statistic (8.1) should be better than using (8.2), because in (8.1) we utilize the available data more exhaustively by using all increments at the lower observation frequency. This intuition is confirmed by Proposition 10.19 in [3] where central limit theorems are developed for both (8.1) and (8.2) and it is shown that (8.1) has a smaller asymptotic variance.* $\qquad\square$

To analyse the limit of $\Psi_{k,T,n}^{J,(l)}$ recall that Corollary 2.2 and Theorem 2.40(a) yield $V^{(l)}(g_4, \pi_n)_T \overset{\mathbb{P}}{\longrightarrow} B^{(l)}(g_4)_T$ respectively $V^{(l)}(g_4, [k], \pi_n)_T \overset{\mathbb{P}}{\longrightarrow} kB^{(l)}(g_4)_T$. Hence on $\Omega_T^{J,(l)}$ the statistic $\Psi_{k,T,n}^{J,(l)}$ converges to 1. On $\Omega_T^{C,(l)}$ on the other hand we obtain that both numerator and denominator in the definition of $\Psi_{k,T,n}^{J,(l)}$ converge to zero. However, we obtain

$$nV^{(l)}(g_4, \pi_n)_T \mathbb{1}_{\Omega_T^{C,(l)}} \overset{\mathbb{P}}{\longrightarrow} 3 \int_0^T (\sigma_s^{(l)})^4 dG_4^{(l)}(s) \mathbb{1}_{\Omega_T^{C,(l)}}$$

$$\frac{n}{k^2} V^{(l)}(g_4, [k], \pi_n)_T \mathbb{1}_{\Omega_T^{C,(l)}} \overset{\mathbb{P}}{\longrightarrow} 3 \int_0^T (\sigma_s^{(l)})^4 dG_4^{[k],(l)}(s) \mathbb{1}_{\Omega_T^{C,(l)}} \tag{8.3}$$

by Corollary 2.19 and Theorem 2.41(a) if we assume that the functions $G_p^{(l),n}$ and $G_p^{[k],(l),n}$ converge to some functions $G_p^{(l)}$ respectively $G_p^{[k],(l)}$ as $n \to \infty$.

In the following condition we state precise assumptions under which the above considerations are valid.

Condition 8.2. *Let Condition 1.3 be fulfilled. Further let $(\pi_n)_{n \in \mathbb{N}}$ be exogenous, let the process σ fulfil $\int_0^T |\sigma_s| ds > 0$ almost surely and suppose that the functions $G_4^{(l),n}(t)$, $G_4^{[k],(l),n}(t)$ converge pointwise on $[0,T]$ in probability to strictly increasing functions $G_4^{(l)}, G_4^{[k],(l)} : [0,T] \to [0,\infty)$.*

The convergence of $\Psi_{k,T,n}^{J,(l)}$ on $\Omega_T^{J,(l)}$ already follows under Condition 1.3. The remaining assumptions in Condition 8.2 are only needed to prove the convergence of $\Psi_{k,T,n}^{J,(l)}$ on $\Omega_T^{C,(l)}$. The exogeneity and the convergence of $G_4^{(l),n}(t)$, $G_4^{[k],(l),n}(t)$ are required such that (8.3) holds while $\int_0^T |\sigma_s| ds > 0$ and that the function $G_4^{(l)}$ is strictly increasing is needed such that the limit of $nV^{(l)}(g_4, [k], \pi_n)_T$ is never equal to zero.

To shorten the notation for the limit of $\Psi_{k,T,n}^{J,(l)}$ on $\Omega_T^{C,(l)}$ we denote

$$C_T^{J,(l)} = \int_0^T (\sigma_s^{(l)})^4 dG_4^{(l)}(s), \quad C_{k,T}^{J,(l)} = \int_0^T (\sigma_s^{(l)})^4 dG_4^{[k],(l)}(s).$$

Theorem 8.3. *Under Condition 8.2 it holds*

$$\Psi_{k,T,n}^{J,(l)} \xrightarrow{\mathbb{P}} \begin{cases} 1, & on \ \Omega_T^{J,(l)}, \\ \dfrac{kC_{k,T}^{J,(l)}}{C_T^{J,(l)}}, & on \ \Omega_T^{C,(l)}. \end{cases} \tag{8.4}$$

Here, the limit $kC_{k,T}^{J,(l)}/C_T^{J,(l)}$ on $\Omega_T^{C,(l)}$ is always different from 1 if the observation scheme is not too degenerated. This property is illustrated in the following remark.

Remark 8.4. *We obtain*

$$G_4^{(l),n}(t) - G_4^{(l),n}(s) + O(|\pi_n|t) \le kG_4^{[k],(l),n}(t) - kG_4^{[k],(l),n}(s) \le kG_4^{(l),n}(t) - kG_4^{(l),n}(s)$$

for all $t \ge s \ge 0$ from the series of elementary inequalities

$$\sum_{i=1}^k a_i^2 \le \left(\sum_{i=1}^k a_i \right)^2 \le k \sum_{i=1}^k a_i^2 \tag{8.5}$$

which holds for any $a_1, \ldots, a_k \ge 0$, $k \in \mathbb{N}$. Here, the second inequality follows from the Cauchy-Schwarz inequality. Equality in (8.5) holds for $a_1 \ge 0$, $a_2 = \ldots = a_k = 0$ respectively $a_1 = \ldots = a_k > 0$. The relations of $G_4^{(l),n}$ and $G_4^{[k],(l),n}$ are preserved in the limit which yields

$$G_4^{(l)}(t) - G_4^{(l)}(s) \le kG_4^{[k],(l)}(t) - kG_4^{[k],(l)}(s) \le kG_4^{(l)}(t) - kG_4^{(l)}(s)$$

for all $t \ge s \ge 0$. Hence we get $kC_{k,T}^{J,(l)}/C_T^{J,(l)} \in [1,k]$. $\qquad\square$

Based on the fact that $\Psi_{k,T,n}^{J,(l)} \xrightarrow{\mathbb{P}} 1$ on $\Omega_T^{J,(l)}$ and that $\Psi_{k,T,n}^{J,(l)}$ converges on $\Omega_T^{C,(l)}$ to a random variable which is strictly greater than 1 if and only if $kC_{k,T}^{J,(l)} > C_T^{J,(l)}$ by Remark 8.4, we will later construct a test with critical region

$$C_{k,T,n}^{J,(l)} = \{\Psi_{k,T,n}^{J,(l)} > 1 + c_{k,T,n}^{J,(l)}\} \tag{8.6}$$

for an appropriate series of decreasing random positive numbers $c_{k,T,n}^{J,(l)}$, $n \in \mathbb{N}$.

Next, we illustrate the result from Theorem 8.3 by looking at two prominent observation schemes: first Poisson sampling, compare Definition 5.1, which is truly random and irregular and second equidistant sampling for which results exist in the literature and which therefore serves as a kind of benchmark.

Example 8.5. *In the case of Poisson sampling Condition 8.2(ii) is fulfilled. To this end note that $n|\mathcal{I}_{i,k,n}^{(l)}|$ is Gamma-distributed with parameters k and λ_l. Hence* $\mathbb{E}[(n|\mathcal{I}_{i,k,n}^{(l)}|)^2] = k/\lambda_l$, $\mathbb{E}[(n|\mathcal{I}_{i,k,n}^{(l)}|)^2] = k(k+1)/\lambda_l$ *and by Lemma 5.4 we obtain*

$$G_4^{(l)}(t) = \frac{2}{\lambda}t, \quad G_4^{[k],(l)}(t) = \frac{k+1}{k\lambda}t.$$

This yields that the limit under the alternative is equal to $(k+1)/2$. □

Remark 8.6. *In the case of equidistant synchronous observations, i.e. $t_{i,n}^{(l)} = i/n$, it holds $G_4^{(l)}(t) = G_4^{[k],(l)}(t) = t$ which yields*

$$\frac{kC_{k,T}^{J,(l)}}{C_T^{J,(l)}} = k.$$

Hence in this setting $\Psi_{k,T,n}^{J,(l)}$ converges on $\Omega_T^{C,(l)}$ to a known deterministic limit as well which also allows to construct a test using $\Psi_{k,T,n}^{J,(l)}$ for the null hypothesis of no jumps; compare Section 10.3 in [3]. This is not immediately possible in the irregular setting, unless the law of the generating mechanism is known to the statistician. □

As a next step we develop a central limit theorem for the statistic $\Psi_{k,T,n}^{J,(l)}$ under the null hypothesis that jumps are present i.e. we derive a central limit theorem which holds on $\Omega_T^{J,(l)}$. To motivate the structure of the upcoming central limit theorem we consider as in Section 7.2 a toy example of the form

$$X_t^{toy,(l)} = \sigma^{(l)} W_t^{(l)} + \sum_{s \leq t} \Delta N^{(l)}(q)_s.$$

First note that it holds

$$\sqrt{n}(\Psi_{k,T,n}^{J,(l)} - 1) = \frac{\sqrt{n}(V^{(l)}(g_4, [k], \pi_n)_T - kV^{(l)}(g_4, \pi_n)_T)}{kV^{(l)}(g_4, \pi_n)_T} \tag{8.7}$$

and therefore it remains to consider the nominator as the denominator converges on $\Omega_T^{J,(l)}$ to $B^{(l)}(g_4)_T > 0$. Denote by $S_{q,p}^{(l)}$, $p \in \mathbb{N}$, an enumeration of the jump times of $N^{(l)}(q)$. In our toy example we then have on the set where any two jumps of $N^{(l)}(q)$ are further apart than $k|\pi_n|_T$

$$\sqrt{n}(V^{(l)}(g_4, [k], \pi_n)_T - kV^{(l)}(g_4, \pi_n)_T)$$

$$= \sqrt{n}\Big(\sum_{i:t_{i,n}^{(l)} \le T} (\Delta_{i-k,n}^{(l)} X^{toy,(l)} + \ldots + \Delta_{i,n}^{(l)} X^{toy,(l)})^4$$

$$- k \sum_{i:t_{i,n}^{(l)} \le T} (\Delta_{i,n}^{(l)} X^{toy,(l)})^4 \Big)$$

$$= \sqrt{n}\Big(\sum_{p:S_{q,p}^{(l)} \le T} 4(\Delta X_{S_{q,p}^{(l)}}^{toy,(l)})^3 \sum_{j=0}^{k-1} \sigma^{(l)} \Delta_{i_n^{(l)}(S_{q,p}^{(l)})+j,k,n}^{(l)} W^{(l)}$$

$$- k \sum_{p:S_{q,p}^{(l)} \le T} 4(\Delta X_{S_{q,p}^{(l)}}^{toy,(l)})^3 \sigma^{(l)} \Delta_{i_n^{(l)}(S_{q,p}^{(l)}),n}^{(l)} W^{(l)} \Big) + o_{\mathbb{P}}(1)$$

$$= 4 \sum_{p:S_{q,p}^{(l)} \le T} (\Delta X_{S_{q,p}^{(l)}}^{toy,(l)})^3 \sqrt{n} \sum_{j=-(k-1),j\neq0}^{k-1} |k-j|\sigma^{(l)} \Delta_{i_n^{(l)}(S_{q,p}^{(l)})+j,n}^{(l)} W^{(l)} + o_{\mathbb{P}}(1)$$

under suitable conditions. Here, the $o_{\mathbb{P}}(1)$-term contains all sums over terms in which higher powers of the increments $\Delta_{i,n}^{(l)} W^{(l)}$ occur that originate from the expansion of

$$(\Delta_{i,k',n}^{(l)} X^{toy,(l)})^4 = (\sigma^{(l)} \Delta_{i,k',n}^{(l)} W^{(l)} + \Delta_{i,k',n}^{(l)} N^{(l)}(q))^4, \quad k' = 1, k.$$

A detailed discussion of the asymptotically negligible terms will be given in the proofs section. Here, the $\mathcal{S}$-conditional variance of the term

$$\sqrt{n} \sum_{j=-(k-1),j\neq0}^{k-1} |k-j|\Delta_{i_n^{(l)}(S_{q,p}^{(l)})+j,n}^{(l)} W^{(l)}$$

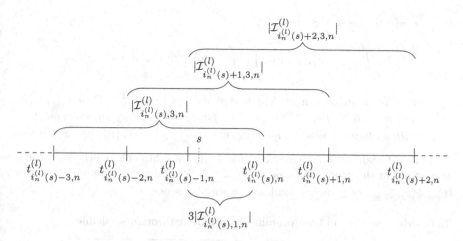

Figure 8.1: Illustrating the origin of $\xi_{k,n,-}^{(l)}(s), \xi_{k,n,+}^{(l)}(s)$ for $k = 3$.

is equal to $\xi_{k,n,-}^{(l)} + \xi_{k,n,+}^{(l)}$ where

$$\xi_{k,n,-}^{(l)}(s) = n \sum_{j=1}^{k-1}(k-j)^2 |\mathcal{I}_{i_n^{(l)}(s)-j,1,n}^{(l)}|,$$

$$\xi_{k,n,+}^{(l)}(s) = n \sum_{j=1}^{k-1}(k-j)^2 |\mathcal{I}_{i_n^{(l)}(s)+j,1,n}^{(l)}|.$$

The form of these terms is illustrated in Figure 8.1.

To obtain a central limit theorem we then need to assume that the random variables $\xi_{k,n,-}^{(l)}(s)$, $\xi_{k,n,+}^{(l)}(s)$ converge in a suitable sense as $n \to \infty$. All assumptions which are required to obtain a central limit theorem for $\Psi_{k,T,n}^{J,(l)}$ are summarized in the following condition.

Condition 8.7. *The process $X^{(l)}$ and the sequence of observation schemes $(\pi_n)_{n \in \mathbb{N}}$ fulfil Condition 8.2. Further the following additional assumptions on the observation schemes hold:*

(i) We have $|\pi_n|_T = o_{\mathbb{P}}(n^{-1/2})$.

(ii) The integral

$$\int_{[0,T]^P} g(x_1, \ldots, x_P) \mathbb{E}\Big[\prod_{p=1}^{P} h_p\big(\xi_{k,n,-}^{(l)}(x_p), \xi_{k,n,+}^{(l)}(x_p)\big)\Big] dx_1 \ldots dx_P$$

converges for $n \to \infty$ *to*

$$\int_{[0,T]^P} g(x_1, \ldots, x_P) \prod_{p=1}^P \int_{\mathbb{R}} h_p(y) \, \Gamma^{J,(l)}(x_p, dy) dx_1 \ldots dx_P$$

for all bounded continuous functions $g \colon \mathbb{R}^P \to \mathbb{R}$, $h_p \colon \mathbb{R}^2 \to \mathbb{R}$, $p = 1, \ldots, P$, *and any* $P \in \mathbb{N}$. *Here* $\Gamma^{J,(l)}(\cdot, dy)$ *is a family of probability measures on* $[0, T]$ *with uniformly bounded first moments and* $\int_0^T \Gamma^{J,(l)}(x, \{(0,0)\}) dx = 0$.

Part (i) of Condition 8.7 guarantees that $|\pi_n|_T$ vanishes sufficiently fast which is required in the proof, while part (ii) of 8.7 yields that the random variables $(\xi_{k,n,-}^{(l)}(s), \xi_{k,n,+}^{(l)}(s))$ converge in law in a suitable sense.

To describe the limit in the upcoming central limit theorem we define

$$\Phi_{k,T}^{J,(l)} = 4 \sum_{p : S_p^{(l)} \leq T} (\Delta X_{S_p^{(l)}}^{(l)})^3 \sqrt{(\sigma_{S_p^{(l)}-})^2 \xi_{k,-}^{(l)}(S_p^{(l)}) + (\sigma_{S_p^{(l)}})^2 \xi_{k,+}^{(l)}(S_p^{(l)})} U_{S_p^{(l)}}^{(l)}.$$

$$(8.8)$$

Here, $(S_p^{(l)})_{p \geq 0}$ denotes an enumeration of the jump times of $X^{(l)}$ and the $\xi_{k,-}^{(l)}(s)$, $\xi_{k,+}^{(l)}(s)$, $s \in [0,T]$, are independent random variables which are distributed according to $\Gamma^{J,(l)}(s, dy)$ and the $U_s^{(l)}$, $s \in [0,T]$, are i.i.d. standard normal distributed random variables. Both the $(\xi_{k,-}^{(l)}(s), \xi_{k,+}^{(l)}(s))$ and the $U_s^{(l)}$ are independent of $X^{(l)}$ and its components and defined on an extended probability space $(\widetilde{\Omega}, \widetilde{\mathcal{F}}, \widetilde{\mathbb{P}})$. Note that $\Phi_{k,T}^{J,(l)}$ is well-defined because the sum in (8.8) is almost surely absolutely convergent and independent of the choice for the enumeration $S_p^{(l)}$, $p \in \mathbb{N}$; compare Proposition 4.1.3 in [30].

Theorem 8.8. *If Condition 8.7 holds, we have the* $\mathcal{X}$*-stable convergence*

$$\sqrt{n}(\Psi_{k,T,n}^{J,(l)} - 1) \xrightarrow{\mathcal{L}-s} \frac{\Phi_{k,T}^{J,(l)}}{kB^{(l)}(g_4)T}$$

$$(8.9)$$

on $\Omega_T^{J,(l)}$.

Here, the limit $\Phi_{k,T}^{J,(l)}/kB_T^{J,(l)}(g_4)$ in (8.9) is by construction defined on the extended probability space $(\widetilde{\Omega}, \widetilde{\mathcal{F}}, \widetilde{\mathbb{P}})$. Further the statement of the $\mathcal{X}$-stable convergence on the set $\Omega_T^{J,(l)}$ means that we have

$$\mathbb{E}\big[g\big(\sqrt{n}(\Psi_{k,T,n}^{J,(l)} - 1)\big)Y \mathbb{1}_{\Omega_T^{J,(l)}}\big] \to \widetilde{\mathbb{E}}\big[g\big(\Phi_{k,T,n}^{J,(l)}/(kB^{(l)}(g_4)T)\big)Y \mathbb{1}_{\Omega_T^{J,(l)}}\big]$$

for all bounded and continuous functions g and all $\mathcal{X}$-measurable bounded random variables Y. For more background information on stable convergence in law see Appendix B.

Example 8.9. *Condition 8.7 is fulfilled in the setting of Poisson sampling discussed in Example 8.5. Part (i) is fulfilled by (5.2) and that part (ii) is fulfilled follows from Lemma 5.10 by choosing $f^{(3)}$ appropriately.* $\qquad\square$

To construct a statistical test for deciding between $\omega \in \Omega^{J,(l)}$ and $\omega \in \Omega^{C,(l)}$ under the null hypothesis that jumps do exist we need to assess the limit distribution in Theorem 8.8. For this purpose we use the methods discussed in Chapter 4. Let $(\widetilde{K}_n)_{n \in \mathbb{N}}$ and $(M_n)_{n \in \mathbb{N}}$ be sequences of natural numbers which tend to infinity and define

$$\hat{\xi}_{k,n,m,-}^{(l)}(s) = n \sum_{j=1}^{k-1} (k-j)^2 |\mathcal{I}_{i_n^{(l)}(s)+V_{n,m}^{(l)}(s)-j,1,n}^{(l)}|,$$

$$\hat{\xi}_{k,n,m,+}^{(l)}(s) = n \sum_{j=1}^{k-1} (k-j)^2 |\mathcal{I}_{i_n^{(l)}(s)+V_{n,m}^{(l)}(s)+j,1,n}^{(l)}|$$

(8.10)

for $m = 1, \ldots, M_n$ where the random variable $V_{n,m}^{(l)}(s)$ attains values in the set $\{-\widetilde{K}_n, \ldots, \widetilde{K}_n\}$ with probabilities

$$\mathbb{P}\big(V_{n,m}^{(l)}(s) = \tilde{k}\big|S\big) = |\mathcal{I}_{i_n^{(l)}(s)+\tilde{k},n}^{(l)}|\Big(\sum_{\tilde{k}'=-\widetilde{K}_n}^{\widetilde{K}_n} |\mathcal{I}_{i_n^{(l)}(s)+\tilde{k}',n}^{(l)}|\Big)^{-1},$$

(8.11)

$\tilde{k} \in \{-\widetilde{K}_n, \ldots, \widetilde{K}_n\}$. Here, $(\hat{\xi}_{k,n,m,-}^{(l)}(s), \hat{\xi}_{k,n,m,-}^{(l)}(s))$ is chosen from the

$$(\xi_{k,n,-}^{(l)}(t_{i_n(s)+\tilde{k},n}), \xi_{k,n,-}^{(l)}(t_{i_n(s)+\tilde{k},n})), \quad \tilde{k} = -\widetilde{K}_n, \ldots, \widetilde{K}_n,$$

which make up the $2\widetilde{K}_n + 1$ different realizations of $(\xi_{k,n,-}^{(l)}(t), \xi_{k,n,-}^{(l)}(t))$ which lie "closest" to s, with probability proportional to the interval length $|\mathcal{I}_{i_n^{(l)}(s)+\tilde{k},n}^{(l)}|$. This corresponds to the probability with which a random variable which is uniformly distributed on the union of the intervals $\mathcal{I}_{i_n^{(l)}(s)+\tilde{k},n}^{(l)}$, $\tilde{k} = -\widetilde{K}_n, \ldots, \widetilde{K}n$, but otherwise independent from the observation scheme would fall into the interval $\mathcal{I}_{i_n(s)+\tilde{k},n}^{(l)}$. Due to the structure of the predictable compensator ν the jump times $S_p^{(l)}$ of the Itô semimartingale $X^{(l)}$ are also evenly distributed in time. This explains why we choose such a random variable $V_{n,m}^{(l)}(s)$ for the estimation of the law of $(\hat{\xi}_{k,n,m,-}^{(l)}(S_p^{(l)}), \hat{\xi}_{k,n,m,-}^{(l)}(S_p^{(l)}))$. The random variables $V_{n,m}^{(l)}(s)$ and

$\hat{\xi}_{k,n,m,-}^{(l)}(S_p^{(l)})$, $\hat{\xi}_{k,n,m,-}^{(l)}(S_p^{(l)})$) can be defined on the extended probability space $(\widetilde{\Omega}, \widetilde{\mathcal{F}}, \widetilde{\mathbb{P}})$.

Using the estimators (8.10) for realizations of $\xi_k^{(l)}(s)$ we build the following estimators for realizations of $\Phi_{k,T}^{J,(l)}$

$$\widehat{\Phi}_{k,T,n,m}^{J,(l)} = 4 \sum_{i:t_{i,n}^{(l)} \leq T} \left(\Delta_{i,n}^{(l)} X^{(l)}\right)^3 \mathbb{1}_{\{|\Delta_{i,n}^{(l)} X^{(l)}| > \beta |\mathcal{I}_{i,n}^{(l)}|^\varpi\}}$$

$$\times \sqrt{(\tilde{\sigma}_n^{(l)}(t_{i,n}^{(l)},-))^2 \hat{\xi}_{k,n,m,-}^{(l)}(t_{i,n}^{(l)}) + (\tilde{\sigma}_n^{(l)}(t_{i,n}^{(l)},+))^2 \hat{\xi}_{k,n,m,+}^{(l)}(t_{i,n}^{(l)})} U_{n,i,m}^{(l)},$$

$$(8.12)$$

$m = 1, \ldots, M_n$, where $\beta > 0$ and $\varpi \in (0, 1/2)$. Here an increment which is large compared to a given threshold, compare Section 2.3, is identified as a jump and the local volatility is approximated using the estimators $\tilde{\sigma}_n^{(l)}(s,-)$, $\tilde{\sigma}_n^{(l)}(s,+)$ defined in Chapter 6. Further the $U_{n,i,m}^{(l)}$ are i.i.d. standard normal distributed random variables which are independent of $\mathcal{F}$ and can be defined on the extended probability space $(\widetilde{\Omega}, \widetilde{\mathcal{F}}, \widetilde{\mathbb{P}})$ as well.

We then denote by

$$\widehat{Q}_{k,T,n}^{J,(l)}(\alpha) = \widehat{Q}_\alpha\left(\{\widehat{\Phi}_{k,T,n,m}^{J,(l)} | m = 1, \ldots, M_n\}\right) \qquad (8.13)$$

the $\lfloor \alpha M_n \rfloor$-th largest element of the set $\{\widehat{\Phi}_{k,T,n,m}^{J,(l)} | m = 1, \ldots, M_n\}$.

We will see that $\widehat{Q}_{k,T,n}^{J,(l)}(\alpha)$ converges on $\Omega_T^{J,(l)}$ under appropriate conditions to the $\mathcal{X}$-conditional α-quantile $Q_{k,T}^{J,(l)}(\alpha)$ of $F_{k,T}^{J,(l)}$ which is defined as the (under the upcoming condition unique) $\mathcal{X}$-measurable $[-\infty, \infty]$-valued random variable $Q_{k,T}^{J,(l)}(\alpha)$ fulfilling

$$\widetilde{\mathbb{P}}(\Phi_{k,T}^{J,(l)} \leq Q_{k,T}^{J,(l)}(\alpha) | \mathcal{X})(\omega) = \alpha, \quad \omega \in \Omega_T^{J,(l)},$$

and for completeness we set $\left(Q_{k,T}^{J,(l)}(\alpha)\right)(\omega) = 0$, $\omega \in (\Omega_T^{J,(l)})^c$ such that $Q_{k,T}^{J,(l)}(\alpha)$ is well defined on all of Ω. Such a random variable $Q_{k,T}^{J,(l)}(\alpha)$ exists if Condition 8.10 is fulfilled because the $\mathcal{X}$-conditional distribution of $\Phi_{k,T}^{J,(l)}$ will be almost surely continuous on $\Omega_T^{J,(l)}$ under Condition 8.10.

Condition 8.10. *Assume that Condition 8.7 is fulfilled and suppose that the set $\{s \in [0,T] : \sigma_s^{(l)} = 0\}$ is almost surely a Lebesgue null set. Further, let the sequence $(b_n)_{n \in \mathbb{N}}$ fulfil $|\pi_n|_T/b_n \xrightarrow{\mathbb{P}} 0$ and suppose that $(\widetilde{K}_n)_{n \in \mathbb{N}}$ and $(M_n)_{n \in \mathbb{N}}$ are sequences of natural numbers converging to infinity with $\widetilde{K}_n/n \to 0$. Additionally,*

(i) it holds

$$\widetilde{\mathbb{P}}\Big(\big|\widehat{\mathbb{P}}((\widehat{\xi}^{(l)}_{k,n,1,-}(s_p),\widehat{\xi}^{(l)}_{k,n,1,+}(s_p)) \le x_p,\ p=1,\ldots,P|\mathcal{S})$$

$$-\prod_{p=1}^{P}\widetilde{\mathbb{P}}((\xi^{(l)}_{k,-}(s_p),\xi^{(l)}_{k,+}(s_p)) \le x_p)\big| > \varepsilon\Big) \to 0$$

as $n \to \infty$, for all $\varepsilon > 0$ and any $x = (x_1,\ldots,x_P) \in \mathbb{R}^{2\times P}$, $P \in \mathbb{N}$, and $s_p \in (0,T)$, $p = 1,\ldots,P$.

(ii) The volatility process $\sigma^{(l)}$ is itself an Itô semimartingale, i.e. a process of the form (1.1).

(iii) On $\Omega^{C,(l)}_T$ we have $kC^{J,(l)}_{k,T} > C^{J,(l)}_T$ almost surely.

Part (i) of Condition 8.10 guarantees that the bootstrapped realizations

$$(\widehat{\xi}^{(l)}_{k,n,m,-}(s),\widehat{\xi}^{(l)}_{k,n,m,+}(s))$$

consistently estimate the distribution of $(\xi^{(l)}_{k,-}(s),\xi^{(l)}_{k,+}(s))$ and thereby that the quantity $\widehat{Q}^{J,(l)}_{k,T,n}(\alpha)$ yields a valid estimator for $Q^{J,(l)}_{k,T}(\alpha)$ on $\Omega^{J,(l)}_T$. Part (ii) is needed for the convergence of the volatility estimators $\tilde{\sigma}^{(l)}_n(S^{(l)}_p,-)$, $\tilde{\sigma}^{(l)}_n(S^{(l)}_p,+)$ for jump times $S^{(l)}_p$, and part (iii) guarantees that $\Psi^{J,(l)}_{k,T,n}$ converges under the alternative to a value different from 1, which is the limit under the null hypothesis.

Theorem 8.11. *If Condition 8.10 is fulfilled, the test defined in (8.6) with*

$$c^{J,(l)}_{k,T,n}(\alpha) = \frac{\widehat{Q}^{J,(l)}_{k,T,n}(1-\alpha)}{\sqrt{nk}V^{(l)}(g_4,\pi_n)_T}, \quad \alpha \in [0,1],$$

has asymptotic level α in the sense that we have

$$\widetilde{\mathbb{P}}(\Psi^{J,(l)}_{k,T,n} > 1 + c^{J,(l)}_{k,T,n}(\alpha)|F^{J,(l)}) \to \alpha, \quad \alpha \in [0,1], \tag{8.14}$$

for all $F^{J,(l)} \subset \Omega^{J,(l)}_T$ with $\mathbb{P}(F^{J,(l)}) > 0$.
The test is consistent in the sense that we have

$$\widetilde{\mathbb{P}}(\Psi^{J,(l)}_{k,T,n} > 1 + c^{J,(l)}_{k,T,n}(\alpha)|F^{C,(l)}) \to 1, \quad \alpha \in (0,1], \tag{8.15}$$

for all $F^{C,(l)} \subset \Omega^{C,(l)}_T$ with $\mathbb{P}(F^{C,(l)}) > 0$.

Note that to carry out the test introduced in Theorem 8.11 the unobservable variable n is not explicitly needed, even though $\sqrt{n}$ occurs in the definition of $c_{k,T,n}^{J,(l)}(\alpha)$. This factor actually cancels in the definition of $c_{k,T,n}^{J,(l)}(\alpha)$ because it also enters $\widehat{Q}_{k,T,n}^{J,(l)}(1-\alpha)$ as a linear factor. What remains is the dependence of b_n and $\widetilde{K}_n$ on n, though, but for these auxiliary variables only a rough idea of the magnitude of n usually is sufficient. Similar observations hold for all tests constructed later on in this chapter and in Chapter 9 as well.

The simulation results in Section 8.2 show that the convergence in (8.14) is rather slow, because certain terms in $\sqrt{n}(\Psi_{k,T,n}^{J,(l)}-1)$ which vanish in the limit contribute significantly in the small sample. Our goal is to diminish this effect by including estimates for those terms in the testing procedure. The asymptotically vanishing terms stem from the continuous part which is mostly captured in the small increments. To estimate their contribution we define

$$A_{k,T,n}^{J,(l)} = n \sum_{i \geq k : t_{i,n}^{(l)} \leq T} (\Delta_{i,k,n}^{(l)} X)^4 \mathbb{1}_{\{|\Delta_{i,k,n}^{(l)} X^{(l)}| \leq \beta |\mathcal{I}_{i,k,n}^{(l)}|^{\varpi}\}}$$

$$- kn \sum_{i \geq 1 : t_{i,n}^{(l)} \leq T} (\Delta_{i,n}^{(l)} X^{(l)})^4 \mathbb{1}_{\{|\Delta_{i,n}^{(l)} X^{(l)}| \leq \beta |\mathcal{I}_{i,n}^{(l)}|^{\varpi}\}}.$$

using the same β, ϖ as in (8.12). We then define for $\rho \in (0,1)$ the adjusted estimator

$$\widetilde{\Psi}_{k,T,n}^{J,(l)}(\rho) = \Psi_{k,T,n}^{J,(l)} - \rho \frac{n^{-1} A_{k,T,n}^{J,(l)}}{kV^{(l)}(g_4, \pi_n)_T}$$

where we partially correct for the contribution of the asymptotically vanishing terms.

Corollary 8.12. *Let $\rho \in (0,1)$. If Condition 8.10 is fulfilled, it holds with the notation from Theorem 8.11*

$$\widetilde{\mathbb{P}}\big(\widetilde{\Psi}_{k,T,n}^{J,(l)}(\rho) > 1 + c_{k,T,n}^{J,(l)}(\alpha)\big|F^{J,(l)}\big) \to \alpha, \quad \alpha \in [0,1], \qquad (8.16)$$

for all $F^{J,(l)} \subset \Omega_T^{J,(l)}$ with $\mathbb{P}(F^{J,(l)}) > 0$ and

$$\widetilde{\mathbb{P}}\big(\widetilde{\Psi}_{k,T,n}^{J,(l)}(\rho) > 1 + c_{k,T,n}^{J,(l)}(\alpha)\big|F^{C,(l)}\big) \to 1, \quad \alpha \in (0,1], \qquad (8.17)$$

for all $F^{C,(l)} \subset \Omega_T^{C,(l)}$ with $\mathbb{P}(F^{C,(l)}) > 0$.

The closer ρ is to 1 the faster is the convergence in (8.16), but also the slower is the convergence in (8.17). Hence an optimal ρ should be chosen somewhere in between. Our simulation results in Section 8.2 show that it is possible to pick a ρ very close to 1 without significantly worsening the power compared to the test from Theorem 8.11.

Example 8.13. *The assumptions on the observation scheme in Condition 8.10 are fulfilled in the Poisson setting. Part (iii) is fulfilled as shown in Example 8.5 and that part (i) is fulfilled follows from Lemma 5.12.* □

In fact for our testing procedure to work in the Poisson setting we do not need the weighting from (8.11). All intervals could also be picked with equal probability. This is due to the fact that the lengths

$$\left(n|\mathcal{I}^{(l)}_{i_n^{(l)}(s)+V_{n,m}^{(l)}(s)+j,n}|\right)_{j=-(k-1),\ldots,-1,1,\ldots,k-1}$$

of intervals to the left and to the right of $\mathcal{I}^{(l)}_{i_n^{(l)}(s)+V_{n,m}^{(l)}(s),n}$ and hence the variables $\hat{\xi}^{(l)}_{k,n,m,-}(s)$, $\hat{\xi}^{(l)}_{k,n,m,+}(s)$ are (asymptotically) independent of $n|\mathcal{I}^{(l)}_{i_n^{(l)}(s)+V_{n,m}^{(l)}(s),n}|$. However, the weighting is important if the interval lengths of consecutive intervals are dependent as illustrated in the following example.

Example 8.14. *Define an observation scheme by*

$$t^{(l)}_{2i,n} = 2i/n, \quad t^{(l)}_{2i+1,n} = (2i+1+\alpha)/n,$$

$i \in \mathbb{N}_0$, *with* $\alpha \in (0,1)$; *compare Example 3 in [8]. Let us consider the case* $k=2$. *The observation scheme is illustrated in Figure 8.2. It can be easily checked that Condition 8.2 holds with* $G_4^{(l)}(t) = (1+\alpha^2)t$ *and* $G_4^{[2],(l)}(t) = t$. *Further it can be shown similarly as in [8] that Condition 8.7 is fulfilled with* $\Gamma^{J,(l)}$ *defined via*

$$\Gamma^{J,(l)}(s,\{(1+\alpha,1+\alpha)\}) = \frac{1-\alpha}{2}, \quad \Gamma^{J,(l)}(s,\{(1-\alpha,1-\alpha)\}) = \frac{1+\alpha}{2}$$

for all $s > 0$. *Hence in order for the distribution of* $\hat{\xi}^{(l)}_{k,n,1}(s)$ *to approximate* $\Gamma^{J,(l)}(s,\cdot)$ *the variable* $i_n^{(l)}(s)+V_{n,m}^{(l)}(s)$ *has to pick the intervals of length* $(1+\alpha)/n$ *with higher probability than those with length* $(1-\alpha)/n$, *because it holds*

$$n|\mathcal{I}^{(l)}_{i_n^{(l)}(s)+V_{n,m}^{(l)}(s),n}| = 1+\alpha \Rightarrow \hat{\xi}^{(l)}_{2,n,m,-}(s) = \hat{\xi}^{(l)}_{2,n,m,+}(s) = 1-\alpha,$$

$$n|\mathcal{I}^{(l)}_{i_n^{(l)}(s)+V_{n,m}^{(l)}(s),n}| = 1-\alpha \Rightarrow \hat{\xi}^{(l)}_{2,n,m,-}(s) = \hat{\xi}^{(l)}_{2,n,m,+}(s) = 1+\alpha.$$

□

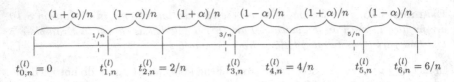

Figure 8.2: The sampling scheme from Example 8.14

8.2 Simulation Results

To verify the effectivity and to study the finite sample properties of the developed tests we apply them on simulated data. In our simulation the observation times originate from a Poisson process with intensity n which corresponds to $\lambda_l = 1$ in Example 8.5 and yields on average n observations in the time interval $[0, 1]$. We simulate from the model

$$dX_t^{(l)} = X_t^{(l)} \sigma dW_t^{(l)} + \alpha \int_{\mathbb{R}} X_{t-}^{(l)} x \mu^{(l)}(dt, dx) \qquad (8.18)$$

where $X_0^{(l)} = 1$ and the Poisson measure $\mu^{(l)}$ has the predictable compensator

$$\nu^{(l)}(dt, dx) = \kappa \frac{\mathbb{1}_{[-h, -l] \cup [l, h]}(x)}{2(h - l)} dt dx.$$

This model is a one-dimensional version of the model used in a similar simulation study in Section 6 of [32].

We consider the parameter settings displayed in Table 8.1 with $\sigma^2 = 8 \times 10^{-5}$ in all cases. We choose $n = 100$, $n = 1,600$ and $n = 25,600$. In a trading day of

Table 8.1: Parameter settings for the simulation.

Case	α	κ	l	h
I-j	0.01	1	0.05	0.7484
II-j	0.01	5	0.05	0.3187
III-j	0.01	25	0.05	0.1238
Cont	0.00			

6.5 hours, this corresponds to observing $X^{(l)}$ on average every 4 minutes, every 15 seconds and every second. The cases I-j to III-j correspond to the presence of jumps of diminishing size. When there are smaller jumps we choose a situation where there are more jumps such that the overall contribution of the jumps to the quadratic variation is roughly the same in all three cases. The fraction of the quadratic variation which originates from the jumps matches the one estimated in real financial data from [46]. In all three cases where the model allows for jumps we only use paths in the simulation study where jumps were realized. In the fourth case **Cont** we consider a purely continuous model. The parameter values in Table 8.1 are taken from [32].

Regarding the free parameters in our testing procedures we set $\beta = 0.03$ and $\omega = 0.49$. Here, we choose the same values as in [32] and thereby follow their recommendation to pick ϖ close to $1/2$ and β to be about 3 to 4 times the magnitude of σ (which in general is unknown but can easily be estimated from the data). This choice for β is reasonable as increments $\Delta_{i,n}^{(l)} X \approx \Delta_{i,n}^{(l)} C$ where the jump part is negligible are roughly normally distributed with variance $\sigma^2 |\mathcal{I}_{i,n}^{(l)}|$. Thereby these increments are filtered out with high probability as a normal distributed random variable rarely exceeds 3 standard deviations. Further we use $b_n = 1/\sqrt{n}$ for the local interval in the estimation of σ_s and $\widetilde{K}_n = \lfloor \ln(n) \rfloor$, $M_n = \lfloor 10\sqrt{n} \rfloor$ in the simulation of the $\hat{\xi}_{k,n,m,-}^{(l)}(s)$, $\hat{\xi}_{k,n,m,+}^{(l)}(s)$. $b_n = 1/\sqrt{n}$ is chosen in the center of the allowed range between b_n constant and $b_n = O(\log(n)/n)$ which balances the benefits from choosing b_n small (smaller bias in the estimation of σ if σ is not flat) and b_n large (less variance in the estimation of σ). This choice is also close to the optimal one in the sense of Theorem 13.3.3 in [30] for the estimation of spot volatility. $\widetilde{K}_n$ is chosen rather small to keep the computation time low, M_n is chosen to be large enough to justify a reasonable approximation to the theoretical quantiles. In the simulation study the results were very robust to the choice of b_n, $\widetilde{K}_n$, M_n.

We applied the two testing procedures from Theorem 8.11 and Corollary 8.12 for $k = 2, 3, 5$ and the results are displayed in Figures 8.4 and 8.5. In Figure 8.4 the results from Theorem 8.11 are presented. In the left column the empirical rejection rates are plotted against the theoretical value of the test and in the right column we show estimated density plots based on the simulated values of $\Phi_{k,T,n}^{J,(l)}$. In Figure 8.5 we present the results from the test in Corollary 8.12 for $\rho = 0.9$ (for the choice of $\rho = 0.9$ see Figure 8.3) in the same way. The density plots here show the estimated density of $\widetilde{\Phi}_{k,T,n}^{J,(l)}(\rho)$.

In Figure 8.4 we observe in the case **Cont** that the power of the test from Theorem 8.11 is very good for $n = 1,600$ and $n = 25,600$. Further the empirical rejection rates match the asymptotic values rather well for all considered values of k in the cases **I-j** and **II-j** at least for the highest observation frequency corresponding to $n = 25,600$. However in the case **III-j** there is severe over-rejection even for $n = 25,600$. In general we observe over rejection in all cases. The empirical rejection rates match the asymptotic values better in the cases where there are on average larger jumps. Further the results are better the smaller k is. Note that for $n = 100$ the cases III-j and Cont are not distinguishable using our test as the rejection curves for $n = 100$ in those two cases are almost identical. The density plots show the convergence of $\Phi_{k,T,n}^{J,(l)}$ to 1 in the presence of jumps and to $(k + 1)/2$ under the alternative as predicted from Example 8.5.

In Figure 8.5 we immediately see that the observed rejection rates from the test based on Corollary 8.12 match the asymptotic values much better than those from the test based on Theorem 8.11. Hence adjusting the estimator has a huge effect on the finite sample performance of the test. Here, the observed rejection rates match the asymptotic values quite well in the case III-j at least for $n = 25{,}600$ and in the cases I-j and II-j we get already for $n = 1{,}600$ very good results. The results in the case Cont show that the power remains to be very good. The density plots show that under the presence of jumps $\widetilde{\Phi}_{k,T,n}^{J,(l)}(\rho)$ is more centered around 1 than $\Phi_{k,T,n}^{J,(l)}$. Under the alternative $\widetilde{\Phi}_{k,T,n}^{J,(l)}(\rho)$ clusters around the value $1+(1-\rho)(k-1)/2$ which is much closer to 1 than $(k+1)/2$, but the observed values of $\widetilde{\Phi}_{k,T,n}^{J,(l)}(\rho)$ still seem to be large enough such that they can be well distinguished from 1 as can be seen from the high empirical rejection rate in the case Cont.

Figure 8.3 illustrates how the performance of the test from Corollary 8.12 depends on the choice of the parameter ρ. For this purpose we investigate for $k = 2$ the empirical rejection rates in the cases III-j and Cont with $n = 25{,}600$ for the test with level $\alpha = 5\%$ dependent on the choice of ρ. We plot for $\rho \in [0, 1]$ the empirical rejection rate under the null hypothesis in the case III-j which serves as a proxy for the type-I error of the test together with one minus the empirical rejection rate under the alternative hypothesis in the case Cont which serves as a proxy for the type-II error of the test. Finally, we plot the sum of both error proxies to obtain an indicator for the overall performance of the test dependent on the choice of ρ. As expected we observe a decrease in the type-I error as ρ increases and an increase in the type-II error. While we observe an approximately linear decrease in the type-I error, the type-II error is equal to zero until 0.8, then slightly increases and starts to steeply increase at $\rho = 0.9$. In this example, the overall error is minimized for a relatively large value of ρ close to $\rho = 0.9$.

Further, we carried out simulations for the same four parameter settings based on equidistant observation times $t_{i,n} = i/n$. In this specific setting the test from Theorem 8.11 coincides with tests discussed in [2] and in Chapter 10.3 of [3]. The simulation results are presented in Figures 8.6, 8.7 and 8.8 in the same fashion as in Figures 8.4, 8.5 and 8.3 for the irregular observations. We observe that the results both from Theorem 8.11 and Corollary 8.12 based on irregular observations are not significantly worse than those obtained in the simpler setting of equidistant observation times. Especially we can conclude that the adjustment technique introduced for Corollary 8.12 cannot only be used to improve the finite sample performance of our test based on irregular observations but also can be used to improve existing tests in the literature which are based on equidistant observations.

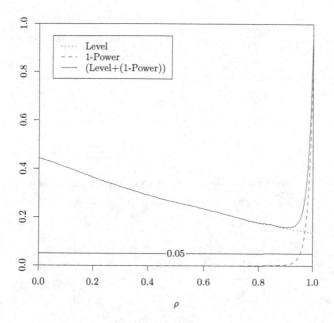

Figure 8.3: This graphic shows for $k = 2$, $\alpha = 5\%$ and $n = 25{,}600$ the empirical rejection rate in the case `Cont` (dotted line) and 1 minus the empirical rejection rate in the case `III-j` (dashed line) from the Monte Carlo simulation based on Corollary 8.12 as a function of $\rho \in [0, 1]$.

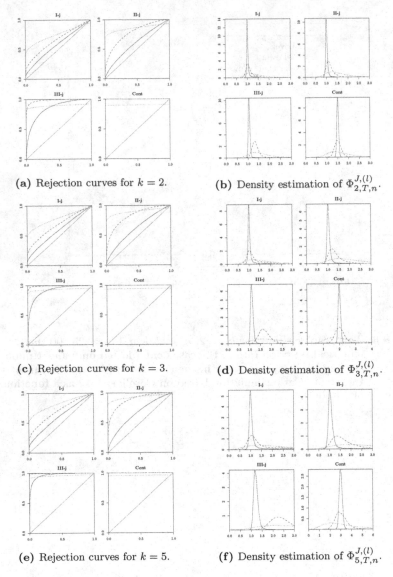

(a) Rejection curves for $k = 2$.

(b) Density estimation of $\Phi_{2,T,n}^{J,(l)}$.

(c) Rejection curves for $k = 3$.

(d) Density estimation of $\Phi_{3,T,n}^{J,(l)}$.

(e) Rejection curves for $k = 5$.

(f) Density estimation of $\Phi_{5,T,n}^{J,(l)}$.

Figure 8.4: These graphics show the simulation results for the test from Theorem 8.11. The dotted lines correspond to $n = 100$, the dashed lines to $n = 1,600$ and the solid lines to $n = 25,600$. In all cases $N = 10,000$ paths were simulated.

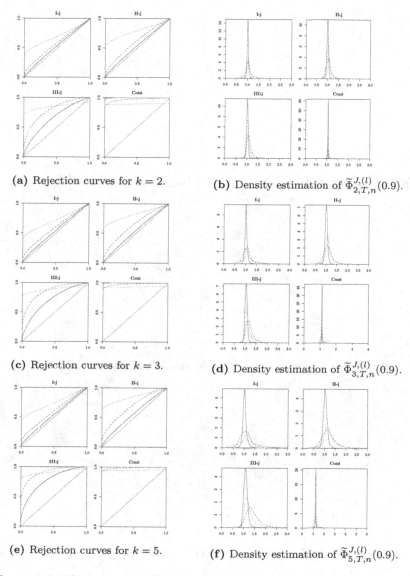

Figure 8.5: These graphics show the simulation results for the test from Corollary 8.12. The dotted lines correspond to $n = 100$, the dashed
lines to $n = 1,600$ and the solid lines to $n = 25,600$. In all cases
$N = 10,000$ paths were simulated.

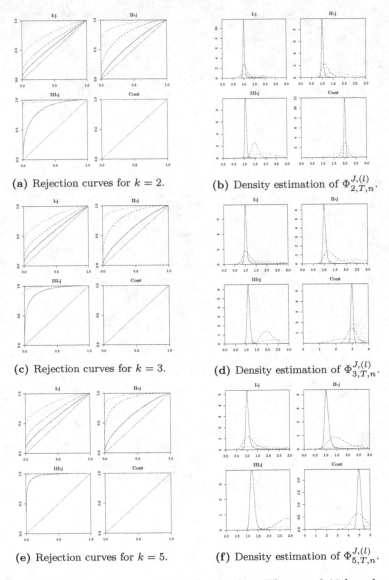

(a) Rejection curves for $k = 2$.

(b) Density estimation of $\Phi_{2,T,n}^{J,(l)}$.

(c) Rejection curves for $k = 3$.

(d) Density estimation of $\Phi_{3,T,n}^{J,(l)}$.

(e) Rejection curves for $k = 5$.

(f) Density estimation of $\Phi_{5,T,n}^{J,(l)}$.

Figure 8.6: Simulation results for the test from Theorem 8.11 based on
equidistant observations $t_{i,n}^{(l)} = i/n$. The dotted lines correspond
to $n = 100$, the dashed lines to $n = 1,600$ and the solid lines to
$n = 25,600$. In all cases $N = 10,000$ paths were simulated.

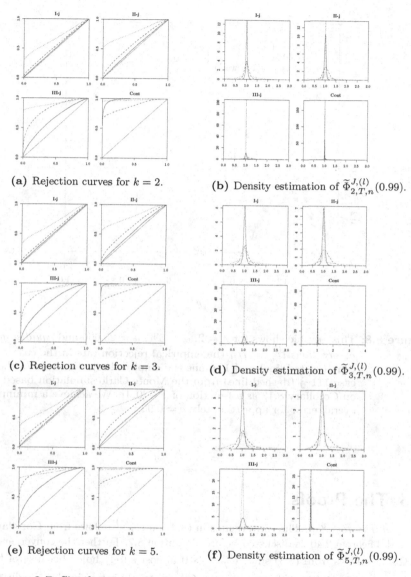

Figure 8.7: Simulation results for the test from Corollary 8.12 based on
equidistant observations $t_{i,n}^{(l)} = i/n$. The dotted lines correspond
to $n = 100$, the dashed lines to $n = 1,600$ and the solid lines to
$n = 25,600$. In all cases $N = 10,000$ paths were simulated.

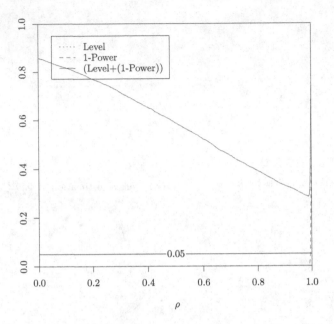

Figure 8.8: This graphic shows for $k = 2$, $\alpha = 5\%$, $n = 25{,}600$ and *equidistant observations* $t_{i,n}^{(l)} = i/n$ the empirical rejection rate in the case Cont (dotted line) and 1 minus the empirical rejection rate in the case III-j (dashed line) from the Monte Carlo simulation based on Corollary 8.12 as a function of $\rho \in [0,1]$. We achieve a minimal overall error for approximately $\rho = 0.99$.

8.3 The Proofs

Proof of Theorem 8.3. The convergence in (8.4) on $\Omega_T^{J,(l)}$ follows from Corollary 2.2 and Theorem 2.40 as discussed before Condition 8.2. Further the convergence on $\Omega_T^{C,(l)}$ follows from (8.3) since $C_T^{J,(l)} > 0$ almost surely holds because of by $\int_0^T |\sigma_s^{(l)}|ds > 0$ almost surely and because $G_4^{(l)}$ is strictly increasing by Condition 8.2. To prove (8.3) note that it holds $X_t^{(l)}\mathbb{1}_{\Omega_T^{C,(l)}} = (B_t^{(l)} + C_t^{(l)})\mathbb{1}_{\Omega_T^{C,(l)}}$ with

$$B_t^{(l)} = \int_0^t \Big(b_s^{(l)} - \int_{\mathbb{R}^2} \delta^{(l)}(s,z) \mathbb{1}_{\{\|\delta(s,z)\| \le 1\}} \lambda(dz) \Big) ds \qquad (8.19)$$

since $\int_0^T \int \delta^{(l)}(s,z) \mu(ds,dz) \equiv 0$ on $\Omega_T^{C,(l)}$. (8.3) then follows by applying Corollary 2.19 and Theorem 2.41(a) to the process $B_t^{(l)} + C_t^{(l)}$. $\qquad \square$

Proof of Theorem 8.8. We prove

$$\sqrt{n} \big(V^{(l)}(g_4, [k], \pi_n)_T - k V^{(l)}(g_4, \pi_n)_T \big) \xrightarrow{\mathcal{L}-s} \Phi_{k,T}^{J,(l)}. \qquad (8.20)$$

The central limit theorem (8.9) then follows easily from (8.20), $V^{(l)}(g_4, \pi_n)_T \xrightarrow{\mathbb{P}} B^{(l)}(g_4)_T$, Proposition B.7(i) and Lemma B.5.

Step 1. Using the discretized versions $\tilde{\sigma}^{(l)}(r,q)$ and $\widetilde{C}^{(l)}(r,q)$ of $\sigma^{(l)}$ and $C^{(l)}$ introduced in Step 1 in the proof of Theorem 3.2 we define

$$R^{(l)}(n,q,r) = 4\sqrt{n} \sum_{i:t_{i,n}^{(l)} \le T} (\Delta_{i,n}^{(l)} N^{(l)}(q))^3 \sum_{j=1}^{k-1} (k-j)(\Delta_{i+j,n}^{(l)} \widetilde{C}^{(l)}(r,q) + \Delta_{i-j,n}^{(l)} \widetilde{C}^{(l)}(r,q)).$$

and show

$$\lim_{q \to \infty} \limsup_{r \to \infty} \limsup_{n \to \infty} \mathbb{P}\big(\big| \sqrt{n}\big(V^{(l)}(g_4, [k], \pi_n)_T - k V^{(l)}(g_4, \pi_n)_T \big)$$

$$- R^{(l)}(n,q,r) \big| > \varepsilon \big) = 0 \quad (8.21)$$

for all $\varepsilon > 0$. As the proof of (8.21) is rather lengthy and technical we will first discuss the complete rough structure of the proof of Theorem 8.8 and present the proof of (8.21) afterwards.

Step 2. Next we prove

$$R^{(l)}(n,q,r) \xrightarrow{\mathcal{L}-s} \Phi_{k,T}^{J,(l)}(q,r) := 4 \sum_{S_{q,p}^{(l)} \le T} (\Delta N^{(l)}(q)_{S_{q,p}^{(l)}})^3$$

$$\times \big((\tilde{\sigma}^{(l)}(r,q)_{S_{q,p}^{(l)}-})^2 \xi_{k,-}^{(l)}(S_{q,p}^{(l)}) + (\tilde{\sigma}^{(l)}(r,q)_{S_{q,p}^{(l)}})^2 \xi_{k,+}^{(l)}(S_{q,p}^{(l)}) \big)^{1/2} U_{S_{q,p}^{(l)}}^{(l)} \quad (8.22)$$

as $n \to \infty$ for all $q > 0$ and $r \in \mathbb{N}$.

To this end note that on the set $\Omega^{(l)}(n,q,r)$ where two different jumps of $N^{(l)}(q)$ are further apart than $|\pi_n|_T$ and any jump time of $N^{(l)}(q)$ is further away from $j2^{-r}$ than $k|\pi_n|_T$ for any $j \in \{1, \ldots, \lfloor T2^r \rfloor\}$ it holds

$$R^{(l)}(n,q,r) \mathbb{1}_{\Omega^{(l)}(n,q,r)} = 4 \sum_{S_{q,p}^{(l)} \le T} (\Delta N^{(l)}(q)_{S_{q,p}^{(l)}})^3 \sqrt{n} \sum_{j=1}^{k-1}(k-j)$$

$$\times (\tilde{\sigma}^{(l)}(r,q)_{S_{q,p}^{(l)}} \Delta_{i_n^{(l)}(S_{q,p}^{(l)})-j,n}^{(l)} \overline{W}^{(l)} + \breve{\sigma}^{(l)}(r,q)_{S_{q,p}^{(l)}} \Delta_{i_n^{(l)}(S_{q,p}^{(l)})+j,n}^{(l)} \overline{W}^{(l)}) \mathbb{1}_{\Omega^{(l)}(n,q,r)},$$

compare (3.23) for the definition of $\overline{W}$. Comparing the proof of (3.25) we find that Condition 8.7(ii) yields the $\mathcal{X}$-stable convergence

$$\left(\left(\sqrt{n}\sum_{j=1}^{k-1}(k-j)\Delta^{(l)}_{i_n^{(l)}(S_{q,p}^{(l)})-j,n}\overline{W}^{(l)},\sqrt{n}\sum_{j=1}^{k-1}(k-j)\Delta^{(l)}_{i_n^{(l)}(S_{q,p}^{(l)})+j,n}\overline{W}^{(l)}\right)_{S_{q,p}^{(l)}\leq T}\right)$$

$$\xrightarrow{\mathcal{L}-s}\left(\left(\sqrt{\xi_{k,-}^{(l)}(S_{q,p}^{(l)})}U^{(l)}_{S_{q,p,-}^{(l)}},\sqrt{\xi_{k,+}^{(l)}(S_{q,p}^{(l)})}U^{(l)}_{S_{q,p,+}^{(l)}}\right)_{S_{q,p}^{(l)}\leq T}\right)$$

for standard normally distributed random variables $(U^{(l)}_{s,-},U^{(l)}_{s,+})$ which are independent of $\mathcal{F}$ and of the $\xi_{k,-}^{(l)}(s),\xi_{k,+}^{(l)}(s)$. Using this stable convergence, Lemma B.3 and the continuous mapping theorem for stable convergence stated in Lemma B.5 we then obtain

$$4\sum_{S_{q,p}^{(l)}\leq T}(\Delta N^{(l)}(q)_{S_{q,p}^{(l)}})^3\sqrt{n}\sum_{j=1}^{k-1}(k-j)$$

$$\times(\tilde{\sigma}^{(l)}(r,q)_{S_{q,p-}^{(l)}}\Delta^{(l)}_{i_n^{(l)}(S_{q,p}^{(l)})-j,n}\overline{W}^{(l)}+\tilde{\sigma}^{(l)}(r,q)_{S_{q,p}^{(l)}}\Delta^{(l)}_{i_n^{(l)}(S_{q,p}^{(l)})+j,n}\overline{W}^{(l)})$$

$$\xrightarrow{\mathcal{L}-s}\Phi_{k,T}^{J,(l)}(q,r).$$

Because of $\mathbb{P}(\Omega^{(l)}(n,q,r))\to 1$ as $n\to\infty$ for any $q,r>0$ this convergence yields (8.22).

Step 3. Finally we need to show

$$\lim_{q\to\infty}\limsup_{r\to\infty}\widetilde{\mathbb{P}}(|\Phi_{k,T}^{J,(l)}-\Phi_{k,T}^{J,(l)}(q,r)|>\varepsilon)=0,\qquad(8.23)$$

for all $\varepsilon>0$. (8.23) can be proven using the fact that $\Gamma^{J,(l)}(\cdot,dy)$ has uniformly bounded first moments together with the boundedness of the jump sizes of $X^{(l)}$ respectively $N^{(l)}(q)$.

Step 4. Finally by combining (8.21)–(8.23) and using Lemma B.6 we obtain (8.20). $\qquad\square$

Proof of (8.21). To simplify notation we set $\Delta^{(l)}_{i,k',n}X=0$, $\mathcal{I}^{(l)}_{i,k',n}=\emptyset$, $k'=1,k$, whenever $t_{i,n}^{(l)}>T$ in this proof.

Step 1. We first prove

$$\limsup_{n\to\infty}\mathbb{P}(|\sqrt{n}(V^{(l)}(g_4,[k],\pi_n)_T-kV^{(l)}(g_4,\pi_n)_T)-R^{(l)}(n)|>\varepsilon)=0\quad(8.24)$$

where

$$R^{(l)}(n)=\sqrt{n}\sum_{i:t_{i,n}^{(l)}\leq T}4(\Delta^{(l)}_{i,n}X^{(l)})^3\sum_{j=1}^{k-1}(k-j)(\Delta^{(l)}_{i-j,n}X^{(l)}+\Delta^{(l)}_{i+j,n}X^{(l)}).$$

Using the identity

$$(x_1 + \ldots + x_k)^4 = \sum_{i=1}^{k}(x_i)^4 + 4\sum_{i=1}^{k}\sum_{j\neq i}(x_i)^3 x_j + 6\sum_{i=1}^{k}\sum_{j>i}(x_i)^2(x_j)^2$$

$$+ 12\sum_{i=1}^{k}\sum_{j\neq i}\sum_{j'>j}(x_i)^2 x_j x_{j'} + 24\sum_{i=1}^{k}\sum_{i'>i}\sum_{j>i'}\sum_{j'>j} x_i x_{i'} x_j x_{j'}$$

$$(8.25)$$

which is a specific case of the multinomial theorem, we derive

$$\sqrt{n}(V^{(l)}(g_4, [k], \pi_n)_T - kV^{(l)}(g_4, \pi_n)_T)$$

$$= \sqrt{n}\sum_{i:t_{i,n}^{(l)}\leq T}\sum_{j=1}^{k-1}(k-j)\big(4(\Delta_{i,n}^{(l)}X^{(l)})^3\Delta_{i-j,n}^{(l)}X^{(l)} + 4(\Delta_{i,n}^{(l)}X^{(l)})^3\Delta_{i+j,n}^{(l)}X^{(l)}\big)$$

$$+ O_{\mathbb{P}}\Big(\sqrt{n}\sum_{i:t_{i,n}^{(l)}\leq T}\sum_{j=1}^{k-1}K(\Delta_{i,n}^{(l)}X^{(l)})^2(\Delta_{i+j,n}^{(l)}X^{(l)})^2\Big) + O_{\mathbb{P}}\big(\sqrt{n}|\pi_n|_T\big) \quad (8.26)$$

where we used the inequalities $|x_i x_i'| \leq (x_i)^2 + (x_{i'})^2$, $|x_j x_{j'}| \leq (x_j)^2 + (x_{j'})^2$ to include terms with powers $(2,1,1,0)$ and $(1,1,1,1)$ from (8.25) into the first summand of the last line of (8.26). The last term in (8.26) is due to boundary effects at T.

By iterated expectations and inequality (1.12) we get

$$\sqrt{n}\mathbb{E}\Big[\sum_{i:t_{i,n}^{(l)}\leq T}\sum_{j=1}^{k-1}K(\Delta_{i,n}^{(l)}X^{(l)})^2(\Delta_{i+j,n}^{(l)}X^{(l)})^2\Big|\mathcal{S}\Big] \leq \sqrt{n}KT|\pi_n|_T.$$

Hence this quantity and the $O_{\mathbb{P}}(\sqrt{n}|\pi_n|_T)$-term vanish for $n \to \infty$ by Condition 8.7(i). This observation yields (8.24).

Step 2. Next we prove

$$\lim_{q\to\infty}\limsup_{n\to\infty}\mathbb{P}(|R^{(l)}(n) - R^{(l)}(n,q)| > \varepsilon) \to 0 \quad (8.27)$$

where

$$R^{(l)}(n,q) = \sqrt{n}\sum_{i:t_{i,n}^{(l)}\leq T}4(\Delta_{i,n}^{(l)}N^{(l)}(q))^3\sum_{j=1}^{k-1}(k-j)\big(\Delta_{i-j,n}^{(l)}C^{(l)} + \Delta_{i+j,n}^{(l)}C^{(l)}\big).$$

Therefore we first consider

$$\lim_{q\to\infty} \limsup_{n\to\infty} \mathbb{P}(|R^{(l)}(n) - \widetilde{R}^{(l)}(n,q)| > \varepsilon) \to 0 \qquad (8.28)$$

for all $\varepsilon > 0$ with

$$\widetilde{R}^{(l)}(n,q) = \sqrt{n} \sum_{i:t_{i,n}^{(l)} \leq T} 4(\Delta_{i,n} N^{(l)}(q))^3 \sum_{j=1}^{k-1} (k-j)(\Delta_{i-j,n}^{(l)} X^{(l)} + \Delta_{i+j,n}^{(l)} X^{(l)})$$

Using

$$|a^3 - b^3| = |a - b||a^2 + ab + b^2| \leq \frac{3}{2}|a - b|(a^2 + b^2)$$

we derive

$$|R^{(l)}(n) - \widetilde{R}^{(l)}(n,q)|$$
$$\leq \sqrt{n}K \sum_{i:t_{i,n}^{(l)} \leq T} |\Delta_{i,n}^{(l)} X^{(l)} - \Delta_{i,n}^{(l)} N^{(l)}(q)|((\Delta_{i,n}^{(l)} X^{(l)})^2 + (\Delta_{i,n}^{(l)} N^{(l)}(q))^2)$$

$$\times \sum_{j=1}^{k-1} |\Delta_{i-j,n}^{(l)} X^{(l)} + \Delta_{i+j,n}^{(l)} X^{(l)}|. \qquad (8.29)$$

The $\mathcal{S}$-conditional expectation of (8.29) is using iterated expectations, the Hölder inequality with $p_1 = 3$, $p_2 = 3/2$ on $\Delta_{i,n}^{(l)}(B^{(l)}(q)+C^{(l)})((\Delta_{i,n}^{(l)} X^{(l)})^2+(\Delta_{i,n}^{(l)} N^{(l)}(q))^2)$, the Cauchy-Schwarz inequality on $\Delta_{i,n}^{(l)} M^{(l)}(q)((\Delta_{i,n}^{(l)} X^{(l)})^2 + (\Delta_{i,n}^{(l)} N^{(l)}(q))^2)$ and Lemma 1.4 bounded by

$$\sqrt{n}K \sum_{i:t_{i,n}^{(l)} \leq T} \left[(K_q|\mathcal{I}_{i,n}^{(l)}|^3 + |\mathcal{I}_{i,n}^{(l)}|^{3/2})^{1/3}(|\mathcal{I}_{i,n}^{(l)}| + K|\mathcal{I}_{i,n}^{(l)}| + K_q|\mathcal{I}_{i,n}^{(l)}|^3)^{2/3} \right.$$
$$\left. + (e_q|\mathcal{I}_{i,n}^{(l)}|)^{1/2}(|\mathcal{I}_{i,n}^{(l)}| + K|\mathcal{I}_{i,n}^{(l)}| + K_q|\mathcal{I}_{i,n}^{(l)}|^4)^{1/2} \right]$$
$$\times \sum_{j=-(k-1),j\neq 0}^{k-1} |\mathcal{I}_{i+j,n}^{(l)}|^{1/2}$$
$$\leq K(K_q(|\pi_n|_T)^{1/6} + (e_q + K_q(|\pi_n|_T)^3)^{1/2})$$
$$\times \sqrt{n} \sum_{i:t_{i,n}^{(l)} \leq T} |\mathcal{I}_{i,n}^{(l)}| \sum_{j=-(k-1),j\neq 0}^{k-1} |\mathcal{I}_{i+j,n}^{(l)}|^{1/2}$$
$$\leq K(K_q(|\pi_n|_T)^{1/6} + (e_q)^{1/2})\sqrt{n} \sum_{i:t_{i,n}^{(l)} \leq T} |\mathcal{I}_{i,k,n}^{(l)}||\mathcal{I}_{i,k,n}^{(l)}|^{1/2}$$

$$\leq K(K_q(|\pi_n|_T)^{1/6} + (e_q)^{1/2})\Big(nT \sum_{i:t_{i,n}^{(l)} \leq T} |\mathcal{I}_{i,k,n}^{(l)}|^2\Big)^{1/2} \qquad (8.30)$$

where we used $|\mathcal{I}_{i,n}^{(l)}||\mathcal{I}_{i+j,n}^{(l)}|^{1/2} \leq |\mathcal{I}_{\max\{i,i+j\},k,n}^{(l)}|^{3/2}$ for $|j| \leq k - 1$ and the Cauchy-Schwarz inequality for sums to obtain the last two inequalities. The last bound vanishes as first $n \to \infty$ and then $q \to \infty$ because of $G_4^{[k],(l)}(T) = O_{\mathbb{P}}(1)$ by Condition 8.2. Hence by Lemma 2.15 we have proved (8.28).

Further we prove

$$\lim_{q\to\infty} \limsup_{n\to\infty} \mathbb{P}(|R^{(l)}(n,q) - \widetilde{R}^{(l)}(n,q))| > \varepsilon) \to 0 \qquad (8.31)$$

for all $\varepsilon > 0$. Denote by $\Omega^{(l)}(n,q)$ the set where two jumps of $N^{(l)}(q)$ are further apart than $k|\pi_n|_T$. By iterated expectations and Lemma 1.4 it holds

$$\mathbb{E}[|R^{(l)}(n,q) - \widetilde{R}^{(l)}(n,q))|\mathbb{1}_{\Omega^{(l)}(n,q)}|\mathcal{S}]$$

$$\leq K\sqrt{n} \sum_{i:t_{i,n}^{(l)} \leq T} \mathbb{E}\big[|\Delta_{i,n}^{(l)} N^{(l)}(q)|^3 \sum_{j=-(k-1),j\neq 0}^{k-1} |\Delta_{i+j,n}^{(l)}(B^{(l)}(q) + M^{(l)}(q))||\mathcal{S}\big]$$

$$\leq K\sqrt{n} \sum_{i:t_{i,n}^{(l)} \leq T} (|\mathcal{I}_{i,n}^{(l)}| + K_q|\mathcal{I}_{i,n}^{(l)}|^3) \sum_{j=-(k-1),j\neq 0}^{k-1} (K_q|\mathcal{I}_{i,n}^{(l)}|^2 + e_q|\mathcal{I}_{i,n}^{(l)}|)^{1/2}$$

$$\leq K(1 + K_q(|\pi_n|_T)^2)(K_q|\pi_n|_T + e_q)^{1/2}\Big(nT \sum_{i:t_{i,n}^{(l)} \leq T} |\mathcal{I}_{i,k,n}^{(l)}|^2\Big)^{1/2}$$

where the last inequality follows as in (8.30). The last bound vanishes as first $n \to \infty$ and then $q \to \infty$ by Condition 8.2. Hence by Lemma 2.15 we have proved (8.31) because of $\mathbb{P}(\Omega^{(l)}(n,q)) \to 1$ as $n \to \infty$ for any $q > 0$ which together with (8.28) yields (8.27).

Step 3. Finally we consider

$$\lim_{q\to\infty} \limsup_{r\to\infty} \limsup_{n\to\infty} \mathbb{P}(|R^{(l)}(n,q) - R^{(l)}(n,q,r)| > \varepsilon) \to 0. \qquad (8.32)$$

Using iterated expectations, inequality (1.11) and inequality (2.1.34) from [30] we obtain

$$\mathbb{E}[|R^{(l)}(n,q) - R^{(l)}(n,q,r)||\mathcal{S}]$$
$$\leq (K + K_q(|\pi_n|_T)^2)$$
$$\times \mathbb{E}\big[\sqrt{n} \sum_{i:t_{i,n}^{(l)} \leq T} |\mathcal{I}_{i,n}^{(l)}| \sum_{j=-(k-1),j\neq 0}^{k-1} (\int_{t_{i+j-1,n}^{(l)}}^{t_{i+j,n}^{(l)}} |\sigma_s^{(l)} - \tilde{\sigma}^{(l)}(r,q)_s|^2 ds)^{1/2}|\mathcal{S}\big]$$

$$\leq (K + K_q(|\pi_n|_T)^2)\mathbb{E}\big[\sqrt{n} \sum_{i:t_{i,n}^{(l)} \leq T} |\mathcal{I}_{i,n}^{(l)}|(2k-2)^{1/2}$$

$$\times \big(\sum_{j=-(k-1),j\neq 0}^{k-1} \int_{t_{i+j-1,n}^{(l)}}^{t_{i+j,n}^{(l)}} |\sigma_s^{(l)} - \tilde{\sigma}^{(l)}(r,q)_s|^2 ds \big)^{1/2}\big|\mathcal{S}\big].$$

This quantity is using the Cauchy-Schwarz inequality for sums further bounded by

$$(K + K_q(|\pi_n|_T)^2)\sqrt{G_4^{(l),n}(T)}$$

$$\times \mathbb{E}\big[\big((2k-2)^2 \sum_{i:t_{i,n}^{(l)} \leq T} \int_{t_{i-1,n}^{(l)}}^{t_{i,n}^{(l)}} |\sigma_s^{(l)} - \tilde{\sigma}^{(l)}(r,q)_s|^2 ds\big)^{1/2}\big|\mathcal{S}\big]$$

$$= (K + K_q(|\pi_n|_T)^2)\sqrt{G_4^{(l),n}(T)}\mathbb{E}\big[\big(\int_0^T |\sigma_s^{(l)} - \tilde{\sigma}^{(l)}(r,q)_s|^2 ds\big)^{1/2}\big|\mathcal{S}\big]$$

$$\xrightarrow{\mathbb{P}} K\sqrt{G_4^{(l)}(T)}\mathbb{E}\big[\big(\int_0^T |\sigma_s^{(l)} - \tilde{\sigma}^{(l)}(r,q)_s|^2 ds\big)^{1/2}\big|\mathcal{S}\big]$$

where the convergence as $n \to \infty$ follows from Condition 8.2. Here the limit vanishes as $r \to \infty$ for any $q > 0$ since the expectation vanishes as $r \to \infty$ by dominated convergence because $\sigma^{(l)}, \tilde{\sigma}^{(l)}(r,q)$ are bounded by assumption and because $(\tilde{\sigma}^{(l)}(r,q))_{r\in\mathbb{N}}$ is a sequence of right continuous elementary processes approximating $\sigma^{(l)}$. Hence (8.32) follows from Lemma 2.15.

Step 4. Combining (8.24), (8.27) and (8.32) we obtain (8.21).   □

The structure of the upcoming proof of (8.14) in Theorem 8.11 and especially of (8.33) therein is identical to the structure of the proof of Theorem 4.3 and (4.5) in Chapter 4.

Proof of Theorem 8.11. We first discuss the main ideas for the proof while more technical details will be proved afterwards. For proving (8.14) we need to show

$$\widetilde{\mathbb{P}}\big(\sqrt{n}(V^{(l)}(g_4,[k],\pi_n)_T - kV^{(l)}(g_4,\pi_n)_T) > \widehat{Q}_{k,T,n}^{J,(l)}(1-\alpha)\big|F^{J,(l)}\big) \to \alpha$$

which follows from Theorem 8.8 and

$$\lim_{n\to\infty} \widetilde{\mathbb{P}}\big(\{|\widehat{Q}_{k,T,n}^{J,(l)}(\alpha) - Q_{k,T}^{J,(l)}(\alpha)| > \varepsilon\} \cap \Omega_T^{J,(l)}\big) \to 0 \qquad (8.33)$$

for all $\varepsilon > 0$ and any $\alpha \in [0,1]$. For more details compare the proof of Theorem 4.3. Hence it remains to prove (8.33) for which we will give a proof after we discussed the main idea for the proof of (8.15).

For proving (8.15) we observe that $\Phi_{k,T,n}^{J,(l)}$ converges on $\Omega_T^{C,(l)}$ to a limit strictly greater than 1 by Theorem 8.3 and Condition 8.10(ii). Hence it is sufficient to prove

$$c_{k,T,n}^{J,(l)}(\alpha)\mathbf{1}_{\Omega_T^{C,(l)}} = o_{\widetilde{\mathbb{P}}}(1), \quad \alpha > 0, \tag{8.34}$$

to obtain (8.15). The proof of (8.34) will be given after the proof of (8.33). $\qquad\Box$

Similarly as in Chapter 4 we first prove the following proposition needed in the proof of (8.33).

Proposition 8.15. *Suppose Condition 8.10 is fulfilled. Then it holds*

$$\widetilde{\mathbb{P}}(\{|\frac{1}{M_n}\sum_{m=1}^{M_n}\mathbf{1}_{\{\widehat{\Phi}_{k,T,n,m}^{J,(l)}\leq\Upsilon\}} - \widetilde{\mathbb{P}}(\Phi_{k,T}^{J,(l)}\leq\Upsilon|\mathcal{X})| > \varepsilon\}\cap\Omega_T^{J,(l)}) \to 0 \tag{8.35}$$

for any $\mathcal{X}$-measurable random variable Υ and all $\varepsilon > 0$.

Proof. Denote by $S_{q,p}^{(l)}$, $p \in \mathbb{N}$, an increasing sequence of stopping times which exhausts the jump times of $N^{(l)}(q)$ and introduce the notation

$$Y_k^{(l)}(P,n,m) = \sum_{p=1}^{P}(\Delta_{i_n^{(l)}(S_{q,p}^{(l)}),n}X)^3\mathbf{1}_{\{|\Delta_{i_n^{(l)}(S_{q,p}^{(l)}),n}X|>\beta|\mathcal{I}_{i_n^{(l)}(S_{q,p}^{(l)}),n}|^{\varpi}\}}$$

$$\times ((\tilde{\sigma}_n^{(l)}(t_{i_n^{(l)}(S_{q,p}^{(l)}),n},-))^2\hat{\xi}_{k,n,m,-}^{(l)}(t_{i_n^{(l)}(S_{q,p}^{(l)}),n})$$

$$+ (\tilde{\sigma}_n^{(l)}(t_{i_n(S_{q,p}^{(l)}),n},+))^2\hat{\xi}_{k,n,m,+}^{(l)}(t_{i_n^{(l)}(S_{q,p}^{(l)}),n}))^{1/2}U_{n,i_n(S_{q,p}^{(l)}),m}^{(l)}\mathbf{1}_{\{S_{q,p}^{(l)}\leq T\}},$$

$$Y_k^{(l)}(P) = \sum_{p=1}^{P}(\Delta X_{S_p}^{(l)})^3((\sigma_{S_{q,p}^{(l)}-}^{(l)})^2\xi_{k,-}^{(l)}(S_{q,p}^{(l)}) + (\sigma_{S_{q,p}^{(l)}}^{(l)})^2\xi_{k,+}^{(l)}(S_{q,p}^{(l)}))^{1/2}U_{S_{q,p}^{(l)}}^{(l)}$$

$$\times \mathbf{1}_{\{S_{q,p}^{(l)}\leq T\}}.$$

Step 1. By specifying $A_{n,p}, A_p, \widehat{Z}_{n,m}^p(s), Z^p(s)$ and the function φ in a similar fashion like in Step 1 in the proof of Proposition 4.10 we obtain from Lemma 4.9, Condition 8.7 and Corollary 6.4 the convergence

$$\widetilde{\mathbb{P}}(\{|\frac{1}{M_n}\sum_{m=1}^{M_n}\mathbf{1}_{\{Y_k^{(l)}(P,n,m)\leq\Upsilon\}} - \widetilde{\mathbb{P}}(Y_k^{(l)}(P)\leq\Upsilon|\mathcal{X})| > \varepsilon\}\cap\Omega_T^{J,(l)}) \to 0 \tag{8.36}$$

as $n \to \infty$ for any fixed P.

Step 2. Next we show

$$\lim_{P\to\infty}\limsup_{n\to\infty}\widetilde{\mathbb{P}}(\{|\frac{1}{M_n}\sum_{m=1}^{M_n}(\mathbb{1}_{\{Y_k^{(l)}(P,n,m)\leq\Upsilon\}}-\mathbb{1}_{\{\hat\Phi_{k,T,n,m}^{J,(l)}\leq\Upsilon\}})|>\varepsilon\}\cap\Omega_T^{J,(l)})=0$$

(8.37)

for all $\varepsilon>0$ which follows from

$$\lim_{P\to\infty}\limsup_{n\to\infty}\frac{1}{M_n}\sum_{m=1}^{M_n}\widetilde{\mathbb{P}}(|Y_k^{(l)}(P,n,m)-\hat\Phi_{k,T,n,m}^{J,(l)}|>\varepsilon)=0\qquad(8.38)$$

for all $\varepsilon>0$; compare Step 4 in the proof of Proposition 4.10.

On the set $\Omega^{(l)}(q,P,n)$ on which there are at most P jumps of $N^{(l)}(q)$ in $[0,T]$ and two different jumps of $N^{(l)}(q)$ are further apart than $|\pi_n|_T$ it holds

$$|Y_k^{(l)}(P,n,m)-\hat\Phi_{k,T,n,m}^{J,(l)}|\mathbb{1}_{\Omega^{(l)}(q,P,n)}$$

$$\leq\sum_{i:t_{i,n}^{(l)}\leq T}|\Delta_{i,n}^{(l)}(X^{(l)}-N^{(l)}(q))|^3\mathbb{1}_{\{|\Delta_{i,n}^{(l)}X^{(l)}|>\beta|\mathcal{I}_{i,n}^{(l)}|^\varpi\}}(\hat\xi_{k,n,m}^{(l)}(i))^{1/2}$$

$$\times\Big(\frac{2}{b_n}\sum_{j\neq i:\mathcal{I}_{j,n}^{(l)}\subset[t_{i,n}^{(l)}-b_n,t_{i,n}^{(l)}+b_n]}(\Delta_{j,n}^{(l)}X^{(l)})^2\Big)^{1/2}|U_{n,i,m}^{(l)}|$$

$$\leq K\sum_{i:t_{i,n}^{(l)}\leq T}(|\Delta_{i,n}^{(l)}(B^{(l)}(q)+C^{(l)})|^3+|\Delta_{i,n}^{(l)}M^{(l)}(q)|^3)(\hat\xi_{k,n,m}^{(l)}(i))^{1/2}$$

$$\times\Big(\frac{1}{b_n}\sum_{j\neq i:\mathcal{I}_{j,n}^{(l)}\subset[t_{i,n}^{(l)}-b_n,t_{i,n}^{(l)}+b_n]}(\Delta_{j,n}^{(l)}X^{(l)})^2\Big)^{1/2}|U_{n,i,m}^{(l)}|\qquad(8.39)$$

with

$$\hat\xi_{k,n,m}^{(l)}(i)=\hat\xi_{k,n,m,-}^{(l)}(t_{i,n}^{(l)})+\hat\xi_{k,n,m,+}^{(l)}(t_{i,n}^{(l)}).$$

For the continuous parts in (8.39) we get using the Cauchy-Schwarz inequality, Lemma 1.4 and $\hat\xi_{k,n,m,-}^{(l)}(t_{i,n}),\hat\xi_{k,n,m,+}^{(l)}(t_{i,n})\leq nK|\pi_n|_T$

$$\mathbb{E}\Big[\sum_{i:t_{i,n}^{(l)}\leq T}|\Delta_{i,n}^{(l)}(B^{(l)}(q)+C^{(l)})|^3\Big(\frac{1}{b_n}\sum_{j\neq i:\mathcal{I}_{j,n}^{(l)}\subset[t_{i,n}^{(l)}-b_n,t_{i,n}^{(l)}+b_n]}(\Delta_{j,n}^{(l)}X^{(l)})^2\Big)^{1/2}$$

$$\times(\hat\xi_{k,n,m,-}^{(l)}(t_{i,n}^{(l)})+\hat\xi_{k,n,m,+}^{(l)}(t_{i,n}^{(l)}))^{1/2}|U_{n,i,m}^{(l)}|\,\big|\mathcal{S}\Big]$$

$$\leq K\sum_{i:t_{i,n}^{(l)}\leq T}[(K_q|\mathcal{I}_{i,n}^{(l)}|^6+|\mathcal{I}_{i,n}^{(l)}|^3)\frac{2b_n}{b_n}n|\pi_n|_T]^{1/2}$$

$$=K\sqrt{n}|\pi_n|_T(K_q(|\pi_n|_T)^3+1)^{1/2}T$$

which vanishes as $n \to \infty$ by Condition 8.7(i).

Furthermore we get for the remaining terms containing $|\Delta_{i,n}^{(l)} M^{(l)}(q)|^3$ in (8.39) using inequalities (1.10) and (1.12)

$$
\mathbb{E}\Big[\sum_{i:t_{i,n}^{(l)} \leq T} |\Delta_{i,n}^{(l)} M^{(l)}(q)|^3 \Big(\frac{1}{b_n} \sum_{j \neq i: \mathcal{I}_{j,n}^{(l)} \subset [t_{i,n}^{(l)} - b_n, t_{i,n}^{(l)} + b_n]} (\Delta_{j,n}^{(l)} X^{(l)})^2\Big)^{1/2}
$$
$$
\times (\hat{\xi}_{k,n,m}^{(l)}(i))^{1/2} |U_{n,i,m}^{(l)}| \Big| \mathcal{S}\Big]
$$
$$
\leq K \sum_{i:t_{i,n}^{(l)} \leq T} \Big(\mathbb{E}\Big[\mathbb{E}[|\Delta_{i,n}^{(l)} M^{(l)}(q)|^3 | \mathcal{S}, \mathcal{F}_{t_{i-1,n}^{(l)}}]
$$
$$
\times \Big(\frac{1}{b_n} \sum_{j:\mathcal{I}_{j,n}^{(l)} \subset [t_{i,n}^{(l)} - b_n, t_{i,n}^{(l)})} (\Delta_{j,n} X^{(l)})^2\Big)^{1/2} \Big| \mathcal{S}\Big]
$$
$$
+ \mathbb{E}\Big[|\Delta_{i,n}^{(l)} M^{(l)}(q)|^3 \mathbb{E}\Big[\Big(\frac{1}{b_n} \sum_{j:\mathcal{I}_{j,n}^{(l)} \subset [t_{i,n}^{(l)}, t_{i,n}^{(l)} + b_n]} (\Delta_{j,n}^{(l)} X^{(l)})^2\Big)^{1/2} \Big| \mathcal{S}, \mathcal{F}_{t_{i,n}^{(l)}}\Big] \Big| \mathcal{S}\Big]\Big)
$$
$$
\times \mathbb{E}[(\hat{\xi}_{k,n,m}^{(l)}(i))^{1/2} | \mathcal{S}]
$$
$$
\leq K e_q \sum_{i:t_{i,n} \leq T} \Big(|\mathcal{I}_{i,n}^{(l)}| \Big(\mathbb{E}\Big[\frac{1}{b_n} \sum_{j:\mathcal{I}_{j,n}^{(l)} \subset [t_{i,n}^{(l)} - b_n, t_{i,n}^{(l)})} (\Delta_{j,n}^{(l)} X^{(l)})^2 \Big| \mathcal{S}\Big]\Big)^{1/2}
$$
$$
+ \mathbb{E}\Big[|\Delta_{i,n}^{(l)} M^{(l)}(q)|^3 \Big(\mathbb{E}\Big[\frac{1}{b_n} \sum_{j:\mathcal{I}_{j,n}^{(l)} \subset [t_{i,n}^{(l)}, t_{i,n}^{(l)} + b_n]} (\Delta_{j,n}^{(l)} X^{(l)})^2 \Big| \mathcal{S}, \mathcal{F}_{t_{i,n}^{(l)}}\Big]\Big)^{1/2} \Big| \mathcal{S}\Big]\Big)
$$
$$
\times (\mathbb{E}[\hat{\xi}_{k,n,m}^{(l)}(i) | \mathcal{S}])^{1/2}
$$
$$
\leq K e_q \sum_{i:t_{i,n}^{(l)} \leq T} |\mathcal{I}_{i,n}^{(l)}| \Big(K \frac{b_n}{b_n}\Big)^{1/2} \Big(\sum_{k=-\tilde{K}_n}^{\tilde{K}_n} \Big(n \sum_{j=1}^{k-1} (k-j)^2 (|\mathcal{I}_{i+\tilde{k}-j,n}^{(l)}| + |\mathcal{I}_{i+\tilde{k}+j,n}^{(l)}|)\Big)^{1/2}
$$
$$
\times |\mathcal{I}_{i+\tilde{k},n}^{(l)}| \Big(\sum_{\tilde{k}'=-\tilde{K}_n}^{\tilde{K}_n} |\mathcal{I}_{i+\tilde{k}',n}^{(l)}|\Big)^{-1}\Big)
$$
$$
\leq K e_q \sqrt{n} \sum_{i:t_{i,n}^{(l)} \leq T} |\mathcal{I}_{i,n}^{(l)}| \sum_{\tilde{k}=-\tilde{K}_n}^{\tilde{K}_n} |\mathcal{I}_{i+\tilde{k},n}^{(l)}| \Big(\sum_{\tilde{k}'=-\tilde{K}_n}^{\tilde{K}_n} |\mathcal{I}_{i+\tilde{k}',n}^{(l)}|\Big)^{-1} \sum_{j=-(k-1)}^{k-1} |\mathcal{I}_{i+\tilde{k}+j,n}^{(l)}|^{1/2}
$$
$$
= K e_q \sqrt{n} \sum_{i:t_{i,n}^{(l)} \leq T} |\mathcal{I}_{i,n}^{(l)}| \sum_{j=-(k-1)}^{k-1} |\mathcal{I}_{i+j,n}^{(l)}|^{1/2} \sum_{\tilde{k}=-\tilde{K}_n}^{\tilde{K}_n} |\mathcal{I}_{i+\tilde{k},n}^{(l)}| \Big(\sum_{\tilde{k}'=-\tilde{K}_n}^{\tilde{K}_n} |\mathcal{I}_{i+\tilde{k}+\tilde{k}',n}^{(l)}|\Big)^{-1}
$$
$$
+ O_{\mathbb{P}}(\sqrt{n}|\pi_n|_T) \tag{8.40}
$$

where we changed the index $i \to i + \tilde{k}$ to derive the last equality. Then (8.40) vanishes as in (8.30) because of

$$\sum_{\tilde{k}=-\tilde{K}_n}^{\tilde{K}_n} |\mathcal{I}_{i+\tilde{k},n}| \Big(\sum_{\tilde{k}'=-\tilde{K}_n}^{\tilde{K}_n} |\mathcal{I}_{i+\tilde{k}+\tilde{k}',n}| \Big)^{-1} \leq 2$$

which we obtain as in (4.23). Hence by Lemma 2.15 we get (8.38) because of $\mathbb{P}(\Omega^{(l)}(q,P,n)) \to \infty$ as $P,n \to \infty$ for any $q > 0$.

Step 3. Finally we show

$$\widetilde{\mathbb{P}}(Y_k^{(l)}(P) \leq \Upsilon | \mathcal{X}) \mathbb{1}_{\Omega_T^{J,(l)}} \xrightarrow{\text{P}} \widetilde{\mathbb{P}}(\Phi_{k,T}^{J,(l)} \leq \Upsilon | \mathcal{X}) \mathbb{1}_{\Omega_T^{J,(l)}} \qquad (8.41)$$

as $P \to \infty$. We have $Y_k^{(l)}(P) \xrightarrow{\text{P}} \Phi_{k,T}^{J,(l)}$ as $P \to \infty$ and it can be shown that this convergence yields (8.41) since the $\mathcal{X}$-conditional distribution of $\Phi_{k,T}^{J,(l)}$ is by Condition 8.7 almost surely continuous; compare Step 3 in the proof of Proposition 4.10 where a similar claim is proven in more detail.

Step 4. Combining (8.36), (8.37) and (8.41) yields (8.35). $\qquad\square$

Proof of (8.33). We have

$$\widetilde{\mathbb{P}}(\{\widehat{Q}_{k,T,n}^{J,(l)}(\alpha) > Q_{k,T}^{J,(l)}(\alpha) + \varepsilon\} \cap \Omega_T^{J,(l)})$$

$$= \widetilde{\mathbb{P}}(\{\frac{1}{M_n} \sum_{m=1}^{M_n} \mathbb{1}_{\{\widehat{\Phi}_{k,T,n,m}^{J,(l)} > Q_{k,T}^{J,(l)}(\alpha)+\varepsilon\}} > \frac{M_n - (\lfloor \alpha M_n \rfloor - 1)}{M_n}\} \cap \Omega_T^{J,(l)})$$

$$\leq \widetilde{\mathbb{P}}(\{\frac{1}{M_n} \sum_{m=1}^{M_n} \mathbb{1}_{\{\widehat{\Phi}_{k,T,n,m}^{J,(l)} > Q_{k,T}^{J,(l)}(\alpha)+\varepsilon\}} - \Upsilon(\alpha,\varepsilon) > (1-\alpha) - \Upsilon(\alpha,\varepsilon)\} \cap \Omega_T^{J,(l)})$$

with $\Upsilon(\alpha,\varepsilon) = \widetilde{\mathbb{P}}(\Phi_{k,T}^{J,(l)} > Q_{k,T}^{J,(l)}(\alpha) + \varepsilon | \mathcal{X})$. As the $\mathcal{X}$-conditional distribution of $\Phi_{k,T}^{J,(l)}$ is almost surely continuous on $\Omega_T^{J,(l)}$ by Condition 8.7, we have $\Upsilon(\alpha,\varepsilon) < 1-\alpha$ for almost all $\omega \in \Omega_T^{J,(l)}$. From (8.35) we then obtain

$$\widetilde{\mathbb{P}}(\{\widehat{Q}_{k,T,n}^{J,(l)} > Q_{k,T}^{J,(l)}(\alpha) + \varepsilon\} \cap \Omega_T^{J,(l)}) \to 0 \qquad (8.42)$$

because

$$\frac{1}{M_n} \sum_{m=1}^{M_n} \mathbb{1}_{\{\widehat{\Phi}_{k,T,n,m}^{J,(l)} > Q_{k,T}^{J,(l)}(\alpha)+\varepsilon\}} - \Upsilon(\alpha,\varepsilon)$$

converges on $\Omega_T^{J,(l)}$ in probability to zero by (8.35). Analogously we derive

$$\widetilde{\mathbb{P}}(\{\widehat{Q}_{k,T,n}^{J,(l)} < Q_{k,T}^{J,(l)}(\alpha) - \varepsilon\} \cap \Omega_T^{J,(l)}) \to 0$$

which together with (8.42) yields (8.33). $\qquad\square$

Proof of (8.34). Note that it holds

$$c_{k,T,n}^{J,(l)}(\alpha) = \frac{\sqrt{n}\widehat{Q}_{k,T,n}^{J,(l)}(1-\alpha)}{nkV^{(l)}(g_4,\pi_n)_T} \tag{8.43}$$

where the denominator in (8.43) converges to $kC_T^{J,(l)} > 0$ on $\Omega_T^{C,(l)}$ as shown in the proof of Theorem 8.3. Hence it remains to show that the numerator $\sqrt{n}\widehat{Q}_{k,T,n}^{J,(l)}(1-\alpha)$ is $o_{\widetilde{\mathbb{P}}}(1)$ on $\Omega_T^{C,(l)}$. We have

$$\sqrt{n}\widehat{Q}_{n,T}^{J,(l)}(1-\alpha) \leq \frac{\sqrt{n}}{\lfloor \alpha M_n \rfloor} \sum_{m=1}^{M_n} |\widehat{\Phi}_{k,T,n,m}^{J,(l)}|. \tag{8.44}$$

Further we get using

$$\mathbb{P}(|\Delta_{i,n}^{(l)}(B^{(l)} + C^{(l)})|^p > \beta^p |\mathcal{I}_{i,n}^{(l)}|^{p\varpi}|\mathcal{S}) \leq \frac{K|\mathcal{I}_{i,n}^{(l)}|^p + K_p|\mathcal{I}_{i,n}^{(l)}|^{p/2}}{\beta^p|\mathcal{I}_{i,n}^{(l)}|^{p\varpi}}$$

$$\leq K_p |\mathcal{I}_{i,n}^{(l)}|^{p(1/2-\varpi)}$$

for $p \geq 1$ where the process $(B_t^{(l)})_{t\geq 0}$ is defined as in (8.19), the elementary inequality $\sqrt{a+b} \leq \sqrt{a} + \sqrt{b}$ which holds for all $a, b \geq 0$, iterated expectations and twice the Cauchy-Schwarz inequality

$$\mathbb{E}\big[\sqrt{n}|\widehat{\Phi}_{k,T,n,m}^{J,(l)}|\mathbb{1}_{\Omega_T^{C,(l)}}|\mathcal{S}\big]$$

$$\leq \sqrt{n} \sum_{i:t_{i,n}^{(l)}\leq T} \mathbb{E}\big[|\Delta_{i,n}^{(l)}(B^{(l)} + C^{(l)})|^3 \mathbb{1}_{\{|\Delta_{i,n}^{(l)}(B^{(l)}+C^{(l)})|>\beta|\mathcal{I}_{i,n}^{(l)}|^\varpi\}}$$

$$\times ((\tilde{\sigma}_n^{(l)}(t_{i,n}^{(l)},-))^2\tilde{\xi}_{k,n,m,-}^{(l)}(t_{i,n}^{(l)}) + (\tilde{\sigma}_n^{(l)}(t_{i,n}^{(l)},+))^2\tilde{\xi}_{k,n,m,+}^{(l)}(t_{i,n}^{(l)}))^{1/2}|\mathcal{S}\big]$$

$$\leq \sqrt{n} \sum_{i:t_{i,n}^{(l)}\leq T} \big(\mathbb{E}[(\Delta_{i,n}^{(l)}(B^{(l)} + C^{(l)}))^{12}|\mathcal{S}]$$

$$\times \mathbb{P}(|\Delta_{i,n}^{(l)}(B^{(l)} + C^{(l)})|^p > \beta^p|\mathcal{I}_{i,n}^{(l)}|^{p\varpi}|\mathcal{S}))^{1/4}$$

$$\times \big(\mathbb{E}[(\tilde{\sigma}_n^{(l)}(t_{i,n}^{(l)},-))^2 + (\tilde{\sigma}_n^{(l)}(t_{i,n}^{(l)},+))^2|\mathcal{S}]\big)^{1/2}K\sqrt{n|\pi_n|_T}$$

$$\leq K_p n\sqrt{|\pi_n|_T} \sum_{i:t_{i,n}^{(l)}\leq T} (K_q|\mathcal{I}_{i,n}^{(l)}|^{12} + |\mathcal{I}_{i,n}^{(l)}|^6)^{1/4}|\mathcal{I}_{i,n}^{(l)}|^{p(1-2\varpi)/4}(4b_n/b_n)^{1/2}$$

$$\leq K_p\sqrt{n}(|\pi_n|_T)^{1/2+p(1-2\varpi)/4}\sqrt{n} \sum_{i:t_{i,n}^{(l)}\leq T} |\mathcal{I}_{i,n}^{(l)}|^{3/2}.$$

If we pick p such that $p(1 - 2\varpi)/4 \geq 1/2$ the final bound vanishes as $n \to \infty$ by Condition 8.7(i) and 8.2(ii); compare (8.30). Hence by Lemma 2.15 we obtain $\sqrt{n}\widehat{Q}_{k,T,n}^{J,(l)}(1 - \alpha) = o_{\widehat{\mathbb{P}}}(1)$. $\qquad\square$

Proof of Corollary 8.12. First observe that following the proof of Theorem 8.8 it holds $n^{-1/2}A_{k,T,n}^{J,(l)} \overset{\mathrm{P}}{\longrightarrow} 0$ as $n \to 0$ from which we obtain

$$\sqrt{n}\rho \frac{n^{-1}A_{k,T,n}^{J,(l)}}{kV^{(l)}(g_4,\pi_n)_T} \mathbf{1}_{\Omega_T^{J,(l)}} = \rho \frac{n^{-1/2}A_{k,T,n}^{J,(l)}}{kV^{(l)}(g_4,\pi_n)_T} \mathbf{1}_{\Omega_T^{J,(l)}} \overset{\mathrm{P}}{\longrightarrow} 0. \qquad (8.45)$$

Further using Corollary 2.38(a) and Theorem 2.41(a) we obtain

$$A_{k,T,n}^{J,(l)} \mathbf{1}_{\Omega_T^{C,(l)}} \overset{\mathrm{P}}{\longrightarrow} (k^2 C_{k,T}^{J,(l)} - kC_T^{J,(l)})\mathbf{1}_{\Omega_T^{C,(l)}}$$

which yields

$$\rho \frac{A_{k,T,n}^{J,(l)}}{knV^{(l)}(g_4,\pi_n)_T} \mathbf{1}_{\Omega_T^{C,(l)}} \overset{\mathrm{P}}{\longrightarrow} \rho \frac{k^2 C_{k,T}^{J,(l)} - kC_T^{J,(l)}}{kC_T^{J,(l)}} \mathbf{1}_{\Omega_T^{C,(l)}}. \qquad (8.46)$$

The convergence (8.45) then yields together with Theorem 8.8, Proposition B.7(i) and Lemma B.5 the $\mathcal{X}$-stable convergence

$$\sqrt{n}(\widetilde{\Phi}_{k,T,n}^{J,(l)}(\rho) - 1) = \sqrt{n}\left(\Phi_{k,T,n}^{J,(l)} - \rho \frac{n^{-1}A_{k,T,n}^{J,(l)}}{kV^{(l)}(g_4,\pi_n)_T} - 1\right) \overset{\mathcal{L}-s}{\longrightarrow} \Phi_{k,T}^{J,(l)}$$

on $\Omega_T^{J,(l)}$ and hence (8.16) follows as in the proof of (8.14).

From (8.46) and Theorem 8.3 we derive

$$\left(\Phi_{k,T,n}^{J,(l)} - \rho \frac{n^{-1}A_{k,T,n}^{J,(l)}}{kV^{(l)}(g_4,\pi_n)_T} - 1\right)\mathbf{1}_{\Omega_T^{J,(l)}} \overset{\mathrm{P}}{\longrightarrow} (1 - \rho)\frac{k^2 C_{k,T}^{J,(l)} - kC_T^{J,(l)}}{kC_T^{J,(l)}} \mathbf{1}_{\Omega_T^{C,(l)}}$$

which is almost surely larger than 0 by Condition 8.10(ii) and hence (8.17) follows as in the proof of (8.15). $\qquad\square$

9 Testing for the Presence of Common Jumps

At the beginning of Chapter 8 we discussed that when modelling a univariate process in continuous time one has to decide whether to incorporate a jump component or not. The same problem occurs in a multivariate setting where multiple processes should be modelled at once. In the situation where a multivariate model with jumps has to be specified, not only the individual jump components have to be characterized but also the dependence structure of the individual jump components has to be modelled. Within the set of models where the jump components of different processes are dependent "the easiest ones to tackle are those for which the various components do not jump together"([32], Section 1. Introduction). Further, in the situation where the processes model stock prices, common jumps of multiple processes play an important role in portfolio management. In fact, systemic jumps that affect the whole market are common jumps and have a significant effect on the portfolio value while on the other hand idiosyncratic jumps only play a minor role in a sufficiently diversified portfolio. Among other things, those two reasons motivate the need for formal tests that allow to decide whether common jumps in multiple processes do occur or not.

In the following, we restrict ourselves to the bivariate situation as the question whether multiple processes jump together can be reduced to the question whether any two of these processes jump together. Like in Chapter 8 we are in general not able to answer the above question completely in the high-frequency framework because we only observe a single realization of the processes. Therefore we will develop tests based on asynchronous observations of these two processes that allow to decide *whether common jumps in the observed paths of two processes* $X_t^{(l)}(\omega)$, $t \in [0, T]$, *are present.* So mathematically we are looking for a test based on the observations $X_{t_{i,n}^{(l)}}^{(l)}(\omega)$, $i \in \mathbb{N}_0$, $l = 1, 2$, which allows to decide to which of the following two subsets of Ω

$$\Omega_T^{CoJ} = \{\exists t \in [0, T] : \Delta X_t^{(1)} \Delta X_t^{(2)} \neq 0\},$$
$$\Omega_T^{nCoJ} = (\Omega_T^{CoJ})^c = \{\Delta X_t^{(1)} \Delta X_t^{(2)} = 0 \ \forall t \in [0, T]\}$$

ω belongs. In Section 9.1 we construct a test which works under the null hypothesis that no common jumps exist, i.e. $\omega \in \Omega_T^{nCoJ}$, and in Section 9.2 we will look at

© Springer Fachmedien Wiesbaden GmbH, part of Springer Nature 2019
O. Martin, *High-Frequency Statistics with Asynchronous and Irregular Data*,
Mathematische Optimierung und Wirtschaftsmathematik | Mathematical Optimization
and Economathematics, https://doi.org/10.1007/978-3-658-28418-3_9

two tests which work under the null hypothesis that common jumps do exist, i.e. $\omega \in \Omega_T^{CoJ}$. All tests are based on the observation that by Theorem 2.3 the non-normalized functional $V(\overline{f}_{(p_1,p_2)}, \pi_n)_T$ with $\overline{f}_{(p_1,p_2)}(x,y) = |x|^{p_1}|y|^{p_2}$ converges for $p_1 \wedge p_2 \geq 2$ to a strictly positive value on Ω_T^{CoJ} and to zero on Ω_T^{nCoJ}. The upcoming tests can be understood as generalizations of two tests for common jumps based on synchronous and equidistant observations $t_{i,n}^{(l)} = i/n$ from [32].

9.1 Null Hypothesis of Disjoint Jumps

In this section we develop a statistical test which allows to decide whether common jumps are present in the paths of two stochastic processes under the null hypothesis that *no common jumps are present*.

9.1.1 Theoretical Results

In the following we will work with $V(\overline{f}_{(p_1,p_2)}, \pi_n)_T$ for $p_1 = p_2 = 2$. The function $\overline{f}_{(2,2)}(x,y) = f_{(2,2)}(x,y) = x^2 y^2$ is within the set of functions $\{\overline{f}_{(p_1,p_2)}|p_1, p_2 \geq 0\}$ the one with the smallest exponents p_1, p_2 such that $V(\overline{f}_{(p_1,p_2)}, \pi_n)_T$ converges under no additional conditions on the observation scheme; compare Theorem 2.3. Further this function has due to the quadratic structure nice mathematical properties which will be exploited in the proofs.

By Theorem 2.3 the functional $V(f_{(2,2)}, \pi_n)_T$ converges on Ω_T^{CoJ} to a strictly positive value and on Ω_T^{nCoJ} to zero. Thereby using the statistic $V(f_{(2,2)}, \pi_n)_T$ we are asymptotically able to differentiate between Ω_T^{CoJ} and Ω_T^{nCoJ}. However, the limit of $V(f_{(2,2)}, \pi_n)_T$ might also be very close to zero for $\omega \in \Omega_T^{CoJ}$ and the limit itself might be difficult to interpret. Therefore we consider the following statistic

$$\Psi_{T,n}^{DisJ} = \frac{V(f_{(2,2)}, \pi_n)_T}{\sqrt{V^{(1)}(g_4, \pi_n)_T V^{(2)}(g_4, \pi_n)_T}}$$

where we devide $V(f_{(2,2)}, \pi_n)_T$ by the geometric mean of $V^{(l)}(g_4, \pi_n)_T$, $l = 1, 2$. The statistic $\Psi_{T,n}^{DisJ}$ is scale invariant and converges on the set $\Omega_T^{J,(1)} \cap \Omega_T^{J,(2)}$ where both processes $X^{(l)}$, $l = 1, 2$, have at least one jump in $[0, T]$ to

$$\Psi_T^{DisJ} = \frac{B^*(f_{(2,2)})_T}{\sqrt{B^{(1)}(g_4)_T B^{(2)}(g_4)_T}} = \frac{\sum_{s \leq T}(\Delta X_s^{(1)})^2(\Delta X_s^{(2)})^2}{\left(\sum_{s \leq T}(\Delta X_s^{(1)})^4\right)^{1/2}\left(\sum_{s \leq T}(\Delta X_s^{(2)})^4\right)^{1/2}}.$$

The random variable Ψ_T^{DisJ} can be interpreted as a correlation coefficient of the squared jumps of $X^{(1)}$ and $X^{(2)}$ and thereby yields a measure for how big the proportion of common jumps is compared to the amount of total jumps of

$X^{(1)}$ and $X^{(2)}$. In fact by the Cauchy-Schwarz inequality for sums we find that $\Psi_T^{DisJ} \in [0,1]$ where $\Psi_T^{DisJ} = 0$ holds if and only if no common jumps exist and $\Psi_T^{nCoJ} = 1$ holds if and only if there exists some constant $c > 0$ such that

$$(\Delta X_s^{(1)})^2 = c(\Delta X_s^{(2)})^2, \quad s \in [0,T],$$

holds. Hence, the statistic $\Psi_{T,n}^{DisJ}$ does not only help to decide whether common jumps are present or not but its value is also of direct interest as it yields an indicator for how big a role common jumps play in the model.

By using the statistic $\Psi_{T,n}^{DisJ}$ instead of $V(f_{(2,2)}, \pi_n)_T$ we are restricting ourselves to testing $\omega \in \Omega_T^{CoJ}$ against $\omega \in \Omega_T^{DisJ}$ where

$$\Omega_T^{DisJ} = \Omega_T^{nCoJ} \cap \Omega_T^{J,(1)} \cap \Omega_T^{J,(2)}$$

because the limit of Ψ_T^{DisJ} is only defined on $\Omega_T^{J,(1)} \cap \Omega_T^{J,(2)}$. However, this is no real limitation because using the tests from Chapter 8 we are able to decide in advance whether ω is in $\Omega_T^{J,(1)} \cap \Omega_T^{J,(2)}$ or not. Hence by combining the tests from Chapter 8 with the upcoming test we are also able to test $\omega \in \Omega_T^{CoJ}$ against $\omega \in \Omega_T^{nCoJ}$.

Theorem 9.1. *Suppose Condition 1.3 is fulfilled. Then it holds*

$$\Psi_{T,n}^{DisJ} \mathbb{1}_{\Omega_T^{J,(1)} \cap \Omega_T^{J,(2)}} \overset{\mathbb{P}}{\longrightarrow} \Psi_T^{DisJ} \mathbb{1}_{\Omega_T^{J,(1)} \cap \Omega_T^{J,(2)}}. \tag{9.1}$$

As already discussed earlier, Theorem 9.1 yields that $\Psi_{T,n}^{DisJ}$ converges to 0 on the set Ω_T^{DisJ} and to a strictly positive limit on Ω_T^{CoJ}. So a natural test for the null $\omega \in \Omega_T^{DisJ}$ against $\omega \in \Omega_T^{CoJ}$ makes use of a critical region of the form

$$\mathcal{C}_{T,n}^{DisJ} = \{\Phi_{T,n}^{DisJ} > c_{T,n}^{DisJ}\} \tag{9.2}$$

for a suitable, possibly random sequence $(c_{T,n}^{DisJ})_{n \in \mathbb{N}}$. In order to choose $c_{T,n}^{DisJ}$ such that the test has a certain level α we need knowledge of the asymptotic behaviour of $\Phi_{T,n}^{DisJ}$ on Ω_T^{DisJ} which will be developed in form of a central limit theorem in the following.

Analogously as in (8.7) we observe

$$n(\Psi_{T,n}^{CoJ} - 0) = \frac{nV(f_{(2,2)}, \pi_n)_T}{\sqrt{V^{(1)}(g_4, \pi_n)_T V^{(2)}(g_4, \pi_n)_T}}$$

and hence it remains to find a central limit theorem for $V(f_{(2,2)}, \pi_n)_T$ which holds on Ω_T^{DisJ}. To motivate the structure of the upcoming central limit theorem we again use a toy example of the form

$$X_t^{toy} = \sigma W_t + \sum_{s \le t} \Delta N(q)_s, \quad t \ge 0.$$

Then on the subset of Ω^{DisJ} where any two jumps of $N(q)$ are further apart than $2|\pi_n|_T$ it holds

$$nV(f_{(2,2)}, \pi_n)_T = n \sum_{i,j:t_{i,n}^{(1)} \vee t_{j,n}^{(2)} \leq T} (\Delta_{i,n}^{(1)} C^{(1)})^2 (\Delta_{j,n}^{(2)} C^{(2)})^2 \mathbb{1}_{\{\mathcal{I}_{i,n}^{(1)} \cap \mathcal{I}_{j,n}^{(2)} \neq \emptyset\}} \tag{9.3}$$

$$+ \sum_{l=1,2} \sum_{p:S_{q,p}^{(l)} \leq T} (\Delta N^{(l)}(q)_{S_{q,p}^{(l)}})^2 n \sum_{i:t_{i,n}^{(3-l)} \leq T} (\Delta_{i,n}^{(3-l)} C^{toy})^2 \mathbb{1}_{\{\mathcal{I}_{i_n^{(l)}(S_{q,p}^{(l)}),n}^{(l)} \cap \mathcal{I}_{i,n}^{(3-l)} \neq \emptyset\}} \tag{9.4}$$

$$+ \sum_{l=1,2} \sum_{p:S_{q,p}^{(l)} \leq T} 2\Delta N^{(l)}(q)_{S_{q,p}^{(l)}} \Delta_{i_n^{(l)}(S_{q,p}^{(l)}),n}^{(l)} C^{toy}$$

$$\times n \sum_{i:t_{i,n}^{(3-l)} \leq T} (\Delta_{i,n}^{(3-l)} C^{toy})^2 \mathbb{1}_{\{\mathcal{I}_{i_n^{(l)}(S_{q,p}^{(l)}),n}^{(l)} \cap \mathcal{I}_{i,n}^{(3-l)} \neq \emptyset\}} \tag{9.5}$$

where $C_t^{toy} = \sigma W_t$ and $S_{q,p}^{(l)}$, $p \in \mathbb{N}$, denotes an enumeration of the jump times of $N^{(l)}(q)$. Under appropriate assumptions on the observation scheme which yield $|\mathcal{I}_{i,n}^{(l)}| = O_{\mathbb{P}}(n^{-1})$ we then obtain that (9.3) converges in probability by Theorem 2.22, that (9.4) converges stably in law using methods from Section 3.1 which are based on the stable convergence of the expressions

$$n \sum_{i:t_{i,n}^{(3-l)} \leq T} (\Delta_{i,n}^{(3-l)} C^{toy})^2 \mathbb{1}_{\{\mathcal{I}_{i_n^{(l)}(S_{q,p}^{(l)}),n}^{(l)} \cap \mathcal{I}_{i,n}^{(3-l)} \neq \emptyset\}}$$

$$= \sigma^{(3-l)} n \sum_{i:t_{i,n}^{(3-l)} \leq T} (\Delta_{i,n}^{(3-l)} \overline{W}^{(3-l)})^2 \mathbb{1}_{\{\mathcal{I}_{i_n^{(l)}(S_{q,p}^{(l)}),n}^{(l)} \cap \mathcal{I}_{i,n}^{(3-l)} \neq \emptyset\}}, \tag{9.6}$$

compare (3.23) for the definition of $\overline{W}$, and that (9.5) is asymptotically negigible.

Remark 9.2. *Recall that we have already derived a central limit theorem for $V(f_{(2,2)}, \pi_n)_T$ in Section 3.1. In fact Theorem 3.6 yields*

$$\sqrt{n}(V(f_{(2,2)}, \pi_n)_T - B^*(f_{(2,2)})_T) \xrightarrow{\mathcal{L}-s} \Phi_T^{biv}(f_{(2,2)}).$$

However, $\Phi_T^{biv}(f_{(2,2)})$ only depends on common jumps of $X^{(1)}$ and $X^{(2)}$ and in particular it holds $\Phi_T^{biv}(f_{(2,2)}) \mathbb{1}_{\Omega_T^{nCoJ}} \equiv 0$. But a central limit theorem where the asymptotic distribution is singular is useless for statistical purposes. Hence we need to develop a different central limit theorem on Ω_T^{nCoJ}. As $\sqrt{n}V(f_{(2,2)}, \pi_n)_T$ converges on Ω_T^{nCoJ} to zero we need to multiply $V(f_{(2,2)}, \pi_n)_T$ by a factor which increases faster in n than $\sqrt{n}$. In the following, we will see that by using the factor n error terms that are usually dominated by terms involving common jumps contribute to the asymptotic variance. $\qquad\square$

Before we can state the central limit theorem (whose form we motivated by the computations in the toy example above) we need to introduce some notation. Using the process $\overline{W}$ defined in (3.23) we denote

$$\eta_{n,-}^{(l)}(s) = \sum_{j:\mathcal{I}_{j,n}^{(l)}\leq T} (\Delta_{j,n}^{(l)}\overline{W}^{(l)})^2 \mathbb{1}_{\left\{\mathcal{I}_{j,n}^{(l)}\cap\mathcal{I}_{i_n^{(3-l)}(s),n}^{(3-l)} \neq\emptyset\wedge j<i_n^{(l)}(s)\right\}},$$

$$\eta_{n,+}^{(l)}(s) = \sum_{j:\mathcal{I}_{j,n}^{(l)}\leq T} (\Delta_{j,n}^{(l)}\overline{W}^{(l)})^2 \mathbb{1}_{\left\{\mathcal{I}_{j,n}^{(l)}\cap\mathcal{I}_{i_n^{(3-l)}(s),n}^{(3-l)} \neq\emptyset\wedge j>i_n^{(l)}(s)\right\}},$$

and as in (3.6) we denote

$$\delta_{n,-}^{(l)}(s) = s - t_{i_n^{(l)}(s)-1,n}^{(l)}, \qquad \delta_{n,+}^{(l)}(s) = t_{i_n^{(l)}(s),n}^{(l)} - s,$$

for $l = 1,2$ and $s \geq 0$; see Figure 9.1 for an illustration. Using the notation from (2.28) we then obtain the identity

$$\sum_{j:\mathcal{I}_{j,n}^{(l)}\cap\mathcal{I}_{i_n^{(3-l)}(s),n}^{(3-l)}\neq\emptyset} (\Delta_{j,n}^{(l)}\overline{W}^{(l)})^2 = \eta_{n,-}^{(l)}(s) + \left[(\delta_{n,-}^{(l)}(s))^{1/2}((\overline{W}_s^{(l)} - \overline{W}_{\tau_{n,-}^{(l)}(s)}^{(l)})/(\delta_{n,-}^{(l)}(s))^{1/2})\right.$$

$$+ (\delta_{n,+}^{(l)}(s))^{1/2}((\overline{W}_{\tau_{n,+}^{(l)}(s)}^{(l)} - \overline{W}_s^{(l)})/(\delta_{n,+}^{(l)}(s))^{1/2})\bigg]^2 + \eta_{n,+}^{(l)}(s)$$

for the sum of the squared increments of $\overline{W}^{(l)}$ over intervals $\mathcal{I}_{j,n}^{(l)}$ which overlap with the observation interval $\mathcal{I}_{i_n(s),n}^{(3-l)}$ containing s. In the computations using our toy example in (9.6) this sum occured for s replaced with a jump time $S_{q,p}^{(3-l)}$ of $X^{(3-l)}$. We additionally distinguish between increments of $\overline{W}^{(l)}$ before and after s to allow for different volatilities immediately before and after s due to a volatility jump at time s, and we write

$$Z_n^{DisJ,(l)}(s) = \left(n\eta_{n,-}^{(l)}(s), n\eta_{n,+}^{(l)}(s), n\delta_{n,-}^{(l)}(s), n\delta_{n,+}^{(l)}(s)\right)^*$$

to shorten notation.

The following condition sums up the assumptions on the asymptotics of the sequence of observation schemes $(\pi_n)_{n\in\mathbb{N}}$ which are needed to obtain a central limit theorem.

Condition 9.3. *Suppose that Condition 1.3 is fulfilled and that the observation times are exogeneous. Further,*

(i) *the functions $t \mapsto G_{2,2}^n(t)$ and $t \mapsto H_{0,0,4}^n(t)$, see (2.39) for their definition, converge pointwise on $[0,T]$ in probability to increasing continuous functions $G_{2,2}, H_{0,0,4}: [0,\infty) \to [0,\infty)$.*

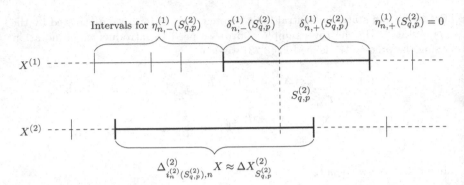

Figure 9.1: Illustration of $Z_n^{DisJ,(1)}(S_{q,p}^{(2)})$ at a jump time $S_{q,p}^{(2)}$ of $X^{(2)}$.

(ii) For all $P_1, P_2 \in \mathbb{N}$ and all bounded continuous functions $g \colon \mathbb{R}^{P_1+P_2} \to \mathbb{R}$ and $h_p^{(l)} \colon \mathbb{R}^4 \to \mathbb{R}$, $p = 1, \ldots, P_l$, $l = 1, 2$, the integral

$$\int_{[0,T]^{k_1+k_2}} g(x_1, \ldots, x_{P_1}, x'_1, \ldots, x'_{P_2}) \mathbb{E}\left[\prod_{p=1}^{P_1} h_p^{(1)}\big(Z_n^{DisJ,(1)}(x_p)\big)\right.$$

$$\left.\times \prod_{p=1}^{P_2} h_p^{(2)}\big(Z_n^{DisJ,(2)}(x'_p)\big)\right] dx_1 \ldots dx_{P_1} dx'_1 \ldots dx'_{P_2}$$

converges to

$$\int_{[0,T]^{k_1+k_2}} g(x_1, \ldots, x_{P_1}, x'_1, \ldots, x'_{P_2}) \prod_{p=1}^{P_1} \int_{\mathbb{R}} h_p^{(1)}(y) \Gamma^{DisJ,(1)}(x_p, dy)$$

$$\times \prod_{p=1}^{P_2} \int_{\mathbb{R}} h_p^{(2)}(y) \Gamma^{DisJ,(2)}(x'_p, dy) dx_1 \ldots dx_{P_1} dx'_1 \ldots dx'_{P_2} \quad (9.7)$$

as $n \to \infty$. Here, $\Gamma^{DisJ,(l)}(\cdot, dy)$, $l = 1, 2$, are families of probability measures on $[0, \infty)^2 \times (0, \infty)^2$ such that the first moments are uniformly bounded.

Condition 9.3(i) is needed to obtain convergence of (9.3) in the motivating example which we will prove using Theorem 2.22. Further, part (ii) is a condition which in similar form already occured in Section 3.1 and which is necessary to obtain stable convergence of the term (9.4). Note that as we are going to derive a central limit theorem restricted to Ω_T^{DisJ} that the random variables $Z_n^{DisJ,(l)}(S_{q,p}^{(l)})$ and $Z_n^{DisJ,(l')}(S_{q,p}^{(l')})$ for $S_{q,p}^{(l)} \neq S_{q,p}^{(l')}$ or $l \neq l'$ are asymptotically $\mathcal{S}$-conditionally independent. This explains the factorization of the expectations with respect to $\Gamma^{DisJ,(l)}$ in (9.7). Here, the form is somewhat simpler compared to Condition 3.5(ii) because we do not have to keep track of the common distribution of $Z_n^{DisJ,(1)}(s)$ and $Z_n^{DisJ,(2)}(s)$.

Next, we define the components occuring in the asymptotic variance of the upcoming central limit theorem. We denote

$$\widetilde{C}_T^{DisJ} = \int_0^T (\sigma_s^{(1)}\sigma_s^{(2)})^2 dG_{2,2}(s) + \int_0^T 2(\rho_s\sigma_s^{(1)}\sigma_s^{(2)})^2 dH_{0,0,4}(s)$$

and

$$\widetilde{D}_T^{DisJ} = \sum_{s \leq T} \left((\Delta X_s^{(1)})^2 R^{DisJ,(2)}(s) + (\Delta X_s^{(2)})^2 R^{DisJ,(1)}(s) \right),$$

where S_p, $p \in \mathbb{N}$, is an enumeration of the jump times of X. Here, $R^{DisJ,(l)}(s)$ is defined by

$$R^{DisJ,(l)}(s) = (\sigma_{s-}^{(2)})^2 \eta_-^{(l)}(s) + \left(\sigma_{s-}^{(l)}(\delta_-^{(l)}(s))^{1/2} U_-^{(l)}(s) + \sigma_s^{(l)}(\delta_+^{(l)}(s))^{1/2} U_+^{(l)}(s) \right)^2$$
$$+ (\sigma_s^{(2)})^2 \eta_+^{(l)}(s), \quad s \in [0,T], \quad l = 1,2,$$

where $Z^{DisJ,(l)}(s) = (\eta_-^{(l)}(s), \eta_+^{(l)}(s), \delta_-^{(l)}(s), \delta_+^{(l)}(s))^*$ are random variables defined on an extended probability space $(\widetilde{\Omega}, \widetilde{\mathcal{F}}, \widetilde{\mathbb{P}})$ which are independent of each other and of $\mathcal{F}$. Their distribution is given by

$$\widetilde{\mathbb{P}}^{Z^{DisJ,(l)}(s)}(dy) = \Gamma^{DisJ,(l)}(s,dy), \quad s \in [0,T].$$

Further, the $U_-^{(l)}(s), U_+^{(l)}(s)$ are i.i.d. $\mathcal{N}(0,1)$ random variables which are independent of all previously introduced random variables and defined on $(\widetilde{\Omega}, \widetilde{\mathcal{F}}, \widetilde{\mathbb{P}})$ as well. The $Z^{DisJ,(l)}(s)$ and $U_-^{(l)}(s)$, $U_+^{(l)}(s)$ are independent of each other and independent for different values s. Note that here we chose $Z^{DisJ,(1)}(s)$ and $Z^{DisJ,(2)}(s)$ to be independent although $Z_n^{DisJ,(1)}(s)$ and $Z_n^{DisJ,(2)}(s)$ are in general not independent. However, as we derive a central limit theorem only on Ω_T^{DisJ} we have $\Delta^{(1)} X_s^{(1)} = 0$ or $\Delta^{(2)} X_s^{(2)} = 0$ for any s. Hence only one of $R^{DisJ,(1)}(s)$, $R^{DisJ,(2)}(s)$ and therefore also only one of $Z^{DisJ,(1)}(s)$, $Z^{DisJ,(2)}(s)$ enters the term $\widetilde{D}_T^{DisJ}$ for any $s \geq 0$. This property allows to conclude that $\widetilde{D}_T^{DisJ}$ does not depend on the common distribution of $Z^{DisJ,(1)}(s)$, $Z^{DisJ,(2)}(s)$ at all which is why we may assume that they are independent for convenience.

Theorem 9.4. *If Condition 9.3 is fulfilled we have the $\mathcal{X}$-stable convergence*

$$n\Psi_{T,n}^{DisJ} \xrightarrow{\mathcal{L}-s} \Phi_T^{DisJ} = \frac{\widetilde{C}_T^{DisJ} + \widetilde{D}_T^{DisJ}}{\sqrt{B^{(1)}(g_4)_T B^{(2)}(g_4)_T}} \quad (9.8)$$

on the set Ω_T^{DisJ}.

For the notion what it means for a sequence of random variables to converge $\mathcal{X}$-stably in law on a subset of Ω compare the paragraph following Theorem 8.8. The term $\widetilde{C}_T^{DisJ}$ in the limit Φ_T^{DisJ} corresponds to (9.3) in the motivating example and is solely due to the continuous martingale part of X. Further, the term $\widetilde{D}_T^{DisJ}$ corresponds to (9.4) and contains the limit of cross terms of the continuous martingale parts and of idiosyncratic jumps.

To illustrate Condition 9.3 and Theorem 9.4 we are going to discuss both of them for the two prominent observation schemes of equidistant observation times and Poisson sampling which feature as the major examples throughout this book.

Example 9.5. *In the setting of equidistant and synchronous observation times* $t_{i,n}^{(l)} = i/n$ *it holds* $|\pi_n|_T = n^{-1}$ *and hence Condition 1.3 is trivially fulfilled. Furthermore, it holds*

$$G_{2,2}^n(t) = H_{0,0,4}^n(t) = n \sum_{i=1}^{\lfloor t/n \rfloor} \left(1/n\right)^2 \to t$$

which yields Condition 9.3(ii). We also have $\eta_{n,-}^{(l)}(s) = \eta_{n,+}^{(l)}(s) = 0$ *and hence for the convergence of* $Z_n^{DisJ,(l)}(s)$ *it remains to consider* $\delta_{n,-}^{(l)}(s)$, $\delta_{n,+}^{(l)}(s)$. *Then Condition 9.3(iii) is fulfilled as shown in Example 3.3 and we obtain*

$$Z^{DisJ,(l)}(s) \sim (0, 0, \kappa^{(l)}(s), 1 - \kappa^{(l)}(s))$$

for i.i.d. $\mathcal{U}[0,1]$*-distributed random variables* $\kappa^{(l)}(s)$, $s \in [0,t]$, $l = 1, 2$.

Hence in the setting of equidistant and synchronous observation times the terms $\widetilde{C}_T^{DisJ}$ *and* $\widetilde{D}_T^{DisJ}$ *in (9.8) are of the form*

$$\widetilde{C}_T^{DisJ} = \int_0^T \left((\sigma_s^{(1)}\sigma_s^{(2)})^2 + 2(\rho_s \sigma_s^{(1)}\sigma_s^{(2)})^2 \right) ds,$$

$$\widetilde{D}_T^{DisJ} = \sum_{s \leq T} \left[(\Delta X_s^{(1)})^2 \left(\sigma_{s-}^{(2)}(\kappa^{(2)}(s))^{1/2} U_-^{(2)}(s) + \sigma_s^{(2)}(1 - \kappa^{(2)}(s))^{1/2} U_+^{(2)}(s) \right)^2 \right.$$
$$\left. + (\Delta X_s^{(2)})^2 \left(\sigma_{s-}^{(1)}(\kappa^{(1)}(s))^{1/2} U_-^{(1)}(s) + \sigma_s^{(1)}(1 - \kappa^{(1)}(s))^{1/2} U_+^{(1)}(s) \right)^2 \right],$$

These terms are identical to the corresponding terms C_T *and* $\widetilde{D}_T$ *in (3.12) and (3.14) of [32], and Theorem 9.4 becomes Theorem 4.1(a) of [32] in this specific setting.* □

Example 9.6. *In the case of Poisson sampling, which is introduced in Definition 5.1, Condition 1.3 is satisfied by (5.2). Further, Condition 9.3(ii) is, as shown in Example 5.7, fulfilled with*

$$G_{2,2}^n(t) \xrightarrow{\mathbb{P}} \left(\frac{2}{\lambda_1} + \frac{2}{\lambda_2} \right)t, \quad H_{0,0,4}^n(t) \xrightarrow{\mathbb{P}} \frac{2}{\lambda_1 + \lambda_2}t.$$

Finally, part (iii) of Condition 9.3 is satisfied by Lemma 5.10 and the discussion following Lemma 5.10 yields that the distributions $\Gamma^{DisJ,(l)}(s,dy)$ do not depend on s and can be constructed as follows: By symmetry we focus on $Z^{DisJ,(1)}(s)$ only. Let $E_{k,-}^{(1)}, E_{k,+}^{(1)} \sim Exp(\lambda_1)$, $k \in \mathbb{N}$, and $E_{-}^{(2)}, E_{+}^{(2)} \sim Exp(\lambda_2)$ be independent exponentially distributed random variables. Then, after rescaling, the lengths of the observation intervals $\mathcal{I}_{i_n^{(l)}(s),n}^{(l)}$, $l = 1,2$, containing s are by the memorylessness of the exponential distribution asymptotically distributed like $E_{1,-}^{(1)} + E_{1,+}^{(1)}$ and $E_{-}^{(2)} + E_{+}^{(2)}$ while the other intervals in $\eta_-^{(1)}(s)$ and $\eta_+^{(1)}(s)$ are $Exp(\lambda_1)$-distributed. Then, if $(U_{k,-})_{k\in\mathbb{N}}$, $(U_{k,+})_{k\in\mathbb{N}}$ are i.i.d. $\mathcal{N}(0,1)$ random variables, it is easy to deduce that

$$Z^{DisJ,(1)}(s) \stackrel{\mathcal{L}}{=} (\eta_-^{(1)}(s), \eta_+^{(1)}(s), E_{1,-}^{(1)}, E_{1,+}^{(1)})^*$$

holds, where we set

$$\eta_-^{(1)}(s) = \sum_{k=2}^{\infty} E_{k,-}^{(1)}(U_{k,-})^2 \mathbf{1}_{\{\sum_{j=1}^{k-1} E_{k,-}^{(1)} < E_-^2\}},$$

$$\eta_+^{(1)}(s) = \sum_{k=2}^{\infty} E_{k,+}^{(1)}(U_{k,+})^2 \mathbf{1}_{\{\sum_{j=1}^{k-1} E_{k,+}^{(1)} < E_+^2\}}.$$

$\square$

To construct a statistical test based on the central limit theorem shown in Theorem 9.4 we need to get hold of the $\mathcal{X}$-conditional distribution of

$$\Phi_T^{DisJ} = \frac{\widetilde{C}_T^{DisJ} + \widetilde{D}_T^{DisJ}}{\sqrt{B^{(1)}(g_4)_T B^{(2)}(g_4)_T}}.$$

Here, the terms $B^{(l)}(g_4)_T$, $l = 1,2$, can be consistently estimated by the functionals $V^{(l)}(g_4, \pi_n)_T$. Further, in the proof of Theorem 9.4 it is shown that the term $\widetilde{C}_T^{DisJ}$ is the limit of

$$n \sum_{i,j: t_{i,n}^{(1)} \vee t_{j,n}^{(2)} \leq T} (\Delta_{i,n}^{(1)} C^{(1)})^2 (\Delta_{j,n}^{(2)} C^{(2)})^2 \mathbf{1}_{\{\mathcal{I}_{i,n}^{(1)} \cap \mathcal{I}_{j,n}^{(2)} \neq \emptyset\}}. \tag{9.9}$$

Hence (9.9) constitutes a natural estimator for $\widetilde{C}_T^{DisJ}$. Unfortunately we only observe increments of X and not of C and therefore the use of (9.9) as an estimator for $\widetilde{C}_T^{DisJ}$ is not feasible. However, in Section 2.3 we have seen that using truncated increments we are able to isolate the contribution of the continuous martingale part C from the increments of X. Motivated by this observation we will use the functional

$$
\overline{V}_-(4, f_{(2,2)}, \pi_n, (\beta, \varpi))_T
$$

$$
= n \sum_{i,j : t_{i,n}^{(1)} \vee t_{j,n}^{(2)} \leq T} \left(\Delta_{i,n}^{(1)} X^{(1)}\right)^2 \left(\Delta_{j,n}^{(2)} X^{(2)}\right)^2 \mathbb{1}_{\{\mathcal{I}_{i,n}^{(1)} \cap \mathcal{I}_{j,n}^{(2)} \neq \emptyset\}}
$$

$$
\times \mathbb{1}_{\{|\Delta_{i,n}^{(1)} X^{(1)}| \leq \beta |\mathcal{I}_{i,n}^{(1)}|^{\varpi}, |\Delta_{j,n}^{(2)} X^{(2)}| \leq \beta |\mathcal{I}_{j,n}^{(2)}|^{\varpi}\}}
$$

for parameters $\beta > 0$ and $\varpi \in (0,1)$. Although $\overline{V}_-(4, f_{(2,2)}, \pi_n, (\beta, \varpi))_T$ converges in general only if the process X is continuous, compare part b) of Theorem 2.38, we will obtain

$$
\overline{V}_-(4, f_{(2,2)}, \pi_n, (\beta, \varpi))_T \mathbb{1}_{\Omega_T^{DisJ}} \xrightarrow{\mathbb{P}} \widetilde{C}_T^{DisJ} \mathbb{1}_{\Omega_T^{DisJ}} \tag{9.10}
$$

restricted on the set Ω_T^{DisJ} also for processes X with jumps. The convergence (9.10) will be formally proven in Section 9.1.3.

To estimate the $\mathcal{X}$-conditional distribution of Φ_T^{DisJ} it then remains to estimate the $\mathcal{X}$-conditional distribution of $\widetilde{D}_T^{DisJ}$. Here, $\widetilde{D}_T^{DisJ}$ is the only term in Φ_T^{DisJ} which is not $\mathcal{X}$-measurable. To estimate the $\mathcal{X}$-conditional distribution of $\widetilde{D}_T^{DisJ}$ we will use the methods introduced in Chapter 4. Let $(K_n)_{n \in \mathbb{N}}$ and $(M_n)_{n \in \mathbb{N}}$ be deterministic sequences of integers which tend to infinity and define

$$
\hat{\eta}_{n,m,-}^{(l)}(s) = n \sum_{i : \mathcal{I}_{i,n}^{(l)} \leq T} |\mathcal{I}_{i,n}^{(l)}| (U_{n,i,m}^{(l)})^2
$$

$$
\times \mathbb{1}_{\{\mathcal{I}_{i,n}^{(l)} \cap \mathcal{I}_{i_n^{(3-l)}(s) + V_{n,m}^{(l,3-l)}(s), n}^{(3-l)} \neq \emptyset \wedge i < i_n^{(l)}(s) + V_{n,m}^{(l,l)}(s)\}},
$$

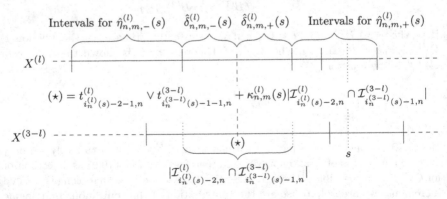

Figure 9.2: Realization of $\widehat{Z}_{n,m}^{DisJ,(l)}(s)$ for $V_{n,m}^{(l,l)} = -2$, $V_{n,m}^{(l,3-l)}(s) = -1$.

$$\hat{\eta}_{n,m,+}^{(l)}(s) = n \sum_{i:\mathcal{I}_{i,n}^{(l)} \leq T} |\mathcal{I}_{i,n}^{(l)}| (U_{n,i,m}^{(l)})^2$$

$$\times \mathbb{1}_{\{\mathcal{I}_{i,n}^{(l)} \cap \mathcal{I}_{i_n^{(3-l)}(s)+V_{n,m}^{(l,3-l)}(s),n}^{(3-l)} \neq \emptyset \wedge i > i_n^{(l)}(s)+V_{n,m}^{(l,l)}(s)\}},$$

$$\hat{\delta}_{n,m,-}^{(l)}(s) = n\Big(\kappa_{n,m}^{(l)}(s)\big|\mathcal{I}_{i_n^{(l)}(s)+V_{n,m}^{(l,l)}(s),n}^{(l)} \cap \mathcal{I}_{i_n^{(3-l)}(s)+V_{n,m}^{(l,3-l)}(s),n}^{(3-l)}\big|$$

$$+ \big(t_{i_n^{(3-l)}(s)+V_{n,m}^{(l,3-l)}(s)-1,n}^{(3-l)} - t_{i_n^{(l)}(s)+V_{n,m}^{(l,l)}(s)-1,n}^{(l)}\big)^+\Big),$$

$$\hat{\delta}_{n,m,+}^{(l)}(s) = n\big|\mathcal{I}_{V_{n,m}^{(l,l)}(s),n}^{(l)}\big| - \hat{\delta}_{n,m,-}^{(l)}(s),$$

$l = 1, 2$. Here, the random variables $U_{n,i,m}^{(l)}$ are i.i.d. $\mathcal{N}(0,1)$-distributed, the $\kappa_{n,m}^{(l)}(s)$ are i.i.d. $\mathcal{U}[0,1]$-distributed random variables and the $V_{n,m}^{(l,l)}(s), V_{n,m}^{(l,3-l)}(s)$ are distributed according to

$$\widetilde{\mathbb{P}}\big((V_{n,m}^{(l,l)}(s), V_{n,m}^{(l,3-l)}(s)) = (k_1, k_2) \big| \mathcal{S}\big) = \big|\mathcal{I}_{i_n^{(l)}(s)+k_1,n}^{(l)} \cap \mathcal{I}_{i_n^{(3-l)}(s)+k_2,n}^{(3-l)}\big|$$

$$\times \Big(\sum_{j_1 \in \mathbb{Z}, |j_2| \leq K_n} \big|\mathcal{I}_{i_n^{(l)}(s)+j_1,n}^{(l)} \cap \mathcal{I}_{i_n^{(3-l)}(s)+j_2,n}^{(3-l)}\big| \Big)^{-1}, \quad (k_1, k_2) \in \mathbb{Z} \times \{-K_n, \ldots, K_n\},$$

and the $(V_{n,m}^{(l,l)}(s), V_{n,m}^{(l,3-l)}(s))$ are $\mathcal{S}$-conditionally independent for different values of $m \in \{1, \ldots, M_n\}$. All newly introduced random variables are defined on the extended probability space $(\widetilde{\Omega}, \widetilde{\mathcal{F}}, \widetilde{\mathbb{P}})$ as well. By construction, the quantity

$$\widehat{Z}_{n,m}^{DisJ,(l)}(s) = (\hat{\eta}_{n,m,-}^{(l)}(s), \hat{\eta}_{n,m,+}^{(l)}(s), \hat{\delta}_{n,m,-}^{(l)}(s), \hat{\delta}_{n,m,+}^{(l)}(s))^*$$

then equals a mixture of the $Z_n^{DisJ,(l)}(u)$ for $u \in [t_{i_n^{(3-l)}(s)-K_n-1,n}^{(3-l)}, t_{i_n^{(3-l)}(s)+K_n,n}^{(3-l)}]$ where the rescaled increments of the Brownian motion $\overline{W}^{(l)}$ in $\eta_{n,-}^{(l)}(u), \eta_{n,+}^{(l)}(u)$ are replaced by independent normally distributed random variables. By the choice of the distribution for $(V_{n,m}^{(l,l)}(s), V_{n,m}^{(l,3-l)}(s))$ and $\kappa_{n,m}^{(l)}(s)$ we obtain further that u in the mixture is chosen uniformly from the interval $[t_{i_n^{(3-l)}(s)-K_n-1,n}^{(3-l)}, t_{i_n^{(3-l)}(s)+K_n,n}^{(3-l)}]$.

Using the estimators $\widehat{Z}_{n,m}^{DisJ,(l)}(s)$ defined above we set for $m = 1, \ldots, M_n$

$$\widehat{D}_{T,n,m}^{DisJ} = \sum_{l=1,2} \sum_{i:t_{i,n}^{(l)} \leq T} (\Delta_{i,n}^{(l)} X^{(l)})^2 \mathbb{1}_{\{|\Delta_{i,n}^{(l)} X^{(l)}| > \beta |\mathcal{I}_{i,n}^{(l)}|^{\varpi}\}} \widehat{R}_{n,m}^{DisJ,(3-l)}(t_{i,n}^{(l)})$$

with

$$\widehat{R}_{n,m}^{DisJ,(l)}(s) = \big(\tilde{\sigma}_n^{(l)}(s,-)\big)^2 \hat{\eta}_{n,m,-}^{(l)}(s) + \big(\tilde{\sigma}_n^{(l)}(s,-)(\hat{\delta}_{n,m,-}^{(l)}(s))^{1/2} U_{n,m,-}^{(l)}(s)$$

$$+ \tilde{\sigma}_n^{(l)}(s,+)(\hat{\delta}_{n,m,+}^{(l)}(s))^{1/2} U_{n,m,+}^{(l)}(s)\big)^2 + \big(\tilde{\sigma}_n^{(l)}(s,+)\big)^2 \hat{\eta}_{n,m,+}^{(l)},$$

$s \in [0,T]$, $l = 1,2$. Hence to obtain $\mathcal{X}$-conditionally independent and identically distributed random variables $\widehat{D}^{DisJ}_{T,n,m}$, $m = 1, \ldots, M_n$, whose $\mathcal{F}$-conditional distribution approximates the $\mathcal{X}$-conditional distribution of $\widetilde{D}^{DisJ}_T$ we estimate jumps of $X^{(l)}$ by unusually large increments, the spot volatility by the estimators $\hat{\sigma}^{(l)}_n(s,-)$, $\hat{\sigma}^{(l)}_n(s,+)$ introduced in Chapter 6 and $Z^{DisJ,(l)}(s)$ by $\widehat{Z}^{DisJ,(l)}_{n,m}(s)$. To estimate quantiles of the $\mathcal{X}$-conditional distribution of $\widetilde{D}^{DisJ}_T$ we define similarly as in (8.13) for $\alpha \in [0,1]$

$$\widehat{Q}^{DisJ}_{T,n}(\alpha) = \widehat{Q}_\alpha(\{\widehat{D}^{DisJ}_{T,n,m} | m = 1, \ldots, M_n\}).$$

We will see that $\widehat{Q}^{DisJ}_{T,n}(\alpha)$ consistently estimates the $\mathcal{X}$-conditional α-quantile $Q^{DisJ}_T(\alpha)$ of $\widetilde{D}^{DisJ}_T$ on Ω^{DisJ}_T. Here, $Q^{DisJ}_T(\alpha) \in [0,\infty]$ is defined as the (under the upcoming condition unique) $\mathcal{X}$-measurable $[0,\infty]$-valued random variable fulfilling

$$\mathbb{P}(\widetilde{D}^{DisJ}_T \le Q^{DisJ}_T(\alpha)|\mathcal{X})(\omega) = \alpha, \quad \omega \in \Omega^{DisJ}_T,$$

and we set $(Q^{DisJ}_T(\alpha))(\omega) = 0$ for $\omega \in (\Omega^{DisJ}_T)^c$ for completeness.

The following condition summarizes all additional assumptions we need in order to obtain an asymptotic test.

Condition 9.7. *Suppose that Condition 9.3 is fulfilled and assume that the set $\{s \in [0,T] : \sigma^{(1)}_s \sigma^{(2)}_s = 0\}$ is almost surely a Lebesgue null set. The sequence $(b_n)_{n\in\mathbb{N}}$ fulfils $b_n \to 0$, $|\pi_n|_T/b_n \xrightarrow{\mathbb{P}} 0$ and $(K_n)_{n\in\mathbb{N}}$, $(M_n)_{n\in\mathbb{N}}$ are sequences of integers converging to infinity with $K_n/n \xrightarrow{\mathbb{P}} 0$. Additionally,*

(i) it holds

$$\widetilde{\mathbb{P}}(|\widetilde{\mathbb{P}}(\widehat{Z}^{DisJ,(l_p)}_{n,1}(s_p) \le x_p, \ p = 1,\ldots,P|\mathcal{S})$$
$$- \widetilde{\mathbb{P}}(Z^{DisJ,(l_p)}(s_p) \le x_p, \ p = 1,\ldots,P)| > \varepsilon) \to 0 \quad (9.11)$$

as $n \to \infty$, for all $\varepsilon > 0$, $P \in \mathbb{N}$, $x = (x_1,\ldots,x_P) \in \mathbb{R}^{4\times P}$, $l_p \in \{1,2\}$ and $s_p \in (0,T)$, $p = 1,\ldots,P$, with $s_p \ne s_{p'}$ for $p \ne p'$.

(ii) The volatility process σ is itself an $\mathbb{R}^{2\times 2}$-valued Itô semimartingale, i.e. a process of similar form as (1.1).

The assumption $\{s \in [0,T] : \sigma^{(1)}_s \sigma^{(2)}_s = 0\}$ ensures in particular that the volatility does not vanish, which yields $\widetilde{D}^{DisJ}_T > 0$ almost surely. Further (9.11) yields that the empirical distribution on the

$$\widehat{Z}^{DisJ,(l)}_{n,m}(s) = (\hat{\eta}^{(l)}_{n,m,-}(s), \eta^{(l)}_{n,m,+}(s), \delta^{(l)}_{n,m,-}(s), \delta^{(l)}_{n,m,+}(s)),$$

$m = 1, \ldots, M_n$, converges to the non-degenerate distribution of $Z^{DisJ,(l)}(s)$ which is essential for the bootstrap method to work. Part (ii) of Condition 9.7 is like in Chapters 4 and 8 needed to ensure the consistency of the estimators $\tilde{\sigma}^{(l)}(s, -)$, $\tilde{\sigma}^{(l)}(s, +)$ for the spot volatilites.

Theorem 9.8. *If Condition 9.7 is satisfied, the test defined in (9.2) with*

$$c_{T,n}^{DisJ}(\alpha) = \frac{\overline{V}_-(4, f_{(2,2)}, \pi_n, (\beta, \varpi))_T + \widehat{Q}_{T,n}^{DisJ}(1 - \alpha)}{n\sqrt{V^{(1)}(g_4, \pi_n)_T V^{(2)}(g_4, \pi_n)_T}}, \quad \alpha \in [0, 1],$$

has asymptotic level α in the sense that we have

$$\widetilde{\mathbb{P}}\left(\Psi_{T,n}^{DisJ} > c_{T,n}^{DisJ}(\alpha)\big| F^{DisJ}\right) \to \alpha \quad \alpha \in [0, 1], \tag{9.12}$$

for all $F^{DisJ} \subset \Omega_T^{DisJ}$ with $\mathbb{P}(F^{DisJ}) > 0$. Because of

$$\widetilde{\mathbb{P}}\left(\Psi_{T,n}^{DisJ} > c_{T,n}^{DisJ}(\alpha)\big| F^{CoJ}\right) \to 1 \quad \alpha \in (0, 1], \tag{9.13}$$

for all $F^{CoJ} \subset \Omega_T^{CoJ}$ with $\mathbb{P}(F^{CoJ}) > 0$ it is consistent as well.

Although n appears in the definition of the critical value $c_{T,n}^{DisJ}(\alpha)$ it is not directly needed for the computation of $c_{T,n}^{DisJ}(\alpha)$ since it also occurs linearly in both $\overline{V}_-(4, f_{(2,2)}, \pi_n, (\beta, \varpi))_T$ and $\widehat{Q}_{T,n}^{DisJ}(1 - \alpha)$. However, it enters indirectly through the choice of b_n, K_n, M_n, but for which usually just a rough idea of the magnitude of n is needed.

Example 9.9. *Condition 9.7(i) is fulfilled in the setting of synchronous equidistant observation times discussed in Example 9.5 where our estimator $\widehat{Q}_{T,n}^{DisJ}(1 - \alpha)$ equals the estimator $Z_n^{(d)}(\alpha)$ defined in (5.10) of [32] for $N_n = M_n$ and any choice of K_n (not necessarily converging to infinity). Note that $K_n \to \infty$ is needed to estimate the distribution of the observation intervals around some time s from the realization of multiple asymptotically similar and independent observation intervals around times s' close to s. Hence, because here all observation intervals are identical the choice $K_n = 1, n \in \mathbb{N}$, is sufficient.* □

Example 9.10. *Condition 9.7 is fulfilled in the Poisson setting discussed in Example 9.6 if we choose $(b_n)_{n \in \mathbb{N}}$ appropriately. Indeed $|\pi_n|_T/b_n \xrightarrow{\mathbb{P}} 0$ follows from (5.2) for every $b_n = O(n^{-\gamma})$ with $\gamma \in (0, 1)$ and that (9.11) holds is shown by Lemma 5.12.* □

9.1.2 Simulation Results

As in Section 8.2 we conduct a simulation study to verify the finite sample properties of the introduced methods. Our benchmark model is the one from Section 6 of [32],

Table 9.1: Parameter settings for the simulation.

Case	ρ	α_1	κ_1	l_1	h_1	α_2	κ_2	l_1	h_1	α_3	κ_3	l_3	h_3
I-j	0.0	0.00				0.00				0.01	1	0.05	0.7484
II-j	0.0	0.00				0.00				0.01	5	0.05	0.3187
III-j	0.0	0.00				0.00				0.01	25	0.05	0.1238
I-m	0.5	0.01	1	0.05	0.7484	0.01	1	0.05	0.7484	0.01	1	0.05	0.7484
II-m	0.5	0.01	5	0.05	0.3187	0.01	5	0.05	0.3187	0.01	5	0.05	0.3187
III-m	0.5	0.01	25	0.05	0.1238	0.01	25	0.05	0.1238	0.01	25	0.05	0.1238
I-d0	0.0	0.01	1	0.05	0.7484	0.01	1	0.05	0.7484				
II-d0	0.0	0.01	5	0.05	0.3187	0.01	5	0.05	0.3187				
III-d0	0.0	0.01	25	0.05	0.1238	0.01	25	0.05	0.1238				
I-d1	1.0	0.01	1	0.05	0.7484	0.01	1	0.05	0.7484				
II-d1	1.0	0.01	5	0.05	0.3187	0.01	5	0.05	0.3187				
III-d1	1.0	0.01	25	0.05	0.1238	0.01	25	0.05	0.1238				

because by using the same configuration as in their paper we are able to directly compare our approach to the case of equidistant and synchronous observations. Further the model can be understood as a natural extension of (8.18) to the bivariate setting. The model for X is given by

$$dX_t^{(1)} = X_t^{(1)} \sigma_1 dW_t^{(1)} + \alpha_1 \int_{\mathbb{R}} X_{t-}^{(1)} x_1 \mu_1(dt, dx_1) + \alpha_3 \int_{\mathbb{R}} X_{t-}^{(1)} x_3 \mu_3(dt, dx_3),$$

$$dX_t^{(2)} = X_t^{(2)} \sigma_2 dW_t^{(2)} + \alpha_2 \int_{\mathbb{R}} X_{t-}^{(2)} x_2 \mu_2(dt, dx_2) + \alpha_3 \int_{\mathbb{R}} X_{t-}^{(2)} x_3 \mu_3(dt, dx_3),$$

where $[W^{(1)}, W^{(2)}]_t = \rho t$ and the Poisson measures μ_i are independent of each other and have predictable compensators ν_i of the form

$$\nu_i(dt, dx_i) = \kappa_i \frac{\mathbb{1}_{[-h_i, -l_i] \cup [l_i, h_i]}(x_i)}{2(h_i - l_i)} dt dx_i$$

where $0 < l_i < h_i$ for $i = 1, 2, 3$, and the initial values are $X_0 = (1, 1)^*$. We consider the same twelve parameter settings which were discussed in [32] of which six allow for common jumps and six do not. In the case where common jumps are possible, we only use the simulated paths which contain common jumps. For the parameters we set $\sigma_1^2 = \sigma_2^2 = 8 \times 10^{-5}$ in all scenarios and choose the parameters for the Poisson measures such that the contribution of the jumps to the total variation remains approximately constant and matches estimations from real financial data; see [25]. The parameter settings are summarized in Table 9.1; compare Table 1 in [32].

To model the observation times we use the Poisson setting discussed in Examples 9.6 and 9.10 for $\lambda_1 = \lambda_2 = 1$ and set $T = 1$ which amounts on average to n observations of each $X^{(1)}$ and $X^{(2)}$ in the interval $[0, T]$. We choose $n = 100$, $n = 400$ and $n = 1,600$ for the simulation. In a trading day of 6.5 hours this

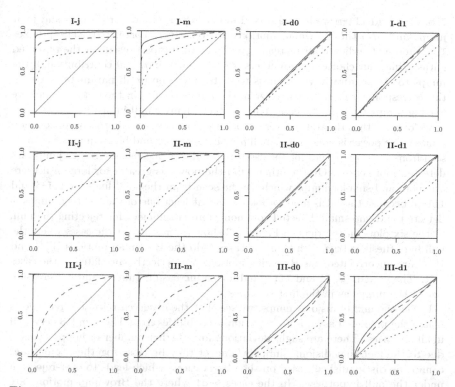

Figure 9.3: Empirical rejection curves from the Monte Carlo simulation for the test derived from Theorem 9.8. The dotted line represents the results for $n = 100$, the dashed line for $n = 400$ and the solid line for $n = 1,600$. In each case $N = 10,000$ paths were simulated.

corresponds to observing $X^{(1)}$ and $X^{(2)}$ on average every 4 minutes, every 1 minute and every 15 seconds. We set $\beta = 0.03$ and $\varpi = 0.49$ for all occuring truncations. We use $b_n = 1/\sqrt{n}$ for the local interval in the estimation of $\sigma_s^{(l)}$ and $K_n = \lfloor \ln(n) \rfloor$, $M_n = n$ in the simulation of the $\widehat{Z}_{n,m}^{(l)}(s)$. For an explanation of this choice of parameters see Section 8.2.

In Figure 9.3 we display the results from the simulation. The plots are constructed as follows: First for different values of α the critical values $c_{T,n}^{DisJ}(\alpha)$ are simulated according to Theorem 9.8. Then we plot the observed rejection frequencies against α.

The six plots on the left show the results for the cases where the alternative of common jumps is true. In the cases I-j, II-j and III-j there exist only joint jumps and the Brownian motions $W^{(1)}$ and $W^{(2)}$ are uncorrelated. In the cases

I-m, II-m and III-m we have a mixed model which allows for disjoint and joint jumps and also the Brownian motions are positively correlated. The prefixes I, II and III indicate an increasing number of jumps present in the observed paths. Since our choice of parameters is such that the overall contribution of the jumps to the quadratic variation is roughly the same in all parameter settings, this corresponds to a decreasing size of the jumps. Hence in the cases I-* we have few big jumps while in the cases III-* we have many small jumps.

We observe that the test has very good power against the alternative of common jumps. The power is greater for small n if there are less and bigger jumps as can be seen from the dotted lines for the cases I-j and I-m because the bigger jumps are detected more easily. On the other hand the power is greater for large n if there are more and smaller jumps which can be seen from the solid lines for III-j and III-m, because then it is more probable that at least one of the common jumps is detected and one small detected common jump is sufficient for rejecting the null.

The six plots on the right in Figure 9.3 show the results for the cases where the null hypothesis is true. While in the cases *-d0 the Brownian motions $W^{(1)}$ and $W^{(2)}$ are uncorrelated, the Brownian motions are perfectly correlated in the cases *-d1. The prefixes I, II and III stand for an increasing number and a decreasing size of the jumps as in the first six cases.

Under the null of disjoint jumps, we see that the observed rejection frequencies match the predicted asymptotic rejection probabilities from Theorem 9.8 very well in all six cases. There are slight deviations for a higher number of jumps. This is due to the fact that disjoint jumps whenever they lie close together, sometimes cannot be distinguished based on the observations which leads to over-rejection under the null hypothesis. In the cases *-d1 where the Brownian motions are perfectly correlated the rejection frequencies are systematically too high for large n. The results are worse than in the cases *-d0.

In general, the results from the Monte Carlo are very similar to the results displayed in Figure 5 (note that the values for n there are 100, 1,600 and 25,600) from [32]. On a closer look, we observe that the power of our test in the asynchronous setting is slightly worse than the power of the test in the equidistant and synchronous setting while under the null hypothesis the rejection levels match the asymptotic levels more closely than in [32]. The loss in power is most pronounced for the smallest observation frequency $n = 100$ and in the range of at most a few percentage points for the more relevant frequency $n = 1,600$. Our results in the cases *-d1 are better than in [32] because the effect of a high correlation in the Brownian motions has less influence on the test statistic due to the asynchronicity. All in all, we conclude that there is no significant drawback of working with asynchronous observations instead of synchronous observations when testing for disjoint jumps in a bivariate process.

9.1.3 The Proofs

Proof of Theorem 9.1. The convergence (9.1) follows from Theorem 2.3, Corollary 2.2 and the continuous mapping theorem for convergence in probability. □

Proof of Theorem 9.4. Step 1. Using Condition 9.3(ii) and Theorem 2.22 we obtain

$$n \sum_{i,j:t_{i,n}^{(1)} \vee t_{j,n}^{(2)} \leq T} (\Delta_{i,n}^{(1)} C^{(1)})^2 (\Delta_{j,n}^{(2)} C^{(2)})^2 \mathbb{1}_{\{\mathcal{I}_{i,n}^{(1)} \cap \mathcal{I}_{j,n}^{(2)} \neq \emptyset\}} \xrightarrow{\mathbb{P}} \widetilde{C}_T^{DisJ}. \quad (9.14)$$

This has been specifically computed in (2.43) in Example 2.23.

Step 2. Recall the discretized versions $\tilde{\sigma}(r,q)$ and $\widetilde{C}(r,q)$ introduced in Step 1 in the proof of Theorem 3.6 and define

$$\widetilde{D}_T^{DisJ}(q,r) = n \sum_{l=1,2} \sum_{i,j:t_{i,n}^{(l)} \vee t_{j,n}^{(3-l)} \leq T} (\Delta_{i,n}^{(l)} N^{(l)}(q))^2 (\Delta_{j,n}^{(3-l)} \widetilde{C}^{(3-l)}(r,q))^2 \mathbb{1}_{\{\mathcal{I}_{i,n}^{(l)} \cap \mathcal{I}_{i,n}^{(3-l)} \neq \emptyset\}}.$$

Let $S_{q,p}^{(l)}$, $p \in \mathbb{N}$, be an enumeration of the jump times of $N^{(l)}(q)$ and denote by $\Omega(n,q,r)$ the subset of Ω_T^{DisJ} where two different jump times $S_{q,p_1}^{(l_1)} \neq S_{q,p_2}^{(l_2)}$ are further apart than $4|\pi_n|_T$ and the jump times $S_{q,p}^{(l)}$ are further away than $2|\pi_n|_T$ from the discontinuities $k/2^r$ of $\sigma(r)$. On this set we get

$$\widetilde{D}_{T,n}^{DisJ}(q,r) \mathbb{1}_{\Omega(n,q,r)} = \sum_{l=1,2} \sum_{p:S_{q,p}^{(l)}} (\Delta N^{(l)}(q)_{S_{q,p}^{(l)}})^2 \widetilde{R}_n^{(3-l)}(S_{q,p}^{(l)}, r, q) \mathbb{1}_{\Omega(n,q,r)}$$

where

$$\widetilde{R}_n^{(l)}(s,r,q) = (\tilde{\sigma}_{s-}^{(l)}(r,q))^2 \eta_{n,-}^{(l)}(s) + \left[\tilde{\sigma}_{s-}^{(l)}(r,q)(\delta_{n,-}^{(l)}(s))^{1/2} U_{n,-}^{(l)}(s) \right.$$
$$\left. + \tilde{\sigma}_s^{(l)}(r,q)(\delta_{n,+}^{(l)}(s))^{1/2} U_{n,+}^{(l)}(s)\right]^2 + (\tilde{\sigma}_s^{(l)}(r,q))^2 \eta_{n,+}^{(l)}(s), \quad s \in [0,T], \quad l=1,2,$$

with $U_{n,-}^{(l)}(s), U_{n,+}^{(l)}(s)$ defined as in (3.40). Comparing the proof of (3.25) we find that Condition 9.3(ii) yields the $\mathcal{X}$-stable convergence

$$\left(\left(\widetilde{R}_n^{(1)}(S_{q,p}^{(2)}, r, q)\right)_{S_{q,p}^{(2)} \leq T}, \left(\widetilde{R}_n^{(2)}(S_{q,p}^{(1)}, r, q)\right)_{S_{q,p}^{(2)} \leq T} \right)$$
$$\xrightarrow{\mathcal{L}-s} \left(\left(\widetilde{R}^{(1)}(S_{q,p}^{(2)}, r, q)\right)_{S_{q,p}^{(2)} \leq T}, \left(\widetilde{R}^{(2)}(S_{q,p}^{(1)}, r, q)\right)_{S_{q,p}^{(2)} \leq T} \right)$$

where

$$\widetilde{R}^{(l)}(s,r,q) = (\tilde{\sigma}_{s-}^{(l)}(r,q))^2 \eta_-^{(l)}(s) + \left[\tilde{\sigma}_{s-}^{(l)}(r,q)(\delta_-^{(l)}(s))^{1/2} U_-^{(l)}(s) \right.$$
$$\left. + \tilde{\sigma}_s^{(l)}(r,q)(\delta_+^{(l)}(s))^{1/2} U_+^{(l)}(s)\right]^2 + (\tilde{\sigma}_s^{(l)}(r,q))^2 \eta_+^{(l)}(s), \quad s \in [0,T], \quad l=1,2.$$

Using this stable convergence, Lemma B.3 and the continuous mapping theorem for stable convergence stated in Lemma B.5 we then obtain

$$\sum_{l=1,2} \sum_{p:S_{q,p}^{(l)}} \left(\Delta N^{(l)}(q)_{S_{q,p}^{(l)}}\right)^2 \widetilde{R}_n^{(3-l)}(S_{q,p}^{(l)}, r, q)$$

$$\stackrel{\mathcal{L}-s}{\longrightarrow} \widetilde{D}_T^{DisJ}(q,r) := \sum_{l=1,2} \sum_{p:S_{q,p}^{(l)}} \left(\Delta N^{(l)}(q)_{S_{q,p}^{(l)}}\right)^2 \widetilde{R}^{(3-l)}(S_{q,p}^{(l)}, r, q).$$

Because of $\mathbb{P}(\Omega(n, q, r)) \to 1$ as $n \to \infty$ for any $q, r > 0$ this convergence yields

$$\widetilde{D}_{T,n}^{DisJ}(q, r) \stackrel{\mathcal{L}-s}{\longrightarrow} \widetilde{D}_T^{DisJ}(q, r) \tag{9.15}$$

for any $q, r > 0$. A more detailed proof of (9.15) can be found in the proof of Proposition A.3 in [36]. In fact the proof of Proposition A.3 in [36] is up to different notation almost identical to the proof of (3.25) which was presented in full detail in Chapter 3.

Step 3. Next we show

$$\lim_{q\to\infty} \limsup_{r\to\infty} \limsup_{n\to\infty} \mathbb{P}\left(\left\{\left|nV(f_{(2,2)}, \pi_n)_T - R(n, q, r)_T\right| > \varepsilon\right\} \cap \Omega_T^{nCoJ}\right) = 0 \tag{9.16}$$

for all $\varepsilon > 0$ where

$$R(n, q, r)_T = n \sum_{i,j:t_{i,n}^{(1)} \vee t_{j,n}^{(2)} \leq T} \left(\left(\Delta_{i,n}^{(1)} C^{(1)}\right)^2 \left(\Delta_{j,n}^{(2)} C^{(2)}\right)^2 + \left(\Delta_{i,n}^{(1)} N^{(1)}(q)\right)^2 \left(\Delta_{j,n}^{(2)} \widetilde{C}^{(2)}(r, q)\right)^2 \right.$$

$$+ \left(\Delta_{i,n}^{(1)} \widetilde{C}^{(1)}(r, q)\right)^2 \left(\Delta_{j,n}^{(2)} N^{(2)}(q)\right)^2\right) \mathbb{1}_{\{\mathcal{I}_{i,n}^{(1)} \cap \mathcal{I}_{j,n}^{(2)} \neq \emptyset\}}.$$

As the proof of (9.16) is rather technical and more involved compared to the other steps it will be given later.

Step 4. Because σ may assumed to be bounded and because the laws of $\Gamma^{DisJ}(\cdot, dy)$ have uniformly bounded first moments we obtain similarly as in the proof of (3.26)

$$\lim_{q\to\infty} \limsup_{r\to\infty} \mathbb{P}\left(\left\{\left|\widetilde{D}_T^{DisJ} - \widetilde{D}_T^{DisJ}(q, r)\right| > \varepsilon\right\} \cap \Omega_T^{DisJ}\right) = 0. \tag{9.17}$$

Step 5. Using (9.14) and (9.15) together with Lemma B.3 and the continuous mapping theorem for stable convergence in law which is stated in Lemma B.5 we obtain

$$R(n, q, r)_T \stackrel{\mathcal{L}-s}{\longrightarrow} \widetilde{C}_T^{DisJ} + \widetilde{D}_T^{DisJ}(q, r). \tag{9.18}$$

Further, combining (9.16), (9.17) and (9.18) and using Lemma (B.6) we obtain

$$nV(f_{(2,2)}, \pi_n)_T \xrightarrow{\mathcal{L}-s} \widetilde{C}_T^{DisJ} + \widetilde{D}_T^{DisJ}$$

which in combination with $V^{(l)}(g_4, \pi_n)_T \xrightarrow{\mathbb{P}} B^{(l)}(g_4)_T$ due to Corollary 2.2 yields (9.8). □

For the proof of (9.16) we need the following Lemma.

Lemma 9.11. *Let Condition 1.3 be satisfied, the processes σ_t, Γ_t be bounded and the observation scheme be exogenous. Then there exists a constant K which is independent of (i,j) such that*

$$\mathbb{E}\Big[(\Delta_{i,n}^{(l)}M(q))^2(\Delta_{j,n}^{(3-l)}M(q'))^2 \mathbb{1}_{\Omega_T^{DisJ}} \big| \sigma(\mathcal{F}_{t_{i,n}^{(1)} \wedge t_{j,n}^{(2)}}, \mathcal{S})\Big] \leq K e_q e_{q'} |\mathcal{I}_{i,n}^{(1)}||\mathcal{I}_{j,n}^{(2)}|.$$
(9.19)

Proof. Arguing similarly as in the proof of Lemma 3.9 it is sufficient to prove (9.19) for $\mathcal{I}_{i,n}^{(1)} = \mathcal{I}_{j,n}^{(2)}$. The claim now follows from (8.17) in [32] which is basically (9.19) for $\mathcal{I}_{i,n}^{(l)} = \mathcal{I}_{j,n}^{(3-l)}$. The generalization to $q \neq q'$ here does not complicate the proof. □

Proof of (9.16). Since γ is bounded by Condition 1.3 we can write

$$X = X_0 + B(q') + C + M(q')$$
(9.20)

on $[0, T]$ for some positive number q' (not necessarily an integer) which yields

$$N(q) = B(q') - B(q) + M(q') - M(q).$$
(9.21)

First we observe that it holds

$$\lim_{r \to \infty} \limsup_{n \to \infty} \mathbb{P}(|R(n,q)_T - R(n,q,r)_T| > \varepsilon) = 0$$
(9.22)

for any $\varepsilon > 0$ and any $q > 0$ where

$$R(n,q)_T = n \sum_{i,j : t_{i,n}^{(1)} \vee t_{j,n}^{(2)} \leq T} \big((\Delta_{i,n}^{(1)}C^{(1)})^2(\Delta_{j,n}^{(2)}C^{(2)})^2 + (\Delta_{i,n}^{(1)}N^{(1)}(q))^2(\Delta_{j,n}^{(2)}C^{(2)})^2$$
$$+ (\Delta_{i,n}^{(1)}C^{(1)})^2(\Delta_{j,n}^{(2)}N^{(2)}(q))^2\big)\mathbb{1}_{\{\mathcal{I}_{i,n}^{(1)} \cap \mathcal{I}_{j,n}^{(2)} \neq \emptyset\}}$$

using similar arguments as for (3.34) in the proof of (3.24).

Next, an application of inequality (2.46) with

$$x_l = \Delta_{i,n}^{(l)}C^{(l)} + \Delta_{i,n}^{(l)}N^{(l)}(q), \quad y_l = \Delta_{i,n}^{(l)}B^{(l)}(q) + \Delta_{i,n}^{(l)}M^{(l)}(q)$$

yields

$$(nV(f_{(2,2)}, \pi_n)_T - \widetilde{R}(n,q)_T) \mathbb{1}_{\Omega(n,q)}$$
$$\leq \theta(\varepsilon) \widetilde{R}(n,q)_T$$
$$+ K_\varepsilon \sum_{l=1,2} \sum_{i,j: t_{i,n}^{(l)} \vee t_{j,n}^{(3-l)} \leq T} \mathbb{1}_{\{\mathcal{I}_{i,n}^{(l)} \cap \mathcal{I}_{j,n}^{(3-l)} \neq \emptyset\}} \left(\Delta_{i,n}^{(l)}(B^{(l)}(q) + M^{(l)}(q))\right)^2$$
$$\times \left(\left(\Delta_{j,n}^{(3-l)}(B^{(3-l)}(q) + M^{(3-l)}(q))\right)^2 + \left(\Delta_{j,n}^{(3-l)}(C^{(3-l)} + N^{(3-l)}(q))\right)^2 \right)$$

where

$$\widetilde{R}(n,q)_T = \sum_{i,j: t_{i,n}^{(1)} \vee t_{j,n}^{(2)} \leq T} \left(\Delta_{i,n}^{(1)}(C^{(1)} + N^{(1)}(q))\right)^2 \left(\Delta_{j,n}^{(2)}(C^{(2)} + N^{(2)}(q))\right)^2$$
$$\times \mathbb{1}_{\{\mathcal{I}_{i,n}^{(1)} \cap \mathcal{I}_{j,n}^{(2)} \neq \emptyset\}}.$$

It holds

$$\widetilde{R}(n,q)_T \leq 4 \sum_{i_1, i_2: t_{i_1,n}^{(1)} \vee t_{i_2,n}^{(2)} \leq T} \prod_{l=1,2} \left((\Delta_{i_l,n}^{(l)} C^{(l)})^2 + (\Delta_{i_l,n}^{(l)} N^{(l)}(q))^2 \right)$$
$$\times \mathbb{1}_{\{\mathcal{I}_{i_1,n}^{(1)} \cap \mathcal{I}_{i_2,n}^{(2)} \neq \emptyset\}}$$

and hence by (9.14), (9.15) and (9.22) the term $\widetilde{R}(n,q)_T$ is bounded in probability for $n, q \to \infty$ which yields that $\theta(\varepsilon) \widetilde{R}(n,q)_T$ vanishes as first $n \to \infty$, then $q \to \infty$ and finally $\varepsilon \to 0$. Further, we also obtain for $l = 1, 2$ using (9.21), Lemma 1.4, Lemma 3.9 and Lemma 9.11

$$\mathbb{E}\Big[K_\varepsilon n \sum_{i,j: t_{i,n}^{(l)} \vee t_{j,n}^{(3-l)} \leq T} \left(\Delta_{i,n}^{(l)} B^{(l)}(q) + M^{(l)}(q)\right)^2 \Big[\left(\Delta_{j,n}^{(3-l)}(B^{(3-l)}(q) + M^{(3-l)}(q))\right)^2$$
$$+ \left(\Delta_{j,n}^{(3-l)} C^{(3-l)} + \Delta_{j,n}^{(3-l)} N^{(3-l)}(q)\right)^2 \Big] \mathbb{1}_{\{\mathcal{I}_{i,n}^{(l)} \cap \mathcal{I}_{i,n}^{(3-l)} \neq \emptyset\}} \mathbb{1}_{\Omega_T^{nCoJ}} \Big| \mathcal{S} \Big]$$
$$\leq K_\varepsilon n \sum_{i,j: t_{i,n}^{(l)} \vee t_{j,n}^{(3-l)} \leq T} \mathbb{1}_{\{\mathcal{I}_{i,n}^{(l)} \cap \mathcal{I}_{i,n}^{(3-l)} \neq \emptyset\}} \left(K_q |\mathcal{I}_{i,n}^{(l)}| + K e_q\right) |\mathcal{I}_{i,n}^{(l)}|$$
$$\times \left(K_q |\mathcal{I}_{j,n}^{(3-l)}| + K e_q + 2K + 8(K_q + K_{q'})|\mathcal{I}_{j,n}^{(3-l)}| + 8K(e_q + e_{q'})\right) |\mathcal{I}_{j,n}^{(3-l)}|$$
$$\leq K_\varepsilon \left(K_q |\pi_n|_T + K e_q\right)\left(K_q |\pi_n|_T + K e_q + K\right) G_{2,2}^n(T),$$

where the latter bound converges to zero for $n \to \infty$ and then $q \to \infty$ for any $\varepsilon > 0$. Hence we obtain

$$\lim_{\varepsilon \to 0} \limsup_{q \to \infty} \limsup_{n \to \infty} \mathbb{P}(\{|nV(f_{(2,2)}, \pi_n)_T - \widetilde{R}(n,q)_T| > \delta\} \cap \Omega_T^{nCoJ}) = 0 \quad (9.23)$$

for any $\delta > 0$ and it remains to show that, restricted to the set Ω_T^{nCoJ}, the quantity $\widetilde{R}(n,q)_T - R(n,q)_T$ vanishes as first $n \to \infty$ and then $q \to \infty$. To this end note that $\widetilde{R}(n,q)_T - R(n,q)_T$ is equal to

$$n \sum_{l=1,2} \sum_{i,j:t_{i,n}^{(l)} \vee t_{j,n}^{(3-l)} \leq T} ((\Delta_{i,n}^{(l)} C^{(l)})^2 + 2(\Delta_{i,n}^{(l)} C^{(l)})(\Delta_{i,n}^{(l)} N^{(l)}(q)) + (\Delta_{i,n}^{(l)} N^{(l)}(q))^2)$$

$$\times 2((\Delta_{j,n}^{(3-l)} C^{(3-l)})(\Delta_{j,n}^{(3-l)} N^{(3-l)}(q))) \mathbb{1}_{\{\mathcal{I}_{i,n}^{(l)} \cap \mathcal{I}_{j,n}^{(3-l)} \neq \emptyset\}}.$$

Here, the sum over terms containing the product $(\Delta_{i,n}^{(l)} N^{(l)}(q))(\Delta_{j,n}^{(3-l)} N^{(3-l)}(q))$ converges to zero because we are on Ω_T^{nCoJ}. For the remaining terms we obtain

$$n \sum_{i,j:t_{i,n}^{(l)} \vee t_{j,n}^{(3-l)} \leq T} (\Delta_{i,n}^{(l)} C^{(l)})^2 (\Delta_{j,n}^{(3-l)} C^{(3-l)})(\Delta_{j,n}^{(3-l)} N^{(3-l)}(q)) \mathbb{1}_{\{\mathcal{I}_{i,n}^{(l)} \cap \mathcal{I}_{j,n}^{(3-l)} \neq \emptyset\}}$$

$$\leq \left(\sup_{j:t_{j,n}^{(3-l)} \leq T} \Delta_{j,n}^{(3-l)} C^{(3-l)} \right) n \sum_{i,j:t_{i,n}^{(l)} \vee t_{j,n}^{(3-l)} \leq T} (\Delta_{i,n}^{(l)} C^{(l)})^2 (\Delta_{j,n}^{(3-l)} N^{(3-l)}(q))$$

$$\times \mathbb{1}_{\{\mathcal{I}_{i,n}^{(l)} \cap \mathcal{I}_{j,n}^{(3-l)} \neq \emptyset\}}$$

where, restricted to Ω_T^{nCoJ}, the right-hand side tends to zero as $n \to \infty$ for all $q > 0$ because the supremum vanishes since C is continuous and because the sum converges stably in law on Ω_T^{nCoJ} to

$$\sum_{S_{q,p}^{(3-l)} \leq T} \Delta N^{(3-l)}(q)_{S_{q,p}^{(3-l)}} R^{(l)}(S_{q,p}^{(3-l)}),$$

where $S_{q,p}^{(3-l)}$, $p \in \mathbb{N}$, denotes an enumeration of the jump times of $N^{(3-l)}(q)$. The stable convergence can be proven similarly as (9.15) and follows from Condition 9.3(ii). $\qquad \square$

Proof of (9.10). Looking at the proof of (9.23), it is enough to show that

$$n \sum_{i,j:t_{i,n}^{(1)} \vee t_{j,n}^{(2)} \leq T} (\Delta_{i,n}^{(1)} C^{(1)} + \Delta_{i,n}^{(1)} N^{(1)}(q))^2 (\Delta_{j,n}^{(2)} C^{(2)} + \Delta_{j,n}^{(2)} N^{(2)}(q))^2$$

$$\times \mathbb{1}_{\{|\Delta_{i,n}^{(1)} X^{(l)}| \leq \beta |\mathcal{I}_{i,n}^{(1)}|^{\varpi} \wedge |\Delta_{j,n}^{(2)} X^{(2)}| \leq \beta |\mathcal{I}_{j,n}^{(2)}|^{\varpi}\}} \mathbb{1}_{\{\mathcal{I}_{i,n}^{(1)} \cap \mathcal{I}_{j,n}^{(2)} \neq \emptyset\}} \qquad (9.24)$$

converges on the set Ω_T^{DisJ} to $\widetilde{C}_T$.

We first deal with the terms in (9.24) involving big jumps. Let $S_{q,p}^{(l)}$, $p \in \mathbb{N}$, denote tan enumeration of the jump times of $N^{(l)}(q)$. Then it holds

$$
\left| n \sum_{l=1,2} \sum_{i,j:t_{i,n}^{(l)} \vee t_{j,n}^{(3-l)} \leq T} \Delta_{i,n}^{(l)} N^{(l)}(q) \big(\Delta_{i,n}^{(l)} (C^{(l)} + N^{(l)}(q)) \right.
$$
$$
\times \big(\Delta_{j,n}^{(3-l)} (C^{(3-l)} + N^{(3-l)}(q)) \big)^2
$$
$$
\left. \times \mathbb{1}_{\{|\Delta_{i,n}^{(l)} X^{(l)}| \leq \beta |\mathcal{I}_{i,n}^{(l)}|^{\varpi} \wedge |\Delta_{j,n}^{(3-l)} X^{(3-l)}| \leq \beta |\mathcal{I}_{j,n}^{(3-l)}|^{\varpi}\}} \mathbb{1}_{\{\mathcal{I}_{i,n}^{(l)} \cap \mathcal{I}_{j,n}^{(3-l)} \neq \emptyset\}} \right| \mathbb{1}_{\Omega(n,q,T)}
$$
$$
\leq n \sum_{l=1,2} \sum_{p:S_{q,p}^{(l)} \leq T} |\Delta N^{(l)}(q)_{S_{q,p}^{(l)}}| \mathbb{1}_{\{|\Delta_{i_n^{(l)}(S_{q,p}^{(l)}),n}^{(l)} X^{(l)}| \leq \beta |\mathcal{I}_{i_n^{(l)}(S_{q,p}^{(l)}),n}^{(l)}|^{\varpi}\}}
$$
$$
\times |\Delta_{i_n^{(l)}(S_{q,p}^{(l)}),n}^{(l)} (C^{(l)} + N^{(l)}(q))| \sum_{j:t_{j,n}^{(3-l)} \leq T} \big(\Delta_{j,n}^{(3-l)} C^{(3-l)} + \Delta_{j,n}^{(3-l)} N^{(3-l)}(q) \big)^2
$$

$$(9.25)$$

where $\Omega(n,q,T)$ denotes the set where any two different jumps of $N(q)$ in $[0,T]$ are further apart than $2|\pi_n|_T$. (9.25) converges in probability to zero as $n \to \infty$ because of

$$
|\Delta_{i_n^{(l)}(S_{q,p}^{(l)}),n}^{(l)} X| \xrightarrow{\mathbb{P}_{\mathcal{X}}} |\Delta N^{(l)}(q)_{S_{q,p}^{(l)}}| > 0, \quad |\mathcal{I}_{i_n^{(l)}(S_{q,p}^{(l)}),n}^{(l)}| \xrightarrow{\mathbb{P}_{\mathcal{X}}} 0
$$

where $\mathbb{P}_{\mathcal{X}}$ denotes convergence in $\mathcal{X}$-conditional probabilities. $\mathbb{P}(\Omega(n,q,T)) \to 1$ as $n \to \infty$ then yields that the terms involving big jumps in (9.24) vanish asymptotically. Hence only the terms involving squared increments of $C^{(1)}, C^{(2)}$ contribute in the limit.

Using (9.14) it is sufficient to show

$$
\widetilde{L}_{T,n} = n \sum_{i,j:t_{i,n}^{(1)} \vee t_{j,n}^{(2)} \leq T} \big(\Delta_{i,n}^{(1)} C^{(1)} \big)^2 \big(\Delta_{j,n}^{(2)} C^{(2)} \big)^2
$$
$$
\times \mathbb{1}_{\{|\Delta_{i,n}^{(1)} X^{(1)}| > \beta |\mathcal{I}_{i,n}^{(1)}|^{\varpi} \vee |\Delta_{j,n}^{(2)} X^{(2)}| > \beta |\mathcal{I}_{j,n}^{(2)}|^{\varpi}\}} \mathbb{1}_{\{\mathcal{I}_{i,n}^{(1)} \cap \mathcal{I}_{j,n}^{(2)} \neq \emptyset\}} \xrightarrow{\mathbb{P}} 0. \quad (9.26)
$$

To this end note that the conditional Markov inequality plus an application of inequality (1.12) give

$$
\mathbb{P}\big(|\Delta_{i,n}^{(l)} X^{(l)}| > \beta |\mathcal{I}_{i,n}^{(l)}|^{\varpi} \big| \mathcal{S} \big) \leq K |\mathcal{I}_{i,n}^{(l)}|^{1-2\varpi}. \quad (9.27)
$$

Then using $\mathbb{1}_{A\vee B} \leq \mathbb{1}_A + \mathbb{1}_B$ for events $A, B \in \mathcal{F}$ and the Cauchy-Schwarz inequality, as well as the inequalities (1.9) and (9.27), we get

$$
\mathbb{E}\big[\widetilde{L}_{T,n}\big|\mathcal{S}\big]
$$

$$
\leq Kn \sum_{i_1,i_2:t^{(1)}_{i_l,n}\vee t^{(2)}_{i_2,n}\leq T} |\mathcal{I}^{(1)}_{i_1,n}||\mathcal{I}^{(2)}_{i_2,n}| \sum_{l=1,2} |\mathcal{I}^{(l)}_{i_l,n}|^{(1-2\varpi)/2}\mathbb{1}_{\{\mathcal{I}^{(1)}_{i,n}\cap\mathcal{I}^{(2)}_{j,n}\neq\emptyset\}}
$$

$$
\leq K(|\pi_n|_T)^{(1-2\varpi)/2}G^n_{2,2}(T), \tag{9.28}
$$

which tends to zero by Condition 1.3 and Condition 9.3(i). This estimate proves (9.26).   □

To prove (9.12) in Theorem 9.8 we first prove a proposition similarly as in Chapters 4, 7 and 8.

Proposition 9.12. *Suppose that Condition 9.7 is satisfied. Then*

$$
\widetilde{\mathbb{P}}\big(\big\{\big|\frac{1}{M_n}\sum_{m=1}^{M_n}\mathbb{1}_{\{\widehat{D}^{DisJ}_{T,n,m}\leq\Upsilon\}} - \widetilde{\mathbb{P}}\big(\widetilde{D}^{DisJ}_T\leq\Upsilon|\mathcal{X}\big)\big| > \varepsilon\big\}\cap\Omega^{DisJ}_T\big)\to 0 \tag{9.29}
$$

as $n\to\infty$ for any $\mathcal{X}$-measurable random variable Υ and all $\varepsilon > 0$.

Proof. Step 1. Denote by $S_{q,p}$, $p\in\mathbb{N}$, an increasing sequence of stopping times which exhausts the jump times of $N(q)$. Recall that on Ω^{DisJ}_T only one component of X jumps at $S_{q,p}$, and we use l_p as the index of the component involving the p-th jump. Therefore, setting

$$
A_{n,p} = \big(\Delta^{(l_p)}_{i^{(l_p)}_n(S_{q,p}),n}X^{(l_p)}\mathbb{1}_{\{|\Delta^{(l_p)}_{i^{(l_p)}_n(S_{q,p}),n}X^{(l_p)}|>\beta|\mathcal{I}^{(l_p)}_{i^{(l_p)}_n(S_{q,p}),n}|^\varpi\}},
$$

$$
\tilde{\sigma}^{(3-l_p)}_n(S_{q,p},-),\tilde{\sigma}^{(3-l_p)}_n(S_{q,p},+)\big)
$$

$$
A_p = \big(\Delta X^{(l_p)}_{S_{q,p}},\sigma^{(3-l_p)}_{S_{q,p}-},\sigma^{(3-l_p)}_{S_{q,p}}\big),\quad \widehat{Z}^p_{n,m} = \widehat{Z}^{DisJ}_{n,m}(S_{q,p}),\quad Z^p = Z^{DisJ}(S_{q,p})
$$

and defining φ via

$$
\varphi\big((A_p,Y^{(l_p)}(S_{q,p}))_{p=1,\dots,P}\big) = \sum_{p=1}^{P}(\Delta X^{(l_p)}_{S_{q,p}})^2 R^{(3-l_p)}(S_{q,p})\mathbb{1}_{\{S_{q,p}\leq T\}},
$$

Lemma 4.9 proves

$$
\widetilde{\mathbb{P}}\big(\big\{\big|\frac{1}{M_n}\sum_{m=1}^{M_n}\mathbb{1}_{\{Y(P,n,m)\leq\Upsilon\}} - \widetilde{\mathbb{P}}\big(Y(P)\leq\Upsilon|\mathcal{X}\big)\big| > \varepsilon\big\}\cap\Omega^{DisJ}_T\big)\to 0 \tag{9.30}
$$

where we used the notation

$$Y(P,n,m) = \sum_{p=1}^{P} \left(\Delta_{i_n^{(l_p)}(S_{q,p}),n}^{(l_p)} X^{(l_p)} \right)^2 \mathbf{1}_{\{|\Delta_{i_n^{(l_p)}(S_{q,p}),n}^{(l_p)} X^{(l_p)}| > \beta |\mathcal{I}_{i_n^{(l_p)}(S_{q,p}),n}^{(l_p)}|^{\varpi}\}}$$
$$\times \widehat{R}_{n,m}^{(3-l_p)}(S_{q,p}) \mathbf{1}_{\{S_{q,p} \le T\}},$$

$$Y(P) = \sum_{p=1}^{P} \left(\Delta X_{S_{q,p}}^{(l_p)} \right)^2 R^{(3-l_p)}(S_{q,p}) \mathbf{1}_{\{S_{q,p} \le T\}}.$$

Step 2. Next, we prove

$$\lim_{P \to \infty} \limsup_{n \to \infty} \frac{1}{M_n} \sum_{m=1}^{M_n} \widetilde{\mathbb{P}} \left(|Y(P,n,m) - \widehat{D}_{T,n,m}^{DisJ}| > \varepsilon \right) = 0 \qquad (9.31)$$

for all $\varepsilon > 0$. Denote by $\Omega(P,q,n)$ the set on which there are at most P jumps of $N(q)$ in $[0,T]$ and two different jumps of $N(q)$ are further apart than $|\pi_n|_T$. Obviously, $\mathbb{P}(\Omega(P,q,n)) \to 1$ for $P,n \to \infty$ and any $q > 0$. On the set $\Omega(P,q,n)$ we have

$$\left| Y(P,n,m) - \widehat{D}_{T,n,m}^{DisJ} \right| \mathbf{1}_{\Omega_T^{DisJ}} \mathbf{1}_{\Omega(P,q,n)}$$

$$\le \sum_{l=1,2} \sum_{t_{i,n}^{(l)} \le T, \nexists p: S_{q,p} \in \mathcal{I}_{i,n}^{(l)}} \left(\Delta_{i,n}^{(l)} B(q) + \Delta_{i,n}^{(l)} C + \Delta_{i,n}^{(l)} M(q) \right)^2$$

$$\times \mathbf{1}_{\{|\Delta_{i,n}^{(l)} X^{(l)}| > \beta |\mathcal{I}_{i,n}^{(l)}|^{\varpi}\}} \widehat{R}_{n,m}^{(3-l)}(t_{i,n}^{(l)})$$

$$\le 2 \sum_{l=1,2} \sum_{t_{i,n}^{(l)} \le T} \left(\Delta_{i,n}^{(l)} B(q) + \Delta_{i,n}^{(l)} C + \Delta_{i,n}^{(l)} M(q) \right)^2 \mathbf{1}_{\{|\Delta_{i,n}^{(l)} X^{(l)}| > \beta |\mathcal{I}_{i,n}^{(l)}|^{\varpi}\}}$$

$$\times \left(\frac{1}{b_n} \sum_{j:\mathcal{I}_{j,n}^{(3-l)} \subset [t_{i,n}^{(l)} - b_n, t_{i,n}^{(l)} + b_n]} \left(\Delta_{j,n}^{(3-l)} X \right)^2 \right) \widehat{\eta}_{n,m}^{(3-l)}(t_{i,n}^{(l)}) \qquad (9.32)$$

where we used the notation

$$\widehat{\eta}_{n,m}^{(3-l)}(t_{i,n}^{(l)})$$
$$= \widehat{\eta}_{n,m,-}^{(3-l)}(s) + \widehat{\delta}_{n,m,-}^{(3-l)}(s) U_{n,m,-}^{(3-l)}(s)^2 + \widehat{\delta}_{n,m,+}^{(3-l)}(s) U_{n,m,+}^{(3-l)}(s)^2 + \widehat{\eta}_{n,m,+}^{(3-l)}(s).$$

We first consider the increments over the overlapping observation intervals in the right-hand side of (9.32). The $\mathcal{F}$-conditional mean of their sum is bounded by

$$\frac{3|\pi_n|_T}{b_n} n \sum_{l=1,2} \sum_{i,j:t_{i,n}^{(l)} \vee t_{j,n}^{(3-l)} \le T} \left(\Delta_{i,n}^{(l)} B(q) + \Delta_{i,n}^{(l)} C + \Delta_{i,n}^{(l)} M(q) \right)^2$$

$$\times \mathbf{1}_{\{|\Delta_{i,n}^{(l)} X^{(l)}| > \beta |\mathcal{I}_{i,n}^{(l)}|^{\varpi}\}} \left(\Delta_{j,n}^{(3-l)} X \right)^2 \mathbf{1}_{\{\mathcal{I}_{i,n}^{(l)} \cap \mathcal{I}_{j,n}^{(3-l)} \ne \emptyset\}}, \qquad (9.33)$$

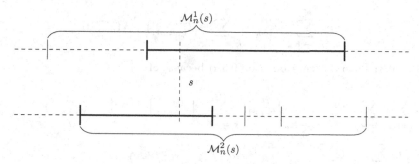

Figure 9.4: Illustration for the variables $\mathcal{M}_n^{(l)}(s)$, $l = 1, 2$.

since with $\mathcal{M}_n^{(l)}(s) = \sum_{i:t_{i,n}^{(l)} \leq T} |\mathcal{I}_{i,n}^{(l)}| \mathbb{1}_{\{\mathcal{I}_{i,n}^{(l)} \cap \mathcal{I}_{i_n^{(3-l)}(s),n}^{(3-l)} \neq \emptyset\}}$ we get

$$\mathbb{E}\big[\hat{\eta}_{n,m}^{(3-l)}(t_{i,n}^{(l)})\big|\mathcal{F}\big] = n \sum_{k_1 \in \mathbb{Z}, |k_2| \leq K_n} |\mathcal{I}_{k_1,n}^{(3-l)} \cap \mathcal{I}_{i+k_2,n}^{(l)}|$$

$$\times \big(\sum_{j_1 \in \mathbb{Z}, |j_2| \leq K_n} |\mathcal{I}_{j_1,n}^{(3-l)} \cap \mathcal{I}_{i+j_2,n}^{(l)}|\big)^{-1} \mathcal{M}_n^{(3-l)}(t_{i+k_2,n}^{(l)})$$

$$= n \sum_{k_2=-K_n}^{K_n} |\mathcal{I}_{i+k_2}^{(l)}| \big(\sum_{j_2=-K_n}^{K_n} |\mathcal{I}_{i+j_2,n}^{(l)}|\big)^{-1} \mathcal{M}_n^{(3-l)}(t_{i+k_2,n}^{(l)})$$

$$\leq \sup_{k=-K_n,\ldots,K_n} n \mathcal{M}_n^{(3-l)}(t_{i+k,n}^{(l)}) \leq 3n|\pi_n|_T. \tag{9.34}$$

Because of Theorem 9.4 the sum in (9.33) is of order n^{-1} on Ω_T^{DisJ}, while $|\pi_n|_T/b_n \xrightarrow{\mathbb{P}} 0$ for $n \to \infty$ by Condition 9.7. Hence, (9.33) vanishes as $n \to \infty$ for any $q > 0$.

Next, we deal with the increments over non-overlapping observation intervals in the right-hand side of (9.32). An upper bound is obtained by taking iterated $\mathcal{S}$-conditional expectations using Lemma 1.4, the Cauchy-Schwarz inequality as in (9.28) and (9.34), and it is given by

$$\sum_{l=1,2} \sum_{t_{i,n}^{(l)} \leq T} \big(K_q |\mathcal{I}_{i,n}^{(l)}|^2 + K|\mathcal{I}_{i,n}^{(l)}|^{1+(1-2\varpi)/2} + Ke_q|\mathcal{I}_{i,n}^{(l)}|\big) \frac{2Kb_n}{b_n}$$

$$\times n \sum_{k=-K_n}^{K_n} |\mathcal{I}_{i+k}^{(l)}| \big(\sum_{k'=-K_n}^{K_n} |\mathcal{I}_{i+k',n}^{(l)}|\big)^{-1} \mathcal{M}_n^{(3-l)}(t_{i+k,n}^{(l)})$$

$$\leq K\big(K_q|\pi_n|_T + (|\pi_n|_T)^{(1-2\varpi)/2} + e_q\big) \sum_{l=1,2} \sum_{t_{i,n}^{(l)} \leq T} |\mathcal{I}_{i,n}^{(l)}|$$

$$\times n \sum_{k=-K_n}^{K_n} |\mathcal{I}_{i+k}^{(l)}| \Big(\sum_{k'=-K_n}^{K_n} |\mathcal{I}_{i+k',n}^{(l)}| \Big)^{-1} \mathcal{M}_n^{(3-l)}(t_{i+k,n}^{(l)}).$$

Now (9.31) follows from Condition 9.3(i) because of

$$n \sum_{l=1,2} \sum_{t_{i,n}^{(l)} \leq T} |\mathcal{I}_{i,n}^{(l)}| \sum_{k=-K_n}^{K_n} |\mathcal{I}_{i+k}^{(l)}| \Big(\sum_{k'=-K_n}^{K_n} |\mathcal{I}_{i+k',n}^{(l)}| \Big)^{-1} \mathcal{M}_n^{(3-l)}(t_{i+k,n}^{(l)})$$

$$= n \sum_{l=1,2} \sum_{i,j : t_{i,n}^{(l)}, t_{j,n}^{(3-l)} \leq T} |\mathcal{I}_{i,n}^{(l)}| |\mathcal{I}_{j,n}^{(3-l)}| \mathbb{1}_{\{\mathcal{I}_{i,n}^{(l)} \cap \mathcal{I}_{j,n}^{(3-l)} \neq \emptyset\}}$$

$$\times \sum_{k=-K_n}^{K_n} |\mathcal{I}_{i+k,n}^{(l)}| \Big(\sum_{m=-K_n}^{K_n} |\mathcal{I}_{i+k+m,n}^{(l)}| \Big)^{-1}$$

$$\leq n \sum_{l=1,2} \sum_{t_{i,n}^{(l)}, t_{j,n}^{(3-l)} \leq T} |\mathcal{I}_{i,n}^{(l)}| |\mathcal{I}_{j,n}^{(3-l)}| \mathbb{1}_{\{\mathcal{I}_{i,n}^{(l)} \cap \mathcal{I}_{j,n}^{(3-l)} \neq \emptyset\}}$$

$$\times \Big(\sum_{k=-K_n}^{0} |\mathcal{I}_{i+k,n}^{(l)}| \Big(\sum_{m=-K_n}^{0} |\mathcal{I}_{i+m,n}^{(l)}| \Big)^{-1} + \sum_{k=0}^{K_n} |\mathcal{I}_{i+k,n}^{(l)}| \Big(\sum_{m=0}^{K_n} |\mathcal{I}_{i+m,n}^{(l)}| \Big)^{-1} \Big)$$

$$\leq 2 G_{2,2}^n(T)$$

where we used the same index change trick as in (4.23).

Step 3. Using dominated convergence, we obtain $Y(P) \xrightarrow{\tilde{\mathbb{P}}} \tilde{D}_T^{DisJ}$ as $P \to \infty$ which allows to deduce

$$\tilde{\mathbb{P}}(Y(P) \leq \Upsilon | \mathcal{X}) \mathbb{1}_{\Omega_T^{DisJ}} \xrightarrow{\tilde{\mathbb{P}}} \tilde{\mathbb{P}}(\tilde{D}_T^{DisJ} \leq \Upsilon | \mathcal{X}) \mathbb{1}_{\Omega_T^{DisJ}} \qquad (9.35)$$

for $P \to \infty$. For details compare Step 3 in the proof of Proposition 4.10.

Step 4. Following the arguments in Step 4 in the proof of Proposition 4.10 the claim (9.29) follows from (9.30), (9.31) and (9.35). $\qquad \square$

Proof of Theorem 9.8. For proving (9.12) it is sufficient to show

$$\tilde{\mathbb{P}}(nV(f, \pi_n)_T > \overline{V}_-(4, f_{(2,2)}, \pi_n, (\beta, \varpi))_T + \hat{Q}_{T,n}^{DisJ}(1-\alpha) | F^{DisJ}) \to \alpha \quad (9.36)$$

for all $F^{DisJ} \subset \Omega_T^{DisJ}$ with $\mathbb{P}(F^{DisJ}) > 0$. To this end, note that we obtain

$$\hat{Q}_{n,T}^{DisJ}(\alpha) \mathbb{1}_{\Omega_T^{DisJ}} \xrightarrow{\tilde{\mathbb{P}}} Q_T^{DisJ}(\alpha) \mathbb{1}_{\Omega_T^{DisJ}} \qquad (9.37)$$

as $n \to \infty$ for each $\alpha \in [0,1]$ from Proposition 9.12; compare Step 1 in the proof of Theorem 4.3.

Then, Theorem 9.4 together with the convergences (9.10) and (9.37) yield using Lemma B.3 the $\mathcal{X}$-stable convergence

$$(nV(f, \pi_n)_T, \overline{V}_-(4, f_{(2,2)}, \pi_n, (\beta, \varpi))_T, \widehat{Q}_{n,T}^{DisJ}(1 - \alpha))$$
$$\xrightarrow{\mathcal{L}-s} (\widetilde{C}_T^{DisJ} + \widetilde{D}_T^{DisJ}, \widetilde{C}_T^{DisJ}, Q_T^{DisJ}(1 - \alpha)).$$

on Ω_T^{DisJ} from which we obtain using Lemma B.5

$$\widetilde{\mathbb{P}}(\{nV(f, \pi_n)_T > \overline{V}_-(4, f_{(2,2)}, \pi_n, (\beta, \varpi))_T + \widehat{Q}_{T,n}^{DisJ}(1 - \alpha)\} \cap F^{DisJ})$$
$$\to \widetilde{\mathbb{P}}(\{\widetilde{D}_T^{DisJ} > Q_T^{DisJ}(1 - \alpha)\} \cap F^{DisJ}) = \alpha\mathbb{P}(F^{DisJ}).$$

Here, the last equality follows from the definition of $Q_T^{DisJ}(\alpha)$. This convergence implies (9.36) and hence (9.12).

The consistency claim (9.13) follows from the fact that $\Psi_{n,T}^{DisJ}$ converges to a strictly positive limit on Ω_T^{CoJ} by Theorem 9.1 while $c_{T,n}^{DisJ}(\alpha) = o_{\widetilde{\mathbb{P}}}(1)$. To see this note that we have

$$c_{T,n}^{DisJ}(\alpha) = \frac{n^{-1}\overline{V}_-(4, f_{(2,2)}, \pi_n, (\beta, \varpi))_T + n^{-1}\widehat{Q}_{T,n}^{DisJ}(1 - \alpha)}{\sqrt{V^{(1)}(g_4, \pi_n)_T V^{(2)}(g_4, \pi_n)_T}}$$

where the denominator converges in probability to a strictly positive value and $n^{-1}\overline{V}_-(4, f_{(2,2)}, \pi_n, (\beta, \varpi))_T$ vanishes as $n \to \infty$ because big common jumps are eventually filtered out due to the indicator and because of (2.14). Further, it holds

$$n^{-1}\widehat{Q}_{T,n}^{DisJ}(1 - \alpha) \leq \frac{n^{-1}}{\lfloor(1 - \alpha)M_n\rfloor} \sum_{m=1}^{M_n} |\widehat{D}_{T,n,m}^{DisJ}|$$

where $\mathbb{E}[|\widehat{D}_{T,n,m}^{DisJ}| \| \mathcal{S}] = O_{\mathbb{P}}(1)$ follows using arguments from Step 2 in the proof of Proposition 9.12. □

9.2 Null Hypothesis of Joint Jumps

In this section we develop a statistical test which allows to decide whether common jumps are present or not in the paths of two stochastic processes under the null hypothesis that *common jumps are present*.

9.2.1 Theoretical Results

Here, like in Section 9.1, we will work with functionals based on the function $f_{(2,2)}$ and we consider a ratio statistic very similar to the one used in Chapter 8 to test for the presence of jumps. In fact the methods presented in this section can be understood as a natural extension of the methods based on the statistic $\Psi_{T,n}^{J,(l)}$ from Chapter 8 and we define

$$\Psi_{k,T,n}^{CoJ} = \frac{V(f_{(2,2)}, [k], \pi_n)_T}{k^2 V(f_{(2,2)}, \pi_n)_T}$$

for natural numbers $k \geq 2$.

Remark 9.13. *In the setting of equidistant observation times* $t_{i,n}^{(l)} = i/n$, $l = 1, 2$, *the statistic* $\Psi_{k,T,n}^{CoJ}$ *is equal to*

$$\frac{\sum_{i,j=k}^{\lfloor nT \rfloor} (X_{i/n}^{(1)} - X_{(i-k)/n}^{(1)})^2 (X_{j/n}^{(2)} - X_{(j-k)/n}^{(2)})^2 \mathbb{1}_{\{((i-k)/n, n] \cap ((j-k)/n, j/n] \neq \emptyset\}}}{k^2 \sum_{i=1}^{\lfloor nT \rfloor} (X_{i/n}^{(1)} - X_{(i-1)/n}^{(1)})^2 (X_{i/n}^{(2)} - X_{(i-1)/n}^{(2)})^2}.$$

In [32] a test for common jumps is constructed based on the statistic

$$\frac{\sum_{i=1}^{\lfloor nT/k \rfloor} (X_{ki/n}^{(1)} - X_{k(i-1)/n}^{(1)})^2 (X_{ki/n}^{(2)} - X_{k(i-1)/n}^{(2)})^2}{\sum_{i=1}^{\lfloor nT \rfloor} (X_{i/n}^{(1)} - X_{(i-1)/n}^{(1)})^2 (X_{i/n}^{(2)} - X_{(i-1)/n}^{(2)})^2} \qquad (9.38)$$

where at the lower observation frequency n/k *only increments over the intervals* $\mathcal{I}_{ki,k,n}^{(l)}$, $l = 1, 2$, *enter the estimation. Further in Section 14.1 of [3] a test for common jumps based on*

$$\frac{\sum_{i=k}^{\lfloor nT \rfloor} (X_{i/n}^{(1)} - X_{(i-k)/n}^{(1)})^2 (X_{i/n}^{(2)} - X_{(i-k)/n}^{(2)})^2}{k \sum_{i=1}^{\lfloor nT \rfloor} (X_{i/n}^{(1)} - X_{(i-1)/n}^{(1)})^2 (X_{i/n}^{(2)} - X_{(i-1)/n}^{(2)})^2} \qquad (9.39)$$

is discussed. As argued in Remark 8.1 it seems advantegeous to use the statistic (9.39) *over* (9.38). *However, in the asynchronous setting it is a priori not clear which observation intervals should be best paired, because there is no one-to-one correspondence of observation intervals in one process to observation intervals in the other process as there is in the synchronous situation. To use the available data as exhaustively as possible therefore here products of squared increments over all overlapping observation intervals at the lower observation frequency are included in the numerator of* $\Psi_{k,T,n}^{CoJ}$. $\qquad \square$

Under the null hypothesis that common jumps are present we obtain the following result from Theorem 2.3 and part b) of Theorem 2.40.

Theorem 9.14. *Suppose Condition 1.3 is fulfilled. Then it holds*

$$\Psi_{k,T,n}^{CoJ} \mathbb{1}_{\Omega_T^{CoJ}} \xrightarrow{\mathbb{P}} \mathbb{1}_{\Omega_T^{CoJ}}. \qquad (9.40)$$

Under the alternative $\omega \in \Omega_T^{nCoJ}$ the asymptotics of $\Psi_{k,T,n}^{CoJ}$ are more complicated. In fact by Theorems 2.3 and 2.40 both numerator and denominator of $\Psi_{k,T,n}^{CoJ}$ vanish on Ω_T^{nCoJ}. As in (8.3) we therefore expand the fraction by n to derive the asymptotics of $\Psi_{k,T,n}^{CoJ}$ on Ω_T^{nCoJ}. As shown in the proof of Theorem 9.4 it holds

$$nV(f_{(2,2)}, \pi_n)_T \mathbb{1}_{\Omega_T^{nCoJ}} \xrightarrow{\mathcal{L}-s} (\widetilde{C}_T^{DisJ} + \widetilde{D}_T^{DisJ}) \mathbb{1}_{\Omega_T^{nCoJ}}$$

and we will obtain a similar result for $n/k^2 V(f_{(2,2)}, [k], \pi_n)_T$. To describe the limit of $n/k^2 V(f_{(2,2)}, [k], \pi_n)_T$ on Ω_T^{nCoJ} and also the dependence structure of the limits of $n/k^2 V(f_{(2,2)}, [k], \pi_n)_T$ and $nV(f_{(2,2)}, \pi_n)_T$ we need to introduce some further notation. Recall the Brownian motion $\overline{W}_t$ defined in (3.23) and set

$$\eta_{k',n,-}^{(l)}(s) = \sum_{j \geq k' : t_{j,n}^{(3-l)} \leq T} \mathbb{1}_{\{s \in \mathcal{I}_{j,k',n}^{(3-l)}\}} \sum_{i:\mathcal{I}_{i,k',n}^{(l)} \cap \mathcal{I}_{j,k',n}^{(3-l)} \neq \emptyset} (\Delta_{i,k',n}^{(l)} \overline{W}^{(l)})^2 \mathbb{1}_{\{i < i_n^{(l)}(s)\}},$$

$$\eta_{k',n,+}^{(l)}(s) = \sum_{j \geq k' : t_{j,n}^{(3-l)} \leq T} \mathbb{1}_{\{s \in \mathcal{I}_{j,k',n}^{(3-l)}\}} \sum_{i:\mathcal{I}_{i,k',n}^{(l)} \cap \mathcal{I}_{j,k',n}^{(3-l)} \neq \emptyset} (\Delta_{i,k',n}^{(l)} \overline{W}^{(l)})^2 \mathbb{1}_{\{i \geq i_n^{(l)}(s)+k'\}},$$

$$Z_{k',n}^{DisJ,(l)}(s) = \left(\eta_{k',n,-}^{(l)}(s), \Delta_{i_n^{(l)}(s)-k'+1,n}^{(l)} \overline{W}^{(l)}, \ldots, \Delta_{i_n^{(l)}(s)-1,n}^{(l)} \overline{W}^{(l)}, s - t_{i_n^{(l)}(s)-1,n}^{(l)}, \right.$$
$$\left. t_{i_n^{(l)}(s),n}^{(l)} - s, \Delta_{i_n^{(l)}(s)+1,n}^{(l)} \overline{W}^{(l)}, \ldots, \Delta_{i_n^{(l)}(s)+k'-1,n}^{(l)} \overline{W}^{(l)}, \eta_{k',n,+}^{(l)}(s) \right)$$

for $k' = 1, k$. Here it holds $Z_{1,n}^{DisJ,(l)}(s) = Z_n^{DisJ,(l)}(s)$ for $Z_n^{DisJ,(l)}(s)$ defined as in Section 9.1. The necessary conditions required to obtain convergence of $\Psi_{k,T,n}^{CoJ}$ on Ω_T^{nCoJ} are summarized in the following condition.

Condition 9.15. *Suppose that Condition 1.3 holds and that σ almost surely fulfils $\int_0^T |\sigma_s^{(1)} \sigma_s^{(2)}| ds > 0$. Further,*

(i) *the functions $G_{2,2}^n(t), G_{2,2}^{[k],n}(t), H_{0,0,4}^n(t), H_{0,0,4}^{[k],n}(t)$, see (2.39) and (2.89) for their definition, converge pointwise on $[0,T]$ in probability to strictly increasing functions $G_{2,2}(t), \widetilde{G}_{2,2}^{[k]}(t), H_{0,0,4}(t), H_{0,0,4}^{[k]}(t): [0,\infty) \to [0,\infty)$.*

(ii) *The integral*

$$\int_{[0,T]^{P_1+P_2}} g(x_1, \ldots, x_{P_1}, x_1', \ldots, x_{P_2}') \mathbb{E}\left[\prod_{p=1}^{P_1} h_p^{(1)}\left(nZ_{1,n}^{DisJ,(1)}(x_p), \frac{n}{k^2} Z_{k,n}^{DisJ,(1)}(x_p)\right) \right.$$

$$\left. \times \prod_{p=1}^{P_2} h_p^{(2)}\left(nZ_{1,n}^{DisJ,(2)}(x_p'), \frac{n}{k^2} Z_{k,n}^{DisJ,(2)}(x_p')\right) \right] dx_1 \ldots dx_{P_1} dx_1' \ldots dx_{P_2}'$$

converges for $n \to \infty$ to

$$\int_{[0,T]^{P_1+P_2}} g(x_1,\ldots,x_{P_1},x_1',\ldots,x_{P_2}') \prod_{p=1}^{P_1} \int_{\mathbb{R}^2} h_p^{(1)}(y)\Gamma^{DisJ,[k],(1)}(x_p,dy)$$

$$\times \prod_{p=1}^{P_2} \int_{\mathbb{R}^2} h_p^{(2)}(y)\Gamma^{DisJ,[k],(2)}(x_p',dy)dx_1\ldots dx_{P_1}dx_1'\ldots dx_{P_2}'$$

for all bounded continuous functions $g\colon \mathbb{R}^{P_1+P_2} \to \mathbb{R}$, $h_p^{(l)}\colon \mathbb{R}^{2(k+1)} \to \mathbb{R}$, $p = 1,\ldots,P_l$, and any $P_l \in \mathbb{N}$, $l = 1,2$. Here $\Gamma^{DisJ,[k],(l)}(\cdot,dy)$, $l = 1,2$, are families of probability measures on $[0,T]$ where

$$(Z_1^{DisJ,(l)}(x), Z_k^{DisJ,(l)}(x)) \sim \Gamma^{DisJ,[k],(l)}(x,dy)$$

has first moments which are uniformly bounded in x. Further the components of $Z_k^{DisJ,(l)}(x)$ which correspond to the $\Delta_{i_n^{(l)}(x)+j,n}^{(l)}$, $\overline{W}^{(l)}$ have uniformly bounded second moments.

In order to describe the limit of $\Psi_{k,T,n}^{CoJ}$ on Ω_T^{nCoJ} we define

$$\widetilde{C}_{k',T}^{DisJ} = \int_0^T (\sigma_s^{(1)}\sigma_s^{(2)})^2 dG_{2,2}^{[k]}(s) + \int_0^T 2(\rho_s\sigma_s^{(1)}\sigma_s^{(2)})^2 dH_{0,0,4}^{[k]}(s), \quad k' = 1,k,$$

$$\widetilde{D}_{k',T}^{DisJ} = \sum_{p:S_p \leq T} ((\Delta X_{S_p}^{(1)})^2 R_{k'}^{(2)}(S_p) + (\Delta X_{S_p}^{(2)})^2 R_{k'}^{(1)}(S_p)), \quad k' = 1,k.$$

Here, we denote by S_p, $p \in \mathbb{N}$, an enumeration of the jump times of X and $R_{k'}^{(l)}(s)$ is defined via

$$R_{k'}^{(l)}(s) = (\sigma_{s-}^{(l)})^2\eta_{k',-}^{(l)}(s) + \sum_{i=1}^{k'} \left(\sigma_{s-}^{(l)} \sum_{j=-k'+i}^{-1} \chi_j + \sigma_{s-}^{(l)}\sqrt{\delta_-(s)}U_-^{(l)}(s)\right.$$

$$\left. + \sigma_s^{(l)}\sqrt{\delta_+(s)}U_+^{(l)}(s) + \sigma_s^{(l)}\sum_{j=1}^{i-1}\chi_j\right)^2 + (\sigma_s^{(l)})^2\eta_{k',+}^{(l)}(s), \quad l = 1,2, \quad k' = 1,k,$$

for random variables

$$(Z_1^{DisJ,(l)}(s), Z_k^{DisJ,(l)}(s)) = ((\eta_{1,-}^{(l)}(s),\delta_-(s),\delta_+(s),\eta_{1,+}^{(l)}(s)),$$

$$(\eta_{k,-}^{(l)}(s),\chi_{-k+1},\ldots,\chi_{-1},\delta_-(s),\delta_+(s),\chi_1,\ldots,\chi_{k-1},\eta_{k,+}^{(l)}(s)))$$

which are defined on an extended probability space $(\widetilde{\Omega}, \widetilde{\mathcal{F}}, \widetilde{\mathbb{P}})$ and whose distribution is given by

$$\widetilde{\mathbb{P}}^{(Z_1^{DisJ,(l)}(s), Z_k^{DisJ,(l)}(s))}(dy) = \Gamma^{DisJ,[k],(l)}(s, dy), \quad l = 1, 2.$$

The random variables $(Z_1^{DisJ,(l)}(s), Z_k^{DisJ,(l)}(s))$, $s \in [0, T]$, are independent of each other and independent of the process X and its components. The random variables $U_-^{(l)}(s), U_+^{(l)}(s)$ are i.i.d. standard normal and defined on $(\widetilde{\Omega}, \widetilde{\mathcal{F}}, \widetilde{\mathbb{P}})$ as well. Furthermore, they are independent of $\mathcal{F}$ and of the $(Z_1^{DisJ,(l)}(s), Z_k^{DisJ,(l)}(s))$. Here, $\widetilde{C}_{1,T}^{DisJ}$ and $\widetilde{D}_{1,T}^{DisJ}$ correspond to the terms $\widetilde{C}_T^{DisJ}$ and $\widetilde{D}_T^{DisJ}$ occuring in Theorem 9.4.

Theorem 9.16. *Under Condition 9.15 we have the $\mathcal{X}$-stable convergence*

$$\Psi_{k,T,n}^{CoJ} \xrightarrow{\mathcal{L}-s} \frac{k\widetilde{C}_{k,T}^{DisJ} + \widetilde{D}_{k,T}^{DisJ}}{\widetilde{C}_{1,T}^{DisJ} + \widetilde{D}_{1,T}^{DisJ}} \tag{9.41}$$

on Ω_T^{nCoJ}.

For the notion what it means for a sequence of random variables to converge $\mathcal{X}$-stably in law on a subset of Ω see the paragraph following Theorem 8.8.

From Theorem 9.14 we conclude that $\Psi_{k,T,n}^{CoJ}$ converges to 1 under the null hypothesis $\omega \in \Omega_T^{CoJ}$ and from Theorem 9.16 we obtain that $\Psi_{k,T,n}^{CoJ}$ converges under the alternative $\omega \in \Omega_T^{nCoJ}$ to a limit which is almost surely different from 1 if $k\widetilde{C}_{k,T}^{DisJ} \neq \widetilde{C}_{1,T}^{DisJ}$, or under mild additional conditions if there is at least one jump in one of the components of X on $[0, T]$. Indeed if we have $k\widetilde{C}_{k,T}^{DisJ} \neq \widetilde{C}_{1,T}^{DisJ}$, $\Psi_{k,T,n}^{CoJ}$ has to be different from 1 because $\widetilde{C}_{k,T}^{DisJ}, \widetilde{C}_{1,T}^{DisJ}$ are $\mathcal{F}$-measurable and $\widetilde{D}_{k,T}^{DisJ}$ and $\widetilde{D}_{1,T}^{DisJ}$ are either zero or their $\mathcal{F}$-conditional distributions admit densities. If $k\widetilde{C}_{k,T}^{DisJ} = \widetilde{C}_{1,T}^{DisJ}$ holds, mild conditions guarantee $\widetilde{D}_{k,T}^{DisJ} \neq \widetilde{D}_{1,T}^{DisJ}$ almost surely which also yields $\Psi_{k,T,n}^{CoJ} \not\to 1$. Hence, similarly as in (8.6), we will construct a test with critical region

$$\mathcal{C}_{k,T,n}^{CoJ} = \left\{ |\Psi_{k,T,n}^{CoJ} - 1| > c_{k,T,n}^{CoJ} \right\} \tag{9.42}$$

for a (possibly random) sequence $(c_{k,T,n}^{CoJ})_{n \in \mathbb{N}}$.

Like in the previous chapters we consider the situation where the observation times are generated by Poisson processes as an example for a random and irregular sampling scheme which fulfils the conditions we need for the testing procedure to work.

Example 9.17. *Condition 9.15 is fulfilled in the setting of Poisson sampling; compare Definition 5.1. Indeed, part (i) of Condition 9.15 is fulfilled because the functions $G^n_{2,2}(t), G^{[k],n}_{2,2}(t), H^n_{0,0,4}(t), H^{[k],n}_{0,0,4}(t)$ converge by Corollaries 5.6 and 5.8 to deterministic strictly increasing and linear functions. That part (ii) holds follows from Lemma 5.10.* □

Next, we derive a central limit theorem for $\Psi^{CoJ}_{k,T,n}$ on Ω^{CoJ}_T. However, before we start this endeavour, we restrict ourselves to the case $k = 2$. Already in this case, the form of the central limit theorem of $\Psi^{CoJ}_{k,T,n}$ is quite complicated. In the general case even the description of the limit would be very tedious due to the fact that we would have to keep track of numerous overlapping and non-overlapping interval parts. Further, from a practitioners perspective considering only $k = 2$ is not really a limitation as the simulation results in Section 8.2 for the test from Chapter 8 indicate that it is optimal to choose k as small as possible anyway.

To state the upcoming central limit theorem for $\Psi^{CoJ}_{k,T,n}$ we need to introduce some notation. To this end we denote by

$$\mathcal{L}^{CoJ,(l)}_n(s) = n|\mathcal{I}^{(l)}_{i^{(l)}_n(s)-1,n}|, \quad \mathcal{R}^{CoJ,(l)}_n(s) = n|\mathcal{I}^{(l)}_{i^{(l)}_n(s)+1,n}|, \quad l = 1,2,$$

$$\mathcal{L}^{CoJ}_n(s) = n|\mathcal{I}^{(1)}_{i^{(1)}_n(s)-1,n} \cap \mathcal{I}^{(2)}_{i^{(2)}_n(s)-1,n}|, \quad \mathcal{R}^{CoJ}_n(s) = n|\mathcal{I}^{(1)}_{i^{(1)}_n(s)+1,n} \cap \mathcal{I}^{(2)}_{i^{(2)}_n(s)+1,n}|$$

lengths of certain intervals around some time s at which a common jump might occur; for an illustration see Figure 9.5. Further we will use the shorthand notation

$$Z^{CoJ}_n(s) = \left(\mathcal{L}^{CoJ,(1)}_n, \mathcal{R}^{CoJ,(1)}_n, \mathcal{L}^{CoJ,(2)}_n, \mathcal{R}^{CoJ,(2)}_n, \mathcal{L}^{CoJ}_n, \mathcal{R}^{CoJ}_n\right)(s).$$

Note that, although they are of similar structure, the random variable $Z^{CoJ}_n(s)$ differs both from $Z^{biv}_n(s)$ and $Z^{COV}_n(s)$ defined in (3.20) repsectively (7.8) and used in Section 3.1 respectively Section 7.2.

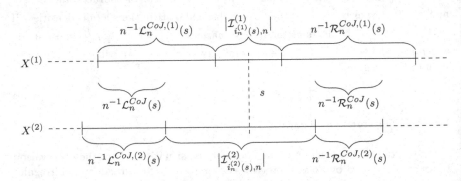

Figure 9.5: Illustration of the components of $Z^{CoJ}_n(s)$.

Limits of the variables $\mathcal{L}_n^{CoJ,(l)}(s), \mathcal{R}_n^{CoJ,(l)}(s), l = 1, 2,$ and $\mathcal{L}_n^{CoJ}(s), \mathcal{R}_n^{CoJ}(s)$ will occur in the central limit theorem for $\Psi_{2,T,n}^{DisJ}$. To ensure convergence of the $Z_n^{CoJ}(s)$ we need to impose the following assumption on the observation scheme.

Condition 9.18. *Assume that Condition 1.3 is fulfilled, that $G_{2,2}^{[2],n} = O_\mathbb{P}(1)$ holds and that we have $\sqrt{n}|\pi_n|_T = O_\mathbb{P}(1)$. Further suppose that the integral*

$$\int_{[0,T]^P} g(x_1, \ldots, x_P)\mathbb{E}\Big[\prod_{p=1}^{P} h_p(Z_n^{CoJ}(x_p))\Big] dx_1 \ldots dx_P$$

converges for $n \to \infty$ to

$$\int_{[0,T]^P} g(x_1, \ldots, x_P) \prod_{p=1}^{P} \int_\mathbb{R} h_p(y)\Gamma^{CoJ}(x_p, dy) dx_1 \ldots dx_P$$

for all bounded continuous functions $g\colon \mathbb{R}^P \to \mathbb{R}$, $h_p\colon \mathbb{R}^6 \to \mathbb{R}$ and any $P \in \mathbb{N}$. Here $\Gamma^{CoJ}(\cdot, dy)$ is a family of probability measures on $[0, T]$ with uniformly bounded first moments.

Using the limit distribution of $Z_n^{CoJ}(s)$ implicitly defined in Condition 9.18 we set

$$\Phi_{2,T}^{CoJ} = 4 \sum_{p:S_p \leq T} \Delta X_{S_p}^{(1)} \Delta X_{S_p}^{(2)}$$

$$\times \Big[\Delta X_{S_p}^{(2)}\Big(\sigma_{S_p-}^{(1)}\sqrt{\mathcal{L}^{CoJ}(S_p)}U_{S_p}^{(1),-} + \sigma_{S_p}^{(1)}\sqrt{\mathcal{R}^{CoJ}(S_p)}U_{S_p}^{(1),+}$$

$$+ \sqrt{(\sigma_{S_p-}^{(1)})^2(\mathcal{L}^{CoJ,(1)} - \mathcal{L}^{CoJ})(S_p) + (\sigma_{S_p}^{(1)})^2(\mathcal{R}^{CoJ,(1)} - \mathcal{R}^{CoJ})(S_p)}U_{S_p}^{(2)}\Big)$$

$$+ \Delta X_{S_p}^{(1)}\Big(\sigma_{S_p-}^{(2)}\rho_{S_p-}\sqrt{\mathcal{L}^{CoJ}(S_p)}U_{S_p}^{(1),-} + \sigma_{S_p}^{(2)}\rho_{S_p}\sqrt{\mathcal{R}^{CoJ}(S_p)}U_{S_p}^{(1),+}$$

$$+ \sqrt{(\sigma_{S_p-}^{(2)})^2(1 - (\rho_{S_p-})^2)\mathcal{L}^{CoJ}(S_p) + (\sigma_{S_p}^{(2)})^2(1 - (\rho_{S_p})^2)\mathcal{R}^{CoJ}(S_p)}U_{S_p}^{(3)}$$

$$+ \sqrt{(\sigma_{S_p-}^{(2)})^2(\mathcal{L}^{CoJ,(2)} - \mathcal{L}^{CoJ})(S_p) + (\sigma_{S_p}^{(2)})^2(\mathcal{R}^{CoJ,(2)} - \mathcal{R}^{CoJ})(S_p)}U_{S_p}^{(4)}\Big)\Big].$$

Here, we denote by S_p, $p \in \mathbb{N}$, an enumeration of the common jump times of $X^{(1)}$ and $X^{(2)}$, the vector

$$(\mathcal{L}^{CoJ,(1)}, \mathcal{R}^{CoJ,(1)}, \mathcal{L}^{CoJ,(2)}, \mathcal{R}^{CoJ(2)}, \mathcal{L}^{CoJ}, \mathcal{R}^{CoJ})(s)$$

is distributed according to $\Gamma^{CoJ}(s, \cdot)$ and the vectors

$$(U_s^{(1),-}, U_s^{(1),+}, U_s^{(2)}, U_s^{(3)}, U_s^{(4)})$$

are standard normally distributed and independent for different values of s. Similarly as for (8.8) we obtain that the infinite sum in the definition of $\Phi_{2,T}^{CoJ}$ has a well-defined limit. Using the variable $\Phi_{2,T}^{CoJ}$ we are able to state the following central limit theorem.

Theorem 9.19. *If Condition 9.18 is fulfilled, we have the $\mathcal{X}$-stable convergence*

$$\sqrt{n}\left(\Psi_{2,T,n}^{CoJ} - 1\right) \xrightarrow{\mathcal{L}-s} \frac{\Phi_{2,T}^{CoJ}}{4B^*(f_{(2,2)})_T} \tag{9.43}$$

on the set Ω_T^{CoJ}.

For the notion what it means for a sequence of random variables to converge $\mathcal{X}$-stably in law on a subset of Ω see the paragraph following Theorem 8.8.

Example 9.20. *Condition 9.18 is fulfilled in the Poisson setting. The property $\sqrt{n}|\pi_n|_T = O_{\mathbb{P}}(1)$ follows from (5.2) and $G_{2,2}^{[2],n} = O_{\mathbb{P}}(1)$ holds true due to Corollary 5.8 as discussed in Example 9.17. Further the convergence of the integrals follows like part (ii) of Condition 9.15 from Lemma 5.10.* □

Example 9.21. *Condition 9.18 is also fulfilled in the setting of equidistant observation times $t_{i,n}^{(1)} = t_{i,n}^{(2)} = i/n$. In that case we have*

$$\left(\mathcal{L}^{CoJ,(1)}, \mathcal{R}^{CoJ,(1)}, \mathcal{L}^{CoJ,(2)}, \mathcal{R}^{CoJ,(2)}, \mathcal{L}^{CoJ}, \mathcal{R}^{CoJ}\right)(s) = (1,1,1,1,1,1)$$

for any $s \in (0,T]$. Hence we get

$$\Phi_{2,T}^{CoJ} = 4\sum_{p:S_p \leq T} \Delta X_{S_p}^{(1)} \Delta X_{S_p}^{(2)} \left[\Delta X_{S_p}^{(2)}\left(\sigma_{S_p-}^{(1)} U_{S_p}^{(1),-} + \sigma_{S_p}^{(1)} U_{S_p}^{(1),+}\right)\right.$$
$$+ \Delta X_{S_p}^{(1)}\left(\sigma_{S_p-}^{(2)} \rho_{S_p-} U_{S_p}^{(1),-} + \sigma_{S_p}^{(2)} \rho_{S_p} U_{S_p}^{(1),+}\right)$$
$$\left. + \sqrt{(\sigma_{S_p-}^{(2)})^2(1-(\rho_{S_p-})^2) + (\sigma_{S_p}^{(2)})^2(1-(\rho_{S_p})^2)} U_{S_p}^{(3)}\right)\right]$$

and $\Phi_{2,T}^{CoJ}$ is $\mathcal{X}$-conditionally normally distributed with mean zero and variance

$$16\sum_{p:S_p \leq T} (\Delta X_{S_p}^{(1)})^2(\Delta X_{S_p}^{(2)})^2 \left[\left(\Delta X_{S_p}^{(2)}\sigma_{S_p-}^{(1)} + \Delta X_{S_p}^{(1)}\sigma_{S_p-}^{(2)} \rho_{S_p-}\right)^2\right.$$
$$+ \left(\Delta X_{S_p}^{(2)}\sigma_{S_p}^{(1)} + \Delta X_{S_p}^{(1)}\sigma_{S_p}^{(2)} \rho_{S_p}\right)^2 + (\Delta X_{S_p}^{(1)})^2\left((\sigma_{S_p-}^{(2)})^2(1-(\rho_{S_p-})^2)\right.$$
$$\left.\left. + (\sigma_{S_p}^{(2)})^2(1-(\rho_{S_p})^2)\right)\right]$$

which is similar to the result in Theorem 4.1(a) of [32]. □

Using the central limit theorem 9.19 and the methods from Chapter 4 we are finally able to construct the statistical test. To this end let $(\widetilde{K}_n)_{n\in\mathbb{N}}$ and $(M_n)_{n\in\mathbb{N}}$ be sequences of natural numbers which tend to infinity and set

$$
\begin{aligned}
\widehat{\mathcal{L}}_{n,m}^{CoJ,(l)}(s) &= n\Big|\mathcal{I}_{i_n^{(l)}(s)+V_{n,m}^{(l)}(s)-1,n}^{(l)}\Big|,\ l=1,2,\\[2mm]
\widehat{\mathcal{R}}_{n,m}^{CoJ,(l)}(s) &= n\Big|\mathcal{I}_{i_n^{(l)}(s)+V_{n,m}^{(l)}(s)+1,n}^{(l)}\Big|,\ l=1,2,\\[2mm]
\widehat{\mathcal{L}}_{n,m}^{CoJ}(s) &= n\Big|\mathcal{I}_{i_n^{(1)}(s)+V_{n,m}^{(1)}(s)-1,n}^{(l)}\cap\mathcal{I}_{i_n^{(2)}(s)+V_{n,m}^{(2)}(s)-1,n}^{(l)}\Big|\\[2mm]
\widehat{\mathcal{R}}_{n,m}^{CoJ}(s) &= n\Big|\mathcal{I}_{i_n^{(1)}(s)+V_{n,m}^{(1)}(s)+1,n}^{(l)}\cap\mathcal{I}_{i_n^{(2)}(s)+V_{n,m}^{(2)}(s)+1,n}^{(l)}\Big|
\end{aligned}
\tag{9.44}
$$

where $V_{n,m}(s)=(V_{n,m}^{(1)}(s),V_{n,m}^{(2)}(s))\in\{-\widetilde{K}_n,\dots,\widetilde{K}_n\}\times\{-\widetilde{K}_n,\dots,\widetilde{K}_n\}$ and

$$
\mathbb{P}\big(V_{n,m}(s)=(\tilde{k}_1,\tilde{k}_2)\big|\mathcal{S}\big)
$$

$$
=\Big|\mathcal{I}_{i_n^{(1)}(s)+\tilde{k}_1,n}^{(1)}\cap\mathcal{I}_{i_n^{(2)}(s)+\tilde{k}_2,n}^{(2)}\Big|\Big(\sum_{\tilde{k}_1',\tilde{k}_2'=-\widetilde{K}_n}^{\widetilde{K}_n}\Big|\mathcal{I}_{i_n^{(1)}(s)+\tilde{k}_1',n}^{(1)}\cap\mathcal{I}_{i_n^{(2)}(s)+\tilde{k}_2',n}^{(2)}\Big|\Big)^{-1}
\tag{9.45}
$$

for $\tilde{k}_1,\tilde{k}_2\in\{-\widetilde{K}_n,\dots,\widetilde{K}_n\}$. Based on (9.44) we define via

$$
\widehat{Z}_{n,m}^{CoJ}(s)=\big(\widehat{\mathcal{L}}_{n,m}^{CoJ,(1)},\widehat{\mathcal{R}}_{n,m}^{CoJ,(1)},\widehat{\mathcal{L}}_{n,m}^{CoJ,(2)},\widehat{\mathcal{R}}_{n,m}^{CoJ,(2)},\widehat{\mathcal{L}}_{n,m}^{CoJ},\widehat{\mathcal{R}}_{n,m}^{CoJ}\big)(s),
$$

$m=1,\dots,M_n$, estimators for realizations of $Z^{CoJ}(s)$. Using the components of $\widehat{Z}_{n,m}^{CoJ}(s)$ and considering increments which are larger than a certain threshold as jumps, compare Section 2.3, we then define

$$
\begin{aligned}
\widehat{\Phi}_{2,T,n,m}^{CoJ} = 4\sum_{i,j:t_{i,n}^{(1)}\vee t_{j,n}^{(2)}\leq T} &\Delta_{i,n}^{(1)}X^{(1)}\Delta_{j,n}^{(2)}X^{(2)}\\
&\times\mathbb{1}_{\{|\Delta_{i,n}^{(1)}X^{(1)}|>\beta|\mathcal{I}_{i,n}^{(1)}|^{\varpi}\}}\mathbb{1}_{\{|\Delta_{j,n}^{(2)}X^{(2)}|>\beta|\mathcal{I}_{j,n}^{(2)}|^{\varpi}\}}\\
&\times\Big[\Delta_{j,n}^{(2)}X^{(2)}\Big(\tilde{\sigma}^{(1)}(t_{i,n}^{(1)},-)\sqrt{\widehat{\mathcal{L}}_{n,m}^{CoJ}(\tau_{i,j}^n)}U_{n,(i,j),m}^{(1,2),-}\\
&\quad+\tilde{\sigma}^{(1)}(t_{i,n}^{(1)},+)\sqrt{\widehat{\mathcal{R}}_{n,m}^{CoJ}(\tau_{i,j}^n)}U_{n,(i,j),m}^{(1,2),+}\\
&\quad+\Big((\tilde{\sigma}^{(1)}(t_{i,n}^{(1)},-))^2(\widehat{\mathcal{L}}_{n,m}^{CoJ,(1)}-\widehat{\mathcal{L}}_{n,m}^{CoJ})(\tau_{i,j}^n)\\
&\qquad+(\tilde{\sigma}^{(1)}(t_{i,n}^{(1)},+))^2(\widehat{\mathcal{R}}_{n,m}^{CoJ,(1)}-\widehat{\mathcal{R}}_{n,m}^{CoJ})(\tau_{i,j}^n)\Big)^{1/2}U_{n,i,m}^{(1)}\Big)\\
&+\Delta_{i,n}^{(1)}X^{(1)}\Big(\tilde{\sigma}^{(2)}(t_{j,n}^{(2)},-)\tilde{\rho}(\tau_{i,j}^n,-)\sqrt{\widehat{\mathcal{L}}_{n,m}^{CoJ}(\tau_{i,j}^n)}U_{n,(i,j),m}^{(1,2),-}
\end{aligned}
$$

$$+ \tilde{\sigma}^{(2)}(t_{j,n}^{(2)}, +)\tilde{\rho}(\tau_{i,j}^n, +)\sqrt{\widehat{\mathcal{R}}_{n,m}^{CoJ}(\tau_{i,j}^n)}U_{n,(i,j),m}^{(1,2),+}$$

$$+ \left((\tilde{\sigma}^{(2)}(t_{j,n}^{(2)}, -))^2(1 - (\tilde{\rho}(\tau_{i,j}^n, -))^2)\widehat{\mathcal{L}}_{n,m}^{CoJ}(\tau_{i,j}^n)\right.$$

$$\left.+ (\tilde{\sigma}^{(2)}(t_{j,n}^{(2)}, +))^2(1 - (\tilde{\rho}(\tau_{i,j}^n, +))^2)\widehat{\mathcal{R}}_{n,m}^{CoJ}(\tau_{i,j}^n)\right)^{1/2}U_{n,j,m}^{(2)}$$

$$+ \left((\tilde{\sigma}^{(2)}(t_{j,n}^{(2)}, -))^2(\widehat{\mathcal{L}}_{n,m}^{CoJ,(2)} - \widehat{\mathcal{L}}_{n,m}^{CoJ})(\tau_{i,j}^n)\right.$$

$$\left.+ (\tilde{\sigma}^{(2)}(t_{j,n}^{(2)}, +))^2(\widehat{\mathcal{R}}_{n,m}^{CoJ,(2)} - \widehat{\mathcal{R}}_{n,m}^{CoJ})(\tau_{i,j}^n)\right)^{1/2}U_{n,j,m}^{(3)}\right)\right]$$

$$\times \mathbb{1}_{\{\mathcal{I}_{i,n}^{(1)} \cap \mathcal{I}_{j,n}^{(2)} \neq \emptyset\}}$$

with $\tau_{i,j}^n = t_{i,n}^{(1)} \wedge t_{j,n}^{(2)}$. Here, we use the estimators $\tilde{\sigma}^{(l)}(s, -)$, $\tilde{\sigma}^{(l)}(s, +)$, $l = 1, 2$, respectively $\tilde{\rho}(s, -)$, $\tilde{\rho}(s, +)$ introduced in Chapter 6 for the estimation of $\sigma_{s-}^{(l)}$, $\sigma_s^{(l)}$, $l = 1, 2$, and ρ_{s-}, ρ_s. The $U_{n,(i,j),m}^{(1,2),-}$, $U_{n,(i,j),m}^{(1,2),+}$, $U_{n,i,m}^{(1)}$, $U_{n,j,m}^{(2)}$, $U_{n,j,m}^{(3)}$ are i.i.d. standard normal distributed random variables and we choose $\beta > 0$ and $\varpi \in (0, 1/2)$. Further, we denote by

$$\widehat{Q}_{2,T,n}^{CoJ}(\alpha) = \widehat{Q}_\alpha(\{|\widehat{\Phi}_{2,T,n,m}^{CoJ}| \big| m = 1, \dots, M_n\})$$

the $\lfloor \alpha M_n \rfloor$-th largest element of the set $\{|\widehat{\Phi}_{2,T,n,m}^{CoJ}| \big| m = 1, \dots, M_n\}$ and we will see that $\widehat{Q}_{2,T,n}^{CoJ}(\alpha)$ converges on Ω_T^{CoJ} under appropriate conditions to the $\mathcal{X}$-conditional α-quantile $Q_{2,T}^{CoJ}(\alpha)$ of $|\Phi_{2,T}^{CoJ}|$ which is defined via

$$\widetilde{\mathbb{P}}(|\Phi_{2,T}^{CoJ}| \leq Q_{2,T}^{CoJ}(\alpha)|\mathcal{X})(\omega) = \alpha, \quad \omega \in \Omega_T^{CoJ},$$

and we set $(Q_{2,T}^{CoJ}(\alpha))(\omega) = 0$, $\omega \in \Omega_T^{nCoJ}$, for completeness. Such a random variable $Q_{2,T}^{CoJ}(\alpha)$ exists because Condition 9.23 will guarantee that the $\mathcal{X}$-conditional distribution of $\Phi_{2,T}^{CoJ}$ is almost surely continuous on Ω_T^{CoJ}.

Remark 9.22. *Note that we could also define the estimator $\widehat{\Phi}_{2,T,n,m}^{CoJ}$ as the sum over all terms where in the product $\Delta_{i,n}^{(1)}X^{(1)}\Delta_{j,n}^{(2)}X^{(2)}$ at least one increment has to be large. This difference has asymptotically no effect because the contribution of idiosyncratic jumps vanishes in the limit. In the corresponding test based on equidistant and synchronous observations from [32] this alternative approach is taken. Although idiosyncratic jumps play no role asymptotically they contribute to the approximation error. Therefore including them in the estimator $\widehat{\Phi}_{2,T,n,m}^{CoJ}$ helps to diminish this approximation error because we estimate part of it. On the contrary, this approach yields that the sum in $\widehat{\Phi}_{2,T,n,m}^{CoJ}$ includes on average much more terms which slows down the numerical implementation of the test. With the simulation study, which will be presented in Section 9.2.2, in mind we therefore* *chose only to include terms where increments of both processes are identified as jumps.* □

The necessary assumptions to obtain a valid statistical test are summarized in the following condition.

Condition 9.23. *The process X and the sequence of observation schemes $(\pi_n)_{n \in \mathbb{N}}$ fulfil Condition 9.15 and 9.18. Further the set $\{s \in [0, T] : \|\sigma_s\| = 0\}$ is almost surely a Lebesgue null set and $\int_0^T \Gamma^{CoJ}(x, \{0\}^6) dx = 0$. $(\widetilde{K}_n)_{n \in \mathbb{N}}$ and $(M_n)_{n \in \mathbb{N}}$ are sequences of natural numbers converging to infinity with $\widetilde{K}_n/n \to 0$ and the sequence $(b_n)_{n \in \mathbb{N}}$ fulfils $|\pi_n|_T/b_n \overset{\mathbb{P}}{\longrightarrow} 0$. Additionally,*

(i) for any $x_p \in \mathbb{R}^6$, $s_p \in (0, T)$, $p = 1, \dots, P$, $P \in \mathbb{N}$, with $s_p \neq s_{p'}$ for $p \neq p'$ it holds

$$\widetilde{\mathbb{P}}\Big(\big|\widetilde{\mathbb{P}}(\widehat{Z}_{n,1}^{CoJ}(s_p) \leq x_p, \ p = 1, \dots, P | \mathcal{S}) - \prod_{p=1}^{P} \widetilde{\mathbb{P}}(Z^{CoJ}(s_p) \leq x_p)\big| > \varepsilon\Big) \to 0$$

as $n \to \infty$ for all $\varepsilon > 0$.

(ii) We have $n^{1/2}(|\pi_n|_T)^{1-\varpi} = o_{\mathbb{P}}(1)$ and $G_{1,2}^n(T) = O_{\mathbb{P}}(1)$, $G_{2,1}^n(T) = O_{\mathbb{P}}(1)$.

(iii) The process σ is a 2×2-dimensional matrix-valued Itô semimartingale.

(iv) It holds $2\widetilde{C}_{2,T}^{DisJ} \neq \widetilde{C}_{1,T}^{DisJ}$ or $\widetilde{D}_{2,T}^{DisJ} \neq \widetilde{D}_{1,T}^{DisJ}$ almost surely.

Part (i) of Condition 9.23 yields that $\widehat{Q}_{2,T,n}^{CoJ}(\alpha)$ consistently estimates $Q_{2,T}^{CoJ}(\alpha)$. Part (ii) is a technical condition which will be needed in the proof of Theorem 9.24. Part (iii) provides that $\tilde{\sigma}_n^{(l)}(S_p, -)$, $\tilde{\sigma}_n^{(l)}(S_p, +)$, $\tilde{\rho}_n(S_p, -)$, $\tilde{\rho}_n(S_p, +)$ are consistent estimators of $\sigma_{S_p-}^{(l)}$, $\sigma_{S_p}^{(l)}$, ρ_{S_p-}, ρ_{S_p}, and part (iv) guarantees that the limit under the alternative is almost surely different from 1. Note also that Condition 9.23(iii) in general does not follow from the convergence of $G_{2,2}^n(T)$ assumed in Condition 9.18.

Theorem 9.24. *Let Condition 9.23 be fulfilled. Then the test defined in (9.42) with*

$$c_{2,T,n}^{CoJ}(\alpha) = \frac{\widehat{Q}_{2,T,n}^{CoJ}(1-\alpha)}{\sqrt{n}4V(f_{(2,2)}, \pi_n)_T}, \quad \alpha \in [0,1], \quad \alpha \in [0,1],$$

has asymptotic level α in the sense that we have

$$\widetilde{\mathbb{P}}\big(|\Psi_{2,T,n}^{CoJ} - 1| > c_{2,T,n}^{CoJ}(\alpha)\big|F^{CoJ}\big) \to \alpha \quad \alpha \in (0,1], \tag{9.46}$$

for all $F^{CoJ} \subset \Omega_T^{CoJ}$ with $\mathbb{P}(F^{CoJ}) > 0$. *Further the test is consistent in the sense that we have*

$$\widetilde{\mathbb{P}}\big(|\Psi_{2,T,n}^{CoJ} - 1| > c_{2,T,n}^{CoJ}(\alpha)\big|F^{nCoJ}\big) \to 1 \qquad (9.47)$$

for all $F^{nCoJ} \subset \Omega_T^{nCoJ}$ with $\mathbb{P}(F^{nCoJ}) > 0$.

To improve the performance of our test in the finite sample we adjust the estimator $\Psi_{2,T,n}^{CoJ}$, similarly as was done in Chapter 8 to obtain the test in Corollary 8.12, by (partially) subtracting terms that contribute in the finite sample but vanish asymptotically. Therefore we define for β, ϖ as above the quantity

$$A_{2,T,n}^{CoJ} = n \sum_{i,j:t_{i,n}^{(1)} \vee t_{j,n}^{(2)} \leq T} (\Delta_{i,2,n}^{(1)} X^{(1)})^2 (\Delta_{j,2,n}^{(2)} X^{(2)})^2 \mathbb{1}_{\{\mathcal{I}_{i,2,n}^{(1)} \cap \mathcal{I}_{j,2,n}^{(2)} \neq \emptyset\}}$$

$$\times \mathbb{1}_{\{|\Delta_{i,2,n}^{(1)} X^{(1)}| \leq \beta|\mathcal{I}_{i,2,n}^{(1)}|^\varpi \vee |\Delta_{j,2,n}^{(2)} X^{(2)}| \leq \beta|\mathcal{I}_{j,2,n}^{(2)}|^\varpi\}}$$

$$- 4n \sum_{i,j:t_{i,n}^{(1)} \vee t_{j,n}^{(2)} \leq T} (\Delta_{i,n}^{(1)} X^{(1)})^2 (\Delta_{j,n}^{(2)} X^{(2)})^2 \mathbb{1}_{\{\mathcal{I}_{i,n}^{(1)} \cap \mathcal{I}_{j,n}^{(2)} \neq \emptyset\}}$$

$$\times \mathbb{1}_{\{|\Delta_{i,n}^{(1)} X^{(1)}| \leq \beta|\mathcal{I}_{i,n}^{(1)}|^\varpi \vee |\Delta_{j,n}^{(2)} X^{(2)}| \leq \beta|\mathcal{I}_{j,n}^{(2)}|^\varpi\}}.$$

Using this expression we then define for $\rho \in (0,1)$ by

$$\widetilde{\Psi}_{2,T,n}^{CoJ}(\rho) = \Psi_{2,T,n}^{CoJ} - \rho \frac{n^{-1} A_{2,T,n}^{CoJ}}{4V(f_{(2,2)}, \pi_n)T}$$

the adjusted estimator.

Corollary 9.25. *Let $\rho \in (0,1)$. If Condition 9.23 is fulfilled, then it holds with the notation of Theorem 9.24*

$$\widetilde{\mathbb{P}}\big(|\widetilde{\Psi}_{2,T,n}^{CoJ}(\rho) - 1| > c_{2,T,n}^{CoJ}(\alpha)\big|F^{CoJ}\big) \to \alpha \quad \alpha \in [0,1], \qquad (9.48)$$

for all $F^{CoJ} \subset \Omega_T^{CoJ}$ with $\mathbb{P}(F^{CoJ}) > 0$ and

$$\widetilde{\mathbb{P}}\big(|\widetilde{\Psi}_{2,T,n}^{CoJ}(\rho) - 1| > c_{2,T,n}^{CoJ}(\alpha)\big|F^{nCoJ}\big) \to 1 \quad \alpha \in (0,1], \qquad (9.49)$$

for all $F^{nCoJ} \subset \Omega_T^{DisJ}$ with $\mathbb{P}(F^{nCoJ}) > 0$.

Example 9.26. *The assumptions on the observation scheme from Condition 9.23 are fulfilled in the setting of Poisson sampling discussed in Examples 9.17 and 9.20. That part (i) of Condition 9.23 holds follows from Lemma 5.12. Part (ii) is fulfilled by (5.2) (note that $1 - \varpi > 1/2$ due to $\varpi \in (0,1/2)$) and Corollary 5.6. The property $\int_0^T \Gamma^{CoJ}(x, \{0\}^6)dx = 0$ follows from the considerations in Remark 5.11* $\qquad \square$

Similarly as discussed in Chapter 8 after Example 8.13 we can omit the weighting in (9.45) for obtaining a working testing procedure also within the bivariate Poisson setting. However, in the setting where both processes $X^{(1)}$ and $X^{(2)}$ are observed at the observation times introduced in Example 8.14 with different $\alpha_1 \neq \alpha_2$ it can be easily verified that the weighting in (9.45) is necessary and leads to a correct estimation of the distribution of $\Phi_{2,T}^{CoJ}$.

9.2.2 Simulation Results

To evaluate the finite sample performance of the tests described in Theorem 9.24 and in Corollary 9.25 we conduct a simulation study. Here, we use the same model and parameter settings as in Section 9.1.2; compare Table 9.1.

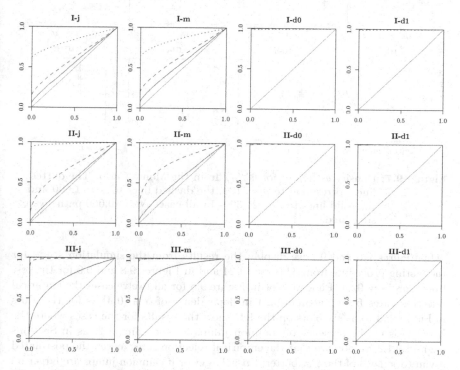

Figure 9.6: Empirical rejection curves from the Monte Carlo simulation for the test derived from Theorem 9.24. The dotted lines represent the results for $n = 100$, the dashed lines for $n = 1,600$ and the solid lines for $n = 25,600$. In each case $N = 10,000$ paths were simulated.

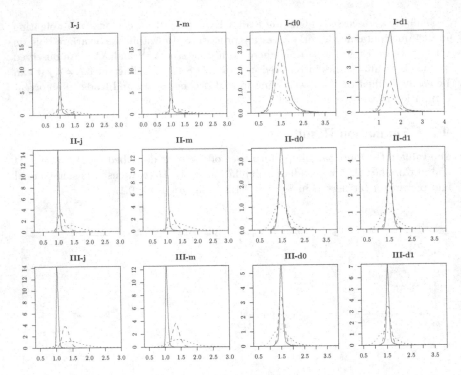

Figure 9.7: Density estimates for $\Phi_{2,T,n}^{CoJ}$ from the Monte Carlo. The dotted
lines correspond to $n = 100$, the dashed lines to $n = 1,600$ and
the solid lines to $n = 25,600$. In all cases $N = 10,000$ paths were
simulated.

In Figures 9.6 and 9.7 we display the results from the simulation study for
the testing procedure from Theorem 9.24 and in Figures 9.8 and 9.9 for the test
from Corollary 9.25. First we plot in Figure 9.6 for all twelve cases the empirical
rejection rates from Theorem 9.24 in dependence of $\alpha \in [0,1]$ as for the plots
in Figure 9.3. The six plots on the left show the results for the cases where the
hypothesis of the presence of common jumps is true. Similarly as in Sections
8.2 and 9.1.2 we observe that the empirical rejection rates match the postulated
asymptotic level of the test better if n is larger or if common jumps are larger on
average. In all six cases where common jumps are present and for all values of
n the test overrejects. This is due to the fact that the asymptotically negligible
terms in $\Psi_{2,T,n}^{CoJ}$ (terms which are contained in $\Psi_{2,T,n}^{CoJ}$ but do not contribute in
the limit) tend to be positive, hence $\Psi_{2,T,n}^{CoJ}$ is on average systematically larger
than 1, see Figure 9.7, which yields the bias. However, at least for $n = 25,600$

the observed rejection rates match the asymptotic level quite well. Only in the cases III-j and III-m, where the jumps are on average very small, the empirical rejection rates are still far higher than the asymptotic level. The results are worse in the mixed model than in the model where there are only common jumps because idiosyncratic jumps contribute to the asymptotically vanishing error. The test has very good power against the alternative of idiosyncratic jumps as can be seen in the six plots on the right-hand side of Figure 9.6 which correspond to the cases where there are no common jumps.

Figure 9.7 shows density estimates for $\Psi_{2,T,n}^{CoJ}$ in all twelve cases. If there are common jumps it is visible in the density plots that $\Psi_{2,T,n}^{CoJ}$ converges to 1 as $n \to \infty$. However for $n = 100$, $n = 1{,}600$ in the cases II-j, II-m and for $n = 100$, $n = 1{,}600$, $n = 25{,}600$ in the cases III-j, III-m the density peaks at a value significantly larger than 1 which corresponds to the overrejection in Figure 9.6. Under the alternative of disjoint jumps $\Psi_{2,T,n}^{CoJ}$ tends to cluster around 1.5 which corresponds to the results obtained in Example 8.5 in the univariate setting.

Our simulation results from Theorem 9.24 are worse than the results in the equidistant setting displayed in Figure 4 of [32] while the power of our test is much better. This effect is partly due to the fact that, contrary to our approach, in [32] idiosyncratic jumps, although their contribution is asymptotically negligible, are included in the estimation of the asymptotic variance in the central limit theorem; compare Remark 9.22. Hence they consistently overestimate the asymptotic variance which yields lower rejection rates. Further, the asymptotically negligible terms in $\Psi_{2,T,n}^{CoJ}$ are larger relative to the asymptotically relevant terms in the asynchronous setting than in the setting of synchronous observation times which also increases the rejection rates in the asynchronous setting.

The test from Corollary 9.25 outperforms the test from Theorem 9.24 in the simulation study. This can be seen in Figures 9.8 and 9.9 which show the results from the Monte Carlo simulation for the test from Corollary 9.25 with $\rho = 0.75$ (for the choice of $\rho = 0.75$ see Figure 9.10) in the same fashion as for Theorem 9.24 in Figures 9.6 and 9.7. In the cases where common jumps are present we observe that the empirical rejection rates match the asymptotic level much better than in Figure 9.6. In Figure 9.8 we see that in the cases I-j, I-m, II-j, II-m we get good results already for $n = 1{,}600$ and in the cases III-j, III-m at least for $n = 25{,}600$. The power of the test from Corollary 9.25 is practically as good as for the test from Theorem 9.24. Hence using the adjusted statistic $\widetilde{\Psi}_{2,T,n}^{CoJ}(\rho)$ instead of $\Psi_{2,T,n}^{CoJ}$ allows to get far better level results while the power of the test remains almost the same.

Figure 9.9 shows that the adjusted estimator $\widetilde{\Psi}_{2,T,n}^{CoJ}(\rho)$ is much more centered around 1 than $\Psi_{2,T,n}^{CoJ}$ if there exist common jumps which is most pronounced in the cases III-j and III-m. On the other hand, $\widetilde{\Psi}_{2,T,n}^{CoJ}(\rho)$ clusters around a value very close to 1 also if there exist no common jumps. However, in all cases *-d0

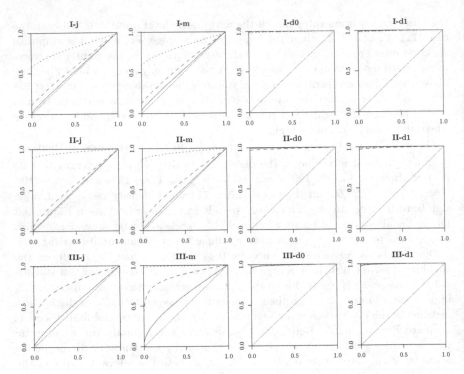

Figure 9.8: Empirical rejection curves from the Monte Carlo simulation for
the test derived from Corollary 9.25. The dotted lines represent
the results for $n = 100$, the dashed lines for $n = 1,600$ and the
solid lines for $n = 25,600$. In each case $N = 10,000$ paths were
simulated.

and *-d1 the peak of the density still occurs at a value which is noticeably larger
than 1.

Figure 9.10 illustrates similarly as Figure 8.3 for the test for jumps the performance of the test from Corollary 9.25 in dependence of ρ. We choose the cases
III-j and III-d0 as representatives for the null hypothesis and the alternative.
As expected the level as well as the power of the test decrease as ρ increases. Here
we get the lowest overall error for a value of ρ close to 0.75.

As demonstrated in Section 8.2 the adjustment method leading to the test
in Corollary 9.25 can also be used to improve the finite sample performance of
existing tests based on equidistant and synchronous observations. To justify this
claim we repeated the above simulation study for observation times given by
$t_{i,n}^{(1)} = t_{i,n}^{(2)} = i/n$. The results are presented in Figures 9.11–9.15 in the same

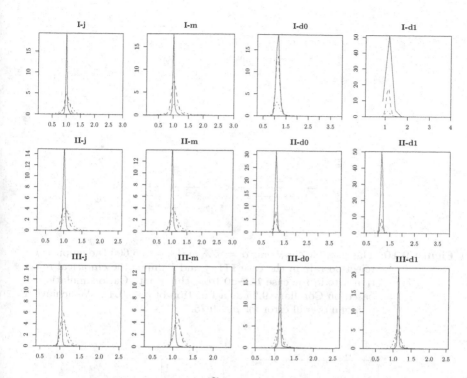

Figure 9.9: Density estimates for $\widetilde{\Psi}^{CoJ}_{2,T,n}(\rho)$ from the Monte Carlo. The dotted lines correspond to $n = 100$, the dashed lines to $n = 1{,}600$ and the solid lines to $n = 25{,}600$. In all cases $N = 10{,}000$ paths were simulated.

fashion as above. Like in the setting of Poisson sampling we observe that the level of the test is much closer to the ideal asymptotic value for the test based on the adjusted statistic $\widetilde{\Psi}^{CoJ}_{2,T,n}(\rho)$ compared to the test based on $\Psi^{CoJ}_{2,T,n}$. Figure 9.15 shows that, for the here chosen parameter settings, the optimal value $\rho \approx 0.79$ under equidistant and synchronous observations is very close to the optimal value $\rho \approx 0.75$ in the Poisson setting.

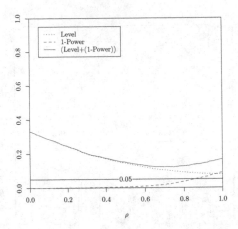

Figure 9.10: This graphic shows for $\alpha = 5\%$ and $n = 25{,}600$ the empirical rejection rate in the case III-j and 1 minus the empirical rejection rate in the case III-d0 from the Monte Carlo simulation based on Corollary 9.25 as a function of $\rho \in [0,1]$. We achieve a minimal overall error for $\rho \approx 0.75$.

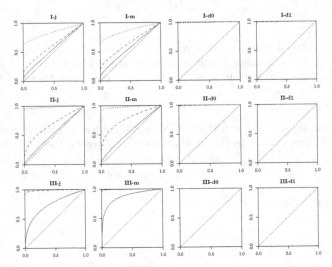

Figure 9.11: Empirical rejection curves as in Figure 9.6 from the Monte Carlo simulation for the test derived from Theorem 9.24 based on *equidistant observations* $t_{i,n}^{(l)} = i/n$.

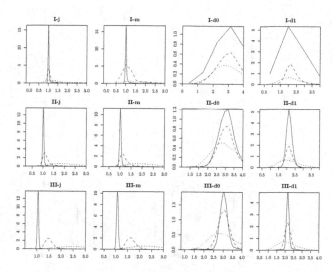

Figure 9.12: Density estimates for $\Psi_{2,T,n}^{CoJ}$ as in Figure 9.7 from the Monte Carlo based on *equidistant observations* $t_{i,n}^{(l)} = i/n$.

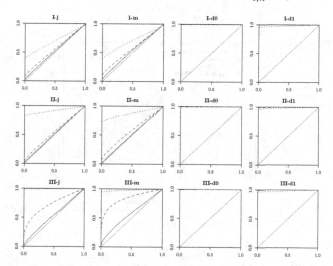

Figure 9.13: Empirical rejection curves as in Figure 9.8 from the Monte Carlo simulation for the test derived from Corollary 9.25 for $\rho = 0.75$ based on *equidistant observations* $t_{i,n}^{(l)} = i/n$.

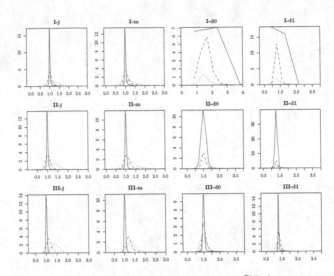

Figure 9.14: Density estimates as in Figure 9.9 for $\widetilde{\Psi}_{2,T,n}^{CoJ}(0.75)$ from the Monte Carlo based on *equidistant observations* $t_{i,n}^{(l)} = i/n$.

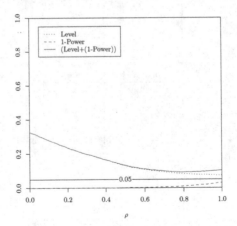

Figure 9.15: This graphic shows for $\alpha = 5\%$, $n = 25{,}600$ and *equidistant observations* $t_{i,n} = i/n$ the empirical rejection rate in the case III-j along with 1 minus the empirical rejection rate in the case III-d0 from the Monte Carlo simulation based on Corollary 9.25 as a function of $\rho \in [0,1]$. We achieve a minimal overall error for $\rho \approx 0.79$.

9.2.3 The Proofs

Proof of Theorem 9.14. By Theorems 2.3 and by Theorem 2.40 b) we obtain $V(f_{(2,2)}, \pi_n)_T \xrightarrow{\mathbb{P}} B^*(f_{(2,2)})_T$ and $V(f_{(2,2)}, [k], \pi_n)_T \xrightarrow{\mathbb{P}} k^2 B^*(f_{(2,2)})_T$. These two convergences and the continuous mapping theorem for convergence in probability yield (9.40). $\qquad\qquad\square$

Proof of Theorem 9.16. It holds

$$\Psi_{k,T,n}^{CoJ} = \frac{n/k^2 V(f_{(2,2)}, [k], \pi_n)_T}{nV(f_{(2,2)}, \pi_n)_T}$$

and the proof of

$$(n/k^2 V(f_{(2,2)}, [k], \pi_n)_T, nV(f_{(2,2)}, \pi_n)_T)$$
$$\xrightarrow{\mathcal{L}-s} (k\widetilde{C}_{k,T}^{DisJ} + \widetilde{D}_{k,T}^{DisJ}, \widetilde{C}_{1,T}^{DisJ} + \widetilde{D}_{1,T}^{DisJ}) \quad (9.50)$$

restricted to Ω_T^{nCoJ} is similar to the proof of Theorem 9.4. For example we obtain

$$\left(n/k^2 \sum_{i,j:t_{i,n}^{(1)} \vee t_{j,n}^{(2)} \leq T} (\Delta_{i,k,n}^{(1)} C^{(1)})^2 (\Delta_{j,k,n}^{(2)} C^{(2)})^2 \mathbb{1}_{\{\mathcal{I}_{i,k,n}^{(1)} \cap \mathcal{I}_{j,k,n}^{(2)} \neq \emptyset\}}, \right.$$
$$\left. n \sum_{i,j:t_{i,n}^{(1)} \vee t_{j,n}^{(2)} \leq T} (\Delta_{i,n}^{(1)} C^{(1)})^2 (\Delta_{j,n}^{(2)} C^{(2)})^2 \mathbb{1}_{\{\mathcal{I}_{i,n}^{(1)} \cap \mathcal{I}_{j,n}^{(2)} \neq \emptyset\}} \right) \xrightarrow{\mathbb{P}} (k\widetilde{C}_{k,T}^{DisJ}, \widetilde{C}_{1,T}^{DisJ})$$

from Theorem 2.22 and Theorem 2.42 b). Using Lemma B.5 we obtain (9.41) from (9.50) because by Condition 9.15 we have $\int_0^T |\sigma_s^{(1)} \sigma_s^{(2)}| ds > 0$ which together with the fact that $G_{2,2}$ is strictly increasing guarantees $\widetilde{C}_{1,T}^{DisJ} + \widetilde{D}_{1,T}^{DisJ} \geq \widetilde{C}_{1,T}^{DisJ} > 0$. $\qquad\square$

Proof of Theorem 9.19. We prove

$$\sqrt{n} \left(V(f_{(2,2)}, [2], \pi_n)_T - 4V(f_{(2,2)}, \pi_n)_T \right) \xrightarrow{\mathcal{L}-s} \Phi_{2,T}^{CoJ} \quad (9.51)$$

from which (9.43) easily follows.

Step 1. Recall the discretized versions $\tilde{\sigma}^{(1)}(r,q)$, $\tilde{\sigma}^{(2)}(r,q)$, $\tilde{\rho}(r,q)$ and $\widetilde{C}(r,q)$ of $\sigma^{(1)}$, $\sigma^{(2)}$, ρ and C defined in Step 1 in the proof of Theorem 3.6. We then show

$$\lim_{q \to \infty} \limsup_{r \to \infty} \limsup_{n \to \infty} \mathbb{P}(|\sqrt{n} \left(V(f_{(2,2)}, [2], \pi_n)_T - 4V(f_{(2,2)}, \pi_n)_T \right) - \widetilde{R}(n,q,r)| > \varepsilon) = 0$$
$$(9.52)$$

for all $\varepsilon > 0$ where

$$\widetilde{R}(n,q,r) = \widetilde{R}^{(1)}(n,q,r) + \widetilde{R}^{(2)}(n,q,r),$$

$$\widetilde{R}^{(l)}(n,q,r) = \sqrt{n} \sum_{i,j \geq 2 : t_{i,n}^{(l)} \wedge t_{j,n}^{(3-l)} \leq T} 2(\Delta_{i-1,1,n}^{(l)} N^{(l)}(q) \Delta_{i,1,n}^{(l)} \widetilde{C}^{(l)}(r,q)$$

$$+ \Delta_{i-1,1,n}^{(l)} \widetilde{C}^{(l)}(r,q) \Delta_{i,1,n}^{(l)} N^{(l)}(q)) \left(\Delta_{j,2,n}^{(3-l)} N^{(3-l)}(q)\right)^2 \mathbb{1}_{\{\mathcal{I}_{i,2,n}^{(l)} \cap \mathcal{I}_{j,2,n}^{(3-l)} \neq \emptyset\}},$$

$l = 1, 2$. As the proof of (9.52) is rather technical it is postponed to after the discussion of the structure of the proof of (9.51).

Step 2. Next, we will show

$$\widetilde{R}(n,q,r) \xrightarrow{\mathcal{L}-s} \Phi_T^{CoJ}(q,r) \tag{9.53}$$

for any $r, q > 0$ where

$$\Phi_T^{CoJ}(q,r) := 4 \sum_{p : S_{q,p} \leq T} \Delta X_{S_{q,p}}^{(1)} \Delta X_{S_{q,p}}^{(2)}$$

$$\times \left[\Delta X_{S_{q,p}}^{(2)} \left(\tilde{\sigma}^{(1)}(r,q)_{S_{q,p}-} \sqrt{\mathcal{L}^{CoJ}(S_{q,p})} U_{S_{q,p}}^{(1),-} \right. \right.$$

$$+ + \tilde{\sigma}^{(1)}(r,q)_{S_{q,p}} \sqrt{\mathcal{R}^{CoJ}(S_{q,p})} U_{S_{q,p}}^{(1),+}$$

$$+ \left[(\tilde{\sigma}^{(1)}(r,q)_{S_{q,p}-})^2 (\mathcal{L}^{CoJ,(1)} - \mathcal{L}^{CoJ})(S_{q,p}) \right.$$

$$\left. (\tilde{\sigma}^{(1)}(r,q)_{S_{q,p}})^2 (\mathcal{R}^{CoJ,(1)} - \mathcal{R}^{CoJ})(S_{q,p}) \right]^{1/2} U_{S_{q,p}}^{(2)} \bigg)$$

$$+ \Delta X_{S_{q,p}}^{(1)} \left(\tilde{\sigma}^{(2)}(r,q)_{S_{q,p}-} \tilde{\rho}(r,q)_{S_{q,p}-} \sqrt{\mathcal{L}^{CoJ}(S_{q,p})} U_{S_p}^{(1),-} \right.$$

$$+ \tilde{\sigma}^{(2)}(r,q)_{S_{q,p}} \tilde{\rho}(r,q)_{S_{q,p}} \sqrt{\mathcal{R}^{CoJ}(S_p)} U_{S_{q,p}}^{(1),+}$$

$$+ \left[(\tilde{\sigma}^{(2)}(r,q)_{S_{q,p}-})^2 (1 - (\tilde{\rho}(r,q)_{S_{q,p}-})^2) \mathcal{L}^{CoJ}(S_{q,p}) \right.$$

$$\left. + (\tilde{\sigma}^{(2)}(r,q)_{S_{q,p}})^2 (1 - (\tilde{\rho}(r,q)_{S_{q,p}})^2) \mathcal{R}^{CoJ}(S_{q,p}) \right]^{1/2} U_{S_{q,p}}^{(3)}$$

$$+ \left[(\tilde{\sigma}^{(2)}(r,q)_{S_{q,p}-})^2 (\mathcal{L}^{CoJ,(2)} - \mathcal{L}^{CoJ})(S_{q,p}) \right.$$

$$\left. \left. + (\tilde{\sigma}^{(2)}(r,q)_{S_{q,p}})^2 (\mathcal{R}^{CoJ,(2)} - \mathcal{R}^{CoJ})(S_{q,p}) \right]^{1/2} U_{S_{q,p}}^{(4)} \bigg) \right].$$

where $S_{q,p}$, $p \in \mathbb{N}$, denotes an enumeration of the jump times of $N(q)$. On the set $\Omega(n,q,r)$ where any two jump times $S_{q,p}, S_{q,p'}$ in $[0, T]$ are further apart than $4|\pi_n|_T$ from each other and from all $j2^{-r}$ with $1 \leq j \leq \lfloor T2^r \rfloor$ we have

$$\widetilde{R}(n,q,r) \mathbb{1}_{\Omega(n,q,r)} = \mathbb{1}_{\Omega(n,q,r)} 4\sqrt{n} \sum_{S_{q,p} \leq T} \Delta N^{(2)}(q)_{S_{q,p}} \Delta N^{(1)}(q)_{S_{q,p}}$$

$$\times \left(\Delta N^{(2)}(q)_{S_p} \left(\tilde{\sigma}^{(1)}_{S_{q,p}-}(r,q) \Delta^{(1)}_{i_n^{(1)}(S_{q,p})-1,n} W^{(1)} + \tilde{\sigma}^{(1)}_{S_{q,p}}(r) \Delta^{(1)}_{i_n^{(1)}(S_{q,p})+1,n} W^{(1)} \right) \right.$$

$$+ \Delta N^{(1)}(q)_{S_{q,p}} \left[\tilde{\sigma}^{(2)}_{S_{q,p}-}(r,q) \right.$$

$$\times \left(\tilde{\rho}_{S_{q,p}-}(r,q) \Delta^{(2)}_{i_n^{(2)}(S_{q,p})-1,n} W^{(1)} \right.$$

$$+ \left. \sqrt{1 - \tilde{\rho}_{S_{q,p}-}(r,q)^2} \Delta^{(2)}_{i_n^{(2)}(S_{q,p})-1,n} W^{(2)} \right)$$

$$+ \tilde{\sigma}^{(2)}_{S_{q,p}}(r,q) \left(\tilde{\rho}_{S_{q,p}}(r,q) \Delta^{(2)}_{i_n^{(2)}(S_{q,p})+1,n} W^{(1)} \right.$$

$$+ \left. \left. \left. \sqrt{1 - (\tilde{\rho}_{S_{q,p}}(r,q))^2} \Delta^{(2)}_{i_n^{(2)}(S_{q,p})+1,n} W^{(2)} \right) \right] \right)$$

where the factor 4 stems from the fact that any jump of $N^{(l)}(q)$ is observed in two consecutive intervals $\mathcal{I}^{(l)}_{i-1,2,n}$, $\mathcal{I}^{(l)}_{i,2,n}$, $l = 1,2$.

Similarly as in the proof of (3.25) it can be shown that Condition 9.18 yields the $\mathcal{X}$-stable convergence

$$\left(\left(\Delta^{(1)}_{i_n^{(1)}(S_{q,p})-1,n} W^{(1)}, \Delta^{(1)}_{i_n^{(1)}(S_{q,p})+1,n} W^{(1)}, \Delta^{(2)}_{i_n^{(2)}(S_{q,p})-1,n} W^{(1)}, \right. \right.$$

$$\left. \Delta^{(2)}_{i_n^{(2)}(S_{q,p})+1,n} W^{(1)}, \Delta^{(2)}_{i_n^{(2)}(S_{q,p})-1,n} W^{(2)}, \Delta^{(2)}_{i_n^{(2)}(S_{q,p})+1,n} W^{(2)} \right)_{S_{q,p} \leq T} \right)$$

$$\xrightarrow{\mathcal{L}-s} \left(\left(\sqrt{\mathcal{L}^{CoJ}(S_{q,p})} U^{(1),-}_{S_{q,p}} + \sqrt{(\mathcal{L}^{CoJ,(1)} - \mathcal{L}^{CoJ})(S_{q,p})} U^{(2),-}_{S_{q,p}}, \right. \right.$$

$$\sqrt{\mathcal{R}^{CoJ}(S_{q,p})} U^{(1),+}_{S_{q,p}} + \sqrt{(\mathcal{R}^{CoJ,(1)} - \mathcal{R}^{CoJ})(S_{q,p})} U^{(2),+}_{S_{q,p}},$$

$$\sqrt{\mathcal{L}^{CoJ}(S_{q,p})} U^{(1),-}_{S_{q,p}} + \sqrt{(\mathcal{L}^{CoJ,(2)} - \mathcal{L}^{CoJ})(S_{q,p})} U^{(3),-}_{S_{q,p}},$$

$$\sqrt{\mathcal{R}^{CoJ}(S_{q,p})} U^{(1),+}_{S_{q,p}} + \sqrt{(\mathcal{R}^{CoJ,(2)} - \mathcal{R}^{CoJ})(S_{q,p})} U^{(3),+}_{S_{q,p}},$$

$$\left. \left. \sqrt{\mathcal{L}^{CoJ,(2)}(S_{q,p})} U^{(4),-}_{S_{q,p}}, \sqrt{\mathcal{R}^{CoJ,(2)}(S_{q,p})} U^{(4),+}_{S_{q,p}} \right)_{S_{q,p} \leq T} \right) \quad (9.54)$$

where $U^{(i),-}_s, U^{(i),+}_s$ for $i = 1, \ldots, 4$ are i.i.d. standard normal distributed random variables which are independent of $\mathcal{F}$ and of the random vectors

$$(\mathcal{L}^{CoJ,(1)}, \mathcal{R}^{CoJ,(1)}, \mathcal{L}^{CoJ,(2)}, \mathcal{R}^{CoJ,(2)}, \mathcal{L}^{CoJ}, \mathcal{R}^{CoJ})(s).$$

Then (9.54) together with Lemma B.3 and the continuous mapping theorem for stable convergence stated in Lemma B.5 yields

$$4\sqrt{n} \sum_{S_{q,p} \in P(q,T)} \Delta N^{(2)}(q)_{S_{q,p}} \Delta N^{(1)}(q)_{S_{q,p}}$$

$$\times \left(\Delta N^{(2)}(q)_{S_p} \left[\tilde{\sigma}^{(1)}_{S_{q,p}-}(r,q) \Delta^{(1)}_{i_n^{(1)}(S_{q,p})-1,n} W^{(1)} \right. \right.$$

$$+ \tilde{\sigma}^{(1)}_{S_{q,p}}(r,q) \Delta^{(1)}_{i^{(1)}_n(S_{q,p})+1,n} W^{(1)} \Big]$$

$$+ \Delta N^{(1)}(q)_{S_{q,p}} \Big[\tilde{\sigma}^{(2)}_{S_{q,p}-}(r,q) \big(\tilde{\rho}_{S_{q,p}-}(r,q) \Delta^{(2)}_{i^{(2)}_n(S_{q,p})-1,n} W^{(1)}$$

$$+ \sqrt{1 - (\tilde{\rho}_{S_{q,p}-}(r,q))^2} \Delta^{(2)}_{i^{(2)}_n(S_{q,p})-1,n} W^{(2)} \big)$$

$$+ \tilde{\sigma}^{(2)}_{S_{q,p}}(r,q) \big(\tilde{\rho}_{S_{q,p}} \Delta^{(2)}_{i^{(2)}_n(S_{q,p})+1,n} W^{(1)}$$

$$+ \sqrt{1 - \tilde{\rho}_{S_{q,p}}(r,q))^2} \Delta^{(2)}_{i^{(2)}_n(S_{q,p})+1,n} W^{(2)} \big) \Big] \Big)$$

$$\xrightarrow{\mathcal{L}-s} \Phi^{CoJ}_T(q,r).$$

Because of $\mathbb{P}(\Omega(n,q,r)) \to 1$ as $n \to \infty$ for any q, r this yields (9.53).

Step 3. Finally we consider

$$\lim_{q\to\infty} \limsup_{r\to\infty} \widetilde{\mathbb{P}}(|\Phi^{CoJ}_T - \Phi^{CoJ}_T(q,r)| > \varepsilon) = 0 \qquad (9.55)$$

for all $\varepsilon > 0$ which can be proven using that the first moments of $\Gamma^{CoJ}(\cdot, dy)$ are uniformly bounded together with the boundedness of the jump sizes of X respectively $N(q)$.

Step 4. Combining (9.52), (9.53) and (9.55) under the use of Lemma B.6 yields (9.51).   □

Proof of (9.52). Step 1. We first prove

$$\lim_{n\to\infty} \mathbb{P}(|\sqrt{n}\,(V(f_{(2,2)},[2],\pi_n)_T - 4V(f_{(2,2)},\pi_n)_T) - R(n)| > \varepsilon) = 0 \quad (9.56)$$

for all $\varepsilon > 0$ where $R(n) = R^{(1)}(n) + R^{(2)}(n)$ and

$$R^{(l)}(n) = \sqrt{n} \sum_{i,j\geq 2: t^{(l)}_{i,n} \wedge t^{(3-l)}_{j,n} \leq T} 2\Delta^{(l)}_{i-1,1,n} X^{(l)} \Delta^{(l)}_{i,1,n} X^{(l)} \big(\Delta^{(3-l)}_{j,2,n} X^{(3-l)}\big)^2 \mathbf{1}_{\{\mathcal{I}^{(l)}_{i,2,n} \cap \mathcal{I}^{(3-l)}_{j,2,n} \neq \emptyset\}},$$

$l = 1, 2$. Therefore note that it holds

$$\sqrt{n}\,\big(V(f_{(2,2)},[2],\pi_n)_T - 4V(f_{(2,2)},\pi_n)_T\big)$$

$$= \sqrt{n} \sum_{i,j\geq 2: t^{(1)}_{i,n} \wedge t^{(2)}_{j,n} \leq T} \Big[\big(\Delta^{(1)}_{i-1,1,n} X^{(1)} + \Delta^{(1)}_{i,1,n} X^{(1)}\big)^2$$

$$\times \big(\Delta^{(2)}_{j-1,1,n} X^{(2)} + \Delta^{(2)}_{j,1,n} X^{(2)}\big)^2$$

$$- \sum_{l_1,l_2=0,1} \big(\Delta^{(1)}_{i-l_1,1,n} X^{(1)}\big)^2 \big(\Delta^{(2)}_{j-l_2,1,n} X^{(2)}\big)^2 \mathbf{1}_{\{\mathcal{I}^{(1)}_{i-l_1,1,n} \cap \mathcal{I}^{(2)}_{j-l_2,1,n} \neq \emptyset\}} \Big]$$

$$\times \mathbf{1}_{\{\mathcal{I}^{(1)}_{i,2,n} \cap \mathcal{I}^{(2)}_{j,2,n} \neq \emptyset\}}$$

$$+ O_{\mathbb{P}}(\sqrt{n}|\pi_n|_T)$$

$$= \sqrt{n} \sum_{i,j \geq 2: t_{i,n}^{(1)} \wedge t_{j,n}^{(2)} \leq T} \Big[2\Delta_{i-1,1,n}^{(1)} X^{(1)} \Delta_{i,1,n}^{(1)} X^{(1)} \big(\Delta_{j-1,1,n}^{(2)} X^{(2)} + \Delta_{j,1,n}^{(2)} X^{(2)} \big)^2$$

$$+ 2\big(\Delta_{i-1,1,n}^{(1)} X^{(1)} + \Delta_{i,1,n}^{(1)} X^{(1)} \big)^2 \Delta_{j-1,1,n}^{(2)} X^{(2)} \Delta_{j,1,n}^{(2)} X^{(2)}$$

$$+ \sum_{l_1,l_2=0,1} \big(\Delta_{i-l_1,1,n}^{(1)} X^{(1)} \big)^2 \big(\Delta_{j-l_2,1,n}^{(2)} X^{(2)} \big)^2 \mathbb{1}_{\{\mathcal{I}_{i-l_1,1,n}^{(1)} \cap \mathcal{I}_{j-l_2,1,n}^{(2)} = \emptyset\}} \Big]$$

$$\times \mathbb{1}_{\{\mathcal{I}_{i,2,n}^{(1)} \cap \mathcal{I}_{j,2,n}^{(2)} \neq \emptyset\}}$$

$$+ O_{\mathbb{P}}(\sqrt{n}|\pi_n|_T), \tag{9.57}$$

where the $O_{\mathbb{P}}(\sqrt{n}|\pi_n|_T)$-terms are due to boundary effects. Because of the indicator $\mathbb{1}_{\{\mathcal{I}_{i-l_1,1,n}^{(1)} \cap \mathcal{I}_{j-l_2,1,n}^{(2)} = \emptyset\}}$ we may use iterated expectations and inequality (1.12) to obtain the following upper bound for the $\mathcal{S}$-conditional expectation of the sum over the terms $(\Delta_{i-l_1,1,n}^{(1)} X^{(1)})^2 (\Delta_{j-l_2,1,n}^{(2)} X^{(2)})^2$ in (9.57)

$$4K\sqrt{n} \sum_{i,j \geq 2: t_{i,n}^{(1)} \wedge t_{j,n}^{(2)} \leq T} |\mathcal{I}_{i,2,n}^{(1)}| |\mathcal{I}_{j,2,n}^{(2)}| \mathbb{1}_{\{\mathcal{I}_{i,2,n}^{(1)} \cap \mathcal{I}_{j,2,n}^{(2)} \neq \emptyset\}} = \frac{32 K G_{2,2}^{[2],n}(T)}{\sqrt{n}}$$

which converges to zero in probability as $n \to \infty$ because of $G_{2,2}^{[2],n}(T) = O_{\mathbb{P}}(1)$ by assumption. Hence we obtain (9.56) by Lemma 2.15.

Step 2. Next, we will show

$$\lim_{q \to \infty} \limsup_{n \to \infty} \mathbb{P}(|R^{(l)}(n) - R^{(l)}(n,q)| > \varepsilon) \to 0 \tag{9.58}$$

for all $\varepsilon > 0$ and $l = 1,2$ with

$$R^{(l)}(n,q) = \sqrt{n} \sum_{i,j \geq 2: t_{i,n}^{(l)} \wedge t_{j,n}^{(3-l)} \leq T} 2\Delta_{i-1,1,n}^{(l)} X^{(l)} \Delta_{i,1,n}^{(l)} X^{(l)} \big(\Delta_{j,2,n}^{(3-l)} N^{(3-l)}(q) \big)^2$$

$$\times \mathbb{1}_{\{\mathcal{I}_{i,2,n}^{(l)} \cap \mathcal{I}_{j,2,n}^{(3-l)} \neq \emptyset\}},$$

$l = 1,2$. Denote by $\Delta_{(j,i),(k_2,k_1),n}^{(3-l,l)} X$ the increment of X over the interval $\mathcal{I}_{i,k_1,n}^{(l)} \cap \mathcal{I}_{j,k_2,n}^{(3-l)}$ and by $\Delta_{(j,i),(k_1,k_2),n}^{(3-l \setminus l)} X$ the increment of X over $\mathcal{I}_{j,k_1,n}^{(3-l)} \setminus \mathcal{I}_{i,k_2,n}^{(l)}$ (which might be the sum of the increments over two separate intervals).

Using the elementary inequality

$$|a^2 - b^2| = (\rho|a+b|)(\rho^{-1}|a-b|) \leq \rho^2(a+b)^2 + \rho^{-2}(a-b)^2$$

which holds for any $a, b \in \mathbb{R}$ and $\rho > 0$ with $a = \Delta_{j,2,n}^{(2)} X^{(2)}$ and $b = \Delta_{j,2,n}^{(2)} N^{(2)}(q)$ yields

$$
\begin{aligned}
\mathbb{E}[&|R^{(l)}(n) - R^{(l)}(n,q)| \| \mathcal{S}] \\
&\leq 2\sqrt{n} \sum_{i,j \geq 2 : t_{i,n}^{(l)} \wedge t_{j,n}^{(3-l)} \leq T} \Big(\rho^2 \mathbb{E}[|\Delta_{i-1,1,n}^{(l)} X^{(l)} \Delta_{i,1,n}^{(l)} X^{(l)}| \\
&\qquad\qquad\qquad \times (\Delta_{j,2,n}^{(3-l)}(X^{(3-l)} + N^{(3-l)}(q)))^2 | \mathcal{S}] \\
&\qquad + \rho^{-2} \mathbb{E}[|\Delta_{i-1,1,n}^{(l)} X^{(l)} \Delta_{i,1,n}^{(l)} X^{(l)}| (\Delta_{j,2,n}^{(3-l)}(X^{(3-l)} - N^{(3-l)}(q)))^2 | \mathcal{S}] \Big) \\
&\qquad \times \mathbb{1}_{\{\mathcal{I}_{i,2,n}^{(l)} \cap \mathcal{I}_{j,2,n}^{(3-l)} \neq \emptyset\}}.
\end{aligned}
\tag{9.59}
$$

Using

$$
|\Delta_{\iota,1,n}^{(l)} X^{(l)}| \leq |\Delta_{(j,\iota),(2,1),n}^{(3-l,l)} X^{(l)}| + |\Delta_{(\iota,j),(1,2),n}^{(l \backslash (3-l))} X^{(l)}|, \quad \iota = i-1, i,
$$

$$
(\Delta_{j,2,n}^{(3-l)} Y)^2 \leq 3 (\Delta_{(j,i),(2,1),n}^{(3-l,l)} Y)^2 + 3 (\Delta_{(j,i-1),(2,1),n}^{(3-l,l)} Y)^2 + 3 (\Delta_{(j,i),(2,2),n}^{(3-l \backslash l)} Y)^2,
$$

for $Y = X^{(3-l)} + N^{(3-l)}(q)$ to treat increments over overlapping and non-overlapping intervals differently we obtain that the first summand in (9.59) is bounded by

$$
\begin{aligned}
6\rho^2 \sqrt{n} \sum_{i,j \geq 2 : t_{i,n}^{(l)} \wedge t_{j,n}^{(3-l)} \leq T} \Big[&\mathbb{E}[|\Delta_{i-1,1,n}^{(l)} X^{(l)} \Delta_{i,1,n}^{(l)} X^{(l)}| \\
&\qquad \times (\Delta_{(j,i),(2,2),n}^{(3-l \backslash l)}(X^{(3-l)} + N^{(3-l)}(q)))^2 | \mathcal{S}] \\
&+ \sum_{\iota = i-1, i} \Big(\mathbb{E}[|\Delta_{2i-1-\iota,1,n}^{(l)} X^{(l)} \Delta_{(j,\iota),(2,1),n}^{(3-l,l)} X^{(l)}| \\
&\qquad \times (\Delta_{(j,\iota),(2,1),n}^{(3-l,l)}(X^{(3-l)} + N^{(3-l)}(q)))^2 | \mathcal{S}] \\
&+ \mathbb{E}[|\Delta_{2i-1-\iota,1,n}^{(l)} X^{(l)} \Delta_{(\iota,j),(1,2),n}^{(l \backslash 3-l)} X^{(l)}| \\
&\qquad \times (\Delta_{(j,\iota),(2,1),n}^{(3-l,l)}(X^{(3-l)} + N^{(3-l)}(q)))^2 | \mathcal{S}] \Big) \Big] \\
&\times \mathbb{1}_{\{\mathcal{I}_{i,2,n}^{(l)} \cap \mathcal{I}_{j,2,n}^{(3-l)} \neq \emptyset\}} \\
\leq K\rho^2 \sqrt{n} \sum_{i,j \geq 2 : t_{i,n}^{(l)} \wedge t_{j,n}^{(3-l)} \leq T} &\Big[(|\mathcal{I}_{i-1,1,n}^{(l)}||\mathcal{I}_{i,1,n}^{(l)}|)^{1/2} (|\mathcal{I}_{j,2,n}^{(3-l)}| + K_q |\mathcal{I}_{j,2,n}^{(3-l)}|^2) \\
&+ \sum_{\iota = i-1, i} \Big(|\mathcal{I}_{2i-1-\iota,1,n}^{(l)}|^{1/2} |\mathcal{I}_{\iota,1,n}^{(l)} \cap \mathcal{I}_{j,2,n}^{(3-l)}|^{1/2}
\end{aligned}
$$

$$\times (|\mathcal{I}_{\iota,1,n}^{(l)} \cap \mathcal{I}_{j,2,n}^{(3-l)}| + K_q|\mathcal{I}_{\iota,1,n}^{(l)} \cap \mathcal{I}_{j,2,n}^{(3-l)}|^4)^{1/2}$$

$$+ (|\mathcal{I}_{i-1,1,n}^{(l)}||\mathcal{I}_{i,1,n}^{(l)}|)^{1/2}(|\mathcal{I}_{\iota,1,n}^{(l)} \cap \mathcal{I}_{j,2,n}^{(3-l)}| + K_q|\mathcal{I}_{\iota,1,n}^{(l)} \cap \mathcal{I}_{j,2,n}^{(3-l)}|^2))\Big]$$

$$\times \mathbf{1}_{\{\mathcal{I}_{i,2,n}^{(l)} \cap \mathcal{I}_{j,2,n}^{(3-l)} \neq \emptyset\}}$$

$$\leq K\rho^2(1 + K_q|\pi_n|_T)G_{2,2}^{[2],n}(T)/\sqrt{n}$$

$$+ K\rho^2(1 + K_q(|\pi_n|_T)^3)^{1/2}\sqrt{n} \sum_{i,j\geq 2: t_{i,n}^{(l)} \wedge t_{j,n}^{(3-l)} \leq T} |\mathcal{I}_{i,2,n}^{(l)}|^{1/2}|\mathcal{I}_{i,2,n}^{(l)} \cap \mathcal{I}_{j,2,n}^{(3-l)}|$$

$$\times \mathbf{1}_{\{\mathcal{I}_{i,2,n}^{(l)} \cap \mathcal{I}_{j,2,n}^{(3-l)} \neq \emptyset\}} \tag{9.60}$$

where we used iterated expectations, the Cauchy-Schwarz inequality for the second summand and Lemma 1.4 repeatedly. The first term in the last bound vanishes as $n \to \infty$ because of $G_{2,2}^{[2],n}(T) = O_{\mathbb{P}}(1)$ while the second term vanishes after letting $n \to \infty$, $q \to \infty$ and then $\rho \to 0$ by $\sqrt{n}|\pi_n|_T = O_{\mathbb{P}}(1)$ and $G_{2,2}^{[2],n}(T) = O_{\mathbb{P}}(1)$ because of

$$\sqrt{n} \sum_{i,j\geq 2: t_{i,n}^{(l)} \wedge t_{j,n}^{(3-l)} \leq T} |\mathcal{I}_{i,2,n}^{(l)}|^{1/2}|\mathcal{I}_{i,2,n}^{(l)} \cap \mathcal{I}_{j,2,n}^{(3-l)}|\mathbf{1}_{\{\mathcal{I}_{i,2,n}^{(l)} \cap \mathcal{I}_{j,2,n}^{(3-l)} \neq \emptyset\}}$$

$$\leq \sqrt{n} \sum_{i\geq 2: t_{i,n}^{(l)} \leq T} |\mathcal{I}_{i,2,n}^{(l)}|^{1/2}|\mathcal{I}_{i,2,n}^{(l)}| \leq \Big(nT \sum_{i\geq 2: t_{i,n}^{(l)} \leq T} |\mathcal{I}_{i,2,n}^{(l)}|^2\Big)^{1/2}$$

$$\leq (2^3 T G_{2,2}^{[2],n}(T))^{1/2} + O_{\mathbb{P}}(\sqrt{n}|\pi_n|_T) \tag{9.61}$$

where the inequality in the second to last line follows from the Cauchy-Schwarz inequality for sums.

Analogously we can bound the second summand in (9.59) by

$$K\rho^{-2}(K_q|\pi_n|_T + 1 + e_q)G_{2,2}^{[2],n}(T)/\sqrt{n}$$

$$+ K\rho^{-2}(K_q(|\pi_n|_T)^3 + |\pi_n|_T + e_q)^{1/2}\Big(nT \sum_{i\geq 2: t_{i,n}^{(l)} \leq T} |\mathcal{I}_{i,2,n}^{(l)}|^2\Big)^{1/2}$$

$$+ O_{\mathbb{P}}(\sqrt{n}|\pi_n|_T)$$

where the first term vanishes as $n \to \infty$ and the second term vanishes as $n \to \infty$ and then $q \to \infty$ for any $\rho > 0$. Hence using Lemma 2.15 we have proved (9.58).

Step 3. Further we will show

$$\lim_{q\to\infty} \limsup_{n\to\infty} \mathbb{P}(|R^{(l)}(n,q) - \widetilde{R}^{(l)}(n,q)| > \varepsilon) \to 0 \tag{9.62}$$

for all $\varepsilon > 0$ and $l = 1, 2$ with

$$\widetilde{R}^{(l)}(n,q) = \sqrt{n} \sum_{i,j\geq 2: t_{i,n}^{(l)}\wedge t_{j,n}^{(3-l)}\leq T} 2(\Delta_{i-1,1,n}^{(l)} N^{(l)}(q)\Delta_{i,1,n}^{(l)} C^{(l)}$$

$$+ \Delta_{i-1,1,n}^{(l)} C^{(l)}\Delta_{i,1,n}^{(l)} N^{(l)}(q))\left(\Delta_{j,2,n}^{(3-l)} N^{(3-l)}(q)\right)^2 \mathbb{1}_{\{\mathcal{I}_{i,2,n}^{(l)}\cap\mathcal{I}_{j,2,n}^{(3-l)}\neq\emptyset\}}.$$

Using

$$\mathbb{E}[\Delta_{(j,\iota),(2,1),n}^{(3-l,l)}(X^{(l)} - N^{(l)}(q))\left(\Delta_{(j,\iota),(2,1),n}^{(3-l,l)} N^{(3-l)}(q)\right)^2 |\mathcal{S}]$$

$$\leq (K_q|\mathcal{I}_{\iota,1,n}^{(l)} \cap \mathcal{I}_{j,2,n}^{(3-l)}|^{1/6} + (e_q)^{1/2})|\mathcal{I}_{\iota,1,n}^{(l)} \cap \mathcal{I}_{j,2,n}^{(3-l)}|$$

which can be derived as in (8.30) we get analogously to (9.60)

$$\mathbb{E}\Big[\sqrt{n} \sum_{i,j\geq 2: t_{i,n}^{(l)}\wedge t_{j,n}^{(3-l)}\leq T} 2\Delta_{i-1,1,n}^{(l)}(X^{(l)} - N^{(l)}(q))\Delta_{i,1,n}^{(l)}(X^{(l)} - N^{(l)}(q))$$

$$\times \left(\Delta_{j,2,n}^{(3-l)} N^{(3-l)}(q)\right)^2 \mathbb{1}_{\{\mathcal{I}_{i,2,n}^{(l)}\cap\mathcal{I}_{j,2,n}^{(3-l)}\neq\emptyset\}}\Big|\mathcal{S}\Big]$$

$$\leq K\sqrt{n} \sum_{i,j\geq 2: t_{i,n}^{(l)}\wedge t_{j,n}^{(3-l)}\leq T} \Big[((K_q|\mathcal{I}_{i-1,1,n}^{(l)}| + 1 + e_q)|\mathcal{I}_{i-1,1,n}^{(l)}|)^{1/2}$$

$$\times ((K_q|\mathcal{I}_{i,1,n}^{(l)}| + 1 + e_q)|\mathcal{I}_{i,1,n}^{(l)}|)^{1/2}(1 + K_q|\mathcal{I}_{j,2,n}^{(3-l)}|)|\mathcal{I}_{j,2,n}^{(3-l)}|$$

$$+ \sum_{\iota=i-1,i} \Big(((K_q|\mathcal{I}_{2i-1-\iota,1,n}^{(l)}| + 1 + e_q)|\mathcal{I}_{2i-1-\iota,1,n}^{(l)}|)^{1/2}$$

$$\times (K_q|\mathcal{I}_{\iota,1,n}^{(l)} \cap \mathcal{I}_{j,2,n}^{(3-l)}|^{1/6} + (e_q)^{1/2})|\mathcal{I}_{\iota,1,n}^{(l)} \cap \mathcal{I}_{j,2,n}^{(3-l)}|$$

$$+ ((K_q|\mathcal{I}_{2i-1-\iota,1,n}^{(l)}| + 1 + e_q)|\mathcal{I}_{2i-1-\iota,1,n}^{(l)}|)^{1/2}((K_q|\mathcal{I}_{\iota,1,n}^{(l)}| + 1 + e_q)|\mathcal{I}_{\iota,1,n}^{(l)}|)^{1/2}$$

$$\times (1 + K_q|\mathcal{I}_{\iota,1,n}^{(l)} \cap \mathcal{I}_{j,2,n}^{(3-l)}|)|\mathcal{I}_{\iota,1,n}^{(l)} \cap \mathcal{I}_{j,2,n}^{(3-l)}|\Big)\Big]\mathbb{1}_{\{\mathcal{I}_{i,2,n}^{(l)}\cap\mathcal{I}_{j,2,n}^{(3-l)}\neq\emptyset\}}$$

$$\leq K(K_q|\pi_n|_T + 1 + e_q)(1 + K_q|\pi_n|_T)G_{2,2}^{[2],n}(T)/\sqrt{n}$$

$$+ (K_q|\pi_n|_T + 1 + e_q)^{1/2}(K_q(|\pi_n|_T)^{1/6} + (e_q)^{1/2})(nT \sum_{i\geq 2: t_{i,n}^{(l)}\leq T} |\mathcal{I}_{i,2,n}^{(l)}|^2)^{1/2}$$

$$\tag{9.63}$$

which vanishes as $n \to \infty$ and then $q \to \infty$. Furthermore we get also analogously to (9.60)

$$\mathbb{E}\Big[\sqrt{n} \sum_{i,j\geq 2: t_{i,n}^{(l)}\wedge t_{j,n}^{(3-l)}\leq T} 2\Delta_{i-1,1,n}^{(l)}(B^{(l)}(q) + M^{(l)}(q))\Delta_{i,1,n}^{(l)}(N^{(l)}(q))$$

$$\times \left(\Delta_{j,2,n}^{(3-l)} N^{(3-l)}(q)\right)^2 \mathbb{1}_{\{\mathcal{I}_{i,2,n}^{(l)} \cap \mathcal{I}_{j,2,n}^{(3-l)} \neq \emptyset\}} \big| \mathcal{S}\big]$$

$$\leq K\sqrt{n} \sum_{i,j \geq 2 : t_{i,n}^{(l)} \wedge t_{j,n}^{(3-l)} \leq T} \Big[\big((K_q|\mathcal{I}_{i-1,1,n}^{(l)}| + e_q)|\mathcal{I}_{i-1,1,n}^{(l)}|\big)^{1/2}$$

$$\times \big((1 + K_q|\mathcal{I}_{i,1,n}^{(l)}|)|\mathcal{I}_{i,1,n}^{(l)}|\big)^{1/2}(1 + K_q|\mathcal{I}_{j,2,n}^{(3-l)}|)|\mathcal{I}_{j,2,n}^{(3-l)}|$$

$$+ \big((K_q|\mathcal{I}_{i-1,1,n}^{(l)}| + e_q)|\mathcal{I}_{i-1,1,n}^{(l)}|\big)^{1/2}\big((1 + K_q|\mathcal{I}_{i,1,n}^{(l)} \cap \mathcal{I}_{j,2,n}^{(3-l)}|)|\mathcal{I}_{i,1,n}^{(l)} \cap \mathcal{I}_{j,2,n}^{(3-l)}|\big)^{1/2}$$

$$\times \big((1 + K_q|\mathcal{I}_{i,1,n}^{(l)} \cap \mathcal{I}_{j,2,n}^{(3-l)}|^3)|\mathcal{I}_{i,1,n}^{(l)} \cap \mathcal{I}_{j,2,n}^{(3-l)}|\big)^{1/2}$$

$$+ \big((K_q|\mathcal{I}_{i-1,1,n}^{(l)} \cap \mathcal{I}_{j,2,n}^{(3-l)}| + e_q)|\mathcal{I}_{i-1,1,n}^{(l)} \cap \mathcal{I}_{j,2,n}^{(3-l)}|\big)^{1/2}\big((1 + K_q|\mathcal{I}_{i,1,n}^{(l)}|)|\mathcal{I}_{i,1,n}^{(l)}|\big)^{1/2}$$

$$\times \big((1 + K_q|\mathcal{I}_{i-1,1,n}^{(l)} \cap \mathcal{I}_{j,2,n}^{(3-l)}|^3)|\mathcal{I}_{i-1,1,n}^{(l)} \cap \mathcal{I}_{j,2,n}^{(3-l)}|\big)^{1/2}$$

$$+ \big((K_q|\mathcal{I}_{i-1,1,n}^{(l)}| + e_q)|\mathcal{I}_{i-1,1,n}^{(l)}|\big)^{1/2}\big((1 + K_q|\mathcal{I}_{i,1,n}^{(l)}|)|\mathcal{I}_{i,1,n}^{(l)}|\big)^{1/2}$$

$$\times (1 + K_q|\mathcal{I}_{i,1,n}^{(l)} \cap \mathcal{I}_{j,2,n}^{(3-l)}|)|\mathcal{I}_{i,1,n}^{(l)} \cap \mathcal{I}_{j,2,n}^{(3-l)}|$$

$$+ \big((K_q|\mathcal{I}_{i-1,1,n}^{(l)}| + e_q)|\mathcal{I}_{i-1,1,n}^{(l)}|\big)^{1/2}\big((1 + K_q|\mathcal{I}_{i,1,n}^{(l)}|)|\mathcal{I}_{i,1,n}^{(l)}|\big)^{1/2}$$

$$\times (1 + K_q|\mathcal{I}_{i-1,1,n}^{(l)} \cap \mathcal{I}_{j,2,n}^{(3-l)}|)|\mathcal{I}_{i-1,1,n}^{(l)} \cap \mathcal{I}_{j,2,n}^{(3-l)}|\Big] \mathbb{1}_{\{\mathcal{I}_{i,2,n}^{(l)} \cap \mathcal{I}_{j,2,n}^{(3-l)} \neq \emptyset\}}$$

$$\leq K(K_q|\pi_n|_T + e_q)^{1/2}(1 + K_q|\pi_n|_T)^{3/2} G_{2,2}^{[2],n}(T)/\sqrt{n}$$

$$+ (K_q|\pi_n|_T + e_q)^{1/2}(1 + K_q|\pi_n|_T)(nT \sum_{i \geq 2 : t_{i,n}^{(l)} \leq T} |\mathcal{I}_{i,2,n}^{(l)}|^2)^{1/2} \qquad (9.64)$$

which vanishes as first $n \to \infty$ and then $q \to \infty$. The same obviously also holds if we switch the roles of $i - 1$ and i. Hence using Lemma 2.15 the estimates (9.63) and (9.64) yield (9.62) because $\Delta_{i-1,1,n}^{(l)} N^{(l)}(q)\Delta_{i,1,n}^{(l)} N^{(l)}(q) = 0$ holds eventually for all i with $t_{i,n}^{(l)} \leq T$ as there are only finitely many big jumps.

Step 4. Finally it remains to show

$$\lim_{q \to \infty} \limsup_{r \to \infty} \limsup_{n \to \infty} \mathbb{P}(|\widetilde{R}(n,q) - \widetilde{R}(n,q,r)| > \varepsilon) = 0 \qquad (9.65)$$

for all $\varepsilon > 0$ with $\widetilde{R}(n,q) = \widetilde{R}^{(1)}(n,q) + \widetilde{R}^{(2)}(n,q)$. However, the proof for (9.65) is identical to the proof of (8.32) because we have

$$|\Delta_{i-\iota,1,n}^{(l)} N^{(l)}(q)|(\Delta_{j,2,n}^{(3-l)} N^{(3-l)}(q))^2 \mathbb{1}_{\{\mathcal{I}_{i,2,n}^{(l)} \cap \mathcal{I}_{j,2,n}^{(3-l)} \neq \emptyset\}} \mathbb{1}_{\Omega(n,q)}$$

$$= |\Delta_{i-\iota,1,n}^{(l)} N^{(l)}(q)|(\Delta_{i-\iota,1,n}^{(3-l)} N^{(3-l)}(q))^2 \mathbb{1}_{\Omega(n,q)} \leq K\|\Delta_{i-\iota,1,n}^{(l)} N(q)\|^3$$

for $\iota = 0, 1$ and $l = 1, 2$ on the set $\Omega(n, q)$ where two different jump times of $N(q)$ are further apart than $4|\pi_n|_T$ and because of $\mathbb{P}(\Omega(n,q)) \to 1$ for any $q > 0$.

Step 5. Combining (9.56), (9.58), (9.62) and (9.65) then yields (9.52). $\qquad \square$

For the proof of (9.46), we require the following Proposition.

Proposition 9.27. *Suppose Condition 9.23 is fulfilled. Then it holds*

$$\widetilde{\mathbb{P}}(\{|\frac{1}{M_n}\sum_{m=1}^{M_n}\mathbb{1}_{\{\widehat{\Phi}_{2,T,n,m}^{CoJ}\le\Upsilon\}}-\widetilde{\mathbb{P}}(\Phi_{2,T}^{CoJ}\le\Upsilon|\mathcal{X})|>\varepsilon\}\cap\Omega_T^{CoJ})\to 0 \quad (9.66)$$

for any $\mathcal{X}$-measurable random variable Υ and all $\varepsilon>0$.

Proof. Step 1. Denote by $S_{q,p}$, $p=\mathbb{N}$, an increasing sequence of stopping times which exhausts the jump times of $N(q)$. Similarly as in the proof of Proposition 8.15 we define

$$Y(P,n,m)=4\sum_{p=1}^{P}\Delta_{i_p,n}^{(1)}X^{(1)}\Delta_{j_p,n}^{(2)}X^{(2)}$$

$$\times\mathbb{1}_{\{|\Delta_{i_p,n}^{(1)}X^{(1)}|>\beta|\mathcal{I}_{i_p,n}^{(1)}|^{\varpi}\}}\mathbb{1}_{\{|\Delta_{j_p,n}^{(2)}X^{(2)}|>\beta|\mathcal{I}_{j_p,n}^{(2)}|^{\varpi}\}}$$

$$\times\Big[\Delta_{j_p,n}^{(2)}X^{(2)}\Big(\tilde{\sigma}^{(1)}(t_{i_p,n}^{(1)},-)\sqrt{\widehat{\mathcal{L}}_{n,m}^{CoJ}(\tau_{i_p,j_p,n})}U_{n,(i_p,j_p),m}^{(1,2),-}$$

$$+\tilde{\sigma}^{(1)}(t_{i_p,n}^{(1)},+)\sqrt{\widehat{\mathcal{R}}_{n,m}^{CoJ}(\tau_{i_p,j_p,n})}U_{n,(i_p,j_p),m}^{(1,2),+}$$

$$+\Big((\tilde{\sigma}^{(1)}(t_{i_p,n}^{(1)},-))^2(\widehat{\mathcal{L}}_{n,m}^{CoJ,(1)}-\widehat{\mathcal{L}}_{n,m}^{CoJ})(\tau_{i_p,j_p,n})$$

$$+(\tilde{\sigma}^{(1)}(t_{i_p,n}^{(1)},+))^2(\widehat{\mathcal{R}}_{n,m}^{CoJ,(1)}-\widehat{\mathcal{R}}_{n,m}^{CoJ})(\tau_{i_p,j_p,n})\Big)^{1/2}U_{n,i_p,m}^{(1)}\Big)$$

$$+\Delta_{i_p,n}^{(1)}X^{(1)}\Big(\tilde{\sigma}^{(2)}(t_{j_p,n}^{(2)},-)\tilde{\rho}(\tau_{i_p,j_p,n},-)\sqrt{\widehat{\mathcal{L}}_{n,m}^{CoJ}(\tau_{i_p,j_p,n})}U_{n,(i_p,j_p),m}^{(1,2),-}$$

$$+\tilde{\sigma}^{(2)}(t_{j_p,n}^{(2)},+)\tilde{\rho}(\tau_{i_p,j_p,n},+)\sqrt{\widehat{\mathcal{R}}_{n,m}^{CoJ}(\tau_{i_p,j_p,n})}U_{n,(i_p,j_p),m}^{(1,2),+}$$

$$+\Big((\tilde{\sigma}^{(2)}(t_{j_p,n}^{(2)},-))^2(1-(\tilde{\rho}(\tau_{i_p,j_p,n},-))^2)\widehat{\mathcal{L}}_{n,m}^{CoJ}(\tau_{i_p,j_p,n})$$

$$+(\tilde{\sigma}^{(2)}(t_{j_p,n}^{(2)},+))^2(1-(\tilde{\rho}(\tau_{i_p,j_p,n},+))^2)\widehat{\mathcal{R}}_{n,m}^{CoJ}(\tau_{i_p,j_p,n})\Big)^{1/2}U_{n,j_p,m}^{(2)}$$

$$+\Big((\tilde{\sigma}^{(2)}(t_{j_p,n}^{(2)},-))^2(\widehat{\mathcal{L}}_{n,m}^{CoJ,(2)}-\widehat{\mathcal{L}}_{n,m}^{CoJ})(\tau_{i_p,j_p,n})$$

$$+(\tilde{\sigma}^{(2)}(t_{j_p,n}^{(2)},+))^2\widehat{\mathcal{R}}_{n,m}^{CoJ,(2)}-\widehat{\mathcal{R}}_{n,m}^{CoJ})(\tau_{i_p,j_p,n})\Big)^{1/2}U_{n,j_p,m}^{(3)}\Big)\Big]$$

$$\times\mathbb{1}_{\{S_{q,p}\le T\}}$$

where $(i_p, j_p) = (i_n^{(1)}(S_{q,p}), i_n^{(2)}(S_{q,p}))$ and further set

$$Y(P) = 4 \sum_{p=1}^{P} \Delta X_{S_{q,p}}^{(1)} \Delta X_{S_{q,p}}^{(2)} \mathbb{1}_{\{S_{q,p} \leq T\}}$$

$$\times \Big[\Delta X_{S_{q,p}}^{(2)} \Big(\sigma_{S_{q,p}-}^{(1)} \sqrt{\mathcal{L}^{CoJ}(S_{q,p})} U_{S_{q,p}}^{(1),-} + \sigma_{S_{q,p}}^{(1)} \sqrt{\mathcal{R}^{CoJ}(S_{q,p})} U_{S_{q,p}}^{(1),+}$$

$$+ \sqrt{(\sigma_{S_{q,p}-}^{(1)})^2 (\mathcal{L}^{CoJ,(1)} - \mathcal{L}^{CoJ})(S_{q,p}) + (\sigma_{S_{q,p}}^{(1)})^2 (\mathcal{R}^{CoJ,(1)} - \mathcal{R}^{CoJ})(S_{q,p})} U_{S_{q,p}}^{(2)} \Big)$$

$$+ \Delta X_{S_{q,p}}^{(1)} \Big(\sigma_{S_{q,p}-}^{(2)} \rho_{S_{q,p}-} \sqrt{\mathcal{L}^{CoJ}(S_{q,p})} U_{S_{q,p}}^{(1),-} + \sigma_{S_{q,p}}^{(2)} \rho_{S_{q,p}} \sqrt{\mathcal{R}^{CoJ}(S_{q,p})} U_{S_{q,p}}^{(1),+}$$

$$+ \sqrt{(\sigma_{S_{q,p}-}^{(2)})^2 (1 - (\rho_{S_{q,p}-})^2) \mathcal{L}^{CoJ}(S_{q,p}) + (\sigma_{S_{q,p}}^{(2)})^2 (1 - (\rho_{S_{q,p}})^2) \mathcal{R}^{CoJ}(S_{q,p})} U_{S_{q,p}}^{(3)}$$

$$+ \sqrt{(\sigma_{S_{q,p}-}^{(2)})^2 (\mathcal{L}^{CoJ,(2)} - \mathcal{L}^{CoJ})(S_{q,p}) + (\sigma_{S_{q,p}}^{(2)})^2 (\mathcal{R}^{CoJ,(2)} - \mathcal{R}^{CoJ})(S_{q,p})} U_{S_{q,p}}^{(4)} \Big) \Big].$$

Using this notation we obtain by applying Lemma (4.9) similarly as in Step 1 in the proof of Proposition 4.11

$$\lim_{n \to \infty} \widetilde{\mathbb{P}}(\{ \big| \frac{1}{M_n} \sum_{m=1}^{M_n} \mathbb{1}_{\{Y(P,n,m) \leq \Upsilon\}} - \widetilde{\mathbb{P}}(Y(P) \leq \Upsilon | \mathcal{X}) \big| > \varepsilon \} \cap \Omega_T^{CoJ}) = 0$$

$$(9.67)$$

for any $P \in \mathbb{N}$. Here, Condition 9.23(iii) is needed such that Corollary 6.4 yields the consistency of the estimators for $\sigma^{(1)}$, $\sigma^{(2)}$ and ρ.

Step 2. Next we prove

$$\lim_{P \to \infty} \limsup_{n \to \infty} \frac{1}{M_n} \sum_{m=1}^{M_n} \widetilde{\mathbb{P}}(|Y(P,n,m) - \widehat{\Phi}_{2,T,n,m}^{CoJ}| > \varepsilon) = 0 \qquad (9.68)$$

for all $\varepsilon > 0$.

On the set $\Omega(q, P, n)$ on which there are at most P jumps of $N(q)$ in $[0, T]$ and on which two different jumps of $N(q)$ are further apart than $2|\pi_n|_T$ it holds

$$\mathbb{E}\big[|Y(P,n,m) - \widehat{\Phi}_{2,T,n,m}^{CoJ}| \mathbb{1}_{\Omega(q,P,n)} \big| \mathcal{F} \big]$$

$$\leq K \sum_{l=1,2} \sum_{i,j : t_{i,n}^{(l)} \vee t_{j,n}^{(3-l)} \leq T} |\Delta_{i,n}^{(l)}(X^{(l)} - N^{(l)}(q))| |\Delta_{j,n}^{(3-l)} X^{(3-l)}|$$

$$\times \Big(|\Delta_{j,n}^{(3-l)} X^{(3-l)}| \tilde{\sigma}^{(l)}(n,i) \mathbb{E}\big[\sqrt{(\widehat{\mathcal{L}}_{n,m}^{CoJ,(l)} + \widehat{\mathcal{R}}_{n,m}^{CoJ,(l)})(t_{i,n}^{(l)})} \big| \mathcal{S} \big]$$

$$+ |\Delta_{i,n}^{(l)}(X^{(l)} - N^{(l)}(q))| \tilde{\sigma}^{(3-l)}(n,j)$$

$$\times \mathbb{E}\big[\sqrt{(\widehat{\mathcal{L}}_{n,m}^{CoJ,(3-l)} + \widehat{\mathcal{R}}_{n,m}^{CoJ,(3-l)})(t_{j,n}^{(3-l)})} \big| \mathcal{S} \big] \Big)$$

$$\times \mathbb{1}_{\{ |\Delta_{i,n}^{(l)} X^{(l)}| > \beta |\mathcal{I}_{i,n}^{(l)}|^{\varpi} \wedge |\Delta_{j,n}^{(3-l)} X^{(3-l)}| > \beta |\mathcal{I}_{j,n}^{(3-l)}|^{\varpi} \}} \mathbb{1}_{\{ \mathcal{I}_{i,n}^{(l)} \cap \mathcal{I}_{j,n}^{(3-l)} \neq \emptyset \}} \qquad (9.69)$$

with

$$\tilde{\sigma}^{(l)}(n,i) = \big((\tilde{\sigma}_n^{(l)}(t_{i,n}^{(l)}, -))^2 + (\tilde{\sigma}_n^{(l)}(t_{i,n}^{(l)}, +))^2\big)^{1/2}.$$

Because of $\mathbb{P}(\Omega(q, P, n)) \to \infty$ as $n, P \to \infty$ for all $q > 0$ it suffices to show that (9.69) vanishes as first $n \to \infty$ and then $q \to \infty$ for proving (9.68).

Reconsidering the notation introduced in Step 2 in the proof of (9.52) we define the following terms

$$Y_{(i,j),n}^{(l)} = \Big(\frac{1}{b_n} \sum_{k \neq i: \mathcal{I}_{k,n}^{(l)} \subset [t_{i,n}^{(l)} - b_n, t_{i,n}^{(l)} + b_n]} (\Delta_{(k,j),(1,1),n}^{(l,3-l)} X^{(l)})^2 \Big)^{1/2},$$

$$\widetilde{Y}_{(i,j),n}^{(l)} = \Big(\frac{1}{b_n} \sum_{k \neq i: \mathcal{I}_{k,n}^{(l)} \subset [t_{i,n}^{(l)} - b_n, t_{i,n}^{(l)} + b_n]} (\Delta_{(k,j),(1,1),n}^{(l \backslash 3-l)} X^{(l)})^2 \Big)^{1/2}, \tag{9.70}$$

$l = 1, 2$. Then the Minkowski inequality yields

$$\tilde{\sigma}^{(l)}(n,i) \leq Y_{(i,j),n}^{(l)} + \widetilde{Y}_{(i,j),n}^{(l)},$$
$$\tilde{\sigma}^{(3-l)}(n,j) \leq Y_{(j,i),n}^{(3-l)} + \widetilde{Y}_{(j,i),n}^{(3-l)} \tag{9.71}$$

which allows to treat the increments in the estimation of $\sigma^{(l)}, \sigma^{(3-l)}$ over intervals which do overlap with $\mathcal{I}_{j,n}^{(3-l)}, \mathcal{I}_{i,n}^{(l)}$ and those which do not, separately.

First we derive using the Cauchy-Schwarz inequality, iterated expectations and Lemma 1.4 the following bound for (9.69) where we replaced the estimators for $\sigma^{(l)}, \sigma^{(3-l)}$ with the first summands from (9.71)

$$K \sum_{i,j: t_{i,n}^{(l)} \vee t_{j,n}^{(3-l)} \leq T} \mathbb{E}\big[|\Delta_{i,n}^{(l)}(X^{(l)} - N^{(l)}(q))| \|\Delta_{j,n}^{(3-l)} X^{(3-l)}|^2 Y_{(i,j),n}^{(l)}$$

$$+ |\Delta_{j,n}^{(3-l)} X^{(3-l)}| |\Delta_{i,n}^{(l)}(X^{(l)} - N^{(l)}(q))|^2 Y_{(j,i),n}^{(3-l)} |\mathcal{S}\big] \sqrt{2n|\pi_n|_T}$$

$$\times \mathbb{1}_{\{\mathcal{I}_{i,n}^{(l)} \cap \mathcal{I}_{j,n}^{(3-l)} \neq \emptyset\}}$$

$$\leq K \sum_{i,j: t_{i,n}^{(l)} \vee t_{j,n}^{(3-l)} \leq T} \big[((K_q |\mathcal{I}_{i,n}^{(l)}|^2 + |\mathcal{I}_{i,n}^{(l)}| + e_q |\mathcal{I}_{i,n}^{(l)}|)(|\mathcal{I}_{j,n}^{(3-l)}|/b_n))^{1/2}$$

$$\times (K|\mathcal{I}_{j,n}^{(3-l)}|)^{1/2}$$

$$+ ((K|\mathcal{I}_{j,n}^{(3-l)}|)(|\mathcal{I}_{i,n}^{(l)}|/b_n))^{1/2} (K_q |\mathcal{I}_{i,n}^{(l)}|^4 + |\mathcal{I}_{i,n}^{(l)}|^2 + e_q |\mathcal{I}_{i,n}^{(l)}|)^{1/2}\big]$$

$$\times \sqrt{n|\pi_n|_T} \mathbb{1}_{\{\mathcal{I}_{i,n}^{(l)} \cap \mathcal{I}_{j,n}^{(3-l)} \neq \emptyset\}}$$

$$\leq K(K_q(|\pi_n|_T + (|\pi_n|_T)^3) + 1 + e_q)^{1/2} (|\pi_n|_T/b_n)^{1/2}$$

$$\times \sqrt{n} \sum_{i,j: t_{i,n}^{(l)} \vee t_{j,n}^{(3-l)} \leq T} \left(|\mathcal{I}_{i,n}^{(l)}|^{1/2} |\mathcal{I}_{j,n}^{(3-l)}| + |\mathcal{I}_{i,n}^{(l)}| |\mathcal{I}_{j,n}^{(3-l)}|^{1/2} \right) \mathbb{1}_{\{\mathcal{I}_{i,n}^{(l)} \cap \mathcal{I}_{j,n}^{(3-l)} \neq \emptyset\}}$$

(9.72)

for $l = 1, 2$ which vanishes as $n \to \infty$ for any $q > 0$ because of Condition 9.23(ii) and because of $|\pi_n|_T / b_n \overset{\mathbb{P}}{\longrightarrow} 0$ by assumption.

Further we obtain the following bound for the $\mathcal{S}$-conditional expectation of (9.69) where we replaced the estimators for $\sigma^{(l)}, \sigma^{(3-l)}$ with the second summands from (9.71) by treating increments over $\mathcal{I}_{i,n}^{(l)} \cap \mathcal{I}_{j,n}^{(3-l)}$ and increments over non-overlapping parts of $\mathcal{I}_{i,n}^{(l)}, \mathcal{I}_{j,n}^{(3-l)}$ differently

$$K \sum_{i,j: t_{i,n}^{(l)} \vee t_{j,n}^{(3-l)} \leq T} \left(\left[\mathbb{E}[|\Delta_{(i,j),(1,1),n}^{(l \backslash 3-l)}(X^{(l)} - N^{(l)}(q))| |\Delta_{j,n}^{(3-l)} X^{(3-l)}|^2 \widetilde{Y}_{(i,j),n}^{(l)} \right. \right.$$

$$+ |\Delta_{(i,j),(1,1),n}^{(l,3-l)}(X^{(l)} - N^{(l)}(q))| |\Delta_{(j,i),(1,1),n}^{(3-l\backslash l)} X^{(3-l)}|^2 \widetilde{Y}_{(j,i),n}^{(l)} |\mathcal{S}]$$

$$+ \mathbb{E}[|\Delta_{(j,i),(1,1),n}^{(3-l\backslash l)} X^{(3-l)}| |\Delta_{i,n}^{(l)}(X^{(l)} - N^{(l)}(q))|^2 \widetilde{Y}_{(i,j),n}^{(3-l)}$$

$$+ |\Delta_{(j,i),(1,1),n}^{(3-l,l)} X^{(3-l)}| |\Delta_{(i,j),(1,1),n}^{(l\backslash 3-l)}(X^{(l)} - N^{(l)}(q))|^2 \widetilde{Y}_{(j,i),n}^{(3-l)} |\mathcal{S}] \left. \right] \sqrt{n} |\pi_n|_T$$

$$+ \mathbb{E}[|\Delta_{(i,j),(1,1),n}^{(l,3-l)}(X^{(l)} - N^{(l)}(q))| |\Delta_{(i,j),(1,1),n}^{(l,3-l)} X^{(3-l)}|^2 \widetilde{Y}_{(i,j),n}^{(l)} |\mathcal{S}]$$

$$\times \mathbb{E}[\sqrt{(\widehat{\mathcal{L}}_{n,m}^{(l)} + \widehat{\mathcal{R}}_{n,m}^{(l)})(t_{i,n}^{(l)} \wedge t_{j,n}^{(3-l)})} |\mathcal{S}]$$

$$+ \mathbb{E}[|\Delta_{(i,j),(1,1),n}^{(l,3-l)} X^{(3-l)}| |\Delta_{(i,j),(1,1),n}^{(l,3-l)}(X^{(l)} - N^{(l)}(q)|^2) \widetilde{Y}_{(j,i),n}^{(3-l)} |\mathcal{S}]$$

$$\times \mathbb{E}[\sqrt{(\widehat{\mathcal{L}}_{n,m}^{(3-l)} + \widehat{\mathcal{R}}_{n,m}^{(3-l)})(t_{i,n}^{(l)} \wedge t_{j,n}^{(3-l)})} |\mathcal{S}] \right) \mathbb{1}_{\{\mathcal{I}_{i,n}^{(l)} \cap \mathcal{I}_{j,n}^{(3-l)} \neq \emptyset\}}$$

$$\leq K \sum_{i,j: t_{i,n}^{(l)} \vee t_{j,n}^{(3-l)} \leq T} \left(\left[(K_q |\mathcal{I}_{i,n}^{(l)}| + 1 + e_q) |\mathcal{I}_{i,n}^{(l)}| \right]^{1/2} |\mathcal{I}_{j,n}^{(3-l)}| \right.$$

$$+ (K_q |\mathcal{I}_{i,n}^{(3-l)}| + 1 + e_q) |\mathcal{I}_{i,n}^{(3-l)}|) |\mathcal{I}_{j,n}^{(l)}|^{1/2} \right] \sqrt{n} |\pi_n|_T \mathbb{1}_{\{\mathcal{I}_{i,n}^{(1)} \cap \mathcal{I}_{j,n}^{(2)} \neq \emptyset\}}$$

$$+ \left[((K_q |\mathcal{I}_{i,n}^{(l)} \cap \mathcal{I}_{j,n}^{(3-l)}|^2 + |\mathcal{I}_{i,n}^{(l)} \cap \mathcal{I}_{j,n}^{(3-l)}|^{1/2} + e_q) |\mathcal{I}_{i,n}^{(l)} \cap \mathcal{I}_{j,n}^{(3-l)}|)^{1/3} \right.$$

$$\times |\mathcal{I}_{i,n}^{(l)} \cap \mathcal{I}_{j,n}^{(3-l)}|^{2/3} \mathbb{E}[\sqrt{(\widehat{\mathcal{L}}_{n,m}^{(l)} + \widehat{\mathcal{R}}_{n,m}^{(l)})(t_{i,n}^{(l)} \wedge t_{j,n}^{(3-l)})} |\mathcal{S}]$$

$$+ |\mathcal{I}_{i,n}^{(l)} \cap \mathcal{I}_{j,n}^{(3-l)}|^{1/2} ((K_q |\mathcal{I}_{i,n}^{(l)} \cap \mathcal{I}_{j,n}^{(3-l)}|^3 + |\mathcal{I}_{i,n}^{(l)} \cap \mathcal{I}_{j,n}^{(3-l)}| + e_q)$$

$$\times |\mathcal{I}_{i,n}^{(l)} \cap \mathcal{I}_{j,n}^{(3-l)}|)^{1/2} \mathbb{E}[\sqrt{(\widehat{\mathcal{L}}_{n,m}^{(3-l)} + \widehat{\mathcal{R}}_{n,m}^{(3-l)})(t_{i,n}^{(l)} \wedge t_{j,n}^{(3-l)})} |\mathcal{S}] \right]). \quad (9.73)$$

Here, we used iterated expectations and

$$\mathbb{E}[\widetilde{Y}_{(i,j),n}^{(l)} |\mathcal{S}] \leq \left(\frac{1}{b_n} \sum_{k \neq i: \mathcal{I}_{k,n}^{(l)} \subset [t_{i,n}^{(l)} - b_n, t_{i,n}^{(l)} + b_n]} K |\mathcal{I}_{k,n}^{(l)}| \right)^{1/2} \leq K$$

along with the similar bound for $\mathbb{E}[\widetilde{Y}^{(3-l)}_{(j,i),n}|\mathcal{S}]$.

The sum over the expression in the first set of square brackets in (9.73) vanishes just like (9.72) while the sum over the expression in the second set of square brackets is bounded by

$$K(K_q(|\pi_n|_T)^{1/2} + e_q)^{1/3} \sum_{i,j:t^{(l)}_{i,n} \vee t^{(3-l)}_{j,n} \leq T} |\mathcal{I}^{(l)}_{i,n} \cap \mathcal{I}^{(3-l)}_{j,n}|$$

$$\times \sum_{\tilde{k}_1,\tilde{k}_2=-\widetilde{K}_n}^{\widetilde{K}_n} |\mathcal{I}^{(l)}_{i+\tilde{k}_1,n} \cap \mathcal{I}^{(3-l)}_{j+\tilde{k}_2,n}| \Big(\sum_{\tilde{k}'_1,\tilde{k}'_2=-\widetilde{K}_n}^{\widetilde{K}_n} |\mathcal{I}^{(l)}_{i+\tilde{k}'_1,n} \cap \mathcal{I}^{(3-l)}_{j+\tilde{k}'_2,n}| \Big)^{-1}$$

$$\times ((n|\mathcal{I}^{(l)}_{i+\tilde{k}_1-1,n}| + n|\mathcal{I}^{(l)}_{i+\tilde{k}_1+1,n}|)^{1/2} + (n|\mathcal{I}^{(3-l)}_{j+\tilde{k}_2-1,n}| + n|\mathcal{I}^{(3-l)}_{j+\tilde{k}_2+1,n}|)^{1/2})$$

$$= K(K_q(|\pi_n|_T)^{1/2} + e_q) \sum_{i,j:t^{(l)}_{i,n} \vee t^{(3-l)}_{j,n} \leq T} |\mathcal{I}^{(l)}_{i,n} \cap \mathcal{I}^{(3-l)}_{j,n}|$$

$$\times ((n|\mathcal{I}^{(l)}_{i-1,n}| + n|\mathcal{I}^{(l)}_{i+1,n}|)^{1/2} + (n|\mathcal{I}^{(3-l)}_{j-1,n}| + n|\mathcal{I}^{(3-l)}_{j+1,n}|)^{1/2})$$

$$\times \sum_{\tilde{k}_1,\tilde{k}_2=-\widetilde{K}_n}^{\widetilde{K}_n} |\mathcal{I}^{(l)}_{i+\tilde{k}_1,n} \cap \mathcal{I}^{(3-l)}_{j+\tilde{k}_2,n}| \Big(\sum_{\tilde{k}'_1,\tilde{k}'_2=-\widetilde{K}_n}^{\widetilde{K}_n} |\mathcal{I}^{(l)}_{i+\tilde{k}_1+\tilde{k}'_1,n} \cap \mathcal{I}^{(3-l)}_{j+\tilde{k}_2+\tilde{k}'_2,n}| \Big)^{-1}$$

$$+ O_\mathbb{P}(\sqrt{n}|\pi_n|_T) \tag{9.74}$$

where the sum in the second to last line is less or equal than 4 which can be shown similarly to (4.23). The $O_\mathbb{P}(\sqrt{n}|\pi_n|_T)$-term is due to boundary effects. Hence (9.74) is bounded by

$$K(K_q(|\pi_n|_T)^{1/2} + e_q)\sqrt{n}\Big(\sum_{i \geq 2:t^{(l)}_{i,n} \leq T} |\mathcal{I}^{(l)}_{i,2,n}|^{3/2} + \sum_{j \geq 2:t^{(3-l)}_{j,n} \leq T} |\mathcal{I}^{(3-l)}_{j,2,n}|^{3/2} \Big)$$

$$+ O_\mathbb{P}(\sqrt{n}|\pi_n|_T)$$

which vanishes as shown in (9.61) for first $n \to \infty$ and then $q \to \infty$ because of $G^{[2],n}_{2,2}(T) = O_\mathbb{P}(1)$.

Step 3. Further we obtain

$$\widetilde{\mathbb{P}}(Y(P) \leq \Upsilon|\mathcal{X})\mathbb{1}_{\Omega^{CoJ}_T} \xrightarrow{\mathbb{P}} \widetilde{\mathbb{P}}(\Phi^{CoJ}_{2,T} \leq \Upsilon|\mathcal{X})\mathbb{1}_{\Omega^{CoJ}_T} \tag{9.75}$$

as in Step 3 in the proof of Proposition 8.15.

Step 4. Finally, (9.66) is obtained from (9.67), (9.68) and (9.75); compare Steps 4 and 5 in the proof of Proposition 4.10. $\qquad\square$

Proof of Theorem 9.24. Analogously as in the proof of (8.33) we obtain

$$\lim_{n\to\infty} \widetilde{\mathbb{P}}(\{|\widehat{Q}^{CoJ}_{2,T,n}(\alpha) - Q^{CoJ}_{2,T}(\alpha)| > \varepsilon\} \cap \Omega^{CoJ}_T) = 0 \tag{9.76}$$

for all $\varepsilon > 0$ and any $\alpha \in [0,1]$ from Proposition 9.27. Then Theorem 9.19 and (9.76) yield, compare Step 2 in the proof of Theorem 4.3,

$$\widetilde{\mathbb{P}}(\sqrt{n}\,|V(f_{(2,2)},[2],\pi_n)_T - 4V(f_{(2,2)},\pi_n)_T| > \widehat{Q}_{2,T,n}^{CoJ}(1-\alpha)\big|F^{CoJ}) \to \alpha$$

from which we immediately conclude (9.46).

For proving (9.47) we observe that $\Psi_{2,T,n}^{CoJ}$ converges on the given F^{nCoJ} to a random variable which is under Condition 9.23(iv) almost surely different from 1 by Theorem 9.16. To see this observe that $\widetilde{C}_{k,T}^{DisJ}, \widetilde{C}_{1,T}^{DisJ}$ are $\mathcal{F}$-measurable while the $\mathcal{F}$-conditional distribution of $\widetilde{D}_{k,T}^{DisJ}, \widetilde{D}_{1,T}^{DisJ}$ is continuous by Condition 9.23. Hence, $k\widetilde{C}_{k,T}^{DisJ} \neq \widetilde{C}_{1,T}^{DisJ}$ or $\widetilde{D}_{k,T}^{DisJ} \neq \widetilde{D}_{1,T}^{DisJ}$ almost surely imply

$$k\widetilde{C}_{k,T}^{DisJ} + \widetilde{D}_{k,T}^{DisJ} \neq \widetilde{C}_{1,T}^{DisJ} + \widetilde{D}_{1,T}^{DisJ} \quad a.s.$$

Then (9.47) follows from

$$c_{2,T,n}^{CoJ}(\alpha)\mathbb{1}_{\Omega_T^{nCoJ}} = o_{\mathbb{P}}(1) \tag{9.77}$$

which we will prove in the following: From the proof of Theorem 9.19 we know that $nV(f_{(2,2)},\pi_n)_T$ converges on Ω_T^{nCoJ} stably in law to a non-negative random variable. Comparing (8.44) in the proof of Theorem 8.11 it then suffices to show

$$\sqrt{n}\widehat{\Phi}_{2,T,n,m}^{CoJ}\mathbb{1}_{\Omega_T^{nCoJ}} = o_{\mathbb{P}}(1) \tag{9.78}$$

uniformly in m for proving (9.77). In order to achieve this goal, observe that it holds

$$\mathbb{E}[\sqrt{n}|\widehat{\Phi}_{2,T,n,m}^{CoJ}|\mathbb{1}_{\Omega_T^{nCoJ}}|\mathcal{F}]$$

$$\leq K\sqrt{n}\sum_{l=1,2}\sum_{i,j:t_{i,n}^{(l)}\vee t_{j,n}^{(3-l)}\leq T}|\Delta_{i,n}^{(l)}X^{(l)}|(\Delta_{j,n}^{(3-l)}X^{(3-l)})^2$$

$$\times|\tilde{\sigma}^{(l)}(t_{i,n}^{(l)},-)+\tilde{\sigma}^{(l)}(t_{i,n}^{(l)},+)|$$

$$\times \mathbb{E}[((\widehat{\mathcal{L}}_{n,m}^{CoJ,(l)}+\widehat{\mathcal{R}}_{n,m}^{CoJ,(l)})(t_{i,n}^{(l)}\wedge t_{j,n}^{(3-l)}))^{1/2}|\mathcal{F}]\mathbb{1}_{\{|\Delta_{i,n}^{(l)}X^{(l)}|>\beta|\mathcal{I}_{i,n}^{(l)}|^{\varpi}\}}$$

$$\times \mathbb{1}_{\{|\Delta_{j,n}^{(3-l)}X^{(3-l)}|>\beta|\mathcal{I}_{j,n}^{(3-l)}|^{\varpi}\}}\mathbb{1}_{\{\mathcal{I}_{i,n}^{(l)}\cap\mathcal{I}_{j,n}^{(3-l)}\neq\emptyset\}}\mathbb{1}_{\Omega_T^{nCoJ}}$$

$$\leq K\sqrt{n}\sum_{l=1,2}\sum_{i,j:t_{i,n}^{(l)}\vee t_{j,n}^{(3-l)}\leq T}|\mathcal{I}_{i,n}^{(l)}|^{-\varpi}(\Delta_{i,n}^{(l)}X^{(l)})^2(\Delta_{j,n}^{(3-l)}X^{(3-l)})^2$$

$$\times|\tilde{\sigma}^{(l)}(t_{i,n}^{(l)},-)+\tilde{\sigma}^{(l)}(t_{i,n}^{(l)},+)|\sqrt{n|\pi_n|_T}\mathbb{1}_{\{\mathcal{I}_{i,n}^{(l)}\cap\mathcal{I}_{j,n}^{(3-l)}\neq\emptyset\}}\mathbb{1}_{\Omega_T^{nCoJ}}. \tag{9.79}$$

Using the notation from (9.70) and the inequalities (9.71) we then obtain

$$(\Delta_{i,n}^{(l)}X^{(l)})^2(\Delta_{j,n}^{(3-l)}X^{(3-l)})^2|\tilde{\sigma}^{(l)}(t_{i,n}^{(l)},-)+\tilde{\sigma}^{(l)}(t_{i,n}^{(l)},+)|\mathbb{1}_{\Omega_T^{nCoJ}}$$

$$\leq 2(\Delta_{i,n}^{(l)}X^{(l)})^2(\Delta_{(j,k),(1,1),n}^{(3-l,l)}X^{(3-l)})^2$$

$$\times\left(\frac{1}{b_n}\sum_{k\neq i:\mathcal{I}_{k,n}^{(l)}\subset[t_{i,n}^{(l)}-b_n,t_{i,n}^{(l)}+b_n]}(\Delta_{k,n}^{(l)}X^{(l)})^2\right)^{1/2}\mathbb{1}_{\Omega_T^{nCoJ}} \qquad (9.80)$$

$$+2(\Delta_{i,n}^{(l)}X^{(l)})^2(\Delta_{(j,k),(1,1),n}^{(3-l\backslash l)}X^{(3-l)})^2$$

$$\left(\frac{1}{b_n}\sum_{k\neq i:\mathcal{I}_{k,n}^{(l)}\subset[t_{i,n}^{(l)}-b_n,t_{i,n}^{(l)}+b_n]}(\Delta_{(k,j),(1,1),n}^{(l,3-l)}X^{(l)})^2\right)^{1/2}\mathbb{1}_{\Omega_T^{nCoJ}}. \quad (9.81)$$

$$+2(\Delta_{i,n}^{(l)}X^{(l)})^2(\Delta_{(j,k),(1,1),n}^{(3-l\backslash l)}X^{(3-l)})^2$$

$$\left(\frac{1}{b_n}\sum_{k\neq i:\mathcal{I}_{k,n}^{(l)}\subset[t_{i,n}^{(l)}-b_n,t_{i,n}^{(l)}+b_n]}(\Delta_{(k,j),(1,1),n}^{(l\backslash 3-l)}X^{(l)})^2\right)^{1/2}\mathbb{1}_{\Omega_T^{nCoJ}}. \quad (9.82)$$

Recall that we may write $X = B(q')+C+M(q')$ for some $q' > 0$ by Condition 1.3; compare (9.20). Then iterated expectations and Lemmata 1.4, 3.30, 9.11 yield that the $\mathcal{S}$-conditional expectation of (9.80) is bounded by $K|\mathcal{I}_{i,n}^{(l)}||\mathcal{I}_{i,n}^{(l)}\cap\mathcal{I}_{j,n}^{(3-l)}|$. Furthermore (9.81) is using iterated expectations, the Cauchy-Schwarz inequality and (1.12) bounded by $K|\mathcal{I}_{i,n}^{(l)}||\mathcal{I}_{j,n}^{(3-l)}\setminus\mathcal{I}_{i,n}^{(l)}|$ and (9.82) is using iterated expectations and inequality (1.12) bounded by $K|\mathcal{I}_{i,n}^{(l)}||\mathcal{I}_{j,n}^{(3-l)}\setminus\mathcal{I}_{i,n}^{(l)}|$ as well. Then alltogether we obtain that the $\mathcal{S}$-conditional expectation of (9.79) is bounded by

$$K\sqrt{n}\sum_{l=1,2}\sum_{i,j:t_{i,n}^{(l)}\vee t_{j,n}^{(3-l)}\leq T}|\mathcal{I}_{i,n}^{(l)}|^{-\varpi}|\mathcal{I}_{i,n}^{(l)}||\mathcal{I}_{j,n}^{(3-l)}|\sqrt{n|\pi_n|_T}\mathbb{1}_{\{\mathcal{I}_{i,n}^{(l)}\cap\mathcal{I}_{j,n}^{(3-l)}\neq\emptyset\}}$$

$$= K\sqrt{n}\sum_{l=1,2}\sum_{i,j:t_{i,n}^{(l)}\vee t_{j,n}^{(3-l)}\leq T}|\mathcal{I}_{i,n}^{(l)}|^{1/2-\varpi}|\mathcal{I}_{i,n}^{(l)}|^{1/2}|\mathcal{I}_{j,n}^{(3-l)}|\sqrt{n|\pi_n|_T}$$

$$\times\mathbb{1}_{\{\mathcal{I}_{i,n}^{(l)}\cap\mathcal{I}_{j,n}^{(3-l)}\neq\emptyset\}}$$

$$\leq Kn^{1/2}(|\pi_n|_T)^{1-\varpi}(G_{1,2}^n(T)+G_{2,1}^n(T))$$

where we used $1/2-\varpi > 0$. Hence (9.78) follows from Condition 9.23(ii) and Lemma 2.15. □

Proof of Corollary 9.25. Note that it holds $n^{-1}A_{2,T,n}^{CoJ} = o_\mathbb{P}(1)$. To see this observe that the second sum in $n^{-1}A_{2,T,n}^{CoJ}$ vanishes as $n\to\infty$ because big common jumps are asymptotically filtered out due to the indicator and because the remaining terms vanish by (2.14) as well. The first sum in $n^{-1}A_{2,T,n}^{CoJ}$ can be discussed similarly. Hence on Ω_T^{CoJ} it holds

$$\sqrt{n}\frac{n^{-1}A_{2,T,n}^{CoJ}}{4V(f_{(2,2)},\pi_n)_T}\mathbb{1}_{\Omega_T^{CoJ}} \xrightarrow{\mathbb{P}} 0$$

and combining this with (9.43) yields the $\mathcal{X}$-stable convergence

$$\sqrt{n}\Big(\Phi_{2,T,n}^{CoJ} - \rho\frac{n^{-1}A_{2,T,n}^{CoJ}}{4V(f_{(2,2)},\pi_n)_T} - 1\Big) \xrightarrow{\mathcal{L}-s} \frac{\Phi_{2,T}^{CoJ}}{4B^*(f_{(2,2)})_T} \tag{9.83}$$

on Ω_T^{CoJ}. Replacing (9.43) with (9.83) in the proof of (9.46) yields (9.48).

Using arguments from the proof of Theorem 9.16 and the proof of (9.10) we obtain

$$A_{2,T,n}^{CoJ}\mathbb{1}_{\Omega_T^{nCoJ}} \xrightarrow{\mathcal{L}-s} \big(4(2\widetilde{C}_{2,T}^{DisJ} - \widetilde{C}_{1,T}^{DisJ}) + 4(\widetilde{D}_{2,T}^{DisJ} - \widetilde{D}_{1,T}^{DisJ})\big)\mathbb{1}_{\Omega_T^{nCoJ}} \tag{9.84}$$

under the alternative $\omega \in \Omega_T^{nCoJ}$. Hence based on Theorem 9.16 and (9.84) we conclude

$$(\Phi_{2,T,n}^{CoJ} - \rho\frac{A_{2,T,n}^{CoJ}}{4nV(f_{(2,2)},\pi_n)_T} - 1)\mathbb{1}_{\Omega_T^{nCoJ}}$$

$$\xrightarrow{\mathcal{L}-s} (1-\rho)\frac{(2\widetilde{C}_{2,T}^{DisJ} - \widetilde{C}_{1,T}^{DisJ}) + (\widetilde{D}_{2,T}^{DisJ} - \widetilde{D}_{1,T}^{DisJ})}{\widetilde{C}_{1,T}^{DisJ} + \widetilde{D}_{1,T}^{DisJ}}\mathbb{1}_{\Omega_T^{nCoJ}}$$

where the limit is almost surely different from zero by Condition 9.23(iv). We then obtain (9.49) as in the proof of Theorem 9.24 because of $c_{2,T,n}^{CoJ}(\alpha)\mathbb{1}_{\Omega_T^{nCoJ}} = o_\mathbb{P}(1)$; compare (9.77). $\square$

Bibliography

[1] Yacin Aït-Sahalia, Jianqing Fan, and Dacheng Xiu. High-frequency estimates with noisy and asynchronous financial data. *Journal of the American Statistical Association*, 492(105):1504–1516, 2010.

[2] Yacine Aït-Sahalia and Jean Jacod. Testing for jumps in a discretely observed process. *The Annals of Statistics*, 37(1):184–222, 2009.

[3] Yacine Aït-Sahalia and Jean Jacod. *High-Frequency Financial Econometrics*. Princeton University Press, 2014. ISBN: 0-69116-143-3.

[4] Torben G. Andersen and Tim Bollerslev. Answering the skeptics: Yes, standard volatility models do provide accurate forecasts. *International Economic Review*, 39(4):885–905, 1998.

[5] Ole Eiler Barndorff-Nielsen, Peter Reinhard Hansen, Asger Lunde, and Neil Shephard. Multivariate realised kernels: Consistent positive semi-definite estimators of the covariation of equity prices with noise and non-synchronous trading. *Journal of Econometrics*, 162(2):149–169, 2011.

[6] Ole Eiler Barndorff-Nielsen and Neil Shephard. Econometric analysis of realized volatility and its use in estimating stochastic volatility models. *Journal of the Royal Statistical Society. Series B (Statistical Methodology)*, 64(2):253–280, 2002.

[7] Markus Bibinger, Nikolaus Hautsch, Peter Malec, and Markus Reiß. Estimating the spot covariation of asset prices - statistical theory and empirical evidence. *To appear in: Journal of Business and Economic Statistics*, 2017.

[8] Markus Bibinger and Mathias Vetter. Estimating the quadratic covariation of an asynchronously observed semimartingale with jumps. *Annals of the Institute of Statistical Mathematics*, 67:707–743, 2015.

[9] Patrick Billingsley. *Convergence of probability measures*. Wiley series in probability and statistics. Wiley, 2nd edition, 1999.

[10] Peter J. Brockwell and Richard A. Davis. *Time Series: Theory and Methods*. Springer Series in Statistics. Springer, 2nd edition, 1991.

© Springer Fachmedien Wiesbaden GmbH, part of Springer Nature 2019
O. Martin, *High-Frequency Statistics with Asynchronous and Irregular Data*,
Mathematische Optimierung und Wirtschaftsmathematik | Mathematical Optimization
and Economathematics, https://doi.org/10.1007/978-3-658-28418-3

[11] Donald Lyman Burkholder. Martingale transforms. *The Annals of Mathematical Statistics*, 37:1494–1504, 1966.

[12] Marek Capinski, Ekkehard Kopp, and Janusz Traple. *Stochastic Calculus for Finance*. Cambridge University Press, 2012.

[13] Freddy Delbaen and Walter Schachermayer. A general version of the fundamental theorem of asset pricing. *Mathematische Annalen*, 3(300):463–520, 1993.

[14] Thomas W. Epps. Comovements in stock prices in the very short run. *Journal of the American Statistical Association*, 74(366):291–298, 1979.

[15] Jianqing Fan and Yazhen Wang. Spot volatility estimation for high-frequency data. *Statistics and its Interface*, 1:279–288, 2008.

[16] Thomas S. Ferguson. *A course in large sample theory*. Chapman & Hall, 1st edition, 1996.

[17] Otto Forster. *Analysis 2 - Differentialrechnung im $\mathbb{R}^n$, gewöhnliche Differentialgleichungen*. Vieweg+Teubner, 8th edition, 2008.

[18] Masaaki Fukasawa and Mathieu Rosenbaum. Central limit theorems for realized volatility under hitting times of an irregular grid. *Stochastic Processes and their Applications*, 122(12):3901–3920, 2012.

[19] Godfrey Harold Hardy, John Edensor Littlewood, and George Pólya. *Inequalities*. Cambridge Mathematical Library. Cambridge University Press, 2nd edition, 1952.

[20] Takaki Hayashi, Jean Jacod, and Nakahiro Yoshida. Irregular sampling and central limit theorems for power variations: the continuous case. *Annales de l'Institut Henri Poincaré Probabilités et Statistiques*, 47(4):1197–1218, 2011.

[21] Takaki Hayashi and Shigeo Kusuoka. Consistent estimation of covariation under nonsychronicity. *Statistical Inference for Stochastic Processes*, 11(1):93–106, 2008.

[22] Takaki Hayashi and Nakahiro Yoshida. On covariance estimation of non-synchronously observed diffusion processes. *Bernoulli*, 11(2):359–379, 2005.

[23] Takaki Hayashi and Nakahiro Yoshida. Asymptotic normality of a covariance estimator for non-synchronously observed processes. *Annals of the Institute of Statistical Mathematics*, 60(2):367–406, 2008.

[24] Takaki Hayashi and Nakahiro Yoshida. Nonsynchronous covarianciation process and limit theorems. *Stochastic Processes and their Applications*, 121:2416–2454, 2011.

[25] Xin Huang and George Tauchen. The relative contribution of jumps to total price variance. *J. Financial Econometrics*, 4:456–499, 2006.

[26] Jean Jacod. On continuous conditional gaussian martingales and stable convergence in law. *Séminaire de Probabilités*, XXXI:232–246, 1997.

[27] Jean Jacod. Asymptotic properties of realized power variations and related functionals of semimartingales. *Stochastic Processes and their Applications*, 118(4):517–559, 2008.

[28] Jean Jacod. Statistics and high frequency data, 2009. Lecture Notes, SEM-STAT Course in La Manga.

[29] Jean Jacod and Philip Protter. Asymptotic error distributions for the Euler method for stochastic differential equations. *The Annals of Probability*, 26:267–307, 1998.

[30] Jean Jacod and Philip Protter. *Discretization of Processes*. Springer, 2012. ISBN: 3-64224-126-3.

[31] Jean Jacod and Albert Shiryaev. *Limit Theorems for Stochastic Processes*. Springer, 2nd edition, 2002. ISBN: 3-540-43932-3.

[32] Jean Jacod and Viktor Todorov. Testing for common arrivals of jumps for discretely observed multidimensional processes. *The Annals of Statistics*, 37(1):1792–1838, 2009.

[33] Achim Klenke. *Probability Theory: A Comprehensive Course*. Springer-Verlag London, 2nd edition, 2014.

[34] Yuta Koike. An estimator for the cumulative co-volatility of asynchronously observed semimartingales with jumps. *Scandinavian Journal of Statistics*, 2(41):460–481, 2014.

[35] Cecilia Mancini and Fabio Gobbi. Identifying the brownian covariation from the co-jumps given discrete observations. *Econometric Theory*, 28(2):249–273, 2012.

[36] Ole Martin and Mathias Vetter. Testing for simultaneous jumps in case of asynchronous observations. *Bernoulli*, 24(4B):3522–3567, 2018.

[37] Ole Martin and Mathias Vetter. Laws of large numbers for Hayashi-Yoshida-type functionals. *Finance and Stochastics*, 23(3):451–500, 2019.

[38] Ole Martin and Mathias Vetter. The null hypothesis of common jumps in case of irregular and asynchronous observations. *Scandinavian Journal of Statistics, to appear,* 2019+.

[39] Per A. Mykland and Lan Zhang. Assessment of uncertainty in high frequency data: The observed asymptotic variance. *Econometrica,* 37(1):197–231, 2017.

[40] Mark Podolskij and Mathias Vetter. Understanding limit theorems for semi-martingales: a short survey. *Statistica Neerlandica,* 64:329–351, 2010.

[41] Philip Protter. *Stochastic Integration and Differential Equations.* Springer, 2nd edition, 2004. ISBN: 978-3-642-05560-7.

[42] Daniel Revuz and Marc Yor. *Continuous Martingales and Brownian Motion.* Springer, 3rd edition, 1999.

[43] Gennady Samorodnitsky and Murad Taqqu. *Stable Non-Gaussian Random Processes: Stochastic Models with Infinite Variance (Stochastic Modeling Series).* Chapman & Hall, 1994.

[44] Carl P. Simon and Lawrence E. Blume. *Mathematics for Economists.* W. W. Norton & Company, 1994.

[45] Roy L. Streit. *Poisson Point Processes: Imaging, Tracking, and Sensing.* Springer, 2010.

[46] Viktor Todorov and George Tauchen. Volatility jumps. *Journal of Business and Economic Statistics,* 29:356–371, 2011.

[47] Mathias Vetter and Tobias Zwingmann. A note on central limit theorems for quadratic variation in case of endogenous observation times. *Electronic Journal of Statistics,* 11(1):963–980, 2017.

Appendix

A Estimates for Itô Semimartingales

The following inequalities allow to bound increments of a local martingale M by increments of its quadratic variation $[M, M]$. The version here is taken from (2.1.34) in [30]. An inequality of that form was first mentioned in [11] for continuous martingales.

Lemma A.1 (Burkholder-Davis-Gundy inequalities).
Let M be a local martingale with $M_0 = 0$ and $S \leq S'$ be two stopping times. Then for $p \geq 1$ there exist constants $0 < c_p < C_p < \infty$ such that it holds

$$c_p \mathbb{E}[([M,M]_{S'}] - [M,M]_S)^{p/2}|\mathcal{F}_S] \leq \mathbb{E}[\sup_{t \in (S,S']} |M_t - M_S|^p|\mathcal{F}_S]$$

$$\leq C_p \mathbb{E}[([M,M]_{S'}] - [M,M]_S)^{p/2}|\mathcal{F}_S]. \quad \text{(A.1)}$$

The constants c_p, C_p are universal i.e. do not depend on the local martingale M or the stopping times S, S'.

Proof. A proof of the inequalities can be found for example in Chapter IV §4 of [42]. $\qquad\square$

Lemma A.1 is used e.g. to derive the bounds for C and $M(q)$ in the proof of Lemma 1.4.

Proof of Lemma 1.4. Using $|\mathbb{1}_{\{\|\delta(s,z)\| \leq 1\}} - \mathbb{1}_{\{\gamma(z) \leq 1/q\}}| \leq \mathbb{1}_{\{\gamma(z) > 1 \wedge 1/q\}}$ it holds

$$\|B(q)_{s+t} - B(q)_s\| \leq \int_s^{s+t} \left(\|b_u\| + \int_{\mathbb{R}^2} \|\delta(s,z)\| \mathbb{1}_{\{\gamma(z) > 1 \wedge 1/q\}} \lambda(dz) \right) du$$

$$\leq \int_s^{s+t} \left(\|b_u\| + \int_{\mathbb{R}^2} \frac{\Gamma_u(\gamma(z))^2}{1 \wedge 1/q} \mathbb{1}_{\{\gamma(z) > 1 \wedge 1/q\}} \lambda(dz) \right) du$$

$$\leq \int_s^{s+t} \left(\|b_u\| + \frac{\Gamma_u}{1 \wedge 1/q} \int_{\mathbb{R}^2} (\gamma(z)^2 \wedge K^2) \lambda(dz) \right) du$$

$$\leq \int_s^{s+t} \left(\|b_u\| + \frac{\Gamma_u(1 \vee K^2)}{1 \wedge 1/q} \int_{\mathbb{R}^2} (\gamma(z)^2 \wedge 1) \lambda(dz) \right) du$$

$$\leq K_q t$$

© Springer Fachmedien Wiesbaden GmbH, part of Springer Nature 2019
O. Martin, *High-Frequency Statistics with Asynchronous and Irregular Data*,
Mathematische Optimierung und Wirtschaftsmathematik | Mathematical Optimization and Economathematics, https://doi.org/10.1007/978-3-658-28418-3

where we used that b_t, Γ_t are bounded by assumption and because $\gamma \leq K$ for some constant $K > 0$ and $\int_{\mathbb{R}^2}(\gamma(z)^2 \wedge 1)\lambda(dz) < \infty$ by Condition 1.3. Taking this inequality to the p-th power yields (1.8).

For $p \geq 1$ the inequality (1.9) follows by the boundedness of σ directly from (2.1.34) in [30]. Using (1.9) for $p = 1$ and the Jensen inequality for the function $x \to x^{1/p}$, $x \geq 0$, $p \in (0, 1)$, we then obtain

$$\mathbb{E}[\|C_{s+t} - C_s\|^p | \mathcal{F}_s] \leq (\mathbb{E}[\|C_{s+t} - C_s\| | \mathcal{F}_s])^p \leq (K_1)^p t^{p/2}$$

which yields (1.9) for $p \in (0, 1)$.

Regarding the inequalities (1.10) and (1.11), Lemma 2.1.5 and Lemma 2.1.7 in [30] yield

$$\mathbb{E}[\|M(q)_{s+t} - M(q)_s\|^p | \mathcal{F}_s] \leq K_p(s\mathbb{E}[\widehat{\delta}_M(p)_{s,t} | \mathcal{F}_s] + s^{p/2}\mathbb{E}[(\widehat{\delta}_M(2)_{s,t})^{p/2} | \mathcal{F}_s]),$$
$$(A.2)$$

$$\mathbb{E}[\|N(q)_{s+t} - N(q)_s\|^{p'} | \mathcal{F}_s] \leq K_{p'}(s\mathbb{E}[\widehat{\delta}_N(p')_{s,t} | \mathcal{F}_s] + s^{p'}\mathbb{E}[(\widehat{\delta}_N(1)_{s,t})^{p'} | \mathcal{F}_s])$$
$$(A.3)$$

for $p \geq 2$ and $p' \geq 1$ with

$$\widehat{\delta}_M(r)_{s,t} = \frac{1}{t}\int_s^{s+t}\int_{\mathbb{R}^2}\|\delta(u,z)\|^r \mathbb{1}_{\{\gamma(z) \leq 1/q\}}\lambda(dz)du, \quad r \geq 2,$$

$$\widehat{\delta}_N(r)_{s,t} = \frac{1}{t}\int_s^{s+t}\int_{\mathbb{R}^2}\|\delta(u,z)\|^r \mathbb{1}_{\{\gamma(z) > 1/q\}}\lambda(dz)du, \quad r \geq 1.$$

We derive for $p \geq 2$

$$\widehat{\delta}_M(p)_{s,t} \leq \frac{1}{t}\int_s^{s+t}\int_{\mathbb{R}^2}((\Gamma_u\gamma(z))^p \mathbb{1}_{\{\gamma(z) \leq 1/q\}}\lambda(dz)du$$

$$\leq K(1/q)^{p-2}\int_{\mathbb{R}^2}(\gamma(z)^2 \wedge 1/q^2)\mathbb{1}_{\{\gamma(z) \leq 1/q\}}\lambda(dz)$$

$$\leq K(1/q)^{p-2}\max\{1, 1/q^2\}\int_{\mathbb{R}^2}(\gamma(z)^2 \wedge 1)\mathbb{1}_{\{\gamma(z) \leq 1/q\}}\lambda(dz)$$

which tends to zero as $q \to \infty$. Hence (A.2) yields (1.10) for all $p \geq 2$ using $t^{p/2} \leq T^{p/2-1}t$. The result for $p \in (0,2)$ then follows using Jensen's inequality as for (1.9).

Further, by Condition 1.3 there exists a constant $K \geq 0$ with $\gamma(z) \leq K$ and by assumption a constant K' with $(\Gamma_u)^{p'} \leq K'$ which yields

$$\widehat{\delta}_N(p')_{s,t} \leq K' \int_{\mathbb{R}^2} (\gamma(z))^{p'} \mathbb{1}_{\{\gamma(z)>1/q\}} \lambda(dz)$$

$$\leq K'K^{p'-2} \int_{\mathbb{R}^2} (\gamma(z)^2 \wedge K^2)\lambda(dz)$$

$$\leq K'K^{p'-2} \max\{1, K^2\} \int_{\mathbb{R}^2} (\gamma(z)^2 \wedge 1)\lambda(dz)$$

for $p' \geq 2$ and

$$\widehat{\delta}_N(p')_{s,t} \leq K' \int_{\mathbb{R}^2} (\gamma(z))^{p'} \mathbb{1}_{\{\gamma(z)>1/q\}} \lambda(dz)$$

$$\leq K'q^{2-p'} \int_{\mathbb{R}^2} (\gamma(z)^2 \wedge K^2)\mathbb{1}_{\{\gamma(z)>1/q\}} \lambda(dz)$$

$$\leq K'q^{2-p'} \max\{1, K^2\} \int_{\mathbb{R}^2} (\gamma(z)^2 \wedge 1)\lambda(dz)$$

for $p' \in [1,2)$. Hence (A.3) yields (1.11).

As γ is bounded by Condition 1.3 there exists a $q' > 0$ with

$$X_t = X_0 + B(q')_t + C_t + M(q')_t.$$

The inequality (1.12) then follows from the elementary inequality

$$\|a + b + c\|^p \leq K_p(\|a\|^p + \|b\|^p + \|c\|^p), \quad a, b, c \in \mathbb{R}^2, \; p \geq 0,$$

and inequalities (1.8)–(1.11). $\qquad\qquad\qquad\qquad\qquad\qquad\qquad\qquad\qquad\square$

Proof of Lemma 1.5. The proof for (1.13) is identical to the proof of (1.8) as all estimates are ω-wise.

As C_t is a local martingale we obtain the following estimate from the upper Burkholder-Davis-Gundy inequality (Lemma A.1)

$$\mathbb{E}[\sup_{t \in (S,S')} \|C_t - C_S\|^p | \mathcal{F}_S]$$

$$\leq K_p \sum_{l=1,2} \mathbb{E}[\sup_{t \in (S,S')} |C_t^{(l)} - C_S^{(l)}|^p | \mathcal{F}_S]$$

$$\leq K_p \sum_{l=1,2} K_p \mathbb{E}[([C^{(l)}, C^{(l)}]_{S'} - [C^{(l)}, C^{(l)}]_S)^{p/2} | \mathcal{F}_S]$$

$$= K_p \sum_{l=1,2} \mathbb{E}[(\int_S^{S'} (\sigma_s^{(l)})^2 ds)^{p/2} | \mathcal{F}_S]$$

$$\leq K_p \sum_{l=1,2} \mathbb{E}[(K(S'-S))^{p/2} | \mathcal{F}_S] \leq K_p \mathbb{E}[(S'-S)^{p/2} | \mathcal{F}_S]$$

which is (1.14).

As M_t is also a local martingale we obtain again using Lemma A.1

$$\mathbb{E}[\sup_{t\in(S,S']} \|M(q)_t - M(q)_S\|^2 | \mathcal{F}_S]$$

$$\leq K \sum_{l=1,2} \mathbb{E}[([M^{(l)}(q), M^{(l)}(q)]_{S'} - [M^{(l)}(q), M^{(l)}(q)]_S) | \mathcal{F}_S]$$

$$= K \sum_{l=1,2} \mathbb{E}[\int_S^{S'} \int_{\mathbb{R}^2} (\delta^{(l)}(s,z))^2 \mathbb{1}_{\{\gamma(z)\leq 1/q\}} \mu(ds,dz) | \mathcal{F}_S]$$

$$\leq 2K\mathbb{E}[\int_S^{S'} \int_{\mathbb{R}^2} \|\delta(s,z)\|^2 \mathbb{1}_{\{\gamma(z)\leq 1/q\}} \mu(ds,dz) | \mathcal{F}_S]$$

$$\leq 2K\mathbb{E}[\int_S^{S'} \int_{\mathbb{R}^2} ((\Gamma_s\gamma(z))^2 \wedge K') \mathbb{1}_{\{\gamma(z)\leq 1/q\}} \mu(ds,dz) | \mathcal{F}_S]$$

$$= 2K\mathbb{E}[\int_S^{S'} \int_{\mathbb{R}^2} ((\Gamma_s\gamma(z))^2 \wedge K') \mathbb{1}_{\{\gamma(z)\leq 1/q\}} \lambda(dz)ds | \mathcal{F}_S]$$

$$\leq 2K \int_{\mathbb{R}^2} ((\gamma(z))^2 \wedge 1) \mathbb{1}_{\{\gamma(z)\leq 1/q\}} \lambda(dz) \mathbb{E}[(S'-S) | \mathcal{F}_S]$$

where we used that the quadratic variation of $M(q)$ equals the sum of squared jumps of $M(q)$, that $(\Gamma_s\gamma(z))^2$ is bounded by some constant K' and that the Lebesgue measure times λ is the predictable compensator of μ. This estimate yields (1.15) with

$$e_q = \int_{\mathbb{R}^2} ((\gamma(z))^2 \wedge 1) \mathbb{1}_{\{\gamma(z)\leq 1/q\}} \lambda(dz).$$

The last inequality (1.16) follows from

$$X_t = X_0 + B(q')_t + C_t + M(q')_t$$

for a sufficiently small q' as the jumps of X are bounded through γ by a constant.

$\square$

B Stable Convergence in Law

In this appendix chapter we give an introduction to the concept of stable convergence in law and a short overview of its basic properties. Stable convergence in law is a slightly stronger mode of convergence of random variables than convergence in law, but still weaker than convergence in probability.[1]

To illustrate the need for a stronger mode of convergence than convergence in law we consider the following generic situation which has already been briefly skteched in the beginning of Chapter 3; compare also Section 2 in [40] and Section 2.2.1 in [30]. Let Y_n, $n \in \mathbb{N}$, be a sequence of random variables which converges in law to a mixed normal distribution i.e. it holds

$$Y_n \xrightarrow{\mathcal{L}} VU \tag{B.1}$$

for independent random variables $V > 0$ and $U \sim \mathcal{N}(0,1)$. In the case where (B.1) depicts a central limit theorem, the asymptotic variance V^2 has to be estimated such that the central limit theorem can be of use for statistical inference. To this end one usually first looks for a sequence of estimators V_n^2, $n \in \mathbb{N}$, which consistently estimates the asymptotic variance V^2, i.e. fulfils $V_n^2 \xrightarrow{\mathbb{P}} V^2$, and after that one tries to deduce the convergence

$$\frac{Y_n}{\sqrt{V_n^2}} \xrightarrow{\mathcal{L}} U \tag{B.2}$$

where all components in the equation, the estimators Y_n, V_n and the asymptotic law of U, are known to the statistician. Hence (B.2) can be used for statistical purposes while (B.1) is infeasible as it contains the unknown variable V. To obtain (B.2) from (B.1) and $V_n^2 \xrightarrow{\mathbb{P}} V^2$ usually the joint convergence

$$(Y_n, V_n^2) \xrightarrow{\mathcal{L}} (UV, V^2) \tag{B.3}$$

is used together with the continuous mapping theorem. However, (B.3) follows from (B.1) and $V_n^2 \xrightarrow{\mathbb{P}} V^2$ in general only if V is deterministic; compare Theorem 6(c) in [16]. If V is non-deterministic this implication is in general false. Hence, some stronger requirement on the convergence of the sequences Y_n and V_n is needed to obtain (B.3), which then using the continuous mapping theorem is

---

[1]Compare the first paragraph on page 4 of [40].

© Springer Fachmedien Wiesbaden GmbH, part of Springer Nature 2019
O. Martin, *High-Frequency Statistics with Asynchronous and Irregular Data*,
Mathematische Optimierung und Wirtschaftsmathematik I Mathematical Optimization
and Economathematics, https://doi.org/10.1007/978-3-658-28418-3

sufficient to get (B.2). This can be achieved if instead of convergence in law of the random variables Y_n to VU in (B.1) we require *stable convergence in law* which is introduced in the following definition; compare Definition 2.1 in [40].

Definition B.1. *Let Y_n, $n \in \mathbb{N}$, be a squence of random variables defined on a probability space $(\Omega, \mathcal{F}, \mathbb{P})$ with values in some Polish space $(E, \mathcal{E})$. Further let $\mathcal{G} \subset \mathcal{F}$ denote some sub-σ-algebra. We say that Y_n converges $\mathcal{G}$-stably in law to some random variable Y, written as $Y_n \xrightarrow{\mathcal{L}-s} Y$, defined on an extended probability space $(\widetilde{\Omega}, \widetilde{\mathcal{F}}, \widetilde{\mathbb{P}})$ if and only if it holds*

$$\mathbb{E}[Zg(Y_n)] \longrightarrow \widetilde{\mathbb{E}}[Zg(Y)] \tag{B.4}$$

for any bounded, continuous function $g\colon E \to \mathbb{R}$ and any bounded real-valued $\mathcal{G}$-measurable random variable Z. □

Here, the notion that $(\widetilde{\Omega}, \widetilde{\mathcal{F}}, \widetilde{\mathbb{P}})$ is an extension of $(\Omega, \mathcal{F}, \mathbb{P})$ means that $(\Omega, \mathcal{F}, \mathbb{P})$ is in some sense contained in $(\widetilde{\Omega}, \widetilde{\mathcal{F}}, \widetilde{\mathbb{P}})$ and all $\mathcal{F}$-measurable random variables have under $\mathbb{P}$ and $\widetilde{\mathbb{P}}$ the same distribution. For a more precise definition see (2.1.26) in [30].

Remark B.2. *As for ordinary convergence in law it is usually sufficient to require that* (B.4) *holds for smaller classes of functions g, i.e. if* (B.4) *holds for all Lipschitz-continuous functions g with compact support it automatically also holds for all bounded, continuous functions.* □

In the context of high-frequency statistics the asymptotic variance very often depends on the observed path $t \mapsto X_t(\omega)$ or its components b, σ and δ, compare Chapers 3 and 7-9. Hence in this field the concept of stable convergence in law is very useful as the sequences Y_n, $n \in \mathbb{N}$, of interest are based on the observed values $X_{t_{i,n}^{(l)}}^{(l)}$ of $X^{(l)}$ on a discrete grid and therefore depend on the observad path as well. Consequently the asymptotic variance is very often random and dependent on the sequence Y_n, $n \in \mathbb{N}$.

The following lemma states two useful equivalent characterizations of stable convergence in law; compare Proposition 2.2 in [40].

Lemma B.3. *Let Y, Y_n, $n \in \mathbb{N}$, be random variables as in Definition B.1. Then the following statements are equivalent.*

(i) The $\mathcal{G}$-stable convergence $Y_n \xrightarrow{\mathcal{L}-s} Y$ holds.

(ii) The convergence in law $(Z, Y_n) \xrightarrow{\mathcal{L}} (Z, Y)$ holds for any $\mathcal{G}$-measurable random variable Z.

(iii) *The* $\mathcal{G}$-*stable convergence* $(Z, Y_n) \xrightarrow{\mathcal{L}-s} (Z, Y)$ *holds for any* $\mathcal{G}$-*measurable random variable* Z.

Proof. The implication (iii)⇒(ii) is obvious. Further (ii) yields that

$$\mathbb{E}[\tilde{g}(Z, Y_n)] \longrightarrow \widetilde{\mathbb{E}}[\tilde{g}(Z, Y)]$$

holds for any continuous and bounded function $\tilde{g}$. Hence (i) follows if we set $\tilde{g}(z, y) = zg(y)$. It remains to prove (i)⇒(iii). To this end we have to show that (i) implies

$$\mathbb{E}[Z'g(Z'', Y_n)] \longrightarrow \widetilde{\mathbb{E}}[Z'g(Z'', Y)] \tag{B.5}$$

for any $\mathcal{G}$-measurable random variables Z', Z'' where Z' is bounded. By a density argument it is sufficient to show (B.5) for all functions of the form $g(z, y) = \mathbb{1}_A(z)\mathbb{1}_B(y)$ and (B.5) then follows from *(i)* if we set $Z = Z'\mathbb{1}_A(Z'')$ in (B.4). $\qquad\square$

The following remark demonstrates why it is necessary to consider an extended probability space $(\widetilde{\Omega}, \widetilde{\mathcal{F}}, \widetilde{\mathbb{P}})$ in Definition B.1 on which the limiting variable Y is defined; compare Lemma 2.3 in [40]. In fact, it turns out that $\mathcal{F}$-stable convergence in law coincides with convergence in probability if we require Y to live on the same probability space $(\Omega, \mathcal{F}, \mathbb{P})$ on which the sequence Y_n is defined. Further for ordinary convergence in law, only the law of the limiting variable Y and not the underlying probability space where the limiting variable Y is defined is important. Contrary we observe that for stable convergence in law also the joint law of the limiting variable Y and $\mathcal{F}$-measurable random variables Z is of importance. Hence Y and these random variables Z have to be defined on a common probability space.

Remark B.4. *Let* Y, Y_n, $n \in \mathbb{N}$, *be random variables as in Definition B.1, assume that the* $\mathcal{F}$-*stable convergence* $Y_n \xrightarrow{\mathcal{L}-s} Y$ *holds and suppose that* Y *is* $\mathcal{F}$-*measurable. Then Lemma B.3 yields* $(Y, Y_n) \xrightarrow{\mathcal{L}} (Y, Y)$ *which by the continuous mapping theorem, compare Theorem 2.7 in [9], implies* $Y_n - Y \xrightarrow{\mathcal{L}} 0$. *Next as convergence in law to a constant is equivalent to convergence in probability to a constant, compare Theorem 6(c) in [16], we get* $Y_n - Y \xrightarrow{\mathbb{P}} 0$. *Hence in the situation where the limit is* $\mathcal{F}$-*measurable* $\mathcal{F}$-*stable convergence in law is equivalent to convergence in probability.* $\qquad\square$

The following lemma generalizes the continuous mapping theorem for convergence in law, compare Theorem 2.7 in [9], to the concept of stable convergence in law.

Lemma B.5. *Let Y, Y_n, $n \in \mathbb{N}$, be random variables as in Definition B.1. If we assume that the $\mathcal{G}$-stable convergence $Y_n \xrightarrow{\mathcal{L}-s} Y$ holds, then we also have the $\mathcal{G}$-stable convergence*

$$g(Y_n) \xrightarrow{\mathcal{L}-s} g(Y) \tag{B.6}$$

for any continuous function $g \colon (E, \mathcal{E}) \to (E', \mathcal{E}')$ for some Polish space $(E', \mathcal{E}')$.

Proof. To prove (B.6) by Definition B.1 we need to show

$$\mathbb{E}[Zh(g(Y_n))] \longrightarrow \widetilde{\mathbb{E}}[Zh(g(Y))] \tag{B.7}$$

for any $\mathcal{G}$-measurable and bounded random variable Z and any continuous function h. However, $Y_n \xrightarrow{\mathcal{L}-s} Y$ immediately yields (B.7) because the function $h \circ g$ is continuous and bounded. $\square$

The next lemma generalizes another well-known result for convergence in law, compare Theorem 6(b) in [16], to the concept of stable convergence in law.

Lemma B.6. *Suppose we are given a sequence of real-valued random variables $(Y_n)_{n \in \mathbb{N}}$ and a double sequence of real-valued random variables $(Z_{n,k})_{n,k \in \mathbb{N}}$ both defined on a probability space $(\Omega, \mathcal{F}, \mathbb{P})$ and real-valued random variables Y, Z_k, $k \in \mathbb{N}$, defined on an extended probability space $(\widetilde{\Omega}, \widetilde{\mathcal{F}}, \widetilde{\mathbb{P}})$. Further assume that it holds*

$$\lim_{k \to \infty} \limsup_{n \to \infty} \mathbb{P}(|Y_n - Z_{n,k}| > \varepsilon) = 0,$$
$$\lim_{k \to \infty} \widetilde{\mathbb{P}}(|Y - Z_k| > \varepsilon) = 0 \tag{B.8}$$

for any $\varepsilon > 0$ and that for some sub-σ-algebra $\mathcal{G} \subset \mathcal{F}$ we have the $\mathcal{G}$-stable convergences

$$Z_{n,k} \xrightarrow{\mathcal{L}-s} Z_k, \quad k \in \mathbb{N}, \tag{B.9}$$

as $n \to \infty$. Then the $\mathcal{G}$-stable convergence $Y_n \xrightarrow{\mathcal{L}-s} Y$ also holds for $n \to \infty$.

Proof. We have to check (B.4). To this end note that it holds

$$|\mathbb{E}[Zg(Y_n)] - \widetilde{\mathbb{E}}[Zg(Y)]| \leq K\widetilde{\mathbb{E}}[|g(Y_n) - g(Y)|]$$
$$\leq K\big(\mathbb{E}[|g(Y_n) - g(Z_{n,k})|] + \widetilde{\mathbb{E}}[|g(Z_{n,k}) - g(Z_k)|] + \widetilde{\mathbb{E}}[|g(Z_k) - g(Y)|]\big).$$

We first conclude $\mathbb{E}[|g(Z_{n,k}) - g(Z_k)|] \to 0$ as $n \to \infty$ for any $k \in \mathbb{N}$ from (B.9). Further

$$\lim_{k \to \infty} \limsup_{n \to \infty} \big(\mathbb{E}[|g(Y_n) - g(Z_{n,k})|] + \widetilde{\mathbb{E}}[|g(Z_k) - g(Y)|]\big)$$

follows from (B.8) for all Lipschitz-continuous functions g. However, it is sufficient to prove (B.4) for such functions only because then by a density argument (B.4) already has to hold for all bounded and continuous functions; compare Remark B.2. □

A standard approximation technique when working with Itô semimartingales is based on the idea to first show desired results for processes with finite jump activity, then let the number of jumps tend to infinity and finally to show that the desired results remain valid in the limit. To verify this method we use Lemma (B.6) in Chapters 3 and 7-9.

The upcoming proposition, compare Proposition 2.5 in [40], finally states that working with stable convergence in law instead of with ordinary convergence in law allows to draw the desired conclusions sketched in the introductory example.

Proposition B.7. *Let Y_n, Y, X be $\mathbb{R}^d$-valued random variables and V_n, V be $\mathbb{R}^{d \times d}$-valued random variables all defined on the extended probability space $(\tilde{\Omega}, \tilde{\mathcal{F}}, \tilde{\mathbb{P}})$ where Y_n, V_n, V are $\mathcal{F}$-measurable. Further let $g \colon \mathbb{R}^d \to \mathbb{R}^{d'}$ be a continuously differentiable function. Then:*

(i) *If the $\mathcal{X}$-stable convergence $Y_n \xrightarrow{\mathcal{L}-s} Y$ holds and we have $V_n \xrightarrow{\mathbb{P}} V$, then it already holds $(V_n, Y_n) \xrightarrow{\mathcal{L}-s} (V, Y)$.*

(ii) *Let $d = 1$ and $Y_n \xrightarrow{\mathcal{L}-s} Y = VU$ for some $U \sim \mathcal{N}(0,1)$ which is independent of $\mathcal{F}$. Further assume $V_n \xrightarrow{\mathbb{P}} V$ with $V_n, V > 0$. Then it holds $Y_n/V_n \xrightarrow{\mathcal{L}-s} U$.*

(iii) *Assume that the random variable Y is $\mathcal{G}$-measurable and that the $\mathcal{G}$-stable convergence $\sqrt{n}(Y_n - Y) \xrightarrow{\mathcal{L}-s} X$ holds. Then we also have*

$$\sqrt{n}(g(Y_n) - g(Y)) \xrightarrow{\mathcal{L}-s} \nabla g(Y) X.$$

Proof. First we will prove part (i): By Lemma B.3 $Y_n \xrightarrow{\mathcal{L}-s} Y$ implies $(V, Y_n) \xrightarrow{\mathcal{L}-s} (V, Y)$. Further we obtain $(V_n, Y_n) - (V, Y_n) \xrightarrow{\mathbb{P}} 0$ and hence Lemma B.6 yields the joint convergence in law $(V_n, Y_n) \xrightarrow{\mathcal{L}-s} (V, Y)$.

The claim (ii) directly follows from (i) using the continuous mapping theorem for stable convergence in law stated in Lemma B.5.

For part (iii) observe that by the mean value theorem we can find ξ_n between Y and Y_n with

$$\sqrt{n}(g(Y_n) - g(Y)) = \nabla g(\xi_n)\sqrt{n}(Y_n - Y).$$

From $Y_n \xrightarrow{\mathcal{L}-s} Y$ we conclude $|Y_n - Y| \xrightarrow{\mathbb{P}} 0$ which yields $\xi_n \xrightarrow{\mathbb{P}} Y$. Consequently by the continuity of g we also obtain $g(\xi_n) \xrightarrow{\mathbb{P}} g(Y)$. Hence, by Lemma B.3 we get $(g(\xi_n), \sqrt{n}(Y_n - Y)) \xrightarrow{\mathcal{L}-s} (g(Y), X)$ and then Lemma B.5 yields the claim. $\square$

The Delta-method for stable convergence in law stated in part (iii) of Proposition B.7 also illustrates the benefits of working with stable convergence in law instead of ordinary weak convergence. Usually, the Delta-method can only be applied if Y is constant. However, using stable convergence we obtain a result also for random $\mathcal{F}$-measurable variables Y; compare [40]. In the setting of high-frequency statistics the occurring limits are very often random; compare e.g. the results in Chapter 2.

Notes. *The above discussion of some elementary properties of stable convergence in law is mostly an adaption of Section 2.1 of [40]. The paper [40] gives an elementary and intuitive introduction to the concept of stable convergence in law and explains its role in the context of high-frequency statistics.*

A more formal discussion of the concept of stable convergence in law and its applications for proving convergences of stochastic processes can be found in [31] and [30].

C Triangular Arrays

In this appendix chapter we state three results which are useful for proving convergence results for random variables which can be represented as sums of martingale differences. To formulate the results we need to introduce the notion of a triangular array.

Definition C.1. *A triangular array with accomodating filtrations is a double sequence of $\mathbb{R}$-valued random variables $(\zeta_k^n)_{k,n \in \mathbb{N}}$ for which filtrations $(\mathcal{G}_k^n)_{k \in \mathbb{N}_0}$, $n \in \mathbb{N}$, exist such that ζ_k^n is $\mathcal{G}_k^n$-measurable.* $\qquad\square$

Let N_n, $n \in \mathbb{N}$, be $(\mathcal{G}_k^n)_{k \in \mathbb{N}_0}$-stopping times. Then for triangular arrays with accomodating filtrations we can obtain simplified conditions for proving

$$\sum_{k=1}^{N_n} \zeta_k^n \xrightarrow{\mathbb{P}} 0$$

based on the conditional expectations $\mathbb{E}[\zeta_k^n | \mathcal{G}_{k-1}^n]$, $\mathbb{E}[(\zeta_k^n)^2 | \mathcal{G}_{k-1}^n]$. In particular the following result holds which is part of Lemma 2.2.11 in [30].

Lemma C.2. *Let $(\zeta_k^n)_{k,n \in \mathbb{N}}$ be a triangular array with accomodating filtrations $(\mathcal{G}_k^n)_{k \in \mathbb{N}_0}$, $n \in \mathbb{N}$, and let N_n, $n \in \mathbb{N}$, be $(\mathcal{G}_k^n)_{k \in \mathbb{N}_0}$-stopping times. Then the two convergences*

$$\sum_{k=1}^{N_n} \mathbb{E}[\zeta_k^n | \mathcal{G}_{k-1}^n] \xrightarrow{\mathbb{P}} 0, \qquad \sum_{k=1}^{N_n} \mathbb{E}[(\zeta_k^n)^2 | \mathcal{G}_{k-1}^n] \xrightarrow{\mathbb{P}} 0,$$

imply

$$\sum_{k=1}^{N_n} \zeta_k^n \xrightarrow{\mathbb{P}} 0.$$

Usually we consider some process $(X_t)_{t \geq 0}$ which is adapted to a filtration $(\mathcal{F}_t)_{t \geq 0}$. Further let $(\tau_{k,n})_{k \in \mathbb{N}_0}$, $n \in \mathbb{N}$, be increasing sequences of stopping times. Then ζ_k^n is some function of the behaviour of the process X in the (random) interval $(\tau_{k-1,n}, \tau_{k-1,n}]$ e.g. $\zeta_k^n = f(X_{\tau_{k,n}} - X_{\tau_{k-1,n}})$ for some deterministic function f. By setting $N_n = \sup\{k : \tau_{k,n} \leq T\}$ we sum up the ζ_k^n up to some time horizon T. Appropriate filtrations are then given by $\mathcal{G}_k^n = \mathcal{F}_{\tau_{k,n}}$. In this specific scenario the

© Springer Fachmedien Wiesbaden GmbH, part of Springer Nature 2019
O. Martin, *High-Frequency Statistics with Asynchronous and Irregular Data*,
Mathematische Optimierung und Wirtschaftsmathematik I Mathematical Optimization
and Economathematics, https://doi.org/10.1007/978-3-658-28418-3

following result holds which even yields u.c.p. convergence, compare page 57 in [41], for the process which we obtain if we vary the time horizon T. This result is part of Lemma 2.2.12 in [30] or Lemma 4.2 in [28].

Lemma C.3. *Let* $(\tau_{k,n})_{k\in\mathbb{N}_0}$, $n \in \mathbb{N}$, *be increasing sequences of stopping times with* $\tau_{0,n} = 0$. *For* $t > 0$ *denote* $N_n(t) = \sup\{k : \tau_{k,n} \le t\}$ *and let* ζ_k^n *be* $\mathcal{G}_{\tau_{k,n}}$*-measurable random variables. Suppose* $(A_t)_{t\ge 0}$ *is a process with continuous paths of finite variation and we have*

$$\sum_{k=1}^{N_n(t)} \mathbb{E}[\zeta_k^n|\mathcal{G}_{\tau_{k-1,n}}] \overset{u.c.p.}{\longrightarrow} A_t, \quad \sum_{k=1}^{N_n(t)} \mathbb{E}[(\zeta_k^n)^2|\mathcal{G}_{\tau_{k-1,n}}] \overset{\mathbb{P}}{\longrightarrow} 0 \ \ \forall t \ge 0,$$

then it also holds $\sum_{k=1}^{N_n(t)} \zeta_k^n \overset{u.c.p.}{\longrightarrow} A_t$.

If the second moments do not vanish as in Lemmata C.2 and C.3 we obtain stable convergence in law of the sum $\sum_{i=1}^{N_n} \zeta_k^n$ and the limit of the sum over the second conditional moments describes the asymptotic variance. The following result is taken from Theorem 2.6 in [40]; compare also Theorem 2.1 in [26].

Proposition C.4. *Let* $(\Omega, (\mathcal{F}_t)_{t\ge 0}, \mathbb{P})$ *be a filtered probability space and let* ζ_k^n *be* $\mathcal{F}_{t_{k,n}}$*-measurable real-valued random variables where* $t_{k,n} = k/n$. *We assume that the* ζ_k^n *are in some sense "fully generated", compare page 6 of [40], by some (one-dimensional) Brownian motion* W. *Futher, suppose that there exist absolutely continuous processes* F, G *and a continuous process* $\widetilde{B}$ *of finite variation such that the following conditions hold*

$$\sum_{k=1}^{\lfloor nt \rfloor} \mathbb{E}[\zeta_k^n|\mathcal{F}_{t_{k-1,n}}] \overset{\mathbb{P}}{\longrightarrow} \widetilde{B}_t, \tag{C.1}$$

$$\sum_{k=1}^{\lfloor nt \rfloor} \left(\mathbb{E}[(\zeta_k^n)^2|\mathcal{F}_{t_{k-1,n}}] - (\mathbb{E}[\zeta_k^n|\mathcal{F}_{t_{k-1,n}}])^2\right) \overset{u.c.p.}{\longrightarrow} F_t = \int_0^t (v_s^2 + w_s^2)ds, \tag{C.2}$$

$$\sum_{k=1}^{\lfloor nt \rfloor} \mathbb{E}[\zeta_k^n(W_{k/n} - W_{(k-1)/n})|\mathcal{F}_{t_{k-1,n}}] \overset{u.c.p.}{\longrightarrow} G_t = \int_0^t v_s^2 ds, \tag{C.3}$$

$$\sum_{k=1}^{\lfloor nt \rfloor} \mathbb{E}[(\zeta_k^n)^2 \mathbf{1}_{\{|\zeta_k^n|>\varepsilon\}}|\mathcal{F}_{t_{k-1,n}}] \overset{u.c.p.}{\longrightarrow} 0, \quad \forall \varepsilon > 0, \tag{C.4}$$

$$\sum_{k=1}^{\lfloor nt \rfloor} \mathbb{E}[\zeta_k^n(\widetilde{M}_{k/n} - \widetilde{M}_{(k-1)/n})|\mathcal{F}_{t_{k-1,n}}] \overset{u.c.p.}{\longrightarrow} 0 \tag{C.5}$$

where $(v_s)_{s\geq 0}$ and $(w_s)_{s\geq 0}$ are predictable processes and the property (C.5) holds for all $(\mathcal{F}_t)_{t\geq 0}$-martingales $\widetilde{M}$ with $[W, \widetilde{M}] \equiv 0$. Then the following $\mathcal{F}$-stable convergence holds

$$\sum_{k=1}^{N_n(t)} \zeta_k^n \xrightarrow{\mathcal{L}_s} \widetilde{B}_t + \int_0^t v_s dW_s + \int_0^t w_s d\widetilde{W}_s, \quad t \geq 0,$$

with $N_n(t) = \lfloor nt \rfloor$ where $(\widetilde{W}_t)_{t\geq 0}$ denotes a standard Brownian motion defined on an extended probability space $(\widetilde{\Omega}, \widetilde{\mathcal{F}}, \widetilde{\mathbb{P}})$ which is independent of $\mathcal{F}$.

Printed in the United States
By Bookmasters